Springer Geology

The book series Springer Geology comprises a broad portfolio of scientific books, aiming at researchers, students, and everyone interested in geology. The series includes peer-reviewed monographs, edited volumes, textbooks, and conference proceedings. It covers the entire research area of geology including, but not limited to, economic geology, mineral resources, historical geology, quantitative geology, structural geology, geomorphology, paleontology, and sedimentology.

More information about this series at http://www.springer.com/series/10172

Vladimir V. Reverdatto · Igor I. Likhanov
Oleg P. Polyansky · Valentin S. Sheplev
Vasiliy Yu. Kolobov

The Nature and Models of Metamorphism

Springer

Vladimir V. Reverdatto
V.S. Sobolev Institute of Geology
and Geophysics
Siberian Branch, Russian
Academy of Sciences
Novosibirsk, Russia

Igor I. Likhanov
V.S. Sobolev Institute of Geology
and Geophysics
Siberian Branch, Russian
Academy of Sciences
Novosibirsk, Russia

Oleg P. Polyansky
V.S. Sobolev Institute of Geology
and Geophysics
Siberian Branch, Russian
Academy of Sciences
Novosibirsk, Russia

Valentin S. Sheplev (Deceased)
V.S. Sobolev Institute of Geology
and Geophysics
Siberian Branch, Russian
Academy of Sciences
Novosibirsk, Russia

Vasiliy Yu. Kolobov
V.S. Sobolev Institute of Geology
and Geophysics
Siberian Branch, Russian
Academy of Sciences
Novosibirsk, Russia

ISSN 2197-9545 ISSN 2197-9553 (electronic)
Springer Geology
ISBN 978-3-030-03028-5 ISBN 978-3-030-03029-2 (eBook)
https://doi.org/10.1007/978-3-030-03029-2

Library of Congress Control Number: 2018961231

Translation from the Russian language edition: *ПРИРОДА И МОДЕЛИ МЕТАМОРФИЗМА* by Ревердатто В.В., © Novosibirsk: Publ. House of Siberian Branch of RAS 2017. All Rights Reserved.

This Springer imprint is published by the registered company Springer Nature Switzerland AG
The registered company address is: Gewerbestrasse 11, 6330 Cham, Switzerland

Preface

Metamorphism is an endogenous transformation of rocks, involving processes of phase transitions in Earth's crust and upper mantle at elevated temperatures (*T*) and pressures (*P*). At the same time, metamorphism involves no chemical changes in the bulk rock composition, with the exception of gain or loss of volatiles, i.e., metamorphism occurs isochemically. If this condition is not fulfilled, and metamorphism is accompanied by a change in the chemical composition of the rock, this process is termed allochemical metamorphism or metasomatism. Metasomatism is typically associated with retrograde alterations. The compositions of coexisting minerals vary with changes in thermodynamic conditions in the Earth's interior as a result of phase transitions, i.e., metamorphic reaction in the presence of an intergranular fluid phase. The fluid is generally released during isochemical metamorphism by dehydration and decarbonatization reactions. The amount of the released metamorphic fluid is small and is controlled by the intrinsic rock permeability that may well be too low, and the composition of the fluid is determined by the composition of the rock. Metamorphic reactions take place in the groundmass or matrix and mineral inclusions. The primary heterogeneity of the protolith is basically preserved during metamorphism in combination with mineralogical, structural, and textural transitions, and within a certain "local" volume neighboring minerals and a fluid in a metamorphic rock represent an association of coexisting minerals in thermodynamic equilibrium which each other, which is known as mineral paragenesis. In a gradient temperature and pressure field, essentially isochemical phase transitions in homogeneous rocks produce a series of "isograds," representing the boundaries between stable mineral assemblages that replace each other. The evolution of the metamorphic *P-T* parameters can be accompanied by the preservation of mineral relics and replacement textures within newly formed minerals. A study of transformations in the mineral assemblage and texture of the rock under favorable conditions makes it possible to trace the change in thermodynamic conditions of metamorphism with time (*t*) and build the so-called *P-T-t* paths of a metamorphic rock.

The P-T conditions and P-T-t paths are directly related to different tectonic and magmatic processes driven by extensive redistribution of mass and heat in the Earth's crust and mantle. These processes destroy the original distribution of matter and temperature and lead to metamorphism. The successive appearance of certain minerals and mineral assemblages in space and time supplemented with geological data can be used as indicators of tectonomagmatic events. This is currently the focus of a great deal of ongoing geological research, which involves the analysis of metamorphism in terms of cause–effect relationships. Our book is an attempt to solve this problem by shedding light on the tectonomagmatic causes, i.e., geological events, as well as the products of metamorphism, i.e., mineral assemblages and textures formed by metamorphic reactions. In order to capture the rigorous reasoning about the problems and for the sake of brevity, we will use mathematical expressions and models that represent the most essential characteristics and qualities of the processes under consideration. Depending on the context, these characteristics may include physical properties of a heterogeneous medium, heat and mass transfer, mechanical impulses and gravity, stress and strain (e.g., for the analysis of tectonomagmatic events), as well as thermodynamic parameters and their variation, duration and conditions of mass transfer, structural and textural parameters, composition, etc. (for describing metamorphism). Depending on the posed questions, we used different mathematical approaches and solutions. The major difficulty in the mathematical modeling of geological phenomena lies in their diversity and complexity, requiring the application of numerical methods. However, the main factors and mechanisms underlying the nature of these phenomena can in most cases be investigated within the context of reasonably simplified models using analytical and numerical solutions. At the same time, instead of considering all the characteristics and parameters of the processes, we only consider the ones that are of primary importance, if the empirical adequacy of the model estimates is generally provided in terms of the most significant signs and qualities of metamorphism. The use of reasonably simplified models makes the result of the interaction of elementary or taken as elementary processes directly visible, thus promoting a deeper understanding of the phenomenon as a whole.

The emphasis in our study is placed on geological causes of metamorphism of rocks in Earth's crust and on mineral transformations in rocks, which is very relevant and deserves special attention. The study of metamorphism helps to understand the thermomechanical state of the lithosphere and its evolution and provides important information on the relationship between metamorphism, tectonic, and magmatic processes in the Earth's history. In this context, particular attention should be paid to crustal deformations caused by extension, collision, and subduction of lithospheric plates, as well as by magma intrusions and diapirism. We do not consider deformation at the level of rock structure or interacting mineral grains, although it plays an important role in the kinetics of metamorphism. Digging deeper into this problem would add another layer of complication to the already complicated task of studying metamorphic reactions. Furthermore, the book does not address in detail other relevant topics such as metamorphism of the mantle rocks, metamorphogenic ore formation, regional metasomatism or metasomatism near intrusive bodies, the nature of protoliths, which are not directly related to the subject area of this book.

The international symposium on the formation of metamorphic minerals and rocks was held at the University of Liverpool in 1964. The main focus was on nucleation and mineral growth, mineral compositions driven by changing pressure and temperature conditions, metamorphic reactions and isograds, the role of tectonic overpressure, etc. Little attention was given to the geological causes of metamorphism, although 20 papers were presented at the symposium. A collection of lectures presented at the symposium was published in 1965 in the Controls of Metamorphism Volume edited by W. S. Pitcher and G. W. Flinn. A Russian version of his book was published by Mir Publishers in 1967, entitled The Nature of Metamorphism under the editorship of V. P. Petrov. Our book has a deceptively similar title, but it addresses not only the factors controlling the formation and evolution of metamorphism minerals and rocks, but also the causes of metamorphism as a geological phenomenon. As mentioned above, this problem is solved using relevant models, which were built and analyzed to greatly improve our understanding.

The book brought to the notice of the readers consists of four unequal parts. Chapter 1 traces the evolution in the understanding of metamorphism. We think that it is important to preface the presentation of the theory with a brief historical essay supplemented with a list of outstanding researchers who contributed substantially to the study of metamorphism in the last century. This essay is an attempt to follow the evolution of various ideas developed within the metamorphic facies concept in terms of the different mineral assemblages. The principle of metamorphic facies in relation to the *P-T*conditions is also used in our book. In Chap. 2, we provide an overview of the most basic tools of metamorphic research—mineral geothermobarometers. Today, this can be considered as the most comprehensive compilation of the geothermometers and geobarometers applied in petrological studies. Chapter 3 of this book considers numerical model-based relationships between metamorphism and geodynamics, discusses tectonomagmatic causes and controls of metamorphism, and makes attempt to link the geological types of metamorphism to the specific *P-T* conditions and *P-T-t* paths. The final Chap. 4 provides a generalized *P-T-t* diagram of the evolution of metamorphic complexes of different geodynamic nature, which are typical for different types of metamorphism. Chapter 4 also provides a model-based characterization of mineral and structural transformations in the rocks, which are controlled by variations in *P-T* conditions, mass transfer, and chemical reactions. The material presented in the book is the result of many years of extensive research, which began in 1970–1974 immediately after the publication of the renowned 4-volume monograph on the facies of metamorphism edited by Academician V. S. Sobolev. Not all four parts cover all our topics in equal detail, because some of them require further investigation. This book is a collective effort, which has been widely discussed and agreed by a group of authors at all stages of the preparation of the manuscript.

The authors of this book are extremely grateful to their colleagues from the Sobolev Institute of Geology and Mineralogy, Siberian Branch, Russian Academy of Sciences for their helpful discussion of the results: A. V. Babichev, N. I. Volkova, G. G. Lepezin, A. D. Nozhkin, V. G. Sverdlova, V. P. Sukhorukov, O. M. Turkina, V. N. Sharapov, A. A. Krylov, P. S. Kozlov (Institute of Geology and Geochemistry, Ural Branch, Russian Academy of Sciences), S. N. Korobeinikov (Institute of Hydrodynamics, Siberian Branch, Russian Academy of Sciences), and A. L. Perchuk (Moscow State University). We extend our particular thanks to Prof. N. V. Sobolev for his valuable comments and suggestions on the manuscript.

The authors find it necessary to acknowledge the outstanding role of Valentin Semenovich Sheplev, Doctor of Chemical Sciences, who passed away too soon and unexpectedly. In this collective work, he studied the processes of mass transfer and chemical reactions accompanying metamorphic and metasomatic alteration of rocks in terms of thermodynamics. His contribution to the development of the theory of metasomatism using mathematical modeling was considerable. It includes the numerical solution of a mathematical model of diffusion bimetasomatism. Using a stationary model, V. S. Sheplev was able to successfully solve mathematical problems and to obtain, for the first time, the analytical expression for all types of metasomatic growth zoning. To select a single solution, he proposed a criterion based on the principle of extremality requiring a minimum amount of thermodynamic data. He was able not only to explain the causes of spatial relations between minerals and mineral assemblages, but to predict the behavior of the systems during chemical interaction. Later, a theory of diffusion metasomatism was transferred by V. S. Sheplev to infiltration metasomatism to first determine in an analytical manner expressions for all characteristics of the process and develop a complete theory of possible infiltration sequences of mineral parageneses. The application of conventional thermobarometry was extended to disequilibrium mineral assemblages, and uncertainties in the equilibrium and disequilibrium thermobarometry were estimated. V. S. Sheplev was widely recognized, both in Russia and Abroad, as a specialist in the field of mass and heat transfer in heterogeneous media where chemical transformations occur. He died too early, but his contribution to the mathematical modeling of chemical and mineral formation processes is a dignified memorial to him.

Novosibirsk, Russia
April 2018

Vladimir V. Reverdatto
Igor I. Likhanov
Oleg P. Polyansky
Valentin S. Sheplev
Vasiliy Yu. Kolobov

About This Book

This book is intended to analyze connections between metamorphism and geodynamics, its tectonomagmatic causes and controls and to outline the geological types of metamorphism with a framework of *P-T* parameters and *P-T-t* paths. Three categories of metamorphism are distinguished based on the magnitude of the heat flux: metamorphism induced by a thermal gradient close to the average ("normal") value, by a higher thermal gradient caused by the supply of additional heat by magmatic intrusions and diapirism, by a lower thermal gradient during the collision of lithospheric plates and crustal blocks. Quantitative methods are widely used in this book to describe metamorphism. The aims of this book are to present mathematical models of metamorphism in the vicinity of magmatic intrusions; to study rifting and diapirism; to characterize mineral transformations in rocks controlled by variations in P-T parameters, mass transfer and chemical reactions; to examine a quasi-stationary model of diffusion-controlled metasomatism to review the zoning in minerals; to develop a new geothermobarometric technique applicable under unsteady equilibrium conditions; to perform a quantitative analysis of mass transfer in the matrix; and to estimate the migration mobility of trace elements during metamorphism.

This book is intended for senior undergraduate and postgraduate students in Earth sciences and also appeals to professional petrologists, mineralogists, and geochemists.

Contents

List of Mineral Abbreviations (Whitney and Evans 2010)

Ab	Albite
Act	Actinolite
Adr	Andradite
Aeg	Aegirine
Ak	Åkermanite
Alm	Almandine
Als	Aluminosilicate
Amp	Amphibole
An	Anorthite
And	Andalusite
Ann	Annite
Arf	Arfvedsonite
Arg	Aragonite
Ath	Anthophyllite
Brc	Brucite
Bt	Biotite
Cal	Calcite
Chl	Chlorite
Clc	Clinochlore
Cld	Chloritoid
Cpx	Clinopyroxene
Crd	Cordierite
Crn	Corundum
Crt	Crossite
Cum	Cummingtonite
Czo	Clinozoisite
Di	Diopside
Dia	Diamond
Dol	Dolomite
Dsp	Diaspore

En	Enstatite
Ep	Epidote
Fa	Fayalite
Fo	Forsterite
Fs	Ferrosilite
Fsp	Feldspar
Ged	Gedrite
Gln	Glaucophane
Gr	Graphite
Grs	Grossular
Grt	Garnet
Hbl	Hornblende
Hc	Hercynite
Hd	Hedenbergite
Hem	Hematite
Hyp	Hypersthene
Ilm	Ilmenite
Jd	Jadeite
Kfs	K-feldspar
Kln	Kaolinite
Ky	Kyanite
Liq	Liquid
Lmt	Laumontite
Lws	Lawsonite
Mgs	Magnesite
Mgt	Magnetite
Mll	Melilite
Mnz	Monazite
Mrg	Margarite
Ms	Muscovite
Mtc	Monticellite
Mw	Merwinite
Nph	Nepheline
Ol	Olivine
Olg	Oligoclase
Omp	Omphacite
Opx	Orthopyroxene
Or	Orthoclase
Per	Periclase
Pg	Paragonite
Ph	Phengite
Phl	Phlogopite
Pl	Plagioclase
Pmp	Pumpellyite
Prh	Prehnite

Prl	Pyrophyllite
Prp	Pyrope
Qz	Quartz
Rt	Rutile
Scp	Scapolite
Sil	Sillimanite
Spl	Spinel
Spn	Sphene (titanite)
Spr	Sapphirine
Sps	Spessartine
Spu	Spurrite
Srp	Serpentine
St	Staurolite
Stp	Stilpnomelane
Tlc	Talc
Tr	Tremolite
Ts	Tschermakite
Ves	Vesuvianite
Wo	Wollastonite
Xtm	Xenotime
Zo	Zoisite

Chapter 1
Evolution in the Understanding of Mineral Transformations and Concept of Metamorphic Facies

1.1 Metamorphic Facies

The basic facts about minerals and rocks have been known to ancient people for a long time, centuries before Christ, although their origin remained enigmatic until the 17th–18th centuries. First, people came to realize, by analogy with modern processes on the Earth's surface, that all sediments accumulate in layers (Nikolas Steno, 1638–1686). The transformations of rocks were explained later by the term "metamorphosis" (Giovanni Arduino, 1713–1795). James Hutton (1726–1797), the father of modern geology, suggested that crystalline schists are derived from clays as a result of mineralogical transformations at high temperatures and pressure in the Earth's interior. He was the first to describe that the rocks surrounding a granitic body at Glen Tilt, Scotland, had suffered marked changes in either color or structure (Playfair 1822). This conclusion gave rise to the doctrine of metamorphism, however, the terms "metamorphism" and "metamorphic rocks" came into common use later (Boue 1820; Lyell 1833). The term "metamorphisme de contact" (contact metamorphism) was coined by Delesse (1857). In the second half of the 19th century, geologists learned that metamorphic rocks differ in the degree of alteration. In 1877, Karl Heinrich Ferdinand Rosenbusch (1836–1914) described a zonal metamorphic aureole in the vicinity of the Barr-Andlau granite pluton in the Vosges (Rosenbusch 1877), and in 1983 Gorge Barrow (1853–1932) first described and mapped zones of regional metamorphism в in the Grampian Highlands of Scotland (Barrow 1893). This was the time when the ideas about two main factors responsible for metamorphism started taking shape: heat of the intruding magma and burial depth. Geologists began to distinguish between regional and contact metamorphism. A special type of metamorphism was found to be related to tectonic processes and deformation imposed during mountain building. During the late 19th—early 20th centuries, these perceptions became embedded in the works of many outstanding geologists and appeared in the textbooks (e.g., Sederholm 1891; Zirkel 1893; Milch 1894; van Hise 1904; others).

V. V. Reverdatto et al., *The Nature and Models of Metamorphism*,
Springer Geology, https://doi.org/10.1007/978-3-030-03029-2_1

The rise of thermodynamics in the 19th century, together with the development of microscopic, chemical, and x-ray analyses for investigation of minerals and rocks have brought significant progress in knowledge of metamorphism. Based on the concept of the equilibrium of heterogeneous system introduced by Josiah Willard Gibbs (1839–1903), European geologists hypothesized in the first half of the 20th century that a great diversity of mineral assemblages is the result of phase transitions and depends on the chemical composition of rocks and *P-T* conditions. The completely general conditions of equilibrium formulated by Gibbs became later the basis for the metamorphic facies concept.

Friedrich Becke (1855–1931) was the first who compared the mineral assemblages and textures of rocks with thermodynamic parameters of metamorphism (Fig. 1.1). He showed that coexisting minerals tend toward a state of chemical equilibrium and formulated the rule according to which mineral-forming reaction involve a decrease of volume with increasing pressure and recognized two depth zones in the Earth's crust: the upper one containing rocks, which are generally a result of hydration and carbonatization processes at moderate temperature and pressure conditions, and the lower one, in which elevated pressures associated with elevated temperatures are accompanied by dehydration and decarbonatization reactions. Becke believed that the preservation of minerals from the lower zone on the Earth's surface can be explained by two kinetic causes: a sharp decrease in the reaction rate with decreasing temperature (Becke 1903).

This idea of a linkage between metamorphism and depth was further elaborated in the works of *Ulrich Grubenmann* (1850–1924), who classified metamorphic rocks into three depth-zones, named epi-, meso-, and kata-zones, in order of

Fig. 1.1 Becke F.

increasing depth of their metamorphism, which differ mostly in the presence of H_2O- and OH-containing minerals. Grubenmann's view was based on the assumption on the apparently direct relations between pressure *P* and temperature *T*, whereas the presence of directed pressure (stress) was implied only in the meso-zone (Grubenmann 1904). When it was first proposed, Grubenmann's classification did not include contact metamorphism, but it was subsequently modified by Paul Niggli (1988–1953) to include rocks in contact aureoles and metasomatic rocks near intrusive bodies irrespective of their actual burial depths, after which the Grubenmann–Niggli depth-zones (Grubenmann and Niggli 1924) were determined by the sequence of metamorphic reactions occurring in response to rising temperature.

Victor Moritz Goldschmidt (1888–1947) by applying an in-depth physicochemical approach heralded a change to modern metamorphic petrology (Fig. 1.2). His work on the Oslo contact hornfelses provided the first example of paragenetic analysis of mineral assemblages (Goldschmidt 1911a). In accordance with Gibbs's doctrine, Goldschmidt discovered systematic relations between rock composition and metamorphic mineral assemblage and showed that these minerals assemblages could be interpreted as having reached thermodynamic equilibrium, which is controlled by external factors, such as temperature and pressure. This correspondence can be defined as the so-called "mineralogical phase rule", a corollary to Gibbs' phase rule: the maximum number of minerals coexisting stably in a system with two or more degrees of freedom is equal to the number of independent (main) chemical components in these phases (Goldschmidt 1911b). Equilibrium during metamorphism was achieved due to mineral transformations in the presence of a saturated pore fluid. Goldschmidt plotted different types of metamorphism on the *P-T* diagram (Goldschmidt 1912), which gave the approximate metamorphic temperatures and pressures estimated on geological and mineralogical grounds. Contact metamorphic rocks represent the minimum pressure conditions, whereas for regional metamorphism, in the pressure range of about 5–12 kbar, he distinguished three zones, but mainly on temperature grounds. At the same time, by assuming CO_2 pressures generally approximating lithostatic pressure, he used the equilibrium Cal + Qz $\leftrightarrows$ Wo + CO_2 calculated from thermodynamic data.

Studying the metamorphic rocks of the Orijärvi region in Finland and comparing them with hornfelses observed by Goldschmidt in the Oslo region, *Pentti Eelis Eskola* (1883–1964) found significant differences in the composition of mineral associations, which he ascribed to different physicochemical conditions of metamorphism (Fig. 1.3). On this basis, Eskola introduced the concept of "metamorphic facies", which denotes a group of rocks metamorphosed under the same pressure and temperature conditions (Eskola 1915). In 1939, he formulated the following definition of the metamorphic facies: "A certain metamorphic facies denotes a group of rocks, which, at identical chemical composition, has an identical mineral content…A unique feature that allows rocks to be attributed to a specific metamorphic facies is that rocks of a given chemical composition have always the same mineralogical composition" (Eskola 1939, p. 339). Based on the mineralogical criteria, Eskola distinguished eight metamorphic facies (Fig. 1.4): sanidinite

Fig. 1.2 Goldschmidt V. M.

Fig. 1.3 Eskola P. E.

(characteristic minerals: sanidine, mullite, tridymite, monticellite, larnite; characteristic assemblage Wo + An), hornfels or pyroxene-hornfels (Di + Hyp, Crd + En, Wo + Di + Pl, Or + And, ± Bt, etc.), greenschist (Ms + Chl + Ab, Bt + Ms, Chl + Ep + Ab, Tlc + Dol, Qz + Dol, serpentine, etc.), epidote-amphibolite

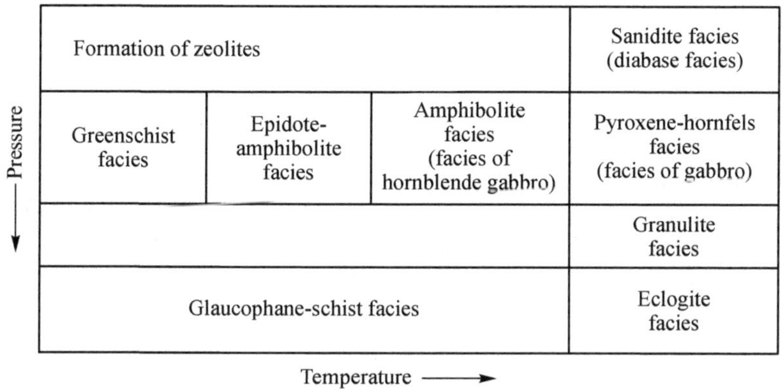

Fig. 1.4 Metamorphic facies and their temperature-pressure relations after Eskola (1939)

(Ab + Ep + Hbl, Ep + Ms + Chl, etc.), amphibolite (Or + Pl, Hbl + Pl + Bt, Di + Hbl, Crd + Ath + Bt, ± St, ± Grt, etc.), granulite (Or + Pl + Grt + Hyp, Pl + Hyp + Cpx, Ky + Sil + Grt + Pl, etc.), glaucophane-schist (glaucophane, jadeite, lawsonite, pumpellyite, garnet, etc.), and eclogite facies (Omp + Prp, ± Ky ± Opx, ± Dia, etc.). The mineral assemblage of the sanidinite facies is stable in xenoliths in volcanic rocks and has crystallized at high temperatures and very low pressures of metamorphism. The hornfels facies is typically developed in the inner zones of contact aureoles under low pressure and high temperature conditions. Rocks of the amphibolite facies typically form in the presence of an aqueous fluid; if they are the result of regional metamorphism they correspond to a higher range of pressures compared to that of hornfelses; rocks of the amphibolite facies occupying the outer zones of contact aureoles usually correspond to a lower range of temperatures. Greenschists are products of regional metamorphism in the upper levels of the Earth's crust at low T and P. The epidote-amphibolite facies covers the transitional temperature range between amphibolite and greenschist facies at the same pressure conditions. The granulite facies forms during regional metamorphism under water-deficient and high-T and high–P conditions. Eclogites are assumed to be products of very deep-seated metamorphism at extreme pressure and temperature which were higher than in the granulite facies. The glaucophane-schist facies correlates with the eclogites facies at least in terms of pressure, but differs by a lower range of temperatures. The main achievement of Eskola is that he rejected the idea of a direct correlation between P and T and proposed a classification of metamorphic rocks in terms of not only temperature but also pressure, thus creating the first scheme of metamorphic facies. Eskola also considered the mechanism of metamorphic reactions and paid special attention to the role of diffusion.

The development of the concept of metamorphic facies has continued in parallel with investigations on progressive metamorphism, which were initiated by Gorge Barrow in the Scottish Highlands (see above). He defined six metamorphic zones based on regular mineralogical changes in the crystalline schists of the Scottish Dalradian: "recrystallization of mica", biotite, garnet, staurolite, kyanite, and sillimanite (Barrow 1893, 1912). The sequence of metamorphic zones mapped by Barrow was correlated with increasing metamorphic temperature and pressure. He assumed that the heat source responsible for the formation of these zones in the Scottish Highlands was provided by deep granitic intrusions. However, this and other ideas were not immediately recognized by the scientific community and it took considerable time for them to be accepted (Oldroyd 1990).

Barrow's method for mapping metamorphic zones was eventually resuscitated by *Cecil Edgar Tilley* (1894–1973) who showed that Dalradian metapelites have almost identical chemical composition and differences in their mineralogy can be attributed to the spatial variation in external physical conditions, i.e., temperature and pressure (Tilley 1925; Elles and Tilley 1930). To generally express the physical implications of metamorphic zoning, Tilley (1924, p. 168) introduced the term "the grade of metamorphism": "Rocks which belong to the same facies can be said to be in the same metamorphic grade, and can be referred to by the terms which I now suggest as isofacial or isogradic. Isograde rocks are those which have originated under closely similar physical conditions of temperature and pressure". He also used the term isograd for a line drawn on the map to mark the outer limit of development of an index mineral (e.g., biotite or garnet) in a rock of some definite composition. The concept of isograd and mapping of metamorphic zones using index minerals became very attractive to many geologists. A number of works which have been published in the 1920s and 1930s were devoted to large-scale mapping of metamorphic zonation within relatively small areas (Vogt 1927; Barth 1936; Billings 1937; Turner 1938; Chapman 1939 and others). Most researchers considered the development of a progressive sequence of metamorphic zone to be associated with the intrusion of granitic magma or depth of the burial or with a combination of these two factors in orogenic areas.

In relation to the facies problem, the question of disequilibrium mineral assemblages in low-grade metamorphic rocks remained a matter of hot debate (Becke 1921; Tilley 1924; Eskola 1939). When these assemblages are observed they are interpreted as disequilibrium by reaction relationships between different generations of minerals assemblages and as reflecting partial phase transitions. Most researchers emphasized that metamorphic reactions are very slow, especially at low temperatures, as opposed to the rapid achievement of equilibrium at higher temperature. Thus, the important conclusion was reached that the sequence in which different minerals appear in time is determined by their textural relationships and depicts the so-called "general path" of metamorphism, i.e., the metamorphic evolution.

In discussing necessary conditions for attaining equilibrium during metamorphism, *Alfred Harker* (1859–1939) proposed the idea that the main factor influencing mineral assemblages in deformed rocks should be directed pressure (shearing stress), P_s. Based on the textural and mineralogical differences between contact and regional metamorphic rocks, he developed the concept of "stress" versus "anti-stress" minerals (Harker 1932). According to Harker, stress minerals were formed under a non-hydrostatic stress regime, whereas only anti-stress minerals were present under a hydrostatic stress regime; the fields of stability on a pressure-temperature diagram are extended by introduction of shearing stress while those of anti-stress minerals are reduced under like conditions. Harker defined chlorite, muscovite, and kyanite, typical products of regional metamorphism, as stress minerals, while cordierite and andalusite, commonly found in contact hornfelses, as anti-stress minerals. The problem was very acute since the origin of some specific types of regional metamorphic zonation could be explained through the action of shearing stress. Nevertheless, Harker's hypothesis has been criticized by Eskola (1939) and other geologists, and the validity of this idea was doubted by later experimental works on the *P-T* stability fields of minerals. At the same time, the discussion on the topic made a great contribution in drawing attention to the fact that stress can affect solubility of a mineral and can also accelerate metamorphic reactions during deformation.

Dmitrii Sergeevich Korzhinskii (1899–1985) provided thermodynamic calculations of a few carbonatization-decarbonatization reactions like those calculated by Goldschmidt (see above) for wollastonite (Fig. 1.5). By assuming that the CO_2 pressure (P_{CO_2}) is equal to the lithostatic pressure, he identified six "depth facies": larnite-merwinite, gehlenite-monticellite, periclase, wollastonite, grossular, and grossular-free (Korzhinskii 1935, 1937, 1940). The calculated equilibria curves sloping toward the coordinate axes define the fields of stability of certain minerals on a *P-T* diagram with increasing pressure (depth). The first three facies were attributed by Korzhinskii to contact metamorphism (at $P_{CO_2} < 1500$ atm) and the latter three to regional metamorphism. He also showed that the equilibrium curves of hydration-dehydration reactions on the *P-T* diagram have a gentler slope than those of the above equilibria in the carbonate system, and therefore, they could be useful for the determination of metamorphic temperatures. The intersection of these lines gives the stability fields of different mineral assemblages in *P* and *T* space and allows in the longer term for the facies scheme to be put on a quantitative basis. Korzhinskii emphasized an important role of variations in the chemistry of minerals in metamorphic rocks, mainly the distribution of Fe and Mg between coexisting phases, and noted that such compositional variations were in accordance with the requirements of the phase rule (Korzhinskii 1936a). In addition, he coined the terms "inert" and "perfectly mobile" (in terms of thermodynamics) components, assuming that only inert components are capable of forming mineral phases while the latter ones may be present in the composition of mineral phases but do not increase the number of coexisting phases (Korzhinskii 1936b). Moreover, he proposed a modification to the phase rule (with respect to the degrees of freedom) for thermodynamically open systems with perfectly mobile components: the maximum number of minerals coexisting in equilibrium in a 2 degree of freedom system is equal to the number of inert components and does not depend on the number of mobile ones. In this modification, the mineralogical phase rule can be applied to infiltration metasomatism.

Fig. 1.5 Korzhinski D. S.

Based on the principles of "local thermodynamic equilibrium" and "differential mobility of components", Korzhinskii formulated a theory of metasomatic zoning. He admitted the widespread occurrence of metamorphism in rocks containing mobile alkalis and a number of other elements (in addition to H_2O and CO_2). Korzhinskii's concept of local and mosaic thermodynamic equilibrium played an essential role in improving our understanding of metamorphism (Korzhinskii 1957, 1973, 1982). He showed that although disequilibrium may exist over the entire metamorphic sequence, the rock may reach a state of equilibrium within each small domain. If changes in temperature, pressure and diffusion of components and in other factors controlling the state of a system occur at a rate slower than that of equilibration of mineral assemblages, then local equilibrium should be attained within each small domain.

Norman Levi Bowen (1887–1956) supported Korzhinskii's conclusions that the intersection of lines representing mineral equilibria gives the stability fields of mineral assemblages in the *P-T* space. Bowen proposed that a petrogenetic grid defined in such a way may furnish basis for facies classification. Based on geological considerations, Bowen deduced a sequence of thirteen mineral equilibria in the system $CaO–MgO–SiO_2–CO_2$, corresponding to thirteen steps in the progressive metamorphism of carbonate rocks. Most of them correspond to equilibria earlier derived by Korzhinskii in determining his depth facies. Each Bowen's step is characterized by the disappearance of its appropriate mineral assemblage or phase, which is stable only below the temperature of the corresponding equilibrium. The disappearance of assemblages at successively higher temperatures occurs in the following order:: Dol + Qz (below step 1), Dol + Tr (below step 2), Cal + Tr + Qz (below step 3), Cal + Tr (below step 4), dolomite (below step 5), Cal + Qz (below step 6), Cal + Fo + Di (below step 7), Cal + Di (below step 8), Cal + Fo (below step 9), Cal + Wo (below step 10), Cal + Ak (below step 11), Spu + Wo (below step 12), Spu + Ak (below step 13). The minerals formed at the various steps with increasing temperature are as follows: tremolite, as an anhydrous phase (step 1), forsterite (step 2), diopside (step 3), Di + Fo (step 4), periclase (step 5), wollastonite (step 6), monticellite (step 7), åkermanite (step 8), Mtc + Per (step 9), spurrite (step 10), merwinite (step 11), larnite (step 12) (Bowen 1940).

Fransis John Turner (1904–1985) elaborated Eskola's concept of metamorphic facies. In two books, the first one published in 1948 (Turner 1948) and the second one written with John Verhoogen (1912–1993) (Turner and Verhoogen 1951), he modified Eskola's scheme of metamorphic facies with a more detailed division into subfacies for some rock compositions (Fig. 1.6). Like Eskola, the facies/subfacies were distinguished by these writers on the basis of mineral assemblages and geological data. At the same time, they emphasized that the identification of critical assemblages that are stable within a relatively narrow range of physical conditions is of critical importance. Following on the work of Korzhinskii (1937), these writers proposed a division of the sanidinite facies for siliceous carbonate rocks into larnite-merwinite-surrite subfacies (at very high temperatures and very low pressures) and monticellite-melilite subfacies (at lower temperatures). Steps 7 and 10 of Bowen were used as subfacies boundaries (Bowen 1940). Another important distinction was made between regional and contact metamorphic rocks. In the amphibolite facies, the basic assemblage in which is hornblende and plagioclase,

Fig. 1.6 Turner F. J.

Turner and Verhoogen originally distinguished the cordierite-anthophyllite, staurolite-kyanite, sillimanite-almandine, and almandine-diopside-hornblende subfacies (based on critical assemblages). The first subfacies was considered as representing a range of low pressures (medium-temperature contact metamorphism), while the remaining were treated as representing a range of elevated pressures (regional metamorphism). The minerals of the staurolite-kyanite subfacies were thought to be formed under the possible influence of shearing stress. The actinolite-epidote-hornfels (low-temperature contact metamorphism) and chloritoid-almandine subfacies were identified in the albite-epidote-amphibolite facies. Turner and Verhoogen proposed to retain the pyroxene-hornfels, granulite, and eclogites facies defined by Eskola and abolish the glaucophane-schist facies based on the assumption that the corresponding rocks were formed from greenschists and albite-epidote amphibolites with the participation of alkaline metasomatism. The greenschist facies was divided into two subfacies: biotite-muscovite and muscovite-chlorite (Turner and Verhoogen 1951); the latter was regarded as representing the lowest-temperature rocks of regional metamorphism. Using mineralogical and petrographic data available in the literature on metamorphic complexes of the world, Turner and Verhoogen were able to characterize facies/subfacies by a definite set of mineral assemblages for different chemical compositions of rocks. Considering changes in the chemical composition of a rock during metamorphism, they draw a distinction between metasomatism and metamorphic differentiation. The former case is referred to the process that involved the chemical alteration in a volume of metamorphic rock due to the introduction and removal of elements by the solutions. The latter case is referred to the selective dissolution of minerals (under the influence of different values of pressure in a pore solution and a solid, different interfacial energy between grains, etc.), differential diffusion of components (over short distances along chemical potential gradients) and the formation ("segregation") of new phases. Types of metasomatism are distinguished according to the name of the added substance, e.g., "alkaline", "calcium", "iron-magnesian-silicate", etc. Metamorphic differentiation was discussed using examples of quartz-kyanite and quartz vein formation, porphyroblast growth, band development, etc., and deformation of rocks was considered as an important contributing factor. Turner and Verhoogen emphasized that the study of structures and textures of metamorphic rock as well as various relics and heterogeneities provides essential information on the type of protoliths and subsequent stages of material transformation (evolution), and helps shed light on the character of deformations accompanying metamorphism.

A few years later, what were previously regarded as subfacies of contact metamorphism (except for high-temperature subfacies) became redefined by *William Sefton Fyfe* (born in 1927) and Turner and Verhoogen (Fyfe et al. 1958; Turner and Verhoogen 1951) as facies and termed the hornblende-hornfels and albite-epidote-hornfels facies (Fig. 1.7). For the first one, the critical assemblages are Qz + Ms + Bt + Crd (+ Pl), Qz + Ms + And + Crd (+ Pl), Di + Pl + Grs+ Qz, Di + Grs + Cal + Qz, Pl + Hbl + Di (+ Qz + Bt), Tlc + Tr (+ Qz), Ath + Crd (+ Qz), Ath + Crd + Spl + Pl, Cal + Di + Fo, Cal + Dol + Fo, Fo + Tlc + Tr + Clc,

Fig. 1.7 Fife W. S.

Ath + Clc, Fo + Tr + Di, Fo + Brc + Spl (+ Di), etc. For the second one, the critical stable assemblages are Qz + Ab + Ms + Bt (+ Crd or And), Qz + Ab + Ep + micas, Ab + Ep + Act + Chl + Qz, Cal + Qz + Tlc, Cal + Tr + Qz, Cal + Spl + Dsp, etc. In metapelites, the boundary between the pyroxene and hornblende hornfels facies is determined by the equilibrium Ms + Qz = Als (And or Sil) + Or + H_2O. In the pyroxene-hornfels facies, the following assemblages were included: Qz + Or + Pl + Als(Al_2SiO_5) + Crd (+ Bt), Qz + Or + Pl + Crd + Hyp, Di + Grs + Wo (+ Ves), Di + Grs + Pl, Pl + Cpx + Hyp + Qz, Cal + Spl + Di + Grs, Cal + Di + Grs + Wo, Cal + Grs + An, Cal + An + Spl + Crn, Cal + Spl + Di + Ol, Cal + Di + Per, Cal + Fo + Per, Hyp + Crd + Spl + An, Spl + An + Hyp + Ol, Spl + An + Cpx + Ol, An + Hyp + Cpx +Ol, Spl + Sil + Crd, Crn + Spl + Crd + An, Spl + Di + An + Grs, etc. Fyfe et al. (1958) and Turner and Verhoogen (1951) added nothing to the earlier division of the sanidinite facies into two subfacies: larnite-merwinite-spurrite and monticellite-melilite. The zeolite facies was established to represent low-temperature regional metamorphic rocks at the boundary with diagenetic processes (at $T \sim 200–300$ °C, $P \sim 2–3$ kbar). The mineral phases in this facies are not fully equilibrated and new phases are represented by zeolite, pumpellyite, prehnite, adular, seladonite, etc. The greenschist facies was divided into three subfacies; the quartz-albite-epidote-almandine subfacies, previously referred to as the chloritoid-almandine subfacies of the albite-epidote-amphibolite facies was added to two earlier subfacies (previously known under other names), quartz-albite-muscovite-chlorite and quartz-albite-epidote-biotite. In the quartz-albite-epidote-almandine subfacies, the stable mineral assemblages are Bt + Ms + Alm + Qz + Ab + ep, Ms + Cld + Alm + Qz + Ab + Ep + Chl, Ms + Cld +

Ky + Qz + Chl, Hbl + Ab +Ep + Alm, Hbl + Ab + Ep + Bt + Qz, Hbl + Chl + Alm, Tlc + Tr + Chl, Cal + Ep + Tr + Qz, etc. In the quartz-albite-muscovite-chlorite subfacies, the stable assemblages are multimineralic and may include, for example, Qz + Ms + Chl + Ab (+ Ep, sometimes +Cld), Cal + Ep +Tr + Qz + Chl, Cal + Dol + Chl + Tr, Ab + Ep + Chl + Act + Spn (+ Stp + Qz), Tlc + Tr + Chl + Qz, Tlc + Mgs + Dol, etc. In the quartz-albite-epidote-biotite subfacies, the critical assemblages are Ms + Cld + Qz (+ Ab + Ep), Bt + Ms + Chl + Qz (+Ab + Ep), Qz +Ab + Bt (+ Ms + Ep), Act + Ep + Ab + Chl + Spn (+ Qz + Bt), Tlc + Act + Chl (+ Qz), Cal + Ep + Tr (+ Qz), etc. *P-T* conditions of the greenschist facies metamorphism were determined in the range of 300–500 °C and 3–8 kbar. The amphibolite facies was renamed as the almandine-amphibolite with *P-T* of 550–750 °C and 3–8 kbar; it was divided into four subfacies: staurolite-almandine, kyanite-almandine-muscovite, sillimanite-almandine-muscovite, and sillimanite-almandine-orthoclase (Turner and Verhoogen 1960). The first one is characterized by the following stable assemblages: Qz + St + Alm + Ms + Bt + Pl, Qz + Ky + St + Ms + Bt + Pl (+ Ep), Hbl + Pl + Alm + Ep (+ Qz + Bt), Cum (or Ath) + Hbl + Alm, Ged + St, Di + Fo + Cal, Cal + Hbl + Ep, etc. In the second subfacies in metapelites, kyanite and almandine are stable instead of staurolite and quartz. In the third subfacies almandine is associated with sillimanite instead of kyanite, because it is assumed that the amphibolite facies is divided into subfacies by the Sil $\rightleftarrows$ Ky equilibrium. The forth subfacies is characterized by the following critical assemblages: Qz + Sil + Alm + Or + Pl + Bt, An + Cpx + Grt + Qz, Cal + Di + Grt + Qz, An + Di +Hbl + Qz, Hbl + Pl + Alm + Qz, Cum (or Ged) + Hbl + Alm (+ Pl), Cum + Tr, Ol + Hbl + Spl, etc. The glaucophane-schist facies developed at low temperatures (300–400 °C) and "very high pressure" was introduced again. This facies is characterized by the following stable assemblages: Ms + Chl + Qz (+ Gln), Qz + Jd (+ Ms + Gln), Qz + Jd + Lws + Gln, Qz + Ms + Stp + Gln, Qz + Ab + Crt, Lws + Pmp + Gln (+ Spn), Alm + Ms + Gln + Lws, Ep + Pmp + Gln (+ Spn), Lws + Jd + Gln, Ab + Ep + Chl + Ms (+ Gln), Cal + Ep + Chl + Gln, Qz +Sps + Stp + Gln, Qz + Crt + Aeg + Sps + Stp, etc. The granulite facies was divided into the pyroxene-granulite and hornblende-granulite subfacies. In the former subfacies, the critical assemblages include Qz + K-Na-Fsp + Grt (+ Pl + Ky or Sil), Qz + K-Na-Fsp + Hyp (+ Grt + Pl), Pl + Hyp + Cpx (+ Qz + K-Na-Fsp), Pl + Hyp + Cpx + Grt, Di + Pl + Cal + Qz, Di + Scp + Cal + Qz, etc. In the latter subfacies, both hornblende and biotite coexist stably with minerals typical of pyroxene granulites, but cordierite may be present as well; in calcareous rocks andesine coexists stably with clinozoisite; the typical rocks are represented by the assemblages Pl + Hbl + Cpx + Hyp, Pl + Hbl + Cpx + Alm; magnesian granulites contain sapphirine in association with gedrite, hypersthene, cordierite, enstatite, and spinel. The granulite facies comprises rocks that formed during regional metamorphism at maximum temperatures of about 700–800 °C and high pressures, which may decrease to several kbar (Turner and Verhoogen 1951). The critical mineral assemblage in eclogites is Omp + Grt, rutile, ilmenite, kyanite, and enstatite are commonly present, whereas diamond is rare. Eclogites are thought to form at a minimum pressure of 10–13 kbar and a temperature of about 700 °C.

Fyfe et al. (1958) in their attempt to associate a facies with a range of *P-T* space, proposed a tentative scheme of metamorphic facies; however, the precise definitions of metamorphic facies boundaries and their slopes remained uncertain. Ten years later, Turner (1968) revised this previous scheme on the basis of experimental data on mineral equilibria. The *P-T* diagram of Turner shown in Fig. 1.8 demonstrates the approximate position of the metamorphic facies without division into subfacies. Major revisions of the previous classification include: the recognition of the prehnite-pumpellyite-metagraywacke facies with the critical assemblages Qz + Ab + Prh + Pmp + Chl + Spn, Qz + Prh + Cal + Pmp, Ab + Pmp + Chl + Spn + Ep + Qz, Qz + Ab + Ms + Pmp + Ep + Stp + Chl + Act, etc.; the recognition of the glaucophane-lawsonite-schist (instead of glaucophane) and amphibolite (instead of almandine-amphibolite) facies, as well as transitional subfacies (mineral assemblages) between greenschist and eclogites facies, etc.

A vast amount of experimental data that became available after the Second World War was used to determine the position of key mineral equilibria representing the boundaries of metamorphic facies and subfacies in *P-T* space. These investigations were carried out mainly at the Carnegie Institution's Geophysical Laboratory, Washington DC, and Pennsylvania State University. Based on the results of experimental studies on the system $MgO–Al_2O_3–SiO_2–H_2O$ at T = 430–990 °C and P = 140–2100 bar, Yoder (1952, 1955) raised a question on the role of volatiles, particularly, water, in metamorphism. He pointed out that the significant change in water content of the system may greatly affect water-containing mineral assemblages, leading to interchangeability of metamorphic facies. However, this conclusion made with regard to a closed system has been criticized by Fyfe et al. (1958) who showed, on geological grounds, that the partial pressure of water (P_{H_2O})

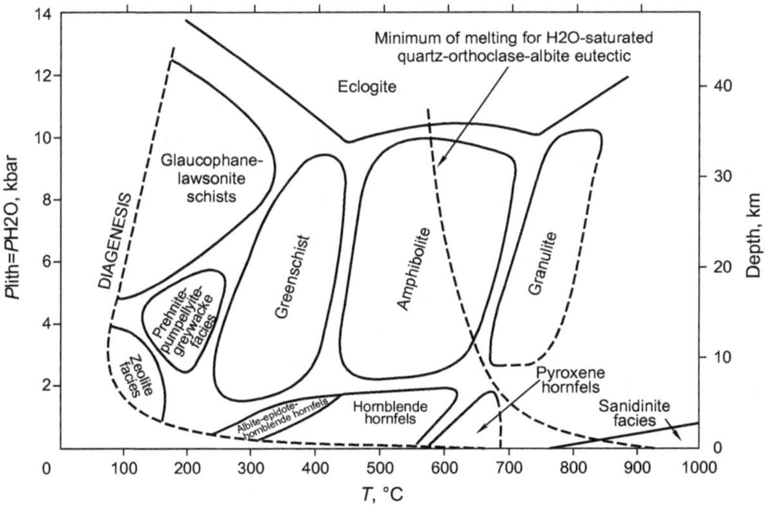

Fig. 1.8 Pressure-temperature fields of metamorphic facies after Turner (1968)

during metamorphism does not depend, in part, on temperature and total fluid pressure ($P_{tot} \approx P_{H_2O} + P_{CO_2}$) and in most cases is associated with the lithostatic pressure (P_{lith}). To a first approximation, it can be assumed that $P_{tot} \approx P_{lith}$. Natural metamorphic rock systems are usually open with regard to water; this is because progressive metamorphism of pelitic rocks is accompanied by notable change in water content. The same conclusion can be drawn with regard to CO_2 in metamorphism of carbonate rocks, when P_{CO2} commonly approaches the difference between P_{tot} and P_{H_2O}. The water phase present in the rocks undergoing metamorphism may be nothing more than a surface film on mineral grains, or the filling of microscopic pores. This surface film is just a few water molecules thick and its thermodynamic properties are different from those of water present in large volumes under the same conditions.

As regards the role of metasomatism, Fyfe, Turner and Verhoogen pointed out that metamorphism is never strictly isochemical, and the rocks must exhibit some (minor) changes in the content of the more mobile components. Mobility of components (after Korzhinskii (1936b)) raises certain complications in applying the phase rule as a method of studying equilibrium in metamorphic assemblages; however, this does not affect the facies concept, though it may necessitate more rigorous definition of some individual facies.

Fyfe et al. (1958) and Turner and Verhoogen (1951) modified the definition of stress in rocks undergoing metamorphism. They defined it as directed pressure (P_s) acting on the chemical potential of the solid and the value of P_s is equal to the arithmetic mean of three principal components of the nonhydrostatic stress which normally causes fracture or deformation is not due to load. Compared to P_{lith} at great depths (>5–10 km), stress may be neglected as a first approximation, although it may be a critical factor at relatively shallow depths; nevertheless, it is assumed that P_s should not exceed 2–3 kbar. Since a rock is generally an aggregate of minerals, the directed pressure (stress) is distributed unevenly and the compressive stress is maximum at grain contacts. Such stressed areas are particularly prone to dissolution, while in other areas the dissolved material may precipitate from a solution that eventually leads to recrystallization of a rock.

According to Fyfe, Turner and Verhoogen, among global-scale causes of metamorphism that may disturb a state of thermodynamic equilibrium in the Earth's crust are thermal effects attributable to igneous intrusions and flows of juvenile fluids, as well as great depths of burial of rocks; deformations were not regarded as the main cause of metamorphism. Fyfe et al. (1958) and Turner (1968) attempted to explain the distribution of metamorphic facies in the crust depending on the differences in geothermal gradient values, fluid pressure (P_{tot}), and lithostatic pressure (P_{lith}). However, this attempt was attended by great difficulties in the interpretation of formation conditions of the glaucophane-schist facies, which were successfully resolved by later workers within the plate tectonics paradigm.

Hans Ramberg (1917–1998) wrote that "rocks formed or recrystallized within a certain *P*, *T*-field, limited by the stability of certain critical minerals of defined composition, belong to the same mineral facies" (Ramberg 1952, p. 136). In his scheme, the following metamorphic facies were identified successively with increasing pressure: greenschist (T < 200 °C), epidote-amphibolite ($T \approx$ 200–500 °C), amphibolite ($T \approx$ 500–700 °C), and granulite (T > 600–700 °C). The eclogites facies ($T \approx$ 300–500 °C) was recognized at pressures exceeding those of the epidote-amphibolite and amphibolite facies. The pyroxene-hornfels facies ($T \approx$ 600–1000 °C) was the only facies representing contact metamorphism. It was assumed that palingenesis and anatexis in metapelites occur at T > 900 °C. In the scarcity of experimental data and based on general thermodynamic considerations, approximate calculations, and Eskola's definitions of metamorphic facies, Ramberg built a series of schematic diagrams illustrating a shift in mineral equilibria and the relative stability of phases of variable composition in metamorphic rocks with changes in pressure and temperature conditions. He explored more than one hundred dehydration and decarbonatization reactions and estimated their role in metamorphism. Ramberg elaborated the idea of the thermodynamic dependence of physical P-T parameters of phase equilibria and variations in the compositions of coexisting solid solutions on variable concentrations of Mg, Fe, Ca, Al, etc. These studies laid the foundation for mineral thermobarometry (Ramberg 1959). Starting with first simple experiments, he roughly calibrated the variation in composition of plagioclase in equilibrium with epidote at various temperatures. However, the thermodynamic parameters of this equilibrium were later found to be incorrect, which eventually resulted in underestimation of temperatures for greenschists and epidote amphibolites in the facies classification. Ramberg believed that the movement of chemical components during metamorphism occurs by diffusion though a fluid film along a grain boundary toward concentration gradients. Using unrealistically high values for the diffusion coefficients at grain boundaries and component concentration gradients, he argued for the large-scale migration of matter (particularly, alkalis) in the lower crust, which should be manifested in metamorphic granitization. However, recent studies could not confirm this assumption. Ramberg believed that the differential stress that drives deformation of rocks undergoing metamorphism may result in segregation of mobile elements, such as Si, Al, K, and Na in areas with less stress (e.g., tensile cracks), leading to metamorphic differentiation.

In the 1950s–1960s, a surge in experimental petrology allowed *Helmut Winkler* (1915–1980) to specify a range of P-T conditions of metamorphic facies in the scheme developed by Fyfe et al. (1958) and Turner and Verhoogen (1951), using his own results (obtained together with his colleagues at the University of Goettingen) and experimental data on the stability of critical mineral assemblages available in the literature (Winkler 1965, 1967) (Fig. 1.9). In his scheme of metamorphic facies plotted in P-T space (Fig. 1.10), Winkler accepted P_{tot} roughly equal to P_{lith}. In the contact metamorphic facies series (up to 2–3 kbar), he recognized the albite-epidote-hornfels, hornblende-hornfels, and K-feldspar-cordierite (formerly designated as pyroxene-hornfels) facies and did not consider the sanidinite facies. The formation of epidote hornfelses, the various classes of which are characterized by the following mineral assemblages: Act + Ep + Ab + Chl ± Bt; Ms + Bt + Ep + Ab ± Chl ± Qz; Prl ± Chl ± Ep ± Ms; Ep + Tr + Chl; Cal ± Ab ±

Fig. 1.9 Winkler H.

Bt ± Ms; Cal + Tr + Ep/Zo + Ab ± Bt ± Tlc etc., occurs at temperatures below 400 °C. The lower stability limit of hornblende hornfelses (with assemblages Hbl + Pl + Ath or Di ± Bt, Bt + Ms + Pl + Crd, And + Bt ± Crd ± Pl ± Ms, Pl + Hbl + Di or Ath ± Bt ± Ms, Cal ± Wo or Pl + Di + Grs ± Bt, Di + Cal + Qz, etc.) is defined at $T = 520–540$ °C and $P = 0.5–2$ kbar. K-Na-feldspar-cordierite hornfelses are characterized by the assemblages Hyp + Di + Pl ± Bt, Or ± Bt + Crd + Pl (free from muscovite), And or Sil ± Crd ± Pl ± Or ± Bt, Pl + Di + Hyp ± Bt ± K-Na-Fsp, Wo or Pl + Di ± Grs ± Bt, etc.; quartz-free, Mg-rich rocks contain periclase and forsterite. The boundary of this facies with the above-mentioned hornblende hornfelses lies at $T = 580–630$ °C. The K-feldspar-cordierite hornfels facies is divided into two subfacies: low-temperature "rhombic amphibole" and high-temperature "rhombic pyroxene".

Following the idea of Turner and Verhoogen (1951), Winkler ascribed the lowest grade metamorphic rocks to the laumontite-prehnite-quartz (formely designated as zeolite) facies. It is characterized by the assemblages Lmt + Prh + Chl + Qz ± Ab, Prh + Cal + Chl + Qz ± Ab, Lmt + Kln + montmorillonite + Chl, etc., while saponite, vermiculite, and seladonite may also be present. The rocks belonging to this facies are stable at the boundary with the diagenesis of sedimentary rocks within a range of temperatures between 200–250 and 350 °C and at pressures up to 5 kbar. The pumpellyite-prehnite-quartz facies is recognized within a narrow temperature range between 350 and 400 °C at the same pressure; it is characterized by the absence of laumontite and the presence of stable assemblages Pmp + Qz + Ab + Chl, Prh + Qz + Chl ± Ab, Pmp + Prh + Qz ± Ab, etc. Epidote, stilpnomelan, and actinolite appear in this facies at the boundary with greenschists.

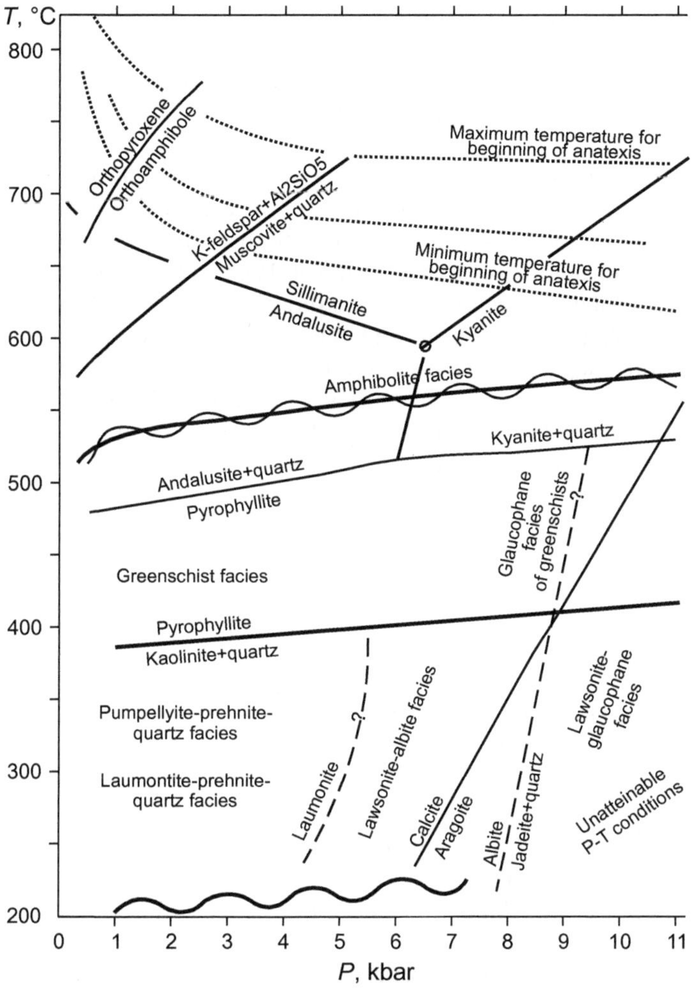

Fig. 1.10 Scheme of metamorphic facies after Winkler (1967) (without contact metamorphism)

At T = 250–400 °C and P greater than 5 kbar, lawsonite is formed at the expense of the anorthite component of plagioclase, heulandite, calcite, and micas or other minerals (Winkler 1965, 1967). This marks the transition to the lawsonite-albite facies, which is characterized by the assemblages Ab + Lws + Chl ± Qz ± Cal ± Ph, Lws + Ab ± Cal ± Arg; crossite, glaucophane, riebeckite, etc. are also present. The transformation of calcite to aragonite at T = 200–400 °C and P = 6–9 kbar marks a definite boundary, which allows a subdivision of the lawsonite-albite facies into two subfacies: with aragonite (without calcite) and with calcite (without aragonite). The appearance of jadeite in association with quartz at the expense of albite occurs at a pressure of about 8 kbar and defines the boundary between the lawsonite-albite and lawsonite-glaucophane facies. Critical mineral assemblages of the lawsonite-glaucophane facies are Gln + Lws ± Jd ± Ab ± Pmp ± Czo or Ep ± Ph ± Spn ± Chl ± Arg ± Grt ± Qz, Crt + Qz ± Ph ± Grt ± Chl ± Arg, Ph + Qz ± Chl ± Gln ± Grt ± montmorillonite, Qz + Jd + Lws ± Gln ± Chl ± Stp, etc. The temperature range for the lawsonite-glaucophane facies is estimated between 230 and 400 °C.

Winkler recognized only two facies of regional metamorphism (greenschist and amphibolite) and divided them into subfacies in accordance with the Barrovian facies series (at elevated pressure of up to 8–9 kbar), which is represented by the kyanite and sillimanite zones in the Grampian Highlands, Scotland, and the facies series of the Abukuma type (at lower pressures of up to 3–3.5 kbar) developed in the Ryoke-Abukuma belt, Japan, and represented by the andalusite-sillimanite zones. The following zonal sequences (based on critical minerals) arranged in order of increasing temperature constitute the Barrovian facies series: Chl → Bt → Grt → St → Ky → Sil; and the Abukuma series: Bt → And → Crd → Sil. Subfacies of the first series are similar to those recognized by Turner and Verhoogen (1951), except for the sillimanite-almandine-muscovite subfacies. The lowest grade quartz-albite-muscovite-chlorite subfacies of the Barrovian-type greenschist facies is characterized by the assemblages Qz + Ms + Chl + Prl ± Ep ± Pg, Prl + Ms + Chl + Cld + Qz, Prl + Czo + Ms + Qz, Cal + Ep/Zo + Chl + Qz + Ms, Cal + Dol + Chl + Qz, Ab + Ep + Chl + Act + Spn + Stp ± Qz; pumpellyite ± epidote, etc. may also be present. In the higher-grade quartz-albite-epidote-biotite subfacies, the assemblages are Ms + Qz + Chl ± Prl ± Pg ± Ep, Ms + Chl + Qz ± Cld ± Ep, Ep + Tr + Chl + Qz ± Ab ± Ms ± Bt, Cal + Tr ± Qz ± Ep, Chl + Act + Ep + Ab + Spn ± Qz ± Bt, Tlc + Act + Chl ± Bt ± Qz, etc. The highest-grade part of the greenschist facies, i.e., quartz-albite-epidote-almandine subfacies, is characterized by the following assemblages: Ms ± Prl (or Ky) ± Cld + Qz + Mg-Chl ± Ep, Ms ± Cld + Alm + Qz ± Chl ± Ep, Ms + Bt + Alm + Qz ± Chl + Ab ± Ep, Ep + Hbl ± Alm ± Bt ± Qz, Cal + Ep + Tr (or Hbl) ± Qz, Hbl + Ep + Ab + Alm + Bt + Qz, Hbl + Alm + Ab ± Mg-Chl ± Tlc, etc. Winkler suggested that the Barrovian-type greenschist facies may also contain microcline (possibly, as rare relics), vesuvianite, margarite, and serpentine.

The Barrovian-type amphibolite facies was designated by Winkler as almandine-amphibolite. Following Turner and Verhoogen (1951), Winkler divided this facies into three subfacies: staurolite-almandine, kyanite-almandine-muscovite, and sillimanite-almandine-orthoclase. The first (relatively low-grade) subfacies is characterized by the mineral assemblages Ky + St + Ms ± Pg + Bt ± Pl + Qz, St + Alm + Ms ± Pg + Bt ± Pl + Qz, Alm + Ms + Bt ± Pl ± Ep + Qz, Pl + Ep + Hbl ± Alm ± Ms ± Qz, Pl + Ep + Hbl + Di ± Ms ± Qz, Cal + Di + Grs ± Qz, Cal +Di + Tr ± Fo, Hbl + Pl + Ep ± Alm ± Bt ± Qz, Hbl + Alm + Cum (or Ath/Ged), etc. At higher temperatures, staurolite commonly disappears in the kyanite-almandine-muscovite subfacies and the following assemblages become stable in metapelites: Ky + Alm + Bt + Qz + Pl ± Ep, Alm + Ms + Bt + Qz + Pl ± Ep, etc. In non-metapelitic rocks, the mineral assemblages present in the kyanite-almandine-muscovite and staurolite-almandine subfacies are identical. In the sillimanite-almandine-orthoclase subfacies, the mineral pair muscovite +

quartz becomes unstable, and mineral assemblages in metapelites are characterized by the appearance of orthoclase; the most typical mineral assemblage is Qz + Sil + Alm + Or ± Pl ± Bt; the other classes of rocks are characterized by the presence of Pl + Hbl + Alm + Qz ± Bt ± Or, Pl + Hbl + Di + Qz ± Bt ± Or, Pl + Grs-Adr + Di + Qz, Cal + Di + Qz ± Grs, Cal + Di + Tr или Fo, Hbl + Pl ± Di ± Qz, Ath/Ged (or Cum) + Hbl + Alm ± Pl, Cum + Tr, Ol + Hbl + Spl, etc.

Critical mineral assemblages of the lowest-grade Abukuma-type quartz-albite-muscovite-biotite-chlorite subfacies are similar to those of the Barrovian-type albite-epidote hornfels and quartz-albite-epidote-biotite subfacies; the biotite-free subfacies of the greenschist facies and assemblages containing stilpnomelane are absent at low pressures. The next (higher-grade) quartz-andalusite- plagioclase-chloriteовая subfacies of the Abukuma-type greenschist facies is characterized by the presence of hornblende in association with oligoclase-andesine, epidote, biotite, and chlorite; the other stable assemblage is andalusite + plagioclase + muscovite + biotite + chlorite.

The Abukuma-type amphibolite facies was designated by Winkler (1965, 1967) as cordierite-amphibolite; it is divided into three subfacies in order of increasing temperature: andalusite-cordierite-muscovite, sillimanite-cordierite-muscovite-almandine, and sillimanite-cordierite-orthoclase-almandine. In the first subfacies, the stable assemblages include Qz + Bt + Pl ± Ms ± Crd ± Sps ± And, Hbl + Pl ± Qz ± Bt ± Di, etc.; on the other hand, staurolite may be present in metapelites of appropriate chemical composition. In the second subfacies, sillimanite will form in metapelites instead of andalusite and almandine may appear in association with cordierite in rocks of appropriate chemical composition. Apart from that, the mineral assemblages of these two subfacies are identical. The third subfacies is devoid of muscovite and contains the following assemblages: Qz + Pl + Or + Bt ± Crd ± Sil ± Alm, Hbl + Pl + Cum + Bt ± Qz, Grs-Adr + Cal + Di ± Wo, etc.

Winkler suggested from the experimental data that the lower boundary of the greenschist facies lies in the temperature range of 370–415 °C, and the transition from greenschist to amphibolite facies takes place at ~550 °C; the temperature at the upper boundary of the amphibolite facies remain uncertain. The granulite and eclogite facies were treated by Winkler as a separate case, being specific within the Earth's crust, and were characterized in accordance with Turner's and Verhoogen's views (1960). In the granulite facies, Winkler established two subfacies: hornblende and pyroxene. In the first subfacies, the diagnostic assemblages are Hyp + Pl + Hbl ± Cpx ± Bt ± Qz ± Ilm, Hbl + Grt + Cpx + Pl ± Bt ± Qz ± Ilm, etc. while in the second one Hyp + Cpx ± Pl ± Qz ± Ilm, Hyp + Grt ± Cpx + Pl ± Qz ± Ilm, etc. Some granulites may contain orthoclase, cordierite, sillimanite or kyanite, zoisite or epidote; carbonate rocks contain calcite, dolomite, and forsterite. Granulites with large contents of K-feldspar are termed "charnokite" and "enderbite"; some of them are of primary magmatic origin. Winkler considered the granulite facies rocks to be formed under H_2O-poor conditions at the temperatures around 800 °C. The eclogites facies (very high P and T), which is characterized by mineral assemblages containing variable amounts of omphacite, garnet, kyanite, zoisite, and hypersthene, was not subdivided by Winkler into subfacies.

Using experimental data, Winkler (1965, 1967) estimated the physical conditions of melting in the pelitic system during metamorphism and showed that the anatexis of rocks may begin in the amphibolite facies. Having emphasized that melting in the crust takes place on a large scale, he denied the granitization hypothesis and claimed that large amounts of granitic melts are formed through anatexis. At the same time, Winkler rejected previous concepts concerning the formation of glaucophane through regional Na–Fe metasomatism. He related unusual conditions of formation of the lawsonite-glaucophane facies ($P > 8$–9 kbar, $T = 230$–400 °C) to the effects of high pressure of the fluid phase liberated during a metamorphic reaction in pore space. As noted above, it is highly unlikely that the fluid pressure is considerably higher than P_{lith} (Fyfe et al. 1958).

Later, Winkler (1974) abolished the concept of metamorphic facies as a generalized characteristic of *P-T* conditions and preferred to employ the concept of isograds. He admitted that results of recent progress in experimental and geological studies of regional and contact metamorphism allow *P-T* conditions to be estimated directly from mineral equilibria and isograds, without using a sequence of facies and subfacies. For practical purposes, he proposed to group isograds into four series: very low-temperature, low-temperature, medium-temperature, and high-temperature. In each series (for $P < 10$–11 kbar), he identified the diagnostic mineral parageneses for different rock compositions (essentially, for metapelites and metabasites). The first series, comprising wairakite- and laumontite-bearing, lawsonite-bearing, glaucophane-lawsonite, and jadeite-quartz mineral assemblages, covers the low temperature range from ~200 to ~350 °C. This series comprises rocks previously assigned by Winkler (1967) to the laumontite-prehnite-quartz, pumpellyite-prehnite-quartz, lawsonite-albite, and lawsonite-albite-glaucophane facies. Fragments of the original rocks are present as relics in this series. The transition between the first and second series is marked by the disappearance of pumpellyite, lawsonite, and then prehnite and the appearance of zoisite or clinozoisite. The first series, like the second one, which covers the temperature range from ~350° to ~500°C, the following minerals typical of the greenschist facies are stable: chlorite, actinolite, white mica, epidote, albite and others. In the first series, glaucophane and aragonite are stable at pressures >5–7 kbar, and jadeite and quartz at $P > 8$–10 kbar. Biotite in association with chlorite, muscovite, and quartzeм appears at the lower boundary of the second series and the assemblage Chl + Zo/Czo ± Act ± Qz is diagnostic in the second series. In metabasites, hornblende appears near the upper boundary of the second series and is, probably, formed by a reaction of actinolite with clinozoisite, chlorite, and quartz; hornblende is commonly found in association with oligoclase. At temperatures between 500 and 550 °C, chlorite with muscovite and quartz, as well as chloritoid disappear from metapelites, while cordierite with biotite, almandine and staurolite appear in various assemblages. In the medium-temperature ($T \approx 500$–600 °C) series, cordierite-bearing assemblages appear in metapelites at $P < 4$ kbar and almandine-bearing assemblages at $P > 4$ kbar; depending on *P-T* conditions rocks may contain andalusite, sillimanite or kyanite. Muscovite + quartz are stable over the entire range of temperatures, but at lower pressures, $P < 2$–3 kbar (during contact metamorphism), they are replaced in metapelites by K-feldspar. Similar to the medium-temperature series, the high-temperature ($T > 600$ °C) series can be subdivided over the entire range of pressures based on the presence (or absence) of cordierite, almandine, and Al_2SiO_5 polymorphs (except for andalusite) in metapelites. Additional evidence for a high temperature range is provided by the absence of primary muscovite, the stability of K-feldspar, the appearance of hypersthene in granulites (in different parageneses with clinopyroxene, feldspars, garnet, biotite, sillimanite, kyanite, and quartz; sometimes in the presence of hornblende, cordierite, sapphirine, spinel, etc.), and anatexis of felsic rocks: the beginning of melting is accepted to occur at $T \approx 680$ °C, $P_{H_2O} \approx 2$ kbar and at $T \approx 620$ °C, $P_{H_2O} \approx 10$ kbar.

Using experimental data on mineral equilibria, isograds in metamorphosed ultramafic and carbonate rocks were grouped by Winkler into series, whose boundaries are not always coincident with series distinguished in metapelites and metabasites. A sequence/combination of isograds in a certain series reflects the evolution of *P-T* conditions of metamorphism. For example, at a low temperature of about 300 °C, the formation of talc from serpentine and quartz takes place in metamorphosed ultramafic rocks (in the system $MgO–SiO_2–H_2O$ at $P < 4$–7 kbar); with increasing temperature forsterite is formed from serpentine + brucite, forsterite + talc are formed from serpentine, anthophyllite is formed from talc + forsterite; at a temperature of about 700 °C, enstatite is formed by a reaction of anthophyllite with forsterite; at 750–780 °C anthophyllite breaks down to enstatite + quartz, etc. (Evans and Trommsdorff 1970; Trommsdorff and Evans 1972). During the metamorphism of siliceous magnesite rocks at 300–500 °C and $P_{H_2O+CO_2} = 2$ kbar (in the system $MgO–SiO_2–H_2O–CO_2$), talc is formed at the expense of quartz, magnesite, and water; a further increase in temperature is accompanied by the formation of forsterite (at 500–550 °C) at the expense of talc and magnesite and the appearance of anthophyllite, which in turn is replaced by enstatite + quartz at 650–760 °C (Johannes 1969).

During the metamorphism of siliceous dolomites, a sequence of 15 mineral parageneses is observed, which include quartz, dolomite (± magnesite), calcite, talc, tremolite, diopside, and forsterite. Some of these parageneses correspond to Korzhinskii's and Bowen's steps (see above). This sequence was investigated experimentally in more detail by Metz and Trommsdorff (1968) within a range of temperatures from ~350 to ~700 °C and pressures from ~1 to ~7 kbar; the position of equilibria depends on *P*, *T* and the amount of CO_2 + H_2O in the fluid. The association of carbonates with talc and tremolite is formed at low temperatures and with forsterite at higher temperatures; at temperatures above 600–800 °C ($X_{CO_2} > 0.1$) dolomite dissociates to form periclase. The characteristic paragenetic association with rare minerals is formed during contact metamorphism of siliceous limestones at low pressures ($P_{tot} < 0.3$–0.5 kbar, $P_{CO_2} > P_{H_2O}$) and high temperatures; the increase of temperature from ~500 to ~1100 °C followed by decarbonatization reactions results in the formation of wollastonite, tilleyite, rankinite, spurrite, larnite, etc. (Zharikov and Shmulovich 1969).

Akiho Miyashiro (1920–2008) introduced several important ideas that furthered understanding of the main causes of metamorphism (Fig. 1.11). His works succeeded in linking metamorphism firmly to geodynamics. Based on Eskola's facies scheme (Eskola 1939), Miyashiro defined three types of metamorphism of low, medium, and high pressures (Miyashiro 1961, 1973). The first type essentially represents contact metamorphism and includes the sanidinite and pyroxene-hornfels facies. It is associated with magmatism in the basement complexes of ancient volcanic arcs and continental margins and is characterized by an elevated geothermal gradient (>25 °C/km). The third type manifested in combination with the eclogites and glaucophane-schist facies is associated with "zones of subsidence" (later called subduction zones) and is characterized by a low geothermal gradient (about or less than 10 °C/km). According to Miyashiro, these types may or may not be spatially associated to form paired metamorphic belts with contrasting tectonic styles (orogeny and subsidence). Classic paired metamorphic belts are found in Japan, e.g., the Ryoke and Sanbagawa, Hida and Sangun, Hidaka and Kamuikotan belts. The second type, transitional in relation to a geothermal gradient (about 20 °C/km) and representing medium-pressure metamorphism, includes four widely distributed facies: greenschist, epidote-amphibolite, amphibolite, and granulite. This type was regarded as being of a dual genetic nature, since the facies of this type may be either representative of paired metamorphic belts or occur independently in folded belts, forming regional metamorphic zones (e.g., in the Barrovian series of Scotland). In addition, Miyashiro (1973) recognized other independent

Fig. 1.11 Miyashiro A.

types of metamorphism: granulite-amphibolite metamorphism in shield regions as ancient cores of continent and ocean-floor metamorphism of ocean-floor basalts.

Next, Miyashiro distinguished five facies series, which supplemented a threefold classification of types of metamorphism based on pressure. For this purpose, he used Eskola's scheme added with two new low-pressure facies proposed by Coombs et al. (1959), Coombs (1961). Miyashiro (1973) illustrated these facies series taking metabasic rocks as an example. He recognized the following facies series with increasing temperature for low-pressure regional metamorphism (series I): zeolite → prehnite-pumpellyite → greenschist → actinolite-plagioclase (not recognized as an independent facies); the same for contact metamorphism: actinolite-plagioclase → amphibolite → pyroxene-hornfels → sanidinite. For regional and contact metamorphism of low but slightly higher (than the previous one) pressures, the following sequence of facies appears with increasing temperature (series II): zeolite → prehnite-pumpellyite → greenschist → amphibolite → granulite. For medium-pressure regional metamorphism (series III): zeolite → prehnite-pumpellyite → greenschist → epidote-amphibolite → amphibolite → granulite. For high-pressure regional metamorphism (series IV): prehnite-pumpellyite → glaucophane schist → greenschist → epidote-amphibolite → amphibolite; for higher-pressure metamorphism (series V): (prehnite-pumpellyite) → glaucophane schist → greenschist; the eclogite facies can be distinguished in some regions.

The concept of Miyashiro that explains the origin of paired metamorphic belts and high-pressure metamorphic rocks as a result of collision of continental and oceanic plates with the involvement of subduction was confirmed in the region at the continent-Pacific boundary (in the circum-Pacific) and became very popular among petrologists, but encountered much difficulties when it was applied to intracontinental folded belts. Therefore, it was assumed that such orogenic belts, often not containing ophiolites and glaucophane schists, may be formed by continent-continent or continent-island-arc collision (Miyashiro 1973; Miyashiro et al. 1982). Miyashiro pointed out that the tectonic sinking (subsidence) of a lithospheric plate was rapid enough to cause a low geothermal gradient during high-pressure metamorphism, as in the case of islands and continents along the Pacific margins. Where sinking was not rapid enough, a medium-pressure complex should be produced (for example, around the Atlantic). According to Myiashiro, exhumation of high-pressure rocks appears to have been associated with thrusting movements. The second type of metamorphism (medium-pressure) not associated with the tectonic sinking was assumed to be related to plutonic magmatism.

In his discussion of the behavior of a fluid during metamorphism, Miyashiro (1973), summing up and giving systematic accounts on this issue, propose four simple models. (1) If the role of CO_2 and other volatiles as well as dissolved components in the fluid is negligibly small, it can be assumed that $P_{lith} \approx P_{tot} \approx P_{H_2O}$. This model is realized, for example, in simple hydrothermal experiments. (2) If pelitic sediments are mixed with carbonate ones, metamorphism should produce a fluid containing H_2O and CO_2; in this case $P_{lith} = P_{tot} > P_{H_2O}$.

(3) If the intergranular spaces and fractures in a rock undergoing metamorphism are mutually connected to each other and even to the surface of the earth, then in such an open system, in the case of a pure aqueous fluid, $P_{lith} > P_{tot} = P_{H_2O}$. This system is both mechanically and thermodynamically unstable. (4) If a fluid in a rock is absent as an independent phase, but can be present as "molecular films" lining the grain boundaries between minerals in a metamorphic rock (Ramberg 1952), then $P_{lith} > P_{H_2O}$.

The rise of temperature during metamorphism should cause progressive dehydration and decarbonatization of minerals and reduce intergranular spaces, thus tending to cut off channels of fluid. As a result, the temperature and pressure of the fluid phase should be equal to those of the adjacent solid phases, and fluid regime should be controlled by models 1 and 2. These conditions are usually realized in progressive regional metamorphism of pelitic and carbonate rocks. In some cases, in high-temperature zone, partial melting of the metapelites could occur. Such a melt phase should absorb a large part of H_2O present in a rock if the system is nearly closed with respect to H_2O. This should result in an abrupt decrease of P_{H_2O}, which may lead to the breakdown of hydrous minerals; this fluid regime was confirmed by observations on the granulite facies rocks. Finally, prolonged metamorphism at relatively high temperatures may cause a virtually complete disappearance of the intergranular fluid phase, though H_2O molecules should still be present in an absorbed state along the intergranular surfaces. These conditions correspond to model 4. The same relation does not hold for open systems where P_{H_2O} should not decrease. If the intergranular fluid liberated by dehydration would migrate through a system of long narrow channels and pores up to the surface of the earth, this situation corresponds to model 3. According to Miyashiro (1973) open-system conditions are realized, for example, in hydrothermal systems of volcanic areas.

Vladimir Stepanovich Sobolev (1908–1982) in his a book "Introduction to the Mineralogy of Silicates" published in 1949, proposed a scheme of high-temperature metamorphic facies (Fig. 1.12). The main difficulty, however, was the lack of relevant experimental data on the *P-T* stability fields of metamorphic minerals, especially, when exploring the role of pressure. For example, to solve the problem concerning the origin of eclogites, he was forced to rely on the notion of stress to link the estimated values of stress during eclogitization derived from geologic data to the values of pressure (CO_2) estimated from carbonate rocks using Korzhinskii's depth-facies. A large role of stress was invoked by Sobolev for the explanation of formation of garnet-bearing amphibolites, hypersthene gneisses, etc. Fifteen years later, following the publication of his book "Introduction to the Mineralogy of Silicates", Sobolev (1964) proposed a new scheme of metamorphic facies, which was based on experimental data on the important phase equilibria that became available at that time. This scheme was fundamentally an expansion of Eskola's (1939) as well as Turner's and Verhoogen's (1960) definitions. In his scheme, calibrated on *T* and *P*, Sobolev did not take into account the stress action, while recognized for the first time the high-pressure rocks as a separate group of facies

Fig. 1.12 Sobolev V. S.

(eclogite, lawsonite-, jadeite-, kyanite-bearing rocks, etc.). The assignment of all kyanite-bearing rocks to high-pressures was based on the erroneous experimental data on the *P-T* position of the Sil ⇆ Ky equilibrium: $P > 10–11$ kbar at $T \approx 500$ °C and $P > 17–18$ kbar at $T \approx 1000$ °C (after Clark et al. 1957; Clark 1961). This scheme (Sobolev 1964) was subsequently revised and became the basis for the compilation of the first map of the metamorphic facies of the USSR (Dobretsov et al. 1966a, b), which was published for a large region such as the USSR (at a scale of 1:7,500,000). The nomenclature of regional metamorphic facies on this map roughly corresponded to that of Eskola's scheme (Eskola 1939); it was also accepted that $P_{tot} \approx P_{lith}$. In order to meet mapping requirement, the facies were distinguished on the basis of the critical mineral assemblages, and boundaries between the individual facies were defined by major reactions in most common rock compositions (types). Four categories of minerals and assemblages were defined in each facies: "diagnostic" minerals and assemblages which are not possible in all other facies, except for this one; "forbidden" minerals and assemblages which are impossible in a particular facies (and, probably, in other facies); "common" minerals and assemblages which are possible in a wide range of compositions at *P-T* conditions of a particular facies and other facies; "exotic" minerals and assemblages which are possible in a particular facies within a narrow range of compositions. It was assumed that each metamorphic facies is characterized to the maximum extent by minerals and mineral assemblages ascribed to the first two categories. In Sobolev's scheme, all regional metamorphic facies were classified into three groups: (A) high-temperature—granulite and amphibolite, (B) medium- and low-temperature—epidote-amphibolite and greenschist, and (C) high-pressure—lawsonite-glaucophane, kyanite schist/gneiss, and eclogite. The boundary between groups A and B was defined by the Sil/And + Kfs + H_2O ⇆ Ms + Qz equilibrium (muscovite + quartz are characteristic only of group B low-temperature facies). The boundary of group C facies is defined by the Sil ⇆ Ky and Ab + Nph ⇆ Jd equilibria, as well as by a transitional *P-T* field, where the basalts were transformed to eclogites due to eclogitization. The "Map of Metamorphic Facies of the USSR" (1966a) clearly illustrated the main regularities of the distribution of regional metamorphism on the Earth's surface and it relationships with geological structures. For example, the occurrence of granulite facies rocks was practically confined to surface outcrops of basement of ancient platforms (shields and massifs). Eclogites, lawsonite- and jadeite-bearing and glaucophane schists in association with ultrabasic rocks were manifested within extended geological structures (in "geosynclinal zones" to use an obsolete term)—metamorphic belts of low-*T*/high-*P* type– along folded platform frames. Areas of disthene schists and gneisses in folded belts at platform boundaries were also attributed to high-pressure metamrophism (Dobretsov et al. 1966a, b). Although this map represented a simplified geological situation, it caught the attention of most geologists and gave the impetus for the further compilation of a series of such maps for the individual regions, countries, and continents under the auspices of the International Geological Society.

In the 1970s, V. Sobolev, together with N. Dobretsov, V. Reverdatto, N. Sobolev, and V. Khlestov published a series of monographs on metamorphic facies (Dobretsov et al. 1972, 1973, 1975; Reverdatto 1973), in which they summarized all data on the world's metamorphic complexes available at that time and discussed the most important theoretical issues. The scheme of metamorphic facies was further developed by these authors (Fig. 1.13). Advances in experimental studies of mineral equilibria made it possible to widely use these data as facies boundaries. The facies was defined as "a *P-T* range of metamorphism bounded by the most important reaction lines", which can be traced "in natural rocks of the most common compositions" (Dobretsov et al. 1972), i.e., in metapelitic, metabasic, and carbonate rocks. The further subdivision of facies into subfacies was based on less significant mineral reactions in rocks of certain compositions. The facies were defined using the earlier established principle of their characterization on the basis of four categories of minerals and assemblages (Dobretsov et al. 1966a, b). However, the meaning and names of categories were modified to some extent: facies were characterized using the "critical" (possible only in a given facies), "typomorphic" (possible in a wide range of rock compositions in a given facies and in rocks of unusual composition in other facies), "common" and "rare" ("exotic") minerals and assemblages. In this scheme (Dobretsov et al. 1972), facies were classified into groups representing different rock pressures: A—facies of contact metamorphism (low pressures), B—facies of common regional metamorphism (medium pressures), C—high-pressure facies, D—ultrahigh-pressure facies (in the mantle). The boundary between groups B and C was defined by the known critical mineral equilibria: Sil ⇆ Ky, Ab + Nph ⇆ Jd, Arg ⇆ Cal, as well as by the stability field of pyrope and a field of complete basalt eclogitization. It is noteworthy that the position of the triple point of Al_2SiO_5 (And–Sil–Ky) was displaced

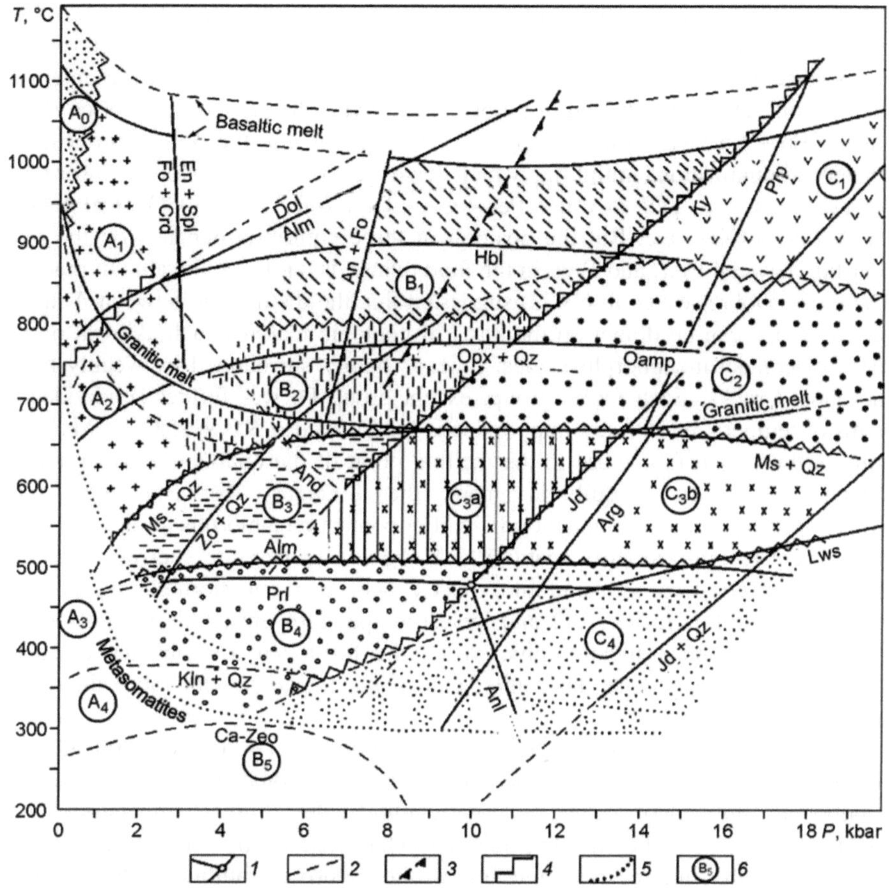

Fig. 1.13 Scheme of metamorphic facies after Dobretsov et al. (1972). 1—lines of mineral equilibria bounding the stability fields of the critical minerals and mineral assemblages (mineral names appear on that side of the line where these minerals are stable) for different values of P_{H_2O}: from 0.3 P_{tot} at high T to 0.9 P_{tot} at low T, 2—the same for inferred equilibria; for basalt melting, the more basic basaltoids are shown as dotted line; for granite melting the solid line is at P_{H_2O} = 0.6 P_{tot} and dotted line at $P_{H_2O} = P_{tot}$, 3—the boundary of the onset of eclogitization for most basaltic rocks, 4—the facies and subfacies boundaries, 5—probable positions of the "kinetic threshold" of metamorphism, 6—letters and other symbols denote fields of individual facies (for further explanations see text)

as compared to that shown in the "Map of Metamorphic Facies of the USSR". On the basis of experimental data from Althaus (1966, 1967), the triple point in the facies scheme was located at $T \approx 600$ °C and $P \approx 6$–7 kbar; the univariant equilibrium Sil = Ky at $T \approx 800$ °C was placed at $P \approx 12$ kbar. According to the most recent estimates, these P values of the triple point of Al_2SiO_5 minerals appear to be overestimated, therefore, not all kyanite-bearing gneisses and schists should be assigned to group C facies. The boundary between A and B facies groups was defined only tentatively, because in the case of metapelitic and metabasic rocks, only univariant assemblages with fixed compositions (primarily, in terms of X_{Fe}) are needed. However, these data were either scarce or lacking at the moment of publication. The only studied equilibrium involving Mg-rich solid phases (Fo + Crd ⇆ En + Spl) was of little help. For carbonate systems, the authors proposed to use decarbonatization equilibria, which have generally steep slopes toward the pressure axis, e.g., Dol ⇆ Cal + Per + CO_2, Di + Cal ⇆ Ak + CO_2, Cal + Fo + Di ⇆ Mtc + CO_2,, etc. But the position of these equilibria depends on the relationship between P_{H_2O} and P_{CO_2} in the fluid, which is hardly achievable in practice. Therefore, Dobretsov et al. (1972) suggested that the distinction between A and B facies groups can be drawn on the basis of geologic features such as local occurrence of contact metamorphism near igneous bodies surrounded by weakly metamorphosed rocks. Each facies group, except group D, was further subdivided in terms of temperature. Of key importance is the Sil/And + Kfs + H_2O ⇆ Ms + Qz equilibrium which falls close to the melting point of the quartz + feldspar + H_2O eutectic; these two equilibria mark the boundary between high- and low-temperature zones of metamorphism. For contact metamorphism, the following facies were recognized (Reverdatto 1973): highest-temperature spurrite-merwinite (formerly referred to as sanidinite) facies—A0, pyroxene-hornfels facies—A1, amphibole-hornfels facies—A2 and lowest-temperature muscovite-hornfels facies —A3. In all contact metamorphic rocks, the forbidden assemblages comprise kyanite, staurolite, jadeite, lawsonite, glaucophane, garnet with more than 20% pyrope, etc.

Facies A0 corresponds to the highest-temperature (~800 to 1200 °C) and lowest pressure (~1 to 300 bar) metamorphism. In carbonate rocks, this facies was subdivided into two subfacies: merwinite-calcite (higher-temperature) and monticellite-spurrite-tilleyite. The facies is characterized by the following diagnostic minerals: larnite, rankinite, merwinite, spurrite, mullite, tridymite, etc.; unstable phases are andalusite, garnets, amphiboles, micas (except phlogopite?), dolomite, Wo + Cal, Qz + Kfs.

Facies A1 corresponds to temperatures of ~700 to 900 °C and pressures from a few hundreds of bars to ~2 kbar; in calcareous rocks two subfacies were recognized: wollastonite-gehlenite-anorthite (higher-temperature) and grossular. In this facies, critical assemblages are Mtc + Mll + Wo + Cal, Mtc + Mll + Grs + Cal, Fo + Crd + Phl, Fe-Crd + Fa + Hc, Grs + Di + Wo + Cal, etc.; forbidden minerals and mineral assemblages are rhombic amphiboles, Ms + Qz, epidote, dolomite, Cal + Qz, etc.; common assemblages are biotite, sillimanite, pyroxenes, hornblende, Or + Crd, Per + Cal, Crn + Or, more rarely Fe-rich garnet, etc.

Facies A2 corresponds to temperatures of ~500 to 550 °C (more often from 600 to ~800 °C) and pressure from a few hundreds of bars to ~2 to 3 kbar. The lower boundary of this facies is defined by the "orthoclase isograds": Ms + Qz ⇆ Or + And/Sil + H_2O and Ms + Bt + Qz ⇆ Or + Crd + H_2O. In the metapelitic rocks, the facies is subdivided into two subfacies: sillimanite (higher-temperature) and andalusite. In the amphibole-hornfels facies, critical minerals and mineral assemblages are Tr + Dol + Di, Tr + Di + Cal, Fo + Dol + Cal, Pl + Hbl + Qz, Di + Fo + Cal + Qz, En + Di + Qz, Per + Cal, etc.; forbidden minerals and mineral assemblages are pyrophyllite, Dol + Qz, Chl + Ms + Qz, epidote, Zo + Qz, staurolite (?), etc.; common minerals and mineral assemblages are hornblende, sillimanite/andalusite, almandine, biotite, Or + And, Or + Crd + Qz, Ged + Crd, An + Wo, diopside, dolomite, Cal + Qz, etc. Facies A3 corresponds to temperatures below ~500 to 600 °C; the minimum temperatures coincide with the kinetic type of metamorphism at low pressures (~350 °C).

Facies A3 corresponds to temperatures below ~500 to 600 °C; the minimum temperatures coincide with the kinetic threshold of metamorphism at low pressures (~350 °C). In facies A3, critical minerals and mineral assemblages are Ms + Qz, Crd + Pl + Bt + Ms, Bt + Crd + Ms + And + Qz, Chl + Bt + Ms + And, Chl + Tr/Act + Tlc (?), serpentine, etc.; forbidden minerals and mineral assemblages are sillimanite, kyanite, anthophyllite, staurolite, chloritoid (?), pyrophyllite, Zo + Qz (?), etc. At a temperature below ~350 °C and low pressures, contact alterations around a magmatic body are assumed to be metasomatic in nature.

Four facies were recognized in medium-pressure regional metamorphic rocks (Dobretsov et al. 1972, 1973): B1—granulite (two-pyroxene gneiss), $T \approx 750$–1000 °C, $P \approx 4$–13 kbar; B2—amphibolite (sillimanite-biotite gneiss), $T \approx 650$–800 °C, P from 2–3 to 8–10 kbar; B3—epidote-amphibolite (andalusite/sillimanite-muscovite schist), $T \approx 500$–650 °C, P from ~2 to 5–7 kbar; B4—greenschist, $T \approx 350$–550 °C, P from ~2 to 7–10 kbar. It was assumed that the pressure of H_2O tend to decrease from B4 (where $P_{H_2O} \approx P_{tot}$) to B1 (where $P_{H_2O} \approx 0.3P_{tot}$), and conversely, P_{CO_2} increases within the corresponding limits. In all medium-pressure facies (as in the case of contact metamorphism), the forbidden minerals include kyanite, jadeite, lawsonite, and garnet with more than 50% pyrope.

Some minerals and mineral assemblages observed in regional metamorphic rocks, for example, in metapelites and metabasites, are stable during contact metamorphism, i.e., they can be used as diagnostic only for temperature conditions and not for pressure (aside from variations in mineral compositions, i.e., shift in mineral equilibria used as boundaries). The diagnostic mineral assemblages in B1 include Hyp + Grt + Crd + Bt + Kfs + Pl, Hyp + Cpx + Bt + Pl + Kfs, Grt + Crd + Sil + Bt + Spl, Hyp with Fe/(Fe + Mg) < 50% + Cpx + Qz, Hyp with Fe/(Fe + Mg) < 20% + Qz, Grt + Cpx + Opx + Hbl + Bt + Pl, Grt + Hbl + Bt + Qz, Cal + Di + Ca-Grt + Wo + Pl, Dol + Ol + Spl + Hbl + Bt, Ol + Di + Cal + Dol, Spr + Opx + Spl + Prg + Phl, etc.; forbidden minerals and mineral assemblages are rhombic amphiboles, cummingtonite, staurolite, muscovite, epidote, andalusite (except for viridine, unless it is a secondary mineral?), biotite with Fe/(Fe + Mg) < 30% + Qz, Kfs + Pl (with less than 20% anorthite) + Qz, etc.; common mineral assemblages are Grt + Sil + Kfs + Crd + Qz, Cpx + Grt + Kfs + Pl + Qz, Ol + Hyp + Qz, Di + Pl + Cal + Qz, etc.; rare assemblages are Crn + Kfs, En + Spr, etc.

Facies B2 is distinguished from facies B1 by equilibrium Opx + Cpx + Kfs + Qz $\leftrightarrows$ Grt + Bt + Amp; this boundary is defined by the disappearance of rhombic amphiboles and biotite-sillimanite assemblage. The most common minerals are various amphiboles, biotite, cordierite, garnet, sillimanite, various plagioclases, K-feldspar, etc. In this facies, the diagnostic assemblages are Bt + Sil + Kfs + Qz, Grt + Spl + St + Sil, etc.; common assemblages are Hbl + Pl ± Cpx ± Bt ± Qz, Hbl + Cpx + Opx, Grt + Hbl + Pl + Qz, Bt + Sil ± Grt ± Kfs ± Crd, Grt + Crd + Bt + Pl, Cal + Qz + Amp + Cpx + Pl ± Kfs ± Bt, Hbl + Cal + Bt + Ep, Sil + Crd + Ged, etc.; forbidden assemblages and minerals are chloritoid, Fe-rich epidote, St + Qz, Ep + Pl with less than 50% anorthite, Dol + Qz, etc.

Facies B3 is distinguished from facies B2 by the stability of muscovite + quartz and staurolite + quartz. In facies B3, the stable minerals include various amphiboles, Fe-rich garnets, diopside, hornblende, anthophyllite, epidote, staurolite, sillimanite/andalusite, chloritoid, chlorite, plagioclases, Bt + Ms + Qz, etc. The diagnostic minerals and mineral assemblages are St + Sil/And + Qz, Grt + Bt + Sil + Ms + Qz, Hbl + Bt + Ep + Qz + Pl containing 10–30% anorthite, etc.; common assemblages are Grt + Crd + Bt + Chl + Ms + Qz, Grt + St + Sil + Bt + Ms + Qz, Hbl + Pl + Alm + Bt + Qz, Amp + Cal + Qz ± Bt, Cal + Tr/Act + Ep ± sodic Pl ± Bt ± Ms ± Chl, Dol + Fo ± Di, Cpx + Amp + Ep + Pl + Cal + Qz, Grt + Cum + Hbl + Bt + Pl + Qz + Mgt, etc.; forbidden minerals and mineral assemblages are Fe-rich chlorites, wollastonite, etc. Facies B4 is characterized by the disappearance of staurolite and the expansion of the hornblende field at the expense of actinolite, epidote, and albite. At the upper boundary of the greenschist facies, the assemblage muscovite + chlorite (and quartz) is replaced by the assemblage biotite + cordierite + andalusite or biotite + garnet + kyanite. The lower boundary of this facies is defined by the absence of kaolinite, diaspore, zeolites, etc.

In facies B4, the stable minerals include various chlorites, white micas, tremolite, actinolite, epidote, talc, chloritoid, stilpnomelane, Ms + Qz, etc. The diagnostic minerals and mineral assemblages in metapelites are Chl + Cal + Qz, Dol + Qz, Prl + Qz, Ms + Chl + Stp + Ab + Qz, Ms + Chl + Ep + Ab + Qz, Ms + Chl + Act + Ab + Qz, etc.; common are Chl + Cum + Ms + Ab + Qz, Chl + Cum + Tlc + Ab + Qz, Chl + Ab + Stp + Qz ± Ms ± Tlc, Cld + Chl + Mgt + Ab + Pg + Qz ± Ms, Cum + Mgt + Tlc ± Ab ± Pg ± Qz, Chl + Mgt + Stp + Ab + Pg + Qz ± Ms, Ab + Pg + Mgt + Hem + Qz ± Ms, Chl + Tlc + Ab + Pg + Stp + Qz ± Ms, Cum + Tlc + Mgt + Qz, Chl + Tlc + Mgt + Stp + Ms + Qz, Prl + Pg + Mgt + Qz ± Ms, Chl + Cld + Mgt + Prl + Pg + Qz, Amp + Cum + Mgt + Tlc + Qz ± Ms, etc.; forbidden are staurolite, cordierite, sillimanite, garnet, aluminous amphiboles, oligoclase, and basic plagioclases, more rarely Mg-rich biotites, etc. In metabasites, the common greenschist mineral assmbelages are Act + Tlc + Cum + Qz, Act + Tlc + Cal + Qz, Act + Ep + Chl + Cal + Qz, Ep + Prh + Grs-Adr + Chl + Srp, Prh + Ep + Cal + Qz, Ep + Als (And-Ky) + Cal + Qz, etc. Dobretsov et al. (1972) noted that the low-temperature part ($T < 300$–350 °C) of facies B4 is bounded by the so-called "zeolite facies" B5 and the field of regional epigenesis, where mineral transformation in the rocks are commonly incomplete due to kinetic factors, which may prevent coexisting minerals and rocks fragments from attaining equilibrium (because of the kinetic threshold of metamorphism). For this reason, the "zeolite facies" was not classified as regional metamorphic facies.

Rocks in group C facies (high-pressure) undergo metamorphism at T = 300–1000 °C and P > 8–14 kbar. There are four facies in this group (Dobretsov et al. 1972, 1975): C1—eclogite, C2—kyanite gneiss and amphibolite, C3—kyanite schist, C4—jadeite-lawsonite-glaucophane (glaucophane schist). The eclogite facies C1 corresponds to the highest-temperature range: T > 800 °C, P > 8–14 kbar. The diagnostic minerals and mineral assemblages are Grt + Omp + Rt, garnet with Fe/(Fe + Mg) < 50%; forbidden are Hyp + Pl, sillimanite, cordierite, amphiboleы; common are garnet, kyanite, jadeite, Ol + Grt, Ky + Cpx, graphite, etc. Facies C2 corresponds in its temperature range to facies B2 (and the lower part of facies B1). The maximum temperature does not exceed 800–900 °C, while the pressure is above 8–12 kbar. The diagnostic minerals and mineral assemblages in metapelites are Ky + Kfs (± basic Pl), Grt + Hbl (karinthin) + Omp, Grt + Ged + Ky, etc.; the most characteristic is the paragenesis Grt + Ky + Bt + Kfs + Pl + Qz; migmatites are widespread; common assemblages are the same as in facies B2; rare assemblages are Opx + Ky ± Spr, Opx + Crn + Spl, Spr + Qz, etc.; forbidden minerals and mineral assemblages include sillimanite, andalusite, Ms + Qz. Cordierite is rarely present in facies C2 while garnet (pyralspite) is more common than in facies B2; chloritoid and some carbonates such as dolomite, siderite, ankerite and others become more abundant in the kyanite gneisses. The kyanite gneiss facies metabasic rocks contain the following characteristic eclogites-like assemblages: Grt + Hbl + Cpx + Opx + Pl + Qz ± Scp. The boundary between the kyanite schist and kyanite gneiss facies is defined by the Ms + Qz ⇆ Ky + Kfs + H_2O equilibrium; for $P_{H_2O} \approx 0.6P_{tot}$, this equilibrium is located at $T \approx 650$ °C. The condition $P_{H_2O} < P_{tot}$ was accounted for by migmatization of metapelites and absorption of water by a melt. For lower metamorphic temperatures, it was assumed that $P_{H_2O} \approx P_{tot}$ (for metapelites). It was noted that the An ⇆ Zo + Sil + Qz equilibrium is located close to a triple point for Al_2SiO_5 (And–Sil–Ky) and that the assemblage kyanite + clinozoisite + quartz under P-T conditions approaching to the triple point should persist at $P_{H_2O} \approx 0.9P_{tot}$. As in the case of epidote in marbles, the stability of staurolite + quartz and chloritoid in metapelites is limited by P-T conditions of the kyanite schist facies. It was shown that the Ca content of garnets in metapelitic rocks increases with increasing pressure. The diagnostic mineral assemblages in facies C3 are Gln + Grt, Ky + St + Qz, Ky + Ms + Qz, etc.; forbidden is Ky + Kfs; common are the same as in facies B3. The main rock-forming minerals in the kyanite schist facies (quartz, biotite, muscovite, garnet, plagioclase, kyanite, chloritoid, chlorite, and epidote) occur in almost all possible combinations of any 4–6 (rarely 7) phases. The associations with chloritoid and biotite are very rare. In The assemblage Act + Bi + Ep + Chl + sodic Pl + Qz ± Grt ± Ms is the most typical of the lowest grade kyanite schist facies under conditions transitional to greenschist facies. Garnet-bearing varieties (Grt + Hbl + Pl ± Bt + Qz) are common in metabasites of facies C3.

Facies C3 in ultrabasic rocks are commonly represented by anthophyllite- and tremolite-bearing varieties. In carbonate rocks of the kyanite schist facies, the typical assemblages such as Dol + Qz и Tr + Cal + Qz ± Bt ± Ms ± Pl ± Ep, etc.

are stable at low pressures. The assemblage Cal + Chl + Qz becomes unstable here and the rocks containing this assemblage should be assigned to greenschist facies. The jadeite-lawsonite-glaucophane (glaucophane schist) facies C4 is stable at T = 300–550 °C, P > 8–10 kbar. The high-temperature boundary of facies C4 (similar to facies B4) was defined by the appearance of hornblende in metabasites and almandine in metapelites along the equilibrium line Prl $\leftrightarrows$ Ky + Qz + H_2O, as well as by the disappearance of lawsonite at P_{tot} > 15 kbar. This facies is characterized by the following diagnostic minerals and mineral assemblages: lawsonite, aragonite, Jd + Qz, glaucophane + zoisite + pumpellyite + epidote, etc.; forbidden minerals and mineral assemblages are the same as in facies B4; common minerals are phengite, Mg-pumpellyite, epidote, glaucophane, chlorites, aegirine-jadeite, etc. The glaucophane-schist facies field can be divided into a series of subfacies, which were named after certain diagnostic assemblages: albite-lawsonite, crossite-actinolite, glaucophane-epidote-chlorite, pumpellyite-lawsonite- glaucophane, epidote-lawsonite-glaucophane, quartz-jadeite-glaucophane, and almandine-lawsonite-glaucophane. The low-temperature boundary of the glaucophane-schist facies is kinetic; it is located at lower temperatures than in the greenschist facies and lies at 250–300 °C. The boundary is defined by certain mineral transformations, e.g., the replacement of Ca-zeolites (laumontite and others) and prehnite by lawsonite and the disappearance of clay minerals such as kaolinite, montmorillonite, etc., as in the greenschist facies.

When considering the origin of facies C4 rocks, the idea that regional Na-metasomatism is the sole cause of glaucophane-producing metamorphism was subjected to criticism. Geochemical studies, including statistical analysis of whole-rock chemical compositions showed that glaucophane schists and eclogites do not differ from basalts of various types. However, some glaucophane-schist facies metabasites show conspicuous enrichment in Na and depletion in K relative to the average basalt composition, which can be explained by their premetamorphic spilitization and the formation of greenstones. Following Miyashiro's arguments (Miyashiro 1961, 1973; Dobretsov et al. 1975) came to the conclusion that glaucophane-producing metamorphism was caused by high pressures at low temperature within narrow tectonic zones (belts) of crustal thickening and subsuidence due to thrusting. Some of these belts developed at the continent—ocean boundary, in particular, along the Pacific margins, whereas others were intracontinental, like the Ural—Tien Shan belt.

At pressures above ~15 to 30 kbar and temperatures above 900–1000 °C, two facies were recognized in facies group D of the upper mantle (Dobretsov et al. 1975): graphite-bearing eclogite facies (in the crust it corresponds to facies C1) and diamond-bearing eclogite facies. The boundary between them is defined by the Gr $\leftrightarrows$ Dia equilibrium, which lies at pressures of 40–50 kbar at $T \approx$ 1200–1400 °C. The graphite-bearing eclogite facies is further divided into three subfacies: garnet peridotite, grospydite, and coesite eclogite. It was assumed that the formation of eclogites by transformation of basalts requires P_{H_2O} much lower than P_{tot}. The pressure range of eclogitization of basalts of variable compositions was accepted to lie between 9 and 15 kbar at ~900 °C and between 11 and 18 kbar at ~1200 °C. It was assumed that eclogites may have directly crystalized from mafic magma at high pressures. The garnet peridotite subfacies is defined by a transition from clino- and orthopyroxene and spinel to garnet and olivine ($P \approx$ 15–21 kbar, $T \approx$ 800–1200 °C); the transition to the grospydite subfacies is marked by the An ⇆ Grs + Ky + Qz equilibrium ($P \approx$ 17–25 kbar, $T \approx$ 800–1200 °C); the quartz-coesite transition ($P \approx$ 32–35 kbar, $T \approx$ 800–1200 °C) defines the boundary to the coesite eclogite subfacies. It was also shown that the characteristic features of the chemical composition of minerals in diamond-bearing eclogite facies are increased contents of Cr in garnet and K in clinopyroxene, etc. The composition of mineral assemblages in mantle-derived graphite-bearing eclogites differs from that of the facies C1 eclogites produced by metamorphism in a crustal sequence.

Dobretsov et al. (1972, 1973, 1975) attempted to further study the causes, conditions, and factors of metamorphism. For example, it was shown that equilibrium within an overall metamorphic sequence cannot be attained due to the persistence of chemical potential gradients because of the presence rocks of different compositions in the metamorphic series. The introduction of the concept of "mosaic" or "local" equilibrium (Korzhinskii 1950, 1957; Thompson 1959) allowed the application of thermodynamic modeling to such systems. However, the local equilibrium volume proved difficult to estimate even at fixed P-T parameters of metamorphism. Therefore, Dobretsov et al. (1972) suggested that the local equilibrium should be regarded as the absence of metastable states, i.e., the absence of minerals and mineral assemblages outside their stability fields. In the later work, Dobretsov et al. (1975) noted that the fluid pressure during metamorphism is approximately equalized with P_{lith}, but in some instances $P_{tot} > P_{lith}$. It was believed that this may have been caused by dehydration/decarbonatization reactions during rapid heating of low-permeability rocks ("autoclave effect"). It is known that directed pressure increases the solubility of solid phases, causes redeposition of matter and reduces the porosity of a rock, which in turn results in the local increase in P_{tot}. The values of the fluid overpressure and stress, as suggested by Dobretsov et al. (1975), could be large enough over a long period of time and, therefore, they, together with thrusting in tectonic zones, played a considerable role in the formation of glaucophane-schist facies rocks. Later studies showed that the tectonic sinking (subsidence) was still critical for this model. Although metamorphic rocks have low permeabilities, fluid volumes cannot be isolated for a geologically reasonable period of time over and P_{tot} must approach P_{lith} due to migration of volatiles. In addition, the increasing pressure would cause the reaction to stop owing to the reverse overstepping of the equilibrium reaction producing fluid to the low-temperature side, i.e., a return to the initial state (and reactants).

Apart from the fluid "overpressure" in combination with thrusting in "zones of deep-seated faults", Dobretsov et al. (1975) proposed alternative three models to explain the formation of metamorphic rocks: "burial metamorphism" of sediments accumulated in troughs with the formation of rocks belonging to facies B4 and B5; zonal "metamorphism associated with deep-seated magmatism and fluid flows in fold belts" (facies B1 + B2 + B3 ± B4 or C2 + C3 ± B4); "Precambrian polyphase metamorphism (polymetamorphism) in shield regions" (facies B1 ± B2 or B2 ± B3). Based on these assumptions, the conclusion was made that high-grade metamorphism can be related only to elevated mantle heat flow. Due to the predominantly conductive nature, the increase in this heat flow can be explained by the rise of hot mantle material; the role of percolating juvenile fluids in heat transfer was regarded to be insignificant. The processes of vertical transport of magmas was considered under near-equilibrium conditions (under conditions of adiabatic cooling) with taking into account their composition, as well as the shape and position of a minimum on melting curves at $P_{H_2O} \leq P_{tot}$ The limit for ascent of melts in equilibrium with the fluid is boiling accompanied by crystallization. Some water-poor ("dry") magmas move upward toward the earth's surface, while others may rise only toward solidus conditions (for the appropriate melt composition and P_{H_2O}) during ascent. Therefore, anatectic granitic melts could not move to quite shallow depths.

The concept of metamorphic formations was introduced as a combination of features that characterize the compositions of the protoliths (original rock), metamorphic conditions (on the facies basis) and tectonics. They were interpreted as "a characteristic paragenesis of metamorphic rocks of a certain composition, which underwent a certain type of metamorphism" (Dobretsov et al. 1975). The metamorphic formations were recognized within shields and median massifs, folded areas of different ages and tectonic zones of deep-seated faults. These studies provide a close link between the metamorphic formations and ore deposits.

Aleksei Aleksandrovich Marakushev (1925–2014) came to the conclusion that the equality of total and lithostatic pressure in the rocks does not reflect the real conditions of metamorphism (Marakushev 1965, 1973) and that the expression $P_{lith} > P_{H_2O(+CO_2)}$ would be more correct (Fig. 1.14). To obtain quantitative estimates of metamorphic conditions, he used thermodynamic calculations of phase equilibria at constant P_{H_2O} and P_{CO_2} in the fluid and recognized metamorphic facies using the following diagrams: $P_{H_2O} - T$, μH_2O (chemical potential of H_2O) $- T$, μCO_2 (chemical potential of CO_2) $- T$, $P - \mu H_2O$, etc. He believed that the use of $P - \mu H_2O$ diagrams, under a set of assumptions, instead of $P_{tot} - T$, would be one of the most powerful tools for analysis. Some cases that might resemble the $P_{tot} - T$ diagrams and roughly correspond to the real paragenetic relations could be chosen from a great variety of diagrams of multi-component systems (with a large number of minerals and negative degrees of freedom) such as Schreinemakers' diagrams (see Korzhinskii 1973). The resemblance arises from the fact that the chemical potential of H_2O in metamorphic rocks (in an open system) is related to dehydration reactions with increasing temperature. By means of the $P_{lith} - T$ diagram for constant $P_{H_2O} = 1$ kbar and $P_{CO_2} = 0.5$ kbar, Marakushev (1973) established the thermodynamic stability areas of mineral associations, which were "chosen as critical ones" and termed "mineral facies": hypersthene-cordierite hornfelses (at $P_{lith} \approx 0$–4 kbar, $T \approx 600$–900 °C), biotite-sillimanite/andalusite gneisses ($P_{lith} \approx 0$–11 kbar, $T > 500$–800 °C), hypersthene-cordierite gneisses ($P_{lith} \approx 3$–11 kbar, $T > 800$ °C), two-pyroxene granulites ($P_{lith} \approx 5$–10 kbar, $T > 800$ °C),

Fig. 1.14 Marakushev A. A.

eclogites (P_{lith} > 5–8 kbar, $T \approx$ 400–900 °C), quartz-muscovite rocks ($P_{lith} \approx$ 0–10 kbar, T > 600 °C), staurolite schists ($P_{lith} \approx$ 0–5 kbar, $T \approx$ 300–600 °C), andalusite/kyanite schists ($P_{lith} \approx$ 0–5 kbar, T > 300 °C), periclase-calcite marbles ($P_{lith} \approx$ 0–3 kbar, T > 700–800 °C), quartz-calcite marbles ($P_{lith} \approx$ 0–8 kbar, T > 500–600 °C), etc. The use of thermodynamic data for end-members of multicomponent solid solutions allowed him to predict stability limits for relevant mineral facies with respect to temperature and pressure (P_{lith}). For some cases, by taking into account the chemical variability of mineral compositions (e.g., their Mg/Fe ratios), Marakushev considered the most important fields of the $P_{lith} - T$ and $P - \mu H_2O$ diagrams in more detail. In particular, as an example, he used a series of Far East metamorphic complexes in garnet-cordierite-bearing rocks, which are stable over a wide range of thermodynamic conditions due to a considerable variation in the chemical composition of coexisting minerals, to establish five depth subfacies (zones): Sutam, Aldan, Khankai, Namdechen, and Primorye. In the first subfacies, coexisting garnet and cordierite have the lowest X_{Fe} values, which is correlated with the greatest depths (P_{lith} > 8–9 kbar). The highest X_{Fe} values for garnet and cordierite in the last subfacies reflect shallow depths and the P-T conditions of contact metamorphism. The identified subfacies overlap the fields of the above-mentioned mineral facies, which were recognized based on other critical assemblages. This feature proved to be very useful for a paragenetic analysis and seems to be a great advantage. The $P_{lith} - T$ diagrams serve to illustrate a specific type of a natural system characterized by a different degree of openness and different relationship between P_{lith} and P_{tot}. Theoretically, these diagrams can be used to tentatively estimate, in the absence or scarcity of experimental data, the effect of changes in the pressure of volatiles on the stability relations of mineral assemblages at

$P_{lith} > P_{H_2O}$, P_{CO_2}. However, the practical use of such diagrams proved to be complicated. The temperature corresponding to the "normal" (average) geothermal gradient was taken as a lower temperature limit for metamorphism in the mineral facies scheme of Marakushev (1973); therefore, high-pressure rocks, which were interpreted by Miyashiro to be formed at a "low" geothermal gradient, were not incorporated in this scheme. Like in the early work of Turner and Verhoogen (1951), Marakushev related the formation of glaucophane schist to low-temperature Na-metasomatism of quartz-chlorite facies rocks. Intense regional alkali (Na-K) metasomatism was implied during metamorphism of pelitic and basic rocks, which was manifested by granitization, spilitization, eclogitization, etc. Marakushev believed that allochemical metamorphism may accompany the evolution of geosynclines, fold belts and a transition from essentially Na- ("pre-granitic") through Na-K to K geochemical regime in ophiolites and associated rocks, which eventually resulted in the formation of migmatites, charnokites, orthoclase gneisses etc. at the orogenic stage. Building on Korzhinskii's ideas, Marakushev (1965, 1973) believed that widespread metamorphism can be assumed in the rocks where alkalis and some other components are perfectly mobile (in addition to H_2O and CO_2). This was in conflict with the idea of metamorphism as essentially an isochemical process (see above).

Leonid Lvovich Perchuk (1933–2009) formulated a general principle of redistribution of chemical elements between coexisting mineral as a function of temperature and pressure (Fig. 1.15). This principle offered an enormous opportunity to

Fig. 1.15 Perchuk L. L.

compile a large-scale inventory of mineralogical indicators of temperature, pressure and fugacity of fluid components during the geochemical processes (Perchuk 1970, 1973). He developed methods for experimental investigation of exchange equilibria of rock-forming minerals of variable composition at high pressures and temperatures. With these advances, geothermobarometry has become a common practice for estimating thermodynamic regimes and major trends in the evolution of metamorphism and magmatism. L. L. Perchuk was the first to introduce the concept of *P-T-t* metamorphic paths and enshrine it in petrology. He developed a petrogenetic model elucidating transformation of the continental crust into oceanic one. In 1987, he conceived the idea of gravitational redistribution of rocks in the Earth's crust that was subjected to high-temperature and ultrahigh-temperature metamorphism. This model was used to explain the widespread occurrence of granulite complexes at the Earth's surface given a relatively constant crustal thickness. L. L. Perchuk conducted extensive studies on crystalline massifs of igneous and metamorphic rocks in many regions of the USSR and in Europe, USA, Japan, Australia, Africa, etc.

This sums up our overview of the concepts of metamorphic facies from the 19th to the late 20th century. The further development of the concept of metamorphic facies has mainly been concentrated on continued refinement and detailed definition of *P-T* stability fields of mineral assemblages in rocks of different composition using thermodynamic data for solid solutions растворов (for example, Berman 1988; Holland and Powel 1990, 1998 and others). Modified petrogenetic grids and various geothermobarometers were applied to estimate metamorphic *P-T* conditions and to constrain the *P-T* evolution of metamorphism.

Generally speaking, advances in the studies of metamorphism raised widespread interest in geodynamics as the driving force for reorganization of the thermomechanical structure of the Earth's crust. Further advances in the theory of plate tectonics provided explanation for crustal extension and/or subsidence, as well as crustal thinning or thickening (during subduction and thrusting). These and other ideas came to dominate the study of metamorphism. The tectonic aspect of metamorphism became incorporated in geodynamic models and the causes of metamorphism are now linked to the dynamic processes and thermal state of the lithosphere. In recent years, using advancements in continuum mechanics (theory of elasticity, rheology, and hydrodynamics), thermal conductivity, and numerical modeling, geodynamics has made enormous progress in a number of areas critical for understanding of the causes of metamorphism. With the main focus on thermal and rheological properties of the lithosphere, many excellent studies were performed to describe the formation of depressions in the Earth's crust in the areas with different tectonic regimes, during large-scale crustal movements accompanied by a pronounced perturbation of the temperature and mass distribution, as well as a thermotectonic regime driving subduction and orogenesis, including magma ascent and different scale deformations. Another important area of investigation is the calculation of *P-T-t* paths using mineralogical data and modeling results, which provide important constraints and inputs for tectonic reconstructions.

1.2 Kinetics of Metamorphism

The kinetics of metamorphic processes involves the *chemical and physical* kinetics. Chemical kinetics is the study of chemical reactions with respect to their mechanisms, dependence of reaction rates on the concentrations of reactants, temperature, etc. Physical kinetics is the study of thermodynamics of nonequilibrium processes operating in a system in case of a small departure from an equilibrium state. This small departure is accompanied by slow changes in the parameters of state (density, temperature, etc.) during diffusion, heat transfer, etc. Nonequilibrium processes in the systems with no heat exchange (adiabatically isolated) are irreversible and occur with the increase in entropy. In the state of equilibrium, the entropy of a system (isolated system) reaches its maximum (in an irreversible process).

As discussed in Sect. 1.1 Metamorphic Facies, the scientists' interest in metamorphism was focused first of all on the study of the diversity of minerals and mineral assemblages, as well as on the clarification of the *P-T* conditions of their formation in the Earth's crust. Little attention was paid to the kinetics of metamorphism, including the problems of mass transfer, crystal growth, mechanisms of mineral transformations, reaction rates, etc. The first systematic research on these problems began some fifty years ago through the pioneering works of Fyfe et al. (1958) and Turner and Verhoogen (1951). These authors emphasized that metamorphic reactions driven by an increase in temperature involve an increase in entropy, i.e., are endothermic; melting is also an endothermic process. However, metastable mineral assemblages, the transformation of which toward stable assemblages was hindered by slow reaction rates at a low temperature. In such a situation, an increase in temperature may induce an exothermic reaction in which pyroxene is replaced by hornblende during alteration of the basalt to amphibolite.

In this connection, it is important to note that metamorphic reactions caused by an increase in pressure favor the formation of mineral assemblages such that the volume of newly-formed mineral assemblages can be smaller than the volume of the initial ones. If entropy of these reactions decreases (as is usually the case), the mineral transformation is accompanied by the release of heat.

These authors called attention to a fundamental problem of metamorphism: why high-temperature metamorphic rocks survive long periods of cooling. This was attributed to the kinetic irreversibility of chemical reactions, which involves the slow rate of transformation of solid phases at low temperatures and loss of water (Fyfe et al. 1958; Turner and Verhoogen 1951). In an open system, water is driven out from the rocks at the high grades of metamorphism, whereas persistence of high-temperature phases would not occur in a closed system. A metamorphic reaction occurring in the presence of water is likely to consist of a number of different steps, each with its own rate; the overall rate is determined by that of the slowest step and depends on the solubility of a substance and end products in a fluid, mass transfer by diffusion, formation of nucleation sites, reaction kinetics on the surface of the growing solid phase, etc. Minerals of complex composition are formed from transitional phases of simple chemical composition with the involvement of a fluid phase containing atoms, ions, and molecules; anionic aluminosilicates species are likely to be present in aqueous solutions. Separation from an aqueous solution of a new phase involves the formation of crystalline embryos and the growth of those embryos. The overall rate of the metamorphic reaction was thought to be controlled by the formation of such nucleation sites. Nucleation requires a signsificant supersaturation, but crystals may continue to grow at much lower supersaturations of reactants diffusing toward the surface of a new phase. The uneven distribution of strain in the mineral aggregate due to the action of stress can accelerate metamorphic reactions; the resultant elastic or plastic deformation might increase the rates of dissolution, diffusion of matter, and growth of solid phases.

Hydration or dehydration reactions have large entropies. The rock systems in which such reactions occur with increasing temperature should remain adjusted with a lag of not more than a few degrees, from a particular equilibrium. Reactions involving only anhydrous phases usually have very small entropies and an overstepping of several hundred degrees may be required. Fyfe et al. (1958) noticed it is unlikely that assemblage characteristic of different facies could develop under identical *P-T* conditions merely because of differences in reaction kinetics. Barring the case of dry rocks or of reactions with very small entropies, the physical factors that affect the reaction kinetics are not very different from those that govern equilibrium, i.e., the physical variables that determine equilibrium are likely to determine kinetic conditions as well. Thus the important conclusion is that progressive metamorphism reflects changing temperature or pressure gradients prevailing during the spatial replacement of mineral assemblages.

Korzhinskii's idea (1957, 1973, 1982) of the mosaic equilibrium system played a fundamental role in the kinetics of metamorphism. He showed (see above) that if the rate at which temperature and pressure or other factors change and the rate of diffusion are slower than the rate of reactions, then equilibrium should be approached in "local" domains of the rock undergoing metamorphism. Korzhinskii (1955) documented a general sequence of the relative mobility of chemical components from bimetasomatic magnesian skarns developed at contacts between dolomites and gneisses in the metamorphic complexes of the Baikal area (from perfectly mobile to inert): H_2O, CO_2, S, SO_3, Cl, K_2O, Na_2O, F, CaO, O_2, Fe, P_2O_5, BaO, MgO, SiO_2, Al_2O_3, and TiO_2. This sequence may be different in different contact rocks, although its general features remain the same under changing conditions.

Dugald Carmichael (born in 1939) was the first to notice that mineral transformations at isograds can be described as metamorphic reactions balanced for the reacting components, except for volatiles (Carmichael 1969) (Fig. 1.16). It was assumed that reactions represent a close approach to thermodynamic equilibrium under corresponding *PT* conditions. The reaction mechanism was considered by Carmichael as a combination of local mineral transformations that occur contemporaneously but are separated in space. Local systems of phase interaction encompass the small volume of rock that is controlled by the limit of migration for aluminum, which is significantly smaller than that of any other component. The exchange of components among the local systems takes place by diffusion, which

Fig. 1.16 Carmichael D.

determines the volume of the net reaction. In other words, the net metamorphic reaction is considered as the sum of metasomatic cation-exchange reactions among the local systems, which proceed without changes in the bulk composition of the rock (except for volatile components). Deducing from the study of thin sections of pelitic schists composed of quartz, felsic plagioclase, muscovite, biotite, garnet, staurolite, kyanite, and sillimanite, Carmichael found that the limit of migration for aluminum, as the least mobile of the major components, is of the order of 0.2 mm, and that for titanium s of the order of 0.5 mm, while that for potassium, sodium, calcium, magnesium, and iron is of the order of 2–4 mm. This determines the total volume where the balanced metamorphic net reaction is inferred to have been in progress. The distances between the local systems are of the order of several millimeters in coarse-grained metapelites. In fact, the local systems are closed to aluminum, and the net reaction is closed to all components, except for the escape of H_2O, which implies that metamorphism is isochemical with respect to the rock as a whole. Later, the ideas of Carmichael became widely accepted. The advent of the electron probe microanalyzer made it possible to calculate the redistribution of material during complex reactions on the basis of the actual composition of reactant and product minerals (for example, Jones 1972).

George Fisher (born in 1937), using petrographic-mineralogical data of Loberg (1963) on Swedish gneisses, examined corona textures in sillimanite-bearing metapelites, containing mineral segregations consisting of andalusite-biotite-quartz cores surrounded by a quartz-microcline mantle, which were formed during metamorphism (Fig. 1.17). It was shown that these segregations were formed by the opposed migration of K on the one hand, of Fe, mg, and Ca, on the other hand,

Fig. 1.17 Fisher G.

depending on their activity gradients (Fisher 1970). Using this and other observations (see above Carmichael 1969) and on the basis of nonequilibrium thermodynamics, Fisher (1973) developed an isothermal-isobaric model of diffusion-controlled metamorphic reactions. Diffusion is thought of as being statistically interpreted transport of particles of substance treated as a continuum.

Fisher's model implicated diffusion of chemical components along the grain boundaries through an intergranular fluid, which is assumed to be in equilibrium with the local mineral assemblages. Diffusion (*D*) of components in the fluid is driven by chemical potential gradients:

$$J_i^D = -\sum_{j=1}^{n} L_{ij}^D \cdot \nabla\mu_j, (i = 1, 2, \ldots n), \quad (1.1)$$

where J_i^D is the flow (diffusion rate) of the *i*th component, L_{ij}^D is the phenomenological coefficient for diffusion of the *i*th component driven by the gradient μ_i, ∇ is the Hamiltonian operator representing the gradient of the parameter, μ_j is the chemical potential of the *j*-th component in a fluid, *n* is the number of components. The diffusion rate can be found by solving a system of linear equations, which is not easily be done due to the lack of knowledge on the actual values of *L* и μ. To overcome this difficulty, Fisher proposed to estimate diffusion rates formally, based on the difference in the concentrations of components in the reactant and product mineral assemblages using the stoichiometric coefficients for the reactions. It was also assumed that diffusion flows during reaction should be proportional to stoichiometric coefficients, and the process should satisfy the condition of a minimum rate of entropy production, equivalent to stable condition. As an example, Fisher calculated the reaction that took place during the formation of the above-mentioned mineral segregations described by Loberg (1963). Based on the data from Carmichael (1969) on the nearly constant-volume and the nearly constant-aluminum postulates for mineral transformations in metapelites, he obtained the equations of relative diffusion rates using stoichiometric coefficients for K_2O, Na_2O, and CaO. Comparison showed that predictions based on these equations agree reasonably with the diffusion rates calculated from bulk mass transfer (in cations per cubic centimeter) between cores and mantles of segregations.

Several years later, Fisher (1977, 1978) expressed the kinetics of metamorphism as a combination of diffusion and chemical reaction (*R*). Therefore, he added the equation of the reaction rate J_r^R to Eq. (1.1):

$$J_r^R = \sum_{p=1}^{q} L_{rp}^R A_p = \sum_{p=1}^{q} L_{rp}^R \sum_{i=1}^{n} v_i^r \mu_i, (r = 1, 2, \ldots q), \quad (1.2)$$

where L_{rp}^R is the phenomenological coefficient for rate of reaction *r*, depending on affinity of reaction A_p; A_p is the chemical affinity of reaction *p*; v_i^r is the stoichiometric coefficient of the *i*-th component in reaction *r*, μ_i is the chemical potential of the *i*-th component. *L* in Eqs. (1.1) and (1.2) is the matrix of phenomenological coefficients, which was determined to relate diffusion, reactions, and corresponding potential. The relationship between diffusion and reaction is supplied by the equation for the conservation of material:

$$\frac{\partial c_i}{\partial t} = -\frac{dJ_i^D}{dx} + \sum_{r=1}^{p} v_i^r J_r^R, \quad (1.3)$$

where c_i is the concentration of component *i* in a fluid; *t* denotes time; *x* is the spatial coordinate. If, as a first approximation, it is believed that $\partial c_i/\partial t \cong 0$ (Fisher and Elliott 1974), then for most metamorphic reactions

$$\frac{dJ_i}{dx} \cong \sum_{r=1}^{p} v_i^r J_r^R. \quad (1.4)$$

A close examination of Eqs. (1.1)–(1.4) shows that in the case of $L_{rp}^R \gg L_{ij}^D$, the net reaction rate is controlled by the rate of diffusion; the reactions between a mineral and a fluid are so rapid that local equilibrium domains develop in rock undergoing metamorphism. Mass transfer through an intergranular fluid between the reactant assemblages is controlled by chemical potential gradients. In accordance with Eq. (1.4), the reactions take place mostly at the grain and assemblage boundaries, where potential gradients change most rapidly. As a result, diffusion-controlled metamorphic process leads to the growth of well-developed mineral zones, with sharp boundaries.

In the case of $L_{ij}^{D} \gg L_{rp}^{R}$ the chemical potential gradients are small over the volume of rock. The overall rate of interaction is controlled by reaction rates, and mineral growth is driven by the difference between the equilibrium value and the actual values of the chemical potentials in each part of the local domain. Because the chemical potential gradients are negligibly small, affinities and reactions rates tend to be uniform over large volumes of rock. The initial matrix grains are most likely to be dissolved during metamorphism; at the same time, the kinetics of nucleation of new phases and variations in the rates of local reactions play an essential role. If $L_{ij}^{D} \gg L_{rp}^{R}$ variations in composition over rock is slight, and the boundaries between mineral segregations may be vague.

By comparing the values of the L_{ij}^{D} and L_{rp}^{R} coefficients, Fisher concluded that because most metamorphic rocks are characterized by textural and structural heterogeneity, sharp variations in composition and the presence of well-developed boundaries between mineral assemblages, which correlate with variations in the chemical potentials of components, the kinetics of metamorphism is mainly controlled by the small rate of diffusion of components through an intergranular space occupied by a fluid phase.

The values for the L_{ij}^{D} and L_{rp}^{R} coefficients in metamorphic rocks were calculated by Fisher (1977, 1978) based on the available experimental determinations. The most typical values are as follows:

- coefficient of grain boundary diffusion (based on experimental determinations in ceramics and metals, supplemented by a set of measurements of natural grain boundary diffusion in calc-silicate nodules):

$$L_{ij}^{D} = \frac{2.54 \times 10^{-12}}{X^{grn}RT} \exp \frac{-8.37 \times 10^{4}}{RT} \ \mathrm{mol^{2}\ J^{-1}\ m^{-1}\ s^{-1}}, \qquad (1.5)$$

  where X^{grn} is the average grain diameter, R is the gas constant, T is the temperature;
- coefficient of lattice diffusion in silicates (estimated from values for diffusion of K_2O in orthoclase):

$$L_{ij}^{D} = \frac{2.47}{RT} \exp \frac{-2.85 \times 10^{5}}{RT} \ \mathrm{mol^{2}\ J^{-1}\ m^{-1}\ s^{-1}}; \qquad (1.6)$$

- coefficient of reaction rate between silicates and aqueous fluid (estimated from values for silicate dissolution reactions at low temperature):

$$L_{rp}^{R} = \frac{4.59 \times 10^{3}}{X^{gb}RT} \exp \frac{-6.28 \times 10^{4}}{RT} \ \mathrm{mol^{2}\ J^{-1}\ m^{-3}\ s^{-1}}, \qquad (1.7)$$

  where X^{gb} is the thickness of the fluid film between mineral grains. The coefficients for metamorphism may range as much as two-three orders of magnitude above and below these values.

Because the dimensionality of L_{ij}^D and L_{rp}^R coefficients may vary, Fisher (1977) introduced the dimensionless relation to compare these coefficients. The criterion for the overall rate of mineral transformation $\frac{2L_{ij}^D}{L_{rp}^R v_i^r x^{gb} \Delta x}$ will be much lower than 1 in diffusion-controlled processes, and much higher than 1 in reaction-controlled processes. The transition from diffusion-controlled to reaction-controlled process takes place when $2L_{ij}^D = L_{rp}^R v_i^R x^{gb} \Delta x$. The transition from grain boundary diffusion to lattice diffusion occurs when $L_{ij}^{D_1} = L_{ij}^{D_2}$, where D_1 is grain boundary diffusion and D_2 is lattice diffusion. If grain boundary diffusion predominates, the criterion $\frac{(L_{ij}^{D_1})}{(L_{ij}^{D_2})} \gg 1$; in the case of $\frac{(L_{ij}^{D_1})}{(L_{ij}^{D_2})} \ll 1$, lattice diffusion prevails. The criterion for diffusion-controlled process is defined by the relation $1 \gg \frac{9x^S \Delta V_i \Delta c_i}{4\Delta x}$, where x^S is the thickness of the reaction zone, ΔV_i is the volume of rock from which one mole of component i was removed, Δc_i is the difference in the concentration along the diffusion distance Δx.

Using the same values as in Eqs. (1.5)–(1.7), it is possible to estimate roughly the scales of mass transfer as a function of temperature and the kinetics of metamorphism. It is evident that the reaction rates control mass transfer within the nuclei. The growth of mineral grains distinguishable on both micro- and macro-scales (ranging in size from ~0.001 to 1 cm) is driven by diffusion along the grain boundaries at temperatures ranging from 300 to 1000 °C; however, a temperature above 1000 °C is needed for any significant lattice diffusion. Similar to his earlier assumption, Fisher (1973) assumed that mineral reactions during metamorphism occur close to a stable state and the diffusion fluxes of components are proportional to stoichiometric coefficients. This enabled a formal determination of the kinetic parameters of multi-component, multi-mineral metamorphic reactions in diffusion-controlled zonal structures (corona-like) by simultaneous solution of Eqs. (1.1) and (1.3). With known chemical compositions and free energies of phases, at constant T, P, L_{ij}^D and under a number of other assumptions, by solving these equation it is possible to obtain a sequence of simultaneously growing zones at contact between reacting (initially nonequilibrium) minerals, and to calculate the diffusion fluxes of material in each zone and the reactions at zone boundaries. This allowed a prediction of the duration of growth of zonal structures, which began to form at a step controlled by the reaction rate and continued into the diffusion-controlled step. Fisher (1978) derived a number of formulas to calculate the duration of growth of zones in spherical segregations and planar reaction zones at contacts between initially nonequilibrium rocks.

It was found that in both cases, diffusion-controlled mineral growth is proportional to the square root of time while reaction-controlled growth is linearly proportional to t. The $\sqrt{t}$ dependence relates a decrease in the chemical potential gradient to increasing distance over which the diffusion transport of the material occurs. Fisher estimated this growth time at ~50,000 years for 0.5 cm spherical segregations in metapelites described by Loberg (1963). Using the isothermal isobaric model and assuming diffusion-controlled growth of mineral segregations,

Fisher (1977) calculated that the reaction zones must not exceed 10–50 cm across even for the estimated growth time of about 100 Myr.

Using his isothermal isobaric diffusion model, Fisher (1978) among other kinetic factors attempted to consider the heat flow to reaction zones and changes in the reaction enthalpies involved in the growth of minerals. As in the case of reaction-controlled growth, the values of chemical potentials of components vary slightly over the volume of rock, and the boundaries between the newly-formed segregations are vague and ill-defined.

By comparing the above-mentioned kinetic mechanisms, Fisher (1978) came to the conclusion that metamorphic structures tend to pass through several successive stages. Spherical structures pass through three stages: (1) an initial reaction-controlled stage, in which growth of nuclei is linearly proportional to time, (2) an intermediate diffusion-controlled stage, in which growth is proportional to the square root of time, and (3) a final heat-flow-controlled stage, with growth of grains proportional to the cube root of time. The growth of planar reaction zones occurs by either a reaction-controlled or heat-flow-controlled mechanism and is linearly proportional to time. Subsequently, most growth will occur by a diffusion-controlled mechanism and tend to be proportional to the square root of time. In general, Fisher (1978) noted that the reaction-controlled stage will end long before the structure is large enough to detect, and that most growth will occur by either a reaction-controlled or heat-flow-controlled mechanism. Increasing distance between the initial nuclei, increasing grain size, decreasing diffusion coefficients, and increasing heat flow rates all tend to favor both diffusion control and heat-flow control.

Using Fisher's model (1970, 1973), Foster (1977) quantitatively tested several local reactions in metapelitic schists, Maine, USA. The pelitic schists consisting of sillimanite, staurolite, garnet, biotite, muscovite, plagioclasea, quartz, and ilmenite were metamorphose at T = 500–650 °C and P = 4–5 kbar. The calculations were made using whole-rock modal compositions of the rock matrix and three types of sillimanite, staurolite, and garnet segregations, and chemical compositions of all minerals constituting rock. The local reactions were not balanced with respect to all components, i.e., were not isochemical, requiring that exchange of components between different segregations must happen. The system approached balance with respect to all components, except H_2O, in the volume of rock equal to ~ 1 cm^3 (for details see Sect. 4.2.2).

Like Fisher, *Ron Vernon* (born in 1935) considered the diffusion-controlled mass transfer as one of the main factors controlling mineral growth during metamorphism (Vernon 1976) (Fig. 1.18). Diffusion in the intergranular fluid is much faster than in the mineral grains. Based on experimental data, Vernon noted that lattice diffusion coefficients of the cations can be estimated at 10^{-9} to 10^{-16} cm^2 s^{-1}. The relative rate of the individual metamorphic reaction going toward a state of lower free energy is determined by the activation energy, which depends on temperature, and the "frequency factor", which indicates the number of reacting particles per unit time. A multistep process appears to be kinetically favorable. For example, the breakdown of talc to form anthophyllite as a reactive intermediate is more favorable

Fig. 1.18 Vernon R.

than the direct transformation of talc into enstatite plus quartz and H_2O (Greenwood 1963). The overall reaction rate is thus determined by the rate of its sluggish step.

The formation of nuclei in a homogeneous material (for example, in a melt or a perfect crystal) is a random process. The interfacial free energy plays a dominant role at the initial step of nucleation. For an embryo to become a stable nucleus it must surmount the energy barrier. Therefore, a reaction tends to run at an appreciable rate only at some temperature above or below the thermodynamic equilibrium temperature. Nucleation rates in solids are reduced with decreasing diffusion coefficients, and the rates of reactions proceeding with an appreciable volume change tend to increase as a result of the increased elastic strain caused by the "room problem" of growing new phases. Nucleation is assisted by a nucleating agent, which lowers the energy barrier, compared with homogeneous nucleation. Such nucleation was termed by Vernon (1976) "heterogeneous nucleation". These reactions require a large amount of overstepping. The lowering of the energy barrier can be caused by strains associated with a volume change, lattice defects, or diffusion occurring more easily in the presence of defects, epitaxy, etc. Further growth of nuclei is controlled by volume diffusion in an intergranular fluid, and the concentration of the diffusing material varies with time.

Zoning in mineral grains reflects the evolution of the chemical composition of the local system (depending on the involvement of various mineral phases under variably thermodynamic conditions of metamorphism) and concentration gradients. Oscillatory zoning possibly may be explained by local oscillations in the supply of components. Diffusion through the mineral lattice affects development of compositional zoning in a growing grain; if the diffusion rate is too rapid, it may result in homogenization of the composition of the solid phase.

For practical purposes, Vernon (1976) recognized two types of mineral equilibria (reactions) in metamorphic rocks: discontinuous and continuous. The first ones are univariant and the second ones are divariant (or multivariant); they differ in the presence of phases of variable composition and may shift their position with a change in thermodynamic parameters. Vernon suggested that the first reactions may be preferable to the second ones for the delineation of isograds. Minor amounts of components (e.g., MnO, Na_2O, Fe_2O_3, etc.) do not cause the formation of new phases; they substitute isomorphically in the other phases and so they may be additional independent variables. Some of these components may increase the stability of the phases, while other reaction products may reduce the stability of the reactant phases.

Vernon (1976) paid particular attention to structures of metamorphic rocks. He noted that changes in grain shape after the reaction are controlled by interfacial energy. This process may occur spontaneously to minimize interfacial energy. Further growth and adjustment of crystalline grains occur in strain-free areas, leading to the formation of stable structures, provided enough heat and time are available for the stability of such structures. Recrystallization may occur in aggregates with small grain sizes and large surface areas, and/or large porosities, and/or deformed and irregular grain shapes, etc. Such recrystallization can reduce the total interfacial free energy by forming grains as large as possible and some low-energy interfaces. In single phase aggregates, the spread of interfacial angles closely approximating 120° is the result of an attempt to fill space with isotropic grains with the lowest surface area. In polyphase aggregates involving anisotropic minerals, the interfacial angles deflect from 120°, depending on the phases present and total interfacial energy. In this case, interfacial free energy is lower compared with that of monomineralic aggregates. The development of porphyroblasts in the early stages of metamorphism indicates a low rate of nucleation and a relatively high rate of growth of crystal faces, but some porphyroblasts may represent an intergrowth of several mineral grains. Interfacial energy controls only the final shape of porphyroblasts. The formation of poikiloblastic porphyroblasts takes place if the grain boundaries move too fast relative to the diffusion rate of the material in small inclusions through the crystalline lattice outside the porphyroblast.

The structures of crystalline rocks are determined by modes of deformation and metamorphic *P-T* conditions. The increase in temperature causes changes in the rheological properties of rocks while the role of creep deformation, including viscous flow of mineral aggregates (e.g., in the granulite facies) increases. Metamorphic reactions involving dehydration and melting bring about a marked increase in the creep rate. At low temperatures and relatively fast strain rates, intracrystalline slip predominates, and work-hardening occurs. The grains appear to have been flattened and show twinning. At higher temperatures and slower strain rates, grain boundaries become weak, dislocation density increases and grain boundary sliding takes place. These processes are accompanied by recrystallization of crystalline materials and recrystallization of complex compounds may lead to changes in the chemical composition (Vernon 1976). Preferred orientations can be produced by oriented gliding of layers in crystal lattices during recrystallization. The processes of metamorphic differentiation may intensify a previous compositional heterogeneity during deformation. This may be caused by different strain rates between adjacent heterogeneous domains, volume change between different domains, a chemical potential gradient caused by a strain difference between the above-mentioned domains, variations in the dissolution kinetics of different minerals, mass transfer, etc. The question about the role of dissolution of crystalline materials in high-strain areas and its precipitation in lower-strain areas remains unanswered. Nevertheless, Vernon suggested that this process may have taken place in the formation of some specific structures such as "pressure shadows" adjacent to stable mineral grains in deformed rocks.

In 1984, *John Walther* (born in 1950) and *Bernard Wood* (born in 1946) published a paper entitled "Rate and Mechanism in Prograde Metamorphism" (Walther and Wood 1984), in which they provided, after having made some assumptions, general estimates of the rates of metamorphic reactions. Based on experimental data for an aqueous fluid at high temperatures, the mineral dissolution and growth rate constants were given approximately by

$$\log K_r = \frac{-2900}{T} - 6.85, \tag{1.8}$$

where K_r is the rate constant expressed in gm atom oxygen $\mathrm{cm}^{-2}\ \mathrm{s}^{-1}$, and T is the absolute temperature; log K_r varies from −11 to −9.5 in the temperature range ~700 to 1070 K. Taking, as an example, a spherical crystal with radius 0.1 cm and a 1 °C overstep of a dehydration reaction which has entropy changes on the order of $\Delta S = 20\ \mathrm{cal\ mol^{-1}\ °C^{-1}}$ at the reaction rate constant $K_r \approx 2.5 \times 10^{-11}$ (at 773 K), the total time taken for dissolution/growth (recrystallization) would be ~875 years. Thus, dissolution and growth of mineral grains which are in contact with each other are relatively rapid even for very small temperature oversteps (disequilibrium). However, transport of reacting material from sites of dissolution to those of nucleation and growth may be slower than the overall reaction rate, thus making it necessary to consider the constraints which transport may impose on the metamorphic process. Mass transfer during metamorphism involves diffusion either through an intergranular fluid or through the disorganized (deformed) grain boundaries (~10^3 Å in thickness).

The fluid flux of material q through a relatively permeable metamorphic rock is given by

$$q = \frac{d^3 l\tau}{12\eta}\left(\frac{\partial P}{\partial Z}\right), \tag{1.9}$$

where d is the width; l is the length (l = *1/R*, where R is the average grain radius); τ is the tortuosity (a dimensionless quantity ranging between ~1 and 0.5) of fracture or intergranular space; η is the fluid viscosity (poise); $(\partial P/\partial Z)$—is the viscous pressure (dyne cm^{-3}); $(\partial P/\partial Z) = (\rho_r - \rho_f)g = 1.9 \times 10^3$, dyne cm^{-3}); where ρ_r and ρ_f are the rock and fluid densities and g is the acceleration due to gravity; fluids in the system H_2O–CO_2 have viscosities on the order of 10^{-3} to 2×10^{-3} poise. A relationship between fluid flux q and fluid velocity V_{fl} (cm s^{-1}) is calculated from

$$Vfl = q/d \cdot l. \tag{1.10}$$

The calculations show that $q \approx 10^{-9}$ g cm^{-2} s^{-1}, if V_{fl} is on the order of a few meters a year and grain boundary width is $d \approx 10^3$ Å. If d would have to be 10 times lower, diffusion through the fluid film would be more efficient than flow transport (infiltration) at fluid fluxes $q \approx 10^{-11}$ to 10^{-12} g cm^{-2} s^{-1}. If d would be 10 times less (up to 10 Å), diffusion through the fluid would be impossible, and in this case, mass transport is controlled by grain boundary diffusion at $q \approx 10^{-15}$ g cm^{-2} s^{-1} or less (Fig. 1.19).

The conclusion made by Walther and Wood is similar to that of Fisher and Vernon, and to the earlier conclusions of Fyfe, Turner and Verhoogen (see above). They showed that diffusion of solutes through a grain boundary film would be the dominant mechanism of transport during metamorphism and that diffusive mass transport of mostly Si and Mg would be a major control on the overall rate of a

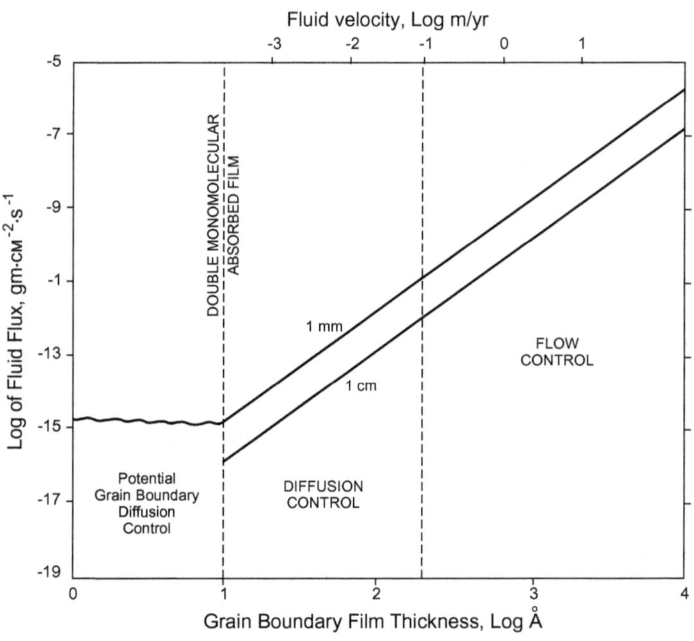

Fig. 1.19 Relationship between the logarithm of fluid flux and grain boundary film thickness for 1 mm and 1 cm average grain size (heavy solid lines labeled 1 mm and 1 cm, respectively). Fluid velocity of the film is shown on top of the diagram. Three regions of controlling transport mechanism are indicated (Walther and Wood 1984)

metamorphic reaction (in metapelites). An increase in the fraction of CO_2 in an aqueous fluid substantially reduces solubilities of the minerals and, thus the concentration of the solutes in the fluid. The isolation of the fluid from the rocks (in fractures) would cause retardation of metamorphic reaction rates or, instead, accelerate them in areas where fluid becomes channeled. This conclusion was substantiated by the results of calculations using data available for the kinetics and model estimates of fluid generation rates during metamorphism (dehydration, decarbonatization). It was also shown that diffusion through the disorganized grain boundary can only be important at temperatures above ~700 °C and where fluid film thickness does not exceed 10 Å, i.e., the amount of fluid is too small. Walther and Wood noted that temperature oversteps (above equilibrium) of metamorphic reactions should rarely exceed a few °C (except for solid-solid reactions). For further discussion on metamorphic reaction rates see Sect. 4.2.3.

In the past decades of the 20th century, several issues, including the role of kinetics in the development of microstructures in rocks, oriented growth of mineral grains and volume changes during metamorphic reactions, the role of the activity of fluid components and temperature oversteps (above equilibrium) during nucleation, etc., all have become the focus for increasing attention. Deformation in metamorphic rocks is considered as a factor considerably affecting not only nucleation and recrystallization of mineral grains, but also the rates of diffusion and mineral transformations. Among other problems of interest are the spatial distribution and size of different mineral grains in rocks depending on metamorphic conditions, whether the spatial distribution of solid phases can be influenced by an initial nucleation alone or also by subsequent retrograde or polymetamorphism. These and other problems were then elaborated in the works of many renowned researchers (Lichtner 1985, 1988; Lasaga 1981, 1986; Ridley 1986; Helgeson and Lichtner 1987; Ortoleva et al. 1987; Schramke et al. 1987; Marsh 1988; Carlson 1989; Joesten 1991a, b; Kerrick et al. 1991; Kretz 1994 and others). Recent years have witnessed a growing interest in experimental studies of the metamorphic reaction kinetics (e.g., Rubie and Thompson 1985)). However, special experimental problems were caused by the need for very long duration experiments at high *P*-*T* conditions and interpretation of experimental results in regard to natural processes. Despite all these difficulties, this appears to be a very important and promising area of investigation, which provides the opportunity to estimate the timescale for metamorphism by applying laboratory-based kinetic parameters.

References

Althaus E (1966) Die Bildung von Pyrophyllit und Andalusit zwischen 2000 und 7000 bar H_2O-Druck. Naturwissenschaften 53:105–106

Althaus E (1967) The triple point andalusite-sillimanite-kyanite: an experimental and petrologic study. Contr Mineral Petrol 16:29–44

Barrow G (1893) On an intrusion of muscovite-biotite gneiss in the south-eastern Highlands of Scotland, and its accompanying metamorphism. Q J Geol Soc Lond 49:330–388

Barrow G (1912) On the geology of the lower Deeside and the southern Highland border. Proc Geol Assoc 23:268–284
Barth TFW (1936) Structural and petrologic studies in Dutchess County, New York. Part II. Geol Soc Am Bull 47:775–850
Becke F (1903) Über Mineralbestand und Struktur der Kristallinin Schiefer. Wien: Denkschriften der Akademie der Wissenschaften, Math-Naturw Kl, Bd 75, S 1–53 (extend German abstract in: Comp Rendu IX Sess Congress Geol International, Vienna, Pt 2, S 553–570)
Becke F (1921) Die optischen Eigenschaften einige Andesine. TMPM 35:31–46
Berman RG (1988) Internally-consistent thermodynamic data for minerals in the system K_2O–Na_2O–CaO–MgO–FeO–Fe_2O_3–Al_2O_3–SiO_2–TiO_2–H_2O–CO_2. J Petrol 29:445–522
Billings MP (1937) Regional metamorphism of the Littelton-Moosilauke area, New Hampshire. Geol Soc Am Bull 48:463–566
Boue A (1820) Essai geologique sur l'Ecosse. Vve Courcier, Paris
Bowen NL (1940) Progressive metamorphism of siliceous limestone and dolomites. J Geol 48:225–274
Carlson WD (1989) The significance of intergranular diffusion to the mechanism and kinetics of porphyroblast crystallization. Contr Mineral Petrol 103:1–24
Carmichael DM (1969) On the mechanism of prograde metamorphic reactions in quartz-bearing pelitic rocks. Contr Mineral Petrol 20:244–267
Chapman CA (1939) Geology of the Mascoma Quadrangle, New Hampshire. Geol Soc Am Bull 50:127–180
Clark SP (1961) A redetermination of equilibrium relations between kyanite and sillimanite. Am J Sci 259:641–650
Clark SP, Robertson EC, Birch F (1957) Experimental determination of kyanite-sillimanite equilibrium relations at high temperature and pressure. Am J Sci 255:628–640
Coombs DS (1961) Some recent work on the lower grades of metamorphism. Australian J Sci 24:203–215
Coombs DS, Ellis AJ, Fyfe WS et al (1959) The zeolite facies, with comments on the interpretation of hydrothermal synthesis. Geochim Cosmochim Acta 17:53–107
Delesse A (1857) Etudes sur le metamorphisme des roches. Annales des Mines, serie 5a 12:89–288, 417–516, 705–772
Dobretsov NL, Reverdatto VV, Sobolev VS et al (1966a) Karta metamorficheskikh fatsiy SSSR. Masshtab 1:7,500,000 (Map of metamorphic facies of the USSR. Scale 1: 7,500,000). Main Department of Geodesy and Cartography of Ministry of Geology of the USSR, Moscow
Dobretsov NL, Reverdatto VV, Sobolev VS et al (1966b) Fatsii regional'nogo metamorfizma SSSR. Ob "yasnitel'naya zapiska k «Karte metamorficheskikh fatsiy SSSR». Masshtab 1:7,500,000 (Facies of regional metamorphism of the USSR. Explanatory note to the Map of metamorphic facies of the USSR. Scale 1:7,500,000). Nauka, Siberian Branch, Novosibirsk
Dobretsov NL, Khlestov VV, Reverdatto VV et al (1972) The facies of metamorphism. Australian National University, Canberra, p 214
Dobretsov NL, Khlestov VV, Sobolev VS (1973) The facies of regional metamorphism at moderate pressure. Australian National University, Canberra, p 236
Dobretsov NL, Sobolev VS, Sobolev NV et al (1975) The facies of regional metamorphism at high pressure. Australian National University, Canberra, p 266
Elles GL, Tilley CE (1930) Metamorphism in relation to structure in the Scottish Highlands. Royal Soc Edinburgh Trans 56:621–646
Eskola P (1915) On the relations between the chemical and mineralogical composition in the metamorphic rocks of the Orijärvi region. Commun Geol Finlande Bull 44:109–145
Eskola P (1939) Die metamorphen Gesteine. In: Barth TFW, Correns CW, Eskola (eds) Die Entstehung der Gesteine. Verlag von Springer, Berlin, pp 263–407
Evans BW, Trommsdorff V (1970) Regional metamorphism of ultramafic rocks in the Central Alps: paragenesis in the system CaO–MgO–SiO_2–H_2O. Schweiz Mineral Petrogr Mitt 50:481–492
Fisher GW (1970) The application of ionic equilibria to metamorphic differentiation: an example. Contrib Mineral Petrol 29:91–103

Fisher GW (1973) Nonequilibrium thermodynamics as a model for diffusion-controlled metamorphic processes. Am J Sci 273:897–924
Fisher GW (1977) Nonequilibrium thermodynamics in metamorphism. In: Fraser DG (ed) Thermodynamics in geology. NATO advanced study institutes series—C, vol 30. Reidel Publ. Co., Dordrecht, pp 381–403
Fisher GW (1978) Rate laws in metamorphism. Geochim Cosmochim Acta 42:1035–1050
Fisher GW, Elliott D (1974) Criteria for quasisteady diffusion and local equilibrium in metamorphism. In: Hofmann et al (eds) Geochemical transport and kinetics. Carnegie institution of Washington, Publ 634, pp 231–241
Foster CT (1977) Mass transfer in sullimanite-bearing pelitic schists near Rangeley, Maine. Am Mineral 62:727–746
Fyfe WS, Turner FJ, Verhoogen J (1958) Metamorphic reactions and metamorphic facies. Geol Soc Am Memoir 73:259
Goldschmidt VM (1911a) Die kontaktmetamorphose im Kristianiagebiet. Videnskapsselskapets Skrifter, Math-NaturV, K1, Bd. 11, 486 S
Goldschmidt VM (1911b) Die Gesetze der Mineralassoziation vom Standpunkt der Phasenregel. Zeits Anorgan Chem 71:313–322
Goldschmidt VM (1912) Die Gesetze der Gesteinsmetamorphose mit Beispielen aus der Geologie des sudlichen Norwegens. Videnskapsselskapets Skrifter, Math-NaturV, K1, N 22, 16 s
Greenwood HJ (1963) The synthesis and stability of anthophyllite. J Petrol 4:317–351
Grubenmann U (1904) Die Kristallinen Schiefer. 1 Allgemeiner Teil. Verlag von G. Borntraeger, Berlin
Grubenmann U, Niggli P (1924) Die Gesteinmetamorphose. 1 Allgemeiner Teil. Verlag von G. Borntraeger, Berlin
Harker A (1932) Metamorphism. A study of the transformations of rock-masses. Methuen and Co. Ltd, London
Helgeson HC, Lichtner PC (1987) Fluid flow and mineral reactions at high temperatures and pressures. J Geol Soc Lond 144:313–326
Holland TJB, Powell R (1990) An enlarged and updated internally consistent thermodyna- mic dataset with uncertainties and correlations: the system K_2O–Na_2O–CaO–MgO–MnO–FeO–Fe_2O_3–Al_2O_3–TiO_2–SiO_2–C–H_2–O_2. J Metamorph Geol 8:89–124
Holland TJB, Powell R (1998) An internally consistent thermodynamic data set for phases of petrological interest. J Metamorph Geol 16:309–343
Joesten R (1991a) Grain-boundary diffusion kinetics in silicate and oxide minerals. In: Ganguly J (ed) Diffusion, atomic ordering and mass transport. Advances in physical geochemistry, vol 8. Springer, New York, pp 345–395
Joesten R (1991b) Local equilibrium in metasomatic processes revisited: diffusion-controlled growth of chert nodule reaction rims in dolomite. Am Mineral 76:743–755
Johannes W (1969) An experimental investigation of the system MgO–SiO_2–H_2O–CO_2. Am J Sci 267:1083–1104
Jones JW (1972) An almandine garnet isograd in the Rogers Pass area, British Columbia: the nature of the reaction and an estimation of the physical conditions during its formation. Contrib Mineral Petrol 37:291–306
Kerrick DM, Lasaga AC, Raeburn SP (1991) Kinetics of heterogeneous reactions. In: Kerrick DM (ed) Contact metamorphism. Reviews in mineralogy, vol 26. Mineralogical Society of America. Book Crafters Inc, Chelsea, pp 583–671
Korzhinskii DS (1935) Termodinamika i geologiya nekotorykh metamorficheskikh re-aktsiy s vydeleniyem gazovoy fazy (Thermodynamics and geology of some metamorphic reactions with the evolution of the gas phase). Zapiski (Proceedings) of VMO 64(1):1–16
Korzhinskii DS (1936a) Parageneticheskiy analiz kvartssoderzhashchikh bednykh kal'tsiyem
kristallicheskikh slantsev arkheyskogo kompleksa Yuzhnogo Pribaykal'ya (Paragenetic analysis of quartz-bearing calcium-poor crystalline schists of the Archean complex of the southern Baikal region). Zapiski (Proceedings) of VMO 65(2):247–280

Korzhinskii DS (1936b) Podvizhnost' i inertnost' komponentov pri metasomatoze (Mobility and immobility of components in metasomatism). Izvestiya AN SSSR, Ser Geol 1:35–60
Korzhinskii DS (1937) Zavisimost' mineraloobrazovaniya ot glubiny (Dependence of mineral formation on depth). Zapiski (Proceedings) of VMO 66(2):369–384
Korzhinskii DS (1940) Факторы минеральных равновесий и минералогические фации глубинности (Factors of mineral equilibria and mineralogical facies of depth). Proc Inst Geol Sci USSR Acad Sci 12(5)
Korzhinskii DS (1950) Faktory ravnovesiya pri metasomatoze (Factors of equilibrium in metasomatism). Izvestiya AN SSSR, Ser Geol 3:21–49
Korzhinskii DS (1955) Ocherk metasomaticheskikh protsessov (An overview of metasomatic processes). In: Betekhtin AG (ed) Key problems in the theory of magmatic mineral deposits, 2nd edn. Izd Acad Nauk SSSR, Moscow, pp 335–456
Korzhinskii DS (1957) Fiziko-khimicheskiye osnovy analiza paragenezisov mineralov (Physico-chemical basis for the analysis of mineral parageneses). Izd Acad Nauk SSSR, Moscow
Korzhinskii DS (1973) Teoreticheskiye osnovy analiza paragenezisov mineralov (Theoretical bases of the analysis of mineral parageneses). Nauka, Moscow
Korzhinskii DS (1982) Teoriya metasomaticheskoy zonal'nosti (The theory of metasomatic zoning). Nauka, Moscow
Kretz R (1994) Metamorphic crystallization. Wiley, Chichester
Lasaga AC (1981) Rate laws of chemical reactions. In: Lasaga AC, Kirkpatrick RJ (eds) Kinetics of geochemical processes. Reviews in mineralogy. Book Crfter Inc, Chelsea, pp 1–68
Lasaga AC (1986) Metamorphic reaction rate laws and development of isograds. Mineral Mag 50:359–373
Lichtner PC (1985) Continuum model for simultaneous chemical reactions and mass transport in hydrothermal systems. Geochim Cosmochim Acta 49:779–800
Lichtner PC (1988) The quasi-stationary state approximation to coupled mass transport and fluid-rock interaction in a porous medium. Geochim Cosmochim Acta 52:143–165
Loberg B (1963) The formation of a flecky gneiss and similar phenomena in relation to the migmatite and vein gneiss problem. Geol Fören Stockholm Förh 85:3–109
Lyell Ch (1833) Principles of geology, being an attempt to explain the former changes of the earth's surface, by reference to causes now in operation, vol 3. John Murray, London
Marakushev AA (1965) Problemy mineralniyh faciy metamorficheskih i metasomaticheskih goniyh porod (Problems of mineral facies of metamorphic and metasomatic rocks). Nauka, Moscow
Marakushev AA (1973) Petrologiya metamorficheskih goniyh porod (Petrology of metamorphic rocks). Moscow University, Moscow
Marsh BD (1988) Crystal size distribution in rocks and the kinetics and dynamics of crystallization. I. Theory. Contrib Mineral Petrol 99:277–291
Metz P, Trommsdorff V (1968) On phase equilibria in metamorphose siliceous dolomites. Contrib Mineral Petrol 18:305–309
Milch L (1894) Beitrage zur Lehre des Regionalmetamorphismus. Neues Jahrb Mineral Geol Paleontol, Beilage-Band 9:101–128
Miyashiro A (1961) Evolution of metamorphic belts. J Petrol 2:277–311
Miyashiro A (1973) Metamorphism and metamorphic belts. Allen and Unwin, London
Miyashiro A, Aki K, Sengör AMC (1982) Orogeny. Wiley, New York
Oldroyd DR (1990) The highlands controversy. Constructing geological knowledge through fieldwork in nineteenth-century Britain. The University of Chicago Press, Chicago
Ortoleva P, Merino E, Moore C et al (1987) Geochemical self-organization. I: Reaction transport feedbacks and modeling approach. Am J Sci 287:979–1007
Perchuk LL (1970) Ravnovesiya porodoobrazuyushih mineralov (Equilibria of rock-forming minerals). Nauka, Moscow
Perchuk LL (1973) Termodynamicheskiy regim glubinnogo petrogenesa (Thermodynamic regime of deep petrogenesis). Nauka, Moscow

Playfair J (1822) Collected works of John Playfair, Esq. with a memoir of the Author, vol 4. Constable and Co, Edinburgh

Ramberg H (1952) The origin of metamorphic and metasomatic rocks. The University of Chicago Press, Chicago

Ramberg H (1959) Evolution of grag folds. Geol Mag 100:97–106

Reverdatto VV (1973) The facies of contact metamorphism. Australian National University Publications no 233, Canberra

Ridley J (1986) Modelling of the relations between reaction enthalpy and the buffering of reaction progress in metamorphism. Mineral Mag 50:375–384

Rosenbusch H (1877) Die Steiger Schiefer und ihre Contactzone an der Granititen von Barr-Andlau und Hohwald. Abhandlungen zur Geol. Specialkarte von Elsass-Lothringen, Band I, Druch und Verlag von R. Schultz and Co., Strassburg

Rubie DC, Thompson AB (1985) Kinetics of metamorphic reactions at elevated temperatures and pressures: an appraisal of available experimental data. In: Thompson AB, Rubie DC (eds) Metamorphic reaction. Kinetics, textures, and deformation. Springer, New York, pp 27–79

Schramke JA, Kerrick DM, Lasaga AC (1987) The reaction muscovite + quartz, andalusite + K-feldspar + water. Part 1. Growth kinetics and mechanism. Am J Sci 287:517–559

Sederholm JJ (1891) Studien uber archaische Eruptivgesteine aus dem sudwestlichen Finnland. Wesen und Ursache der metamorphose. Tschermak's Mineral Petrograph Mitteilungen 12:97–142

Sobolev VS (1964) Fizikochemitcheskie usloviya mineraloobrazovaniya v zemnoi kore i mantii (Physicochemical conditions of mineral formation in the Earth's crust and mantle). Geologiya i Geofisika no 1(7):22

Thompson JP (1959) Local equilibrium in metasomatic processes. In: Abelson PhH (ed) Researches in geochemistry. Wiley, New York, pp 427–457

Tilley CE (1924) The facies classification of metamorphic rocks. Geol Mag 61:167–171

Tilley CE (1925) A preliminary survey of metamorphic zones in the southern Highlands of Scotland. Q J Geol Soc Lond 81:100–112

Trommsdorff V, Evans BW (1972) Progressive metamorphism of antigorite schist in the Bergell tonalite aureole (Italy). Am J Sci 272:423–437

Turner FJ (1938) Progressive regional metamorphism in Southern New Zealand. Geol Mag 75: 160–174

Turner FJ (1948) Mineralogical and structural evolution of the metamorphic rocks, vol 30. Geological Society of America

Turner FJ (1968) Metamorphic petrology. Mineralogical and field aspects. McGraw-Hill, New York

Turner FJ, Verhoogen J (1951, 1960) Igneous and metamorphic petrology. 1st and 2nd edn. McGraw-Hill, New York

Van Hise ChRA (1904) Treatise on Metamorphism. US Geol. Surv Monograph, vol 47. Govt Print Off, Washington

Vernon RH (1976) Metamorphic processes, reactions and microstructure development. George Allen and Unwin, London

Vogt T (1927) Sulitelmafeltets geologi og petrografi. Norges Geol Undersökelse 121:449–531

Walther JV, Wood BJ (1984) Rate and mechanism in prograde metamorphism. Contrib Mineral Petrol 88:246–259

Winkler HGF (1965) Die Genese der metamorphen Gesteine, 1st edn. Springer, Berlin

Winkler HGF (1967) Die Genese der metamorphen Gesteine, 2nd edn. Springer, Berlin

Winkler HGF (1974) Petrogenesis of metamorphic rocks, 3rd edn. Springer, New York

Yoder HS (1952) The $MgO–Al_2O_3–SiO_2–H_2O$ system and the related metamorphic facies. Am J Sci 250-A:569–627

Yoder HS (1955) Role of water in metamorphism. Geol Soc Am Special Paper 62:505–524

Zharikov VA, Shmulovich KI (1969) Visokotemperaturnye mineralnye ravnovesiya v sisteme $CaO-SiO_2-CO_2$ (High-temperature mineral equilibria in the system $CaO-SiO_2-CO_2$). Geokhimiya 9:1039–1056

Zirkel F (1893) Lehrbuch der Petrographie. Erster band. Verlag von W. Engelmann, Leipzig

Chapter 2
Mineral Geothermobarometry

2.1 Concept and General Considerations

The quantification of the pressure-temperature conditions associated with various metamorphic rocks required the development of several targeted methods, which were based on the dependence of the equilibrium constants on pressure and temperature. At given thermodynamic parameters, these constants are fixed by mineral equilibria. Such equilibrium relationships based on the distribution of elements between coexisting minerals were referred to as mineralogical thermometers and barometers, and their application to metamorphic rocks gave rise the subdiscipline of conventional thermobarometry.

The history of mineral geothermobarometry began with the works of Ramberg, Barth and others, who introduced in the mid-20th century the idea of using exchange reaction constants in geologic thermometry. A further contribution to the concept of geothermobarometry was made by Perchuk (1970), who drawing upon Ramberg's ideas, formulated a theory of phase correspondence and performed the calibration of a number of consistent mineral geothermometers. Since Perchuk (1970), geothermobarometry became widely acknowledged in petrological studies as a quantitative method for estimating *P-T* conditions of rock formation. Its further development was largely encouraged in the following ways: to refine the determined reactions and to increase the number of reactions examined. The reliability of the estimated *P-T* conditions depend on many factors, such as high-quality thermodynamic databases, reliable mixing functions, mutual equilibrium between selected mineral pairs. With the advent of models for multicomponent solid solutions and the principle of self-consistency, the accuracy of thermodynamic databases has been considerably improved in recent years, which in turn could not help but increase the number and reliability of calibrated geothermometers and

V. V. Reverdatto et al., *The Nature and Models of Metamorphism*,
Springer Geology, https://doi.org/10.1007/978-3-030-03029-2_2

geobarometers. Therefore, by the end of the last century, the problem related to the absence of a geothermometer or geobarometer for certain rock types was replaced by the selection of the relevant reaction and the most precise calibration of these reactions.

Throughout the nearly 50-year history of application, the method of phase correspondence proved to be efficient for reconstructing thermodynamic parameters of abyssal processes and solving petrological and tectonic problems. The use of mineral equilibria involving multicomponent solid solutions significantly expanded the capabilities of this method. First, the number of independent reactions increased with increasing number of simultaneously determined thermodynamic parameters of mineral formation (given the same phase composition).

Second, this gave the opportunity to independently verify if a mineral assemblage is stable with respect to most key components. This resulted in the generation of an internally consistent system of geothermobarometers (Avchenko 1990; Aranovich 1991), which can be used to determine *P-T* conditions under which mineral parageneses were formed in rocks of variable composition over a wide range of pressures and temperatures. A major step forward in the field of geothermobarometry was in part a result of new instrumentation, including high-resolution in situ trace element microanalysis techniques such as ion microprobe (SIMS) and laser ablation (LA-ICP-MS) analysis. The slower diffusion rates of REEs compared to those of major elements, their contrasting partitioning between metamorphic minerals, and high sensitivity to variations in thermodynamic parameters at low concentrations were taken as a basis for the newly developed REE-based geothermobarometers, which are successfully employed in metamorphic petrology.

The theoretical basis of geothermobarometers based on thermodynamic properties of minerals was discussed in details in the classic work of Wood and Fraser (1976). Each equation for a geothermobarometers consists of two terms. One term is the standard free energy $(\Delta Z^o)_{P,T}$ of the reaction between end-members or stoichiometric relationship between end-member components of a mineral. The other term is the equilibrium constant of the corresponding reaction, which is determined from the compositions of minerals and activity-composition relations for the minerals involved. Conventional geothermobarometry uses the equilibrium thermodynamics of balanced reactions between end-members of minerals, combined with the observed compositions of minerals. Therefore, the correct estimation of *P-T* conditions using conventional geothermobarometry requires equilibrium between coexisting minerals. In order to determine a true mineral paragenesis, certain criteria proposed by Vernon (1977) must be applied. These include common grain contacts between minerals, lack of evidence of one mineral replacing another, finer-grained stable aggregates with sharp boundaries. The agreement (within error) between *P-T* estimates based on two or more independent geothermobarometers provides additional evidence for the judicious choice of reliable equilibria.

A variety of geothermobarometers widely used in metamorphic petrology has been discussed by Essene (1989). Several years later, another set of geothermobarometers applied to specific types of metamorphic rocks has been proposed in the literature: for granulites (Pattison et al. 2003), amphibolites (Aleksandrov 2010), eclogites (Ravna and Paquin 2003), etc. Recent advances in in situ high-precision trace element microanalysis techniques allowed the formulation of new geothermobarometers. In this chapter we provide a compilation of geothermometers and geobarometers most widely applicable in petrological studies (see Table 2.1). Most geothermobarometers in Table 2.1 are based on exchange equilibria and net-transfer reactions. Polymorphic and univariant solid phase transitions and dehydration/decarbonatization reactions in thermobarometry are not listed in Table 2.1, because the equilibrium curves involving hydrous minerals are strongly dependent on the fluid composition and partial pressure of O_2. Solid phase transitions do not involve volatiles. However, their position is strongly affected by trace elements (e.g., Mn in andalusite, Sr in aragonite, etc.). In addition, available experimental data on most solid phase reactions are quite inconsistent. For example, discrepancies between the *P-T* values of the Al_2SiO_5 triple point obtained by Holdaway (1971) and Richardson et al. (1969) are 120 °C and 1.7 kbar. It was also shown that the Al_2SiO_5 triple point assemblage containing all three polymorphs in stable equilibrium is not stable in metapelites of any chemical composition (Pattison 2001; Likhanov and Reverdatto 2013, 2014). Therefore, such assemblages cannot be adequately used in estimations of *P-T* conditions of metamorphism and calibration of geothermobarometers. These reactions can be employed as indicators of the facies series only in combination with other geothermobarometers. For an obvious reason, this table does not include some alternative ways of temperature determination such as illite crystallinity (Thompson and Frey 1984), vitrinite reflectance (Tiechmuller 1987), and conodont color alteration index (Rejebian et al. 1987). These thermometers are based on structural and chemical changes that are not quite apparent. Such thermometers can give only rough estimates of temperature, which depend on mineral transformation rates and are sensitive to changes in thermodynamic conditions, fluid amount and composition, etc.

For a comparative study of different geothermobarometers, we used the results on metapelitic rocks of the Garevka metamorphic complex, Yenisei Ridge (Likhanov et al. 2015). A distinctive feature of these rocks is the presence of relatively large zoned garnet porphyroblasts with three texturally distinct zones (Fig. 2.1a–c) with different proportions of grossular and spessartine components (Fig. 2.1d–f), that show clear evidence of multistage growth. These zones were formed at different stages of metamorphism corresponding to different geotectonic settings Fig. 2.1. The three discrete stages define a counter-clockwise *P-T* path involving initial prograde low pressure heating followed by near isothermal medium-pressure compression and post-peak retrograde decompression and cooling. Since these stages differ significantly in thermodynamic regimes they were used for a comparative analysis of different methods employed in geothermobarometry of metamorphic rocks.

Table 2.1 Summary of geothermobarometers, ranges of application and sources of calibrations

Mineral assemblages	Range of applications	References
Exchange thermometers		
Grt–Bt	GRE + AMP ± GRA	Ferry and Spear (1978), Perchuk and Lavrent'eva (1983), Kleemann and Reinhardt (1994), Holdaway (2000), Kaneko and Miyano (2004), Wu and Zhao (2006), Wu (2017)
Grt–Crd	AMP + GRA	Holdaway and Lee (1977), Perchuk and Lavrent'eva (1983)
Crt–Cpx	GRA + ECL ± AMP	Ganguly (1979), Powell (1985), Pattison and Newton (1989), Ravna (2000b)
Grt–Opx	GRA + ECL	Dahl (1980), Carswell and Harley (1989), Bhattacharya et al. (1991), Berman and Aranovich (1996)
Grt–Hrb	AMP ± GRA	Graham and Powell (1984), Perchuk (1989), Powell (1985), Ravna (2000a)
Grt–Chl	GRE + BLU ± AMP	Dickenson and Hewitt (1986)
Grt–Ol	GRA + ECL	Kawasaki and Matsui (1977), O'Neil and Wood (1979), Wu and Zhao (2007a)
Grt–Phe	BLU + ECL + AMP	Hynes and Forest (1988) Carswell and Harley (1989), Wu and Zhao (2006)
Grt–Ilm	AMP + GRA	Pownceby et al. (1987, 1991)
Grt–Cld	GRE ± AMP	Perchuk (1991)
Grt–Spl	AMP + GRA	Perchuk and Gerya (1989)
Crd–Spl	AMP + GRA	Vielzeuf (1983), Perchuk (1991)
Opx–Bt	GRA	Fonarev and Konilov (1986), Sengupta et al. (1990), Wu et al. (1999)
Opx–Cpx	GRA	Stephenson (1984), Taylor (1998)
Opx–Ilm	GRA	Docka et al. (1986)
Cpx–Ilm	GRA + AMP	Docka et al. (1986)
Ol–Ilm	GRA	Docka et al. (1986)
Ol–Opx	GRA + ECL	Sack (1980), Docka et al. (1986), Carswell and Harley (1989)
Ol–Spl	GRA + ECL	Engi (1983), Wan et al. (2008)
Hbl–Pl	AMP	Blundy and Holland (1990), Perchuk (1991), Holland and Blundy (1994)
Hbl–Bt	AMP	Perchuk and Ryabchikov (1976), Wu et al. (2002)
Chl–Ms	GRE	Kotov (1986)
Chl–Phe	BLU	Vidal and Parra (2000)
Ms–Pl	GRE + AMP	Green and Usdansky (1986)
Ms–Bt	GRE + AMP	Perchuk (1970), Hoisch (1989)
Cld–Bt	GRE + AMP	Perchuk (1991), Vidal et al. (1999)

(continued)

Table 2.1 (continued)

Mineral assemblages	Range of applications	References
Solvus thermometers		
Cpx–Opx	GRA + ECL This thermometer is based on the distribution of Ca and Mg between coexisting pyroxenes	David and Boyd (1966), Saxena (1976), Nickel and Green (1985), Davidson and Lindsley (1985)
Cal–Dol	GRE ± AMP Distribution of Ca and Mg between coexisting carbonates	Goldsmith and Heard (1961), Anovitz and Essene (1987a), Powell et al. (1984)
Pl–Kfs	AMP + GRA Distribution of K and Na between coexisting plagioclase and alkali feldspar	Fuhrman and Lindsley (1988), Elkins and Grove (1990), Kroll et al. (1993), Benisek et al. (2004, 2010)
Ms–Pg	GRE ± AMP Distribution of K and Na between coexisting muscovite and paragonite	Eugster et al. (1972), Chatterjee and Flux (1986), Vance and Holland (1993)
Net transfer equilibria		
Grt–Al_2SiO_5–Pl–Qz (GASP)	AMP + GRA	Ghent (1976), Hodges and Spear (1982), Koziol and Newton (1988), Koziol (1989)
Grt–Al_2SiO_5–Ms–Qz	AMP ± GRA	Hodges and Crowley (1985), Hoisch (1991)
Grt–Al_2SiO_5–Bt–Ms–Qz	AMP ± GRA	Hodges and Crowley (1985), Holdaway et al. (1988), Wu and Zhao (2007b)
Grt–Pl–Bt–Ms	AMP	Ghent and Stout (1981), Hoisch (1990, 1991), Wu (2015)
Grt–Pl–Ms–Qz	AMP	Hodges and Crowley (1985), Hoisch (1991), Wu and Zhao (2006)
Grt–Pl–Bt–Qz	AMP	Wu et al. (2004), Wu and Zhao (2006)
Grt–Pl–Hbl–Qz	BLU + AMP ± GRA	Kohn and Spear (1989), Dale et al. (2000)
Grt–Pl–Ol	GRA	Johnson and Essene (1982), Bohlen et al. (1983a, b), Wu and Zhao (2007a)
Grt–Pl–Opx–Qz	GRA	Perkins and Chipera (1985), Eckert et al. (1991), Faulhaber and Raith (1991), Bhattacharya et al. (1991)
Grt–Pl–Cpx–Qz	GRA	Moecher et al. (1988), Bohlen and Liotta (1986), Anovitz and Essene (1987b), Wu and Zhao (2006)
Grt–Rt–Ilm–Pl–Qz (GRIPS)	AMP + GRA	Bohlen and Liotta (1986), Anovitz and Essene (1987b)
Grt–Al_2SiO_5–Ilm–Rt–Qz (GRAIL)	AMP + GRA	Bohlen et al. (1983b), Essene and Bohlen (1985) Bohlen and Liotta (1986), Koziol and Bohlen (1992)

(continued)

Table 2.1 (continued)

Mineral assemblages	Range of applications	References
Grt–Crd–Sil–Qz	AMP + GRA	Hensen and Green (1973), Holdaway and Lee (1977), Aranovich and Podlesskii (1989)
Grt–Cpx–Opx–Pl–Qz	AMP + GRA	Paria et al. (1988)
Grt–Spl–Sil–Crn	GRA	Shulters and Bohlen (1989)
Grt–Spl–Sil–Qz	GRA	Bohlen et al. (1983b)
Grt–Pl–Wo–Qz	AMP + GRA	Huckenholz et al. (1981)
Cpx–Pl–Qz	BLU + ECL + GRA	Newton (1983), Gasparik (1984), Hemingway et al. (1981), Liou et al. (1987), McCartey and Patino Douce (1998)
Opx–Ol–Qz	GRA	Bohlen and Boettcher (1981), Newton (1983)
Hbl–Pl–Qz	AMP	Bhadra and Bhattacharya (2007)
Grt–Opx	ECL	Carswell and Harley (1989), Brey et al. (2008)
Grt–Cpx–Phe	ECL	Waters and Martin (1993)
Grt–Cpx–Ky–Phe–Qz	BLU + ECL	Holland (1979), Ravna and Terry (2004)
Other major element geothermobarometers		
Phe barometry	GRE + AMP + BLU + ECL This barometer is based on the phengite content of white mica	Powell and Evans (1983), Massonne and Schreyer (1987), Bucher-Nurminen (1987)
Sp–Py–Po barometry	GRE + BLU Fe content of sphalerite coexisting with pyrrhotite and pyrite	Scott (1973, 1976), Lusk and Ford (1978), Jamieson and Crow (1987), Bryndzia et al. (1990)
Mag–Ilm thermometry	AMP + GRA Ti content in magnetite coexisting with ilmenite	Buddington and Lindsley (1964), Stormer (1983), Ghiorso and Sack (1991)
Hbl barometry	AMP Al and/or Na content of hornblende in certain igneous assemblages	Brown (1977), Hollister et al. (1987), Schmidt (1992), Anderson and Smith (1995)
Hbl thermometry	AMP Al and Si contents of hornblende taking into account Fe^{3+} content	Gerya (2002), Zenk and Schulz (2004)
Crd thermometry	AMP + GRA Na content of cordierite	Mirwald et al. (2008)
Na–Ca Amp–Ep–Chl thermometry	AMP + BLU Si, Al, Fe, and Na contents of amphibole and Al and Fe contents of epidote and chlorite	Triboulet (1992)

(continued)

Table 2.1 (continued)

Mineral assemblages	Range of applications	References
Opx–Grt thermometry	GRA + ECL Al content of orthopyroxene coexisting with garnet	Brey and Kohler (1990), Aranovich and Berman (1997), Harley and Motoyoshi (2000)
Ol–Cpx barometry	ECL Ca content of olivine coexisting with clinopyroxene	Kohler and Brey (1990)
Trace element geothermobarometers		
Grt–Xtm thermometry	GRE + AMP This thermometer is based on the Y content of garnet coexisting with xenotime	Pyle and Spear (2000)
Mnz–Grt thermometry	GRE + AMP + GRA Redistribution of HREE и Y between coexisting monazite and garnet	Pyle et al. (2001)
Mnz–Xtm thermometry	GRE + AMP + GRA Mixing in the system (REE,Y) PO_4	Heinrich et al. (1997), Gratz and Heinrich (1997)
Grt thermometry	ECL Ni content of garnet	Griffin et al. (1989), Canil (1999)
Grt barometry	GRE + AMP + GRA Gd/Dy of garnet	Bea et al. (1997)
Opx–Cpx thermometry	GRA + ECL Sc, V, Cr, Mn, Co contents in coexisting pyroxenes	Seitz et al. (1999)
Hbl thermometry	AMP + GRA La/Yb of hornblende	Skublov and Drugova (2003)
Grt–Opx thermometry	GRA Ti content of garnet and orthopyroxene	Kawasaki and Motoyoshi (2007)
Zrn thermometry	GRA Ti content of zircon	Watson et al. (2006), Ferry and Watson (2007), Hofmann et al. (2013)
Rt thermometry	GRA Zr content of rutile	Zack et al. (2004), Tomkins et al. (2007), Hofmann et al. (2013)
Qz thermometry	GRA Ti content of quartz	Wark and Watson (2006), Kawasaki and Osanai (2008)
Qz thermobarometry	GRA Ti content of quartz	Thomas et al. (2010), Huang and Audétat (2012)

Abbreviations for metamorphic facies: *GRE* Greenschist; *BLU* Blueschist; *AMP* Amphibolite; *GRA* Granulite; *ECL* Eclogite. Mineral abbreviations are given at the beginning of the book. Most cited and recent references for each calibration of thermobarometers are given only

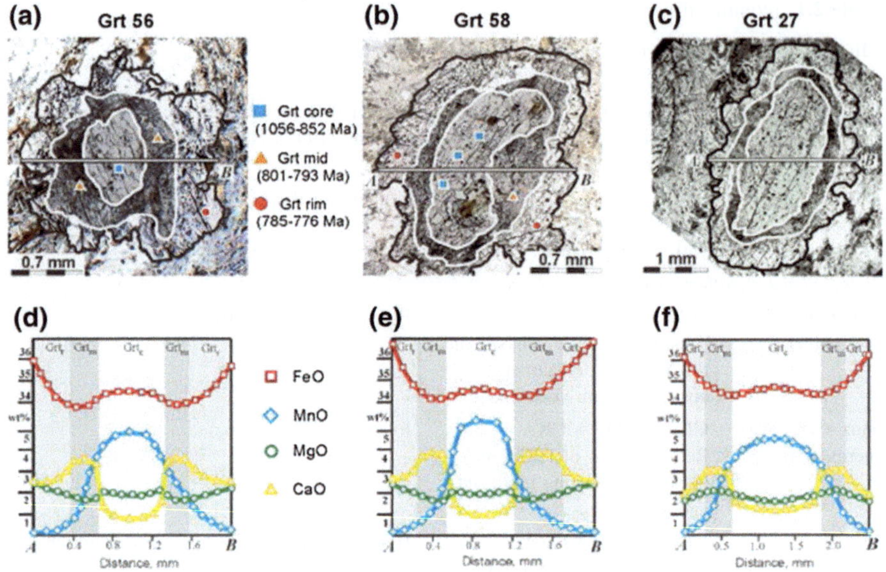

Fig. 2.1 Photomicrographs of samples 56 (**a**), 58 (**b**), and 27 (**c**) showing texture features developed within garnets from pelitic gneisses and schists in the Garevka complex. Compositional profile across a zoned garnet porphyroblasts (sample 56—**d**, sample 58—**e**, sample 27—**f**) with three growth zones is indicated by the light line A-B. Compositionally distinct zones of garnet are indicated by white (core-Grt_c), and different shades of grey (mantle-Grt_m and rim-Grt_r). The locations of dated samples (sample 56—**a**, sample 58—**b**) are given with the age of the dated grain corresponding to the symbol that they are represented by

2.2 Mineral Geothermometers Based on Exchange Reactions

Exchange equilibria involve redistribution of isomorphic components between different minerals or different sites in one mineral. Because the volume changes are very small and the enthalpy effects are significant, the pressure dependence of exchange reactions is not very important. Consequently, they are suitable for metamorphic thermometry. Exchange reactions do not change mineral modes. Only widely used and reliable experimentally-based exchange equilibria are listed in Table 2.1.

Thermometers applicable to metamorphic rocks involve Fe^{2+}–Mg exchange between mineral pairs. Most of the listed equilibria involve garnet. The most reliable and popular geothermometers include Grt–Cpx, Grt–Bt, and Grt–Opx because of the widespread occurrence of these assemblages in rocks (Table 2.1). The assemblage garnet + clinopyroxene is widespread in amphibolite and granulite facies metamorphic rocks, in various eclogtes, blueschists, and garnet peridotites. There are more than 20 calibrations of the Grt–Bt thermometer, which were discussed in detail by Holdaway (2000). The assemblage Grt–Opx is stable over a wide range of temperature (600–1500 °C) and pressure (3–50 kbar), which determines its petrologic significance. A geothermometer calibrated by the solubility of Al_2O_3 in orthopyroxene coexisting with garnet is most useful for estimating equilibration *P* and *T* in ultrahigh-temperature rocks, because other geo geothermometers often yield large underestimates (by 100–150 °C) due to Fe and Mg retrograde exchange by diffusion. At the same time, aluminum is the least mobile among the other components involved in the reaction (Anovitz 1991), and its content in orthopyroxene controlled by the tschermak exchange (Fe, Mg) Si ↔ Al Al is strongly dependent on temperature. In view of the above, the RCLC program applying a correction for retrograde effects was created to estimate *P-T* conditions of formation of ultrahigh-temperature granulites (Pattison et al. 2003). In the absence of any diagnostic indicators of high-temperature metamorphism, the elevated Al_2O_3 content of hypersthene (6–10 wt.%) is thought to be indicative of temperatures above ~900 °C (Harley 2008). The Opx–Cpx, Ol–Opx, and Amp–Pl geothermometers are the most reliable for garnet-free assemblages. The Chl–Ms, Bt–Ms, and Cld–Bt thermometers are used to derive reliable temperature estimates for mineral assemblages of the greenschist facies. The other well-known geothermometers include magnetite-ilmenite and amphibole geothermometers; the first one is based on the Ti content of magnetite coexisting with ilmenite, the second one is based on Al, Si, and Fe^{3+} contents of hornblende.

2.3 Solvus Geothermometers

Solvus thermometers record the exsolution temperature of solid solutions. Two types of component redistribution are used in practice: Ca and Mg in pyroxene and carbonate, as well as K and Na in mica and feldspar (Table 2.1). For the mineral pairs Kfs–Ab and Cal–Dol, these thermometers have a restricted range of applicability and are often useful for estimating the lower temperature limit. This is also true for perthites, in which the exsolution of solid solutions for kinetic reasons might have occurred below a solid solubility limit (Bucher and Grapes 2011). For the pair calcite–dolomite, the early exsolution features are obliterated by recrystallization; for the pair clinopyroxene–orthopyroxene, the Fe^{3+} content has marked effect on the solvus temperature.

2.4 Mineral Geothermobarometers Based on Net-Transfer Reactions

The effect of pressure is significant in "continuous" or sliding equilibria with a variance of ≥ 2. The change in the equilibrium constant of a divariant reaction is accompanied by changes in mineral modes, and the end-member ratios of these minerals are characterized by significant volume changes. Such equilibria with high *dP*/*dT* are sensitive geobarometers, although they all of them are dependent on temperature as well. A good barometer is one with a large volume change (ΔV). Many geobarometers of this type contain garnet and plagioclase; the corresponding equilibria account for grossular component in garnet and anorthite component in plagioclase balanced with different components from coexisting micas, amphiboles, pyroxenes, aluminosilicate, and quartz. These geobarometers may be applied in a wide variety of different rock types. The best known of these is the GASP geobarometer (Grt–Als–Qz–Pl), originally calibrated by Ghent (1976). There are nearly 10 versions of this barometer, which are widely used to obtain pressure estimates for aluminous amphibolite and granulite facies rocks. All of these equilibria depend on the mixing behavior of solid solutions (here garnet) and have limitations with respect to Ca content in garnet and plagioclase. An alternative to this geobarometer is the Grt–Bt–Ms–Pl geobarometer applicable to a wide range of greenschist and amphibolite facies conditions. It probably takes advantage over the GASP geobarometer in pressure estimation because (1) its accuracy does not depend on the number of coexisting Al_2SiO_5 polymorphs and (2) the equilibrium constant is weakly dependent on temperature (Avchenko 1990). However, this barometer is not recommended for rocks containing garnets with less than 3% of grossular component and plagioclase with less than 17% anorthite component.

Another set of geobarometers is based on reactions between the almandine component of garnet and Fe–Ti oxides. The best known of these is the GRAIL (Grt–Rt–Als–Ilm–Qz) and GRIPS (Grt–Rt–Ilm–Pl–Qz) barometers. The Grt–Amp–Pl–Qz geobarometer commonly used to obtain pressure estimates for amphibolites is limited to the following composition of coexisting minerals: $K < 0.4$ a.p.f.u. (atoms per formula unit) and $Na > 0.6$ a.p.f.u. in amphibole; $Ca > 0.15$ a.p.f.u. in plagioclase and $Mn < 0.45$ a.p.f.u. in garnet. Geobarometers based on pyroxene-plagioclase and pyroxene-olivine in equilibrium with and without garnet are used for high-temperature conditions (Table 2.1).

Still another barometer calibrated against the distribution of major elements in minerals is that based on the Fe content of sphalerite in equilibrium with pyrrhotite and pyrite. It is used to obtain pressure estimate for greenschist and blueschist facies rocks. Other practical tools are monomineralic Amp and Ph geobarometers, which provide pressure estimates based on the Al content of hornblende and the amount of phengite component in muscovite. One of the advantages of the phengite geobarometer is that it is applicable to a wide range of *P-T* conditions of metamorphism.

2.5 Mineral Geothermobarometers Based on Trace Element Partitioning

These include two groups of geothermobarometers that are based on the distribution of rare-earth and trace elements. The advantage of trace element geothermobarometry is the slower diffusion rate of tri- and tetravalent cations (Cherniak et al. 2007) that that of major elements (e.g., Ca, Fe, and Mg) and the strong temperature dependence of their partitioning in minerals. Therefore, this type of geothermobarometers appears to offer a considerable potential for the interpretation of metamorphic complexes with a complicated evolutionary history.

Among the REE-based geothermobarometers, the Mnz–Grt and Mnz–Xtm geothermometers were calibrated in the temperature range 400–800 °C and thus can be used to estimate temperatures from greenschist to granulite facies conditions of metamorphism. The Grt–Xtm geothermometer is sensitive to lower-temperature conditions of the garnet and staurolite zones of regional metamorphism (400–500 °C). The amphibole geothermometer, which uses the La/Yb ratio in hornblende can be applied to garnet-free assemblages containing a small proportion of mafic minerals. Application of the Grt geobarometer based on preferential partitioning of trivalent REE in garnet with increasing pressure is limited to the pressure range 4–9 kbar; it cannot be used for higher-pressure complexes.

Another group of geothermometers is useful for the analysis of granulites and pyroxene hornfelses. Thermometry of these rocks using conventional exchange equilibria is complicated by the effect of late retrograde processes, which cause diffusive redistribution of components and/or exsolution of solid solutions. Other REE-based geothermometers include Ti-in-zircon, Zr-in-rutile, and Ti-in-quartz thermometers. The latter (Ti-in-quartz) was also calibrated as a geobarometer. A prerequisite for the application of these tools is the Ti or Zr content of a rock, which favors the appearance of rutile (for Ti-in-zircon and Ti-in-quartz thermometry) or zircon (for Zr-in-rutile thermometry). There is also a thermometer based on Ti partitioning in garnet and orthopyroxene, but it is not widely applicable.

It should be emphasized that in geothermobarometry insufficient attention has been paid to the assessment of uncertainties on *P-T* parameters of metamorphism, although several attempts have been made to account for this, at least for some geothermobarometers. For example, the uncertainties on individual *P-T* estimates calculated using the Grt–Bt (Ferry and Spear 1978) and Grt–Bt–Ms–Pl (Ghent and Stout 1981) geothermometers, and taking account of analytical imprecision of microprobe determinations and errors in a reaction enthalpy do not exceed ±50 °C and ±0.5 kbar (Likhanov et al. 2004a). This is in good agreement with the most widely quoted cited systematic uncertainties in individual geobarometers, including GASP and GRAIL (e.g., Kohn and Spear 1991; Hodges and Crowley 1985; Hodges and McKenna 1987). *P-T* estimates of metamorphism obtained using six Grt–Bt geothermometers and three geobarometers vary significantly between different generations of garnet growth and are similar, within these uncertainties, for the respective zones of garnets collected from three different localities (Table 2.2).

To avoid the random uncertainty contributions to routine calculations, one should use the software packages PET and PTQuick for calculation of *P-T* conditions of the mineral equilibria. The program PET (Petrological Element Tools) developed by Dachs (1998) for MATHEMATICA 5.0 can be downloaded from the site http://www.sbg.ac.at/min/service/petquide.htm. The program PTQuick developed by Dolivo-Dobrovolsky can be downloaded from the site http://www.

Table 2.2 Summary of pressure and temperature conditions calculated with conventional thermobarometry, THERMOCALC and TWQ software for selected samples of Garevka complex

Sample number	*T* (°C)											*P* (kbar)				
	Mineral geothermometers							Thermocalc			TWQ	Geobarometers			Thermocalc	TWQ
	1	2	3	4	5	6	T_{ave}	7	8	9	10	11	12	13	7	10
Yenisei region																
56c	551	553	564	590	574	562	565	576 ± 42	0.96	2.09	543 ± 38	4.97	5.00	5.14	6.3 ± 1.6	4.7 ± 0.6
56m	625	612	616	632	651	614	625	662 ± 65	0.14	2.73	578 ± 69	8.31	7.03	8.16	11.6 ± 2.1	7.9 ± 0.9
56r	478	536	524	521	418	513	500	584 ± 37	0.38	2.63	539 ± 160	4.96	4.59	4.88	6.4 ± 1.9	5.8 ± 2.0
Tis region																
58c	567	566	578	605	601	577	580	543 ± 42	0.73	2.59	564 ± 31	4.81	4.54	4.43	3.1 ± 2.2	4.4 ± 0.4
58m	632	614	618	634	657	615	630	734 ± 66	0.91	2.99	591 ± 74	8.19	6.88	8.17	10.5 ± 3.3	8.2 ± 1.0
58r	483	536	518	524	428	514	500	553 ± 20	0.72	2.19	531 ± 24	5.26	5.31	5.71	8.4 ± 1.4	6.5 ± 2.5
Garevka region																
27c	505	549	545	578	517	560	545	656 ± 18	0.90	2.7	530 ± 33	4.2	4.76	4.5	6.8 ± 0.75	3.9 ± 0.3
27m	758	627	675	681	759	663	690	697 ± 20	0.73	4.0	594 ± 6	9.8	9.61	9.3	11.4 ± 0.9	8.6 ± 0.5
27r	530	567	546	550	546	557	550	657 ± 15	0.75	2.93	535 ± 29	4.9	4.96	4.8	7.9 ± 0.7	4.6 ± 0.3

1—Grt–Bt geothermometer (Ferry and Spear 1978; Hodges and Spear 1982), 2—Grt–Bt geothermometer (Kleemann and Reinhardt 1994), 3—Grt–Bt geothermometer (Perchuk and Lavrent'eva 1983), 4—Grt–Bt geothermometer (Kaneko and Miyano 2004), 5—Grt–Bt geothermometer (Holdaway et al. 1997), 6—Grt–Bt geothermometer (Holdaway 2000), T_{ave}—average temperatures obtained from used exchange thermometers; 7—THERMOCALC calculations (Powell and Holland 1994; Holland and Powell 1998); 8—(σ_{cor}) the correlation coefficient between uncertainties (σT and σP) for a given calculation (σ_{cor}), 9—(σ_{fit}) a measure of the scatter in residuals (the observed minus the calculated values) of the enthalpies and activities normalized by their uncertainties (for further details, see Powell and Holland 1994); 10—TWQ calculations (Berman 1991; Berman and Aranovich 1996); 11—Grt-Bt–Ms–Pl geobarometer (Ghent and Stout 1981; Hodges and Crowley 1985), 12—Grt-Bt–Ms–Pl geobarometer (Hoisch 1990), 13—Grt-Bt–Pl–Qz geobarometer (Wu et al. 2004). The results of the average *P-T* calculations made using the computer software THERMOCALC and TWQ are shown with $\pm 2\sigma$ errors. C, m, and r correspond to core, mantle and rim garnet zones, respectively dimadd.ru/ru/Programs/ptquick. The software PTQuick is used mainly for high-pressure metamorphic rocks while PET is applicable over a wider range of *P*, varying from low to high pressures 69.

2.6 Geothermobarometry Using Multi-equilibria and Internally Consistent Thermodynamic Datasets

Of key importance is the comparison between estimates obtained using various thermobarometers. For example, the Grt–Opx–Pl–Qz (Bhattacharya et al. 1991) and Amp–Grt–Pl (Kohn and Spear 1989) geothermobarometers are often used to estimate *P-T* conditions of granulite and amphibolite facies rocks, respectively. However, what are the grounds for comparing the *P-T* results obtained using different geothermobarometers and to what extent they are consistent with other? In this case, thermobarometry using multi-equilibrium calculations (MET) appear to be particularly useful. This method based on the internally consistent thermodynamic data permits the computation of *P-T* parameters for an independent set of mineral equilibria. The internally consistent thermodynamic dataset is the same for high- and low-temperature equilibria, allowing a correct comparison of the estimated conditions of formation of different metamorphic complexes.

The internally consistent thermodynamic datasets allow application of a large number of equilibria that govern the paragenetic relations within a metamorphic rock. This problem can be solved using the dedicated software programs for thermodynamic data on pure endmembers, the most popular of which are THERMOCALC (Powell and Holland 1994) and TWQ or TWEEQU (Thermobarometry with Estimation of Equilibration State) (Berman 1991). Average *P-T* parameters are computed using the least squares in THERMOCALC and the mathematical method of linear programming in TWQ. This can be illustrated by the results on mineral equilibria calculated in TWQ for the Garevka complex gneisses (Yenisei Ridge), which were expressed as lines for the equilibria intersecting at a certain point on the *P-T* diagram (Fig. 2.2). These results are in good agreement with individual geothermobarometers (Table 2.2). Other computer programs using multi-equilibria thermobarometry include DOMINO/THERIAK (de Capitani and Petrakakis 2010) and PERPLEX (Connolly 1990). They permit the computation of *P-T* for mineral parageneses with a sufficient set of end-member reactions. In other cases, a special option for calculating pressure at a specified *T* and vice versa should be utilized. Nevertheless, the computation of one parameter at a specified second parameter and given the justified equilibrium relationship between the observed mineral compositions should involve a smaller uncertainty. It is possible to account for the effect of each end-member on the computation results and the degree of deviation of entropy and activity of each end-member from the input information (given that equilibrium conditions of *P* and *T* are attained). One apparent advantage of this approach is the mathematical optimization of the results from different types of consistent thermometers and barometers.

2.7 Geothermobarometry Using Zoned Minerals

There are two software packages that are used to model the metamorphic *P-T-t* paths: GEOPATH (Gerya and Perchuk 1990) and PTPATH (Spear 1986; Spear et al. 1991). A detailed description of the operation and routines applied to analyze mineral zoning and reaction textures by the GEOPATH program package can be found in Perchuk et al. (2000). Note that derivation of the *P-T* paths should be controlled using isopleths of mineral compositions calculated for studied mineral assemblages so that the relationship of the *P-T* path with isopleths of the relative amounts of minerals should be consistent with the chemical compositions and zoning of the different textural generations of minerals.

The approach used in the PTPATH package is referred to as the Gibbs method. This differential thermodynamic method uses an analytical formulation of the phase equilibria of a given mineral assemblage in such a way that changes in the composition of coexisting minerals in the assemblage can be monitored as functions of changing *P* and *T*. It is assumed that the phases were in equilibrium during the entire period of mineral growth and zoning. The theoretical basis of this method was described by Spear and Selverstone (1983); the mathematical expressions and the procedure of the thermodynamic calculations were described in detail by Spear (1993). Note that all necessary differential thermodynamic equations are written redardless of the modal contents of the minerals and the bulk compositions of the rocks. Unlike GEOPATH, PTPATH uses all necessary differential thermodynamic equations written regardless of the modal contents of the minerals and the bulk compositions of the rocks. The calculation of metamorphic *P-T* paths by this approach should be based on information on the textural features of the rocks, their mineral assemblages, sequence of mineral reactions, the chemical compositions of minerals, as well as a correlation between the chemical compositions of zoned minerals, and the *P-T* parameters at a certain time. Considering the variance of the system, only key components must be specified to solve equations. The differential calculations are tied to a starting point in *P-T* space where an assemblage and the associated mineral compositions are known. The resolution of this method is within ±10 °C and ±0.1 kbar (Selverstone et al. 1984). The PTPATH packages can be downloaded from the site http://ees2.geo.rpi.edu/MetaPetaRen.

An example illustrating the capabilities of PTPATH (Fig. 2.3) shows the calculated *P-T-t* path for Garevka gneisses of the Yenisei Ridge (Likhanov et al. 2015). The *P-T* diagram shows generalized counter-clockwise *P-T* paths, which are initiated with a low-pressure event followed by a near-isothermal compression and ended with a retrograde decompression. All *P-T* paths have nearly identical slopes for the prograde segment of the metamorphic evolution and differ mainly in the length of the recorded *P-T* history Fig. 2.3. The *P-T* conditions calculated with TWQ (Fig. 2.2) are generally in good agreement with the *P-T* parameters estimated using conventional thermobarometry and *P-T* path calculations (Table 2.2; Fig. 2.3). The slightly higher average *P-T* values calculated using THERMOCALC are broadly consistent with those obtained from conventional geothermobarometry

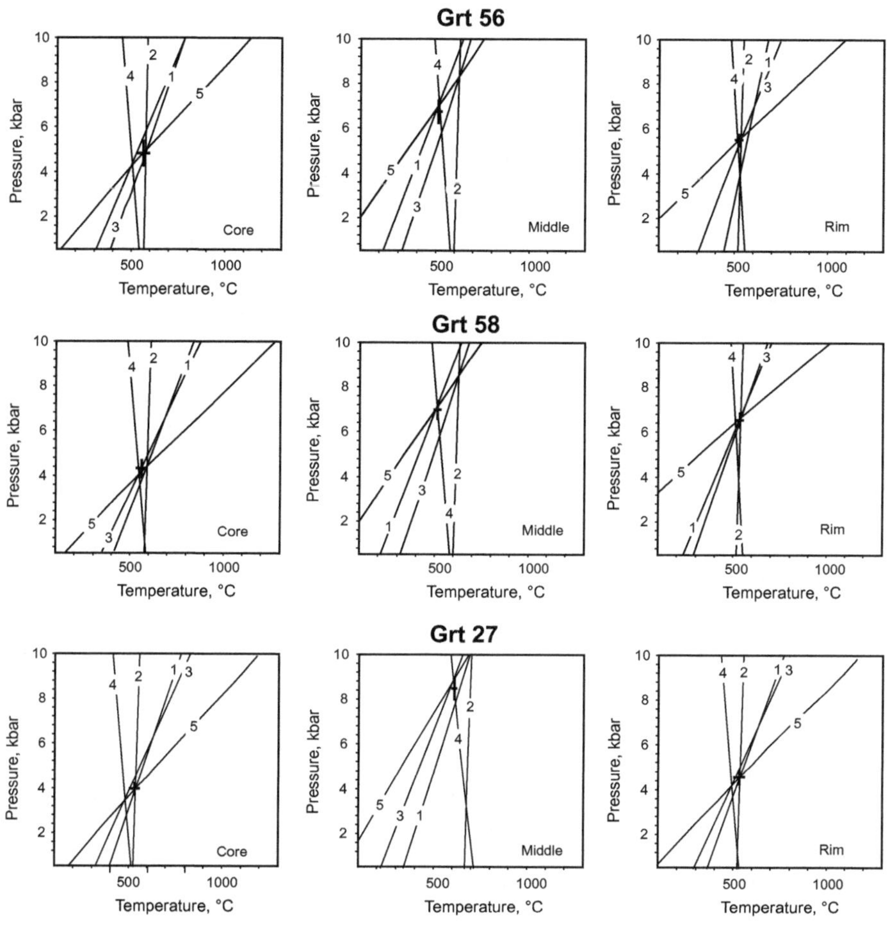

Fig. 2.2 *P-T* estimates based on the TWQ intersections for different stages of garnet evolution. The equilibria for determining (using subprogram INTERSX) average *P-T* results with standard deviations are listed as follows: 1—Prp+Ms=2Ky+Phl+Qz; 2—Phl+Alm=Ann+Prp; 3—Qz+2Ky +Grs=3An; 4—Ms+Grs+Alm=Ann+3An; 5—Prp+Ms+Grs=3An+Phl; with numbers on reaction line corresponding to those in Fig. 2.2

for individual garnet zones and with the overall range of metamorphic pressure and temperature conditions (Table 2.2).

THERMOCALC can also be used for the calculations of *P-T-t* paths from zoned minerals. However, one of the essential systematic problems involved in this package is the determination of the effective bulk composition, a key parameter for constructing phase diagram. This composition may not match the bulk composition of the rock because of the isolation of cores of zoned minerals from the metamorphic reaction. However, this difficulty may be offset by extraction of core composition of a zoned mineral from the overall bulk composition of a rock Core

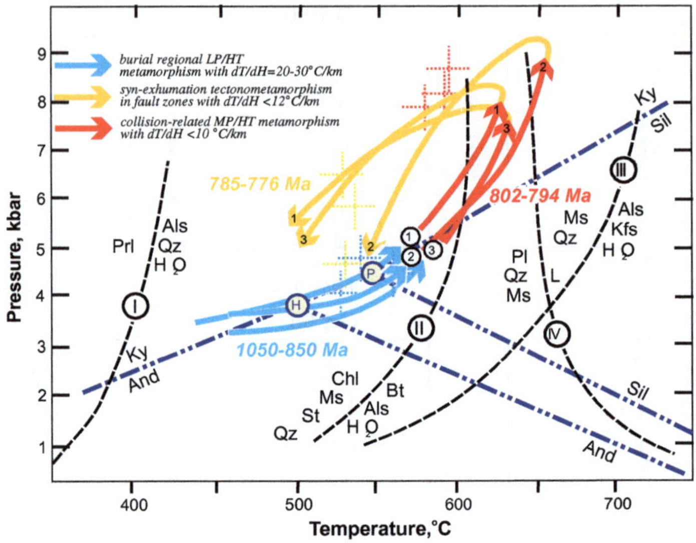

Fig. 2.3 *P-T* diagram showing the generalized counterclockwise *P-T-t* path calculations for metapelites of the different types of metamorphism in the Garevka complex (curve 1—sample 56, curve 2—sample 27 and curve 3—sample 58). The prograde segments of *P-T* trajectories derived from chemical zonation patterns in minerals correspond to the low-pressure regional metamorphism (blue arrows) and medium-pressure collision-related metamorphism (red arrows). The retrograde segment of the *P-T* path for Garevka complex (yellow arrows) reflects the post-collision thrust exhumation of the rocks to upper crustal levels. Each cross labeled by corresponding color for different types of metamorphism indicates inferred peak *P-T* estimates calculated using TWQ. Curve I is the lowest temperature stability of Al_2SiO_5 in aluminous pelites (Haas and Holdaway 1973); curve II shows the upper stability of staurolite + quartz + muscovite + chlorite (Pattison 2001); curve III is muscovite + quartz breakdown (Chatterjee and Johannes 1974); curve IV shows minimum wet melting curve for pelites (Le Breton and Thompson 1988). The coordinates of the aluminum silicate triple point and univariant equilibrium curves of Al_2SiO_5 polymorphs are given after Pattison (1992) (P) and Holdaway (1971) (H)

composition of a zoned mineral, reduced form the whole-rock, is calculated by average cation contents and modal amount of zoned mineral in the thin-sections (Faryad and Chakraborty 2005).

2.8 Geothermobarometry Using P-T Phase Diagrams—Petrogenetic Grids and Pseudosections

P-T phase diagrams are an important tool for analyzing metamorphic processes. The most widely used types of phase diagrams are the *P-T* projections and *P-T* pseudosections. *P-T* projections which Bowen (1940) called "petrogenetic grids" display: (1) thermodynamically stable mineral assemblages in equilibrium with specific *P* and *T* and (2) possible reactions which could take place under conditions of changing thermodynamic parameters. The early versions of petrogenetic grids were based on the observed natural parageneses and used minerals of fixed composition. The second generation of petrogenetic grids accounting for variation in the Fe content of coexisting phases was a combination of petrologic observations and experimental calibrations of a reaction between end-members in the KFASH and KMASH systems. The modern petrogenetic grids are constructed by calculating a series of univariant and divariant reactions using thermodynamic datasets and are developed for the main petrochemical rock types. The most useful types of *P-T* diagrams include (a system is shown in parentheses; *P-T* range of calculation): meta-ultrabasites (NCMASH; 3–22 kbar/700–950 °C) (Schmadicke 2000), metacarbonates (CMSH-CO_2; 0–10 kbar/400–800 °C) (Carmichael 1991), metabasites (NCFMASH; 4–23 kbar/380–620 °C) (Will et al. 1998), metapelites (MnKFMASH; 2–12 kbar/450–700 °C) (Mahar et al. 1997), ultrahigh-temperature granulites (KFMASH; 5–12.5 kbar/840–1000 °C) (Carrington and Harley 1995). Unlike other methods of geothermobarometry, generalized petrogenetic grids give only rough constraints on the *P-T* conditions of metamorphism. However, some inconsistencies remain between petrogenetic grids applied to the same petrogenetic rock type. For example, the assemblage Cld + Bt on the diagrams of Harte and Hudson (1979) is stable at relatively low and high pressures over a narrow temperature range, whereas on the petrogenetic grid of Spear and Cheney (1989) the same assemblage is stable over a wide range of *P-T* conditions. This inconsistency arises from the critical influence of the chemical composition of metapelites, namely, their Fe and Al contents, on the formation of mineral assemblages (Likhanov et al. 2001, 2004b, 2005; Likhanov 1988a, b). These problems may be addressed by using another type of diagrams, the so-called pseudosections. In contrast to petrogenetic grids, pseudosections display the actual stable mineral assemblages for the specific chemical composition of an investigated rock at a specified range of *P-T* conditions. The prefix "pseudo-" refers to the circumstance that not all mineral assemblages calculated for a specified chemical composition are necessarily observed in nature. Pseudosections are calculated using a variety of techniques and computer programs, the most popular of which is THERMOCALC (Powell and Holland 1994). Other programs include DOMINO/THERIAK (de Capitani and Petrakakis 2010), PERPLEX (Connolly 1990), and SELEKTOR-C (Karpov 1981; Karpov et al. 2001), which incorporate minimization of thermodynamic functions.

Compared to other methods, THERMOCALC provides sufficient precision to determine *P-T* parameters. For example, the largest uncertainties in contouring isopleths for the composition of some zoned minerals in pseudosections do not exceed 0.3 kbar and 10 °C (Kelsey 2008). In SELEKTOR-C, of special importance is the quantitative (modal) abundance of a phase calculated from observed chemical compositions of minerals using the service package MS (available from the site http://www.fegi.ru/innov/461-mc). The advantage of this program is that it has two options RESIDUL and ROCK. The first one permits us to evaluate to what degree the calculated modal composition deviates from the actual rock composition. The second one is used for inverse calculations of the chemical composition of rock from the chemical composition of minerals and their modal abundance, and allows an independent control on the calculated modal composition of a rock. An alternative package used to estimate mineral proportions in a rock of different mineralogical composition is MODAN, an interactive computer program (Pactunc 1998). It can be used to trace interactively changing properties of a system with changing temperature, pressure, fluid composition, etc. The fact that with the specified bulk composition at specified *P-T* parameters close to those calculated with thermobarometers it is possible to reconstruct the observed paragenesis including the compositions of minerals involved provides direct evidence that the rock has attained equilibrium corresponding to the minimum of the Gibbs free energy (Avchenko et al. 2009).

As an example, Fig. 2.4 shows a pseudosection (*P-T-X* diagram) for Fe- and Al-rich pelitic gneisses of the Garevka complex of the Yenisei Ridge in the system

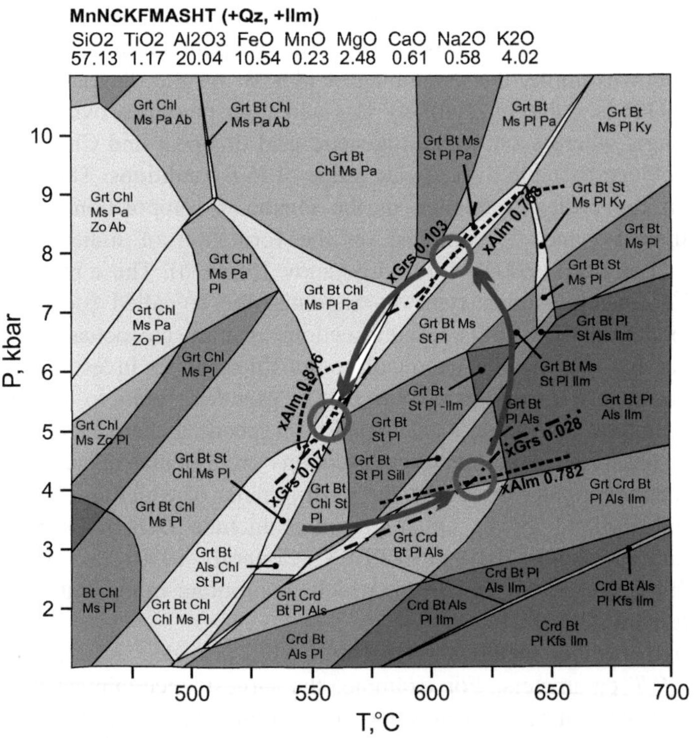

Fig. 2.4 *P-T-X* pseudosection for Fe- and Al-rich gneisses (sample 56) in the system Na_2O–CaO–K_2O–FeO–MgO–MnO–Al_2O_3–SiO_2–TiO_2–H_2O. The pseudosection has been computed at $aH_2O = 1$ using PERPLEX 668 (Connolly 1990, 2005) and the internally consistent thermodynamic dataset 5.5 (Holland and Powell 1998). Inferred peak conditions for different re-equilibrated stages are defined by the intersection of compositional isopleths for the almandine and grossular components (X_{Alm} and X_{Grs}) of garnet. The thick arrows mark the *P-T* paths. The sequence of the inferred mineral assemblages constrains the counterclockwise *P-T* loop

Na_2O–CaO–K_2O–FeO–MgO–MnO–Al_2O_3–SiO_2–TiO_2–H_2O. The temperature and pressure for the mineral equilibria were constructed using the intersection of isopleths of grossular and almandine in different generations of zoned garnet. These results are in accordance with those estimated by other methods of geothermobarometry (Table 2.2; Figs. 2.2, 2.3 and 2.4).

Pseudosections are potentially very useful for geothermometry of low-grade metamorphic rocks (blueschist, zeolite and prehnite-pumpellyite facies/subfacies) due to a lack of well-calibrated mineral indicators for low temperatures. Geothermometry of high-temperature rocks (granulites and pyroxene hornfelses) is complicated by the effect of late retrograde processes, which result in diffusion of components in minerals and exsolution of solid solutions. Traces of the prograde reaction history are seldom preserved in these metamorphic facies rocks. In contrast, medium-temperature facies rocks may better preserve prograde and retrograde events, which can make them most informative for the interpretation of the *P-T* evolution of metamorphic complexes. Application of geobarometers allows us to readily discriminate between low- (zeolite and prehnite-pumpellyite facies/subfacies, hornfelses), medium- (epidote-amphibolite and amphibolite facies), and high-pressure (granulites, blueschists, and eclogites) rocks.

In summary, it is worth noting that in spite of potential pitfalls, all thermobarometric methods are powerful and useful tools for providing detailed information on *P-T* histories. The largest uncertainties on *P-T* estimates arise from the imperfectness of existing solid solution models. It is likely that geothermobarometry based on trace and rare-earth elements has good potential due to the slower diffusion rates of tri- and tetravalent cations as compared to those of major elements and due to the strong temperature-dependence of the partitioning of these cations.

References

Aleksandrov IA (2010) Metamorficheskiye porody amfibolitovoy fatsii Dzhugdzhuro-Stanovoy skladchatoy oblasti: usloviya obrazovaniya i sostav protolitov (Metamorphic rocks of amphibolite facies of the Dzhugdzhuro-Stanovoy folded region: conditions of formation and composition of protolith). Dalnauka, Vladivostok

Anderson JL, Smith DR (1995) The effects of temperature and fO_2 on the Al-in-hornblende barometer. Am Mineral 80:549–559

Anovitz LM (1991) Al zoning in pyroxene and plagioclase: window on the late prograde to early retrograde *P-T* paths in granulite terrains. Am Mineral 76:1328–1343

Anovitz LM, Essene EJ (1987a) Phase equilibria in the system $CaCO_3$–$MgCO_3$–$FeCO_3$. J Petrol 28:389–414

Anovitz LM, Essene EJ (1987b) Compatibility of geobarometers in the system CaO–FeO–Al_2O3–SiO_2–TiO_2 (CFAST): implications for garnet mixing models. J Geol 95:633–645

Aranovich LY (1991) Mineral'nyye ravnovesiya mnogokomponentnykh tverdykh rastvo-rov (Mineral equilibria of multicomponent solid solutions). Nauka, Moscow

Aranovich LY, Podlesskii KK (1989) Geothermobarometry of high-grade metapelites: simultaneously operating reactions. In: Yardley BWD, Daly JS, Cliff RA (eds) Evolution of metamorphic belts, vol 43. Geological Society Special Publications, Blackwell, London, pp 45–61

Aranovich LY, Berman RG (1997) A new garnet-orthopyroxene thermometer based on reversed Al_2O_3 solubility in $FeO–Al_2O_3–SiO_2$ orthopyroxene. Am Mineral 82:345–353

Avchenko OV (1990) Mineral'nyye ravnovesiya v metamorficheskikh porodakh i problemy geobarotermometrii (Mineral equilibria in metamorphic rocks and the problems of geobarothermometry). Nauka, Moscow

Avchenko OV, Chudnenko KV, Aleksandrov IA (2009) Osnovy fiziko-khimicheskogo modelirovaniya mineral'nykh sistem (The principles of physicochemical modeling of mineral systems). Nauka, Moscow

Bea F, Montero P, Garuti G et al (1997) Pressure-dependence of rare earth element distribution in apphibolite- and granulite-grade garnets. A LA-ICP-MS study. Geost Newslett 21:253–270

Benisek A, Kroll H, Cemic L (2004) New developments in two-feldspar thermometry. Am Mineral 89:1496–1504

Benisek A, Dachs E, Kroll H (2010) A ternary feldspar-mixing model based on calorimetric data: development and application. Contrib Mineral Petrol 160:327–337

Berman RG (1991) Thermobarometry using multi-equilibribrium calculations: a new technique, with petrological applications. Can Mineral 29:833–856

Berman RG, Aranovich LY (1996) Optimized standard state and solution properties of minerals. Contrib Mineral Petrol 126:1–24

Bhadra S, Bhattacharya A (2007) The barometer tremolite + tschermakite + 2 albite + 2 pargasite + 8 quartz: constraints from experimental data at unit silica activity, with application to garnet-free natural assemblages. Am Mineral 92:491–502

Bhattacharya A, Krishnakumar KR, Raith M et al (1991) An improved set of a—X parameters for Fe–Mg–Ca garnets and refinements of the orthopyroxene–garnet thermometer and the orthopyroxene-garnet-plagioclase-quartz barometer. J Petrol 32:629–656

Blundy JD, Holland TJB (1990) Calcic amphibole equilibria and new amphibole-plagioclase geothermometer. Contrib Mineral Petrol 104:208–224

Bohlen SR, Boettcher AL (1981) Experimental investigations and geological applications of orthopyroxene geobarometry. Am Mineral 66:951–964

Bohlen SR, Liotta JJ (1986) A barometer for garnet amphibolites and garnet granulites. J Petrol 27:1025–1056

Bohlen SR, Wall VJ, Boettcher AL (1983a) Experimental investigation and application of garnet granulite equilibria. Contrib Mineral Petrol 83:52–61

Bohlen SR, Wall VJ, Boettcher AL (1983b) Experimental investigations and geologic applications of equilibria in the system $FeO–TiO_2–Al_2O_3–SiO_2–H_2O$. Am Mineral 68:1049–1058

Bowen NL (1940) Progressive metamorphism of siliceous limestone and dolomites. J Geol 48:225–274

Brey GP, Kohler T (1990) Geothermobarometry in four-phase lherzolites. II. New thermobarometers, and practical assessment of existing thermobarometers. J Petrol 31:1353–1378

Brey GP, Bulatov VK, Girnis AV et al (2008) Experimental melting of carbonated peridotite at 6–10 GPa. J Petrol 49:797–821

Brown EH (1977) The crossite content of Ca-amphibole as a guide to pressure of metamorphism. J Petrol 18:53–72

Bryndzia LT, Scott SD, Spry PG (1990) Sphalerite and hexagonal pyrrhotite geobarometer: correction in calibration and application. Econ Geol 85:408–411

Bucher K, Grapes R (2011) Petrogenesis of metamorphic rocks, 8th edn. Springer, Berlin, Heidelberg

Bucher-Nurminen KA (1987) A recalibration of the chlorite-biotite-muscovite geobarometer. Contrib Mineral Petrol 96:519–522

Buddington AF, Lindsley DH (1964) Iron-titanium oxide minerals and their synthetic equivalents. J Petrol 5:310–357

Canil D (1999) The Ni-in-garnet geothermometer: calibration at natural abundances. Contrib Mineral Petrol 136:240–246

Carmichael DM (1991) Univariant mixed-volatile reactions: pressure-temperature phase diagrams and reaction isograds. Can Mineral 29:741–754

Carrington DP, Harley SL (1995) Partial melting and phase relations in high-grade metapelites: an experimental petrogenetic grid in the KFMASH system. Contrib Mineral Petrol 120:270–291

Carswell DA, Harley SL (1989) Mineral barometry and thermometry. In: Carswell DA (ed) Eclogites and related rocks. Blackie, Glasgow, pp 83–110

Chatterjee ND, Flux S (1986) Thermodynamic mixing properties of muscovite-paragonite crystalline solutions at high temperatures and pressures, and their geological applications. J Petrol 27:677–693

Chatterjee ND, Johannes WS (1974) Thermal stability and standard thermodynamic properties of synthetic $2M_1$-muscovite, $KAl_2Al_3Si_3O_{10}(OH)_2$. Contrib Mineral Petrol 48:89–114

Cherniak DJ, Manchester J, Watson EB (2007) Zr and Hf diffusion in rutile. Earth Planet Sci Lett 261:267–279

Connolly JAD (1990) Multivariable phase-diagrams—an algorithm based on generalized thermodynamics. Am J Sci 290:666–718

Connolly JAD (2005) Computation of phase equilibria by linear programming: a tool for geodynamic modeling and its application to subduction zone decarbonation. Earth Planet Sci Lett 236:524–541

Dachs E (1998) PET: petrological elementary tools for mathematics. Comput Geosci 24:219–235

Dahl PS (1980) The thermal-compositional dependence of Fe^{2+}–Mg distributions between coexisting garnet and pyroxene: applications to geothermometry. Am Mineral 65:852–866

Dale J, Holland T, Powell R (2000) Hornblende-garnet-plagioclase thermobarometry: a natural assemblage calibration of the thermodynamics of hornblende. Contrib Mineral Petrol 140: 353–362

David BTC, Boyd FR (1966) The join $Mg_2Si_2O_6$–$CaMgSi_2O_6$ at 30 kbar and its application to pyroxene from kimberlites. J Geophys Res 71:3567–3576

Davidson PM, Lindsley DH (1985) Thermodynamic analysis of quadrilateral pyroxenes. Part II: model calibration from experiments and application to geothermometry. Contrib Mineral Petrol 91:390–404

De Capitani C, Petrakakis K (2010) The computation of equilibrium assemblage diagrams with Theriak/Domino software. Am Mineral 95:1006–1016

Dickenson MP, Hewitt D (1986) A garnet-chlotite geothermometer. Geol Soc Am Abstr 18:584

Docka JA, Berg JH, Klewin K (1986) Geothermometry in the Kiglapait aureole. II. Evaluation of exchange thermometry in a well-constrained settings. J Petrol 27:605–626

Eckert JO, Newton RC, Kleppa OJ (1991) The H of reaction and recalibration of garnet-pyroxene-plagioclase-quartz geobarometers in the CMAS system by solution calorimetry. Am Mineral 76:148–160

Elkins LT, Grove TL (1990) Ternary feldspar experiments and thermodynamic models. Am Mineral 75:544–559

Engi M (1983) Equilibria involving Al–Cr spinel: Mg–Fe exchange with olivine. Experiments, thermodynamic analysis, and consequences for geothermometry. Am J Sci 283A:29–71

Essene EJ (1989) The current status of thermobarometry in metamorphic rocks. In: Daly JS, Cliff RA, Yardley BWD (eds) Evolution of metamorphic belts. Geological Society Special Publication, Blackwell, Oxford, pp 1–44

Essene EJ, Bohlen SR (1985) New garnet barometersin the system CaO–FeO–Al_2O_3–SiO_2–TiO_2 (CFAST). EOS Trans Am Geophys Union 66:386

Eugster HP, Albee AL, Bence AE et al (1972) The two-phase region and excess mixing properties of paragonite-muscovite crystalline solutions. J Petrol 13:147–179

Faryad SW, Chakraborty S (2005) Duration of Eo-Alpine metamorphic events obtained from multicomponent diffusion modeling of garnet: a case study from the Eastern Alps. Contrib Mineral Petrol 150:306–318

Faulhaber S, Raith M (1991) Geothermometry and geobarometry of high-grade rocks: a case study on garnet-pyroxene granulites in southern Sri Lanka. Mineral Mag 55:33–56

Ferry JM, Spear FS (1978) Experimental calibration of the partitioning of Fe and Mg between biotite and garnet. Contrib Mineral Petrol 66:113–117

Ferry JM, Watson EB (2007) New thermodynamic models and revised calibrations for the Ti-in-zircon and Zr-in-rutile thermometers. Contrib Mineral Petrol 154:429–437

Fonarev VI, Konilov AN (1986) Experimental study of Fe–Mg distribution between biotite and orthopyroxene at P = 490. Contrib Mineral Petrol 93:227–235

Fuhrman ML, Lindsley DH (1988) Ternary feldspar modeling and thermometry. Am Mineral 73:201–215

Ganguly J (1979) Garnet and clinopyroxene solid solutions and geothermometry based on Fe-Mg distribution coefficient. Geochim Cosmochim Acta 43:1021–1029

Gasparik T (1984) Experimental study of subsolidus phase relations and mixing properties of pyroxene in the system CaO–Al_2O_3–SiO_2. Geochim Cosmochim Acta 48:2537–2546

Gerya TV (2002) P-T-trendy i model' formirovaniya granulitovykh kompleksov dokembriya (P-T trends and model of formation of Precambrian granulite complexes). Doctor of science dissertation, Moscow State University, Moscow

Gerya TV, Perchuk LL (1990) GEOPATH: a new computer program for geothermobarometry and related calculations with the IBM PC computer. In: Abstracts of the 15th general meeting of IMA, Beijing, 28 June–3 July 1990

Ghent ED (1976) Plagioclase-garnet-Al_2SiO_5-quartz: a potential geobarometer-geothrmometer. Am Mineral 61:710–714

Ghent ED, Stout MZ (1981) Geobarometry and geothermometry of plagioclase-biotite-garnet-muscovite assemblages. Contrib Mineral Petrol 76:92–97

Ghiorso MS, Sack RO (1991) Thermochemistry of the oxide minerals. Rev Mineral 25:265–302

Goldsmith JR, Heard HC (1961) Sub-solidus phase relations in the system $CaCO_3$–$MgCO_3$. J Geol 69:45–74

Graham CM, Powell R (1984) A garnet-hornblende geothermometer: calibration, testing, and application to the Pelona Schists, Southern California. J Metamorph Geol 2:13–21

Gratz R, Heinrich W (1997) Monazite-xenotime thermobarometry: experimental calibration of the miscibility gap in the system $CePO_4$–YPO_4. Am Mineral 82:772–780

Green NL, Usdansky SI (1986) Toward a practical plagioclase-muscovite thermometer. Am Mineral 71:1109–1117

Griffin WL, Cousens DR, Ryan CD et al (1989) Ni in chrome pyrope garnets: a new geothermometer. Contrib Mineral Petrol 103:199–202

Haas H, Holdaway MJ (1973) Equilibria in the system Al_2O_3–SiO_2–H_2O involving the stability limits of pyrophyllite, and thermodynamic data of pyrophyllite. Am J Sci 273:348–357

Harley SL (2008) Refining the *P-T* records of UHT crustal metamorphism. J Metamorph Geol 26:125–154

Harley SL, Motoyoshi Y (2000) Al zoning in orthopyroxene in a sapphirine quartzite: evidence for >1120 °C UHT metamorphism in the Napier Complex, Antarctica, and implications for the entropy of sapphirine. Contrib Mineral Petrol 138:293–307

Harte B, Hudson NFC (1979) Pelite facies series and the temperatures and pressures of Dalradian metamorphism in eastern Scotland. In: Harris AL, Holland CH, Leake BE (eds) The caledonides of the British Isles, vol 8. Geologocal Society Special Publication, Blackwell, Oxford, pp 323–337

Heinrich W, Rehs G, Franz G (1997) Monazite-xenotime miscibility gap thermometry. I. An empirical calibration. J Metamorph Geol 15:3–16

Hemingway BS, Krupka KM, Robie RA (1981) Heat capacities of the alkali feldspars between 350 and 1000 K from differential scanning calorimetry, the thermodynamic functions of the alkali feldspars from 298.15 to 1400 K, and the reaction quartz + jadeite = analbite. Am Mineral 66:1202–1215

Hensen BJ, Green DH (1973) Experimental study of the stability of cordierite and garnet in pelitic compositions at high pressures and temperatures. III Synthesis of experimental data and geological application. Contib Mineral Petrol 38:151–166

Hodges KV, Spear FS (1982) Geothermometry, geobarometry and the Al_2SiO_5 triple point at Mt. Moosilauke, New Hampshire. Am Mineral 67:1118–1134

Hodges KV, Crowley PD (1985) Error estimation and empirical geothermobarometry for pelitic system. Am Mineral 70:702–709

Hodges KV, McKenna LW (1987) Realistic propagation of uncertainties in geologic thermobarometry. Am Mineral 72:671–680

Hofmann AE, Baker MB, Eiler JM (2013) An experimental study of Ti and Zr partitioning among zircon, rutile, and granitic melt. Contrib Mineral Petrol 166:235–253

Hoisch TD (1989) A muscovite-biotite geothermometer. Am Mineral 74:565–572

Hoisch TD (1990) Empirical calibration of six geobarometers for the mineral assemblage quartz + muscovite + biotite + plagioclase + garnet. Contrib Mineral Petrol 104:225–234

Hoisch TD (1991) Equilibria within the mineral assemblage quartz + muscovite + biotite + garnet + plagioclase and implications for the mixing properties of octahedrally coordinated cations in muscovite and biotite. Contrib Mineral Petrol 108:43–54

Holdaway MJ (1971) Stability of andalusite and the aluminum silicate phase diagram. Am J Sci 271:97–131

Holdaway MJ (2000) Application of new experimental and garnet Margules data to the garnet-biotite geothermometer. Am Mineral 85:881–892

Holdaway MJ, Lee SM (1977) Fe–Mg cordierite stability in high-grade pelitic rocks based on experimental, theoretical and natural observations. Contrib Mineral Petrol 63:175–198

Holdaway MJ, Dutrow BL, Hinton RW (1988) Devonian and Carboniferous metamorphism in West-Central Maine: the muscovite-almandine geobarometer and the staurolite problem revisited. Am Mineral 73:20–47

Holdaway MJ, Mukhopadhyay B, Dyar MD et al (1997) Garnet-biotite geothermometry revised: new Margules parameters and a natural specimen data set from Maine. Am Mineral 82:582–595

Holland TJB (1979) Experimental determination of the reaction Paragonite = Jadeite + Kyanite + H_2O, and internally consistent thermodynamic data for part of the system Na_2O–Al_2O_3–SiO_2–H_2O, with application to eclogites and blueschists. Contib Mineral Petrol 68:292–301

Holland TJB, Blundy JD (1994) Non-ideal interactions in calcic amphiboles and their bearing on amphibole-plagioclase thermometry. Contrib Mineral Petrol 116:433–447

Holland TJB, Powell R (1998) An internally consistent thermodynamic data set for phases of petrological interest. J Metamorph Geol 16:309–343

Hollister LS, Grissom GC, Peters EK et al (1987) Confirmation of the empirical correlation of Al in hornblende with pressure of solidification of calc-alkaline plutons. Am Mineral 72:231–239

Huang R, Audétat A (2012) The titanium-in-quartz (TitaniQ) thermobarometer: a critical examination and re-calibration. Geochim Cosmochim Acta 84:75–89

Huckenholz HG, Lindhuber W, Fehr KT (1981) Stability of grossular + quartz + wollastonite + anorthite: the effect of andradite and albite. N Jahrb Mineral Abh 142:223–247

Hynes A, Forest RC (1988) Empirical garnet-muscovite geothermometry in low-grade metapelites, Selwyn Range (Canadian Rockies). J Metamorph Geol 6:297–309

Jamieson RA, Crow D (1987) Sphalerite geobarometry in metamorphic terranes: an appraisal with implications for metamorphic pressure in the Otago Schist. J Metamorh Geol 5:87–99

Johnson CA, Essene EJ (1982) The formation of garnet in olivine-bearing metagabbros from the Adirondacks. Contrib Mineral Petrol 81:240–251

Kaneko Y, Miyano T (2004) Recalibration of mutually consistent garnet-biotite and garnet-cordierite geothermometers. Lithos 73:255–269

Karpov IK (1981) Fiziko-khimicheskoye modelirovaniye v geokhimii (Physico-chemical modeling in geochemistry). Nauka, Novosibirsk

Karpov IK, Chudnenko KV, Kulik DA et al (2001) Minimizatsiya energii Gibbsa v geokhimicheskikh sistemakh metodom vypuklogo programmirovaniya (Minimization of Gibbs energy in geochemical systems by the method of convex programming). Geochemistry 39(11):1207–1219

Kawasaki T, Matsui Y (1977) Partitioning of Fe^{2+} and Mg^{2+} between olivine and garnet. Earth Planet Sci Lett 37:159–166

Kawasaki T, Motoyoshi Y (2007) Solubility of TiO_2 in garnet and orthopyroxene: Ti thermometer for ultrahigh-temperature granulites. Short research paper 038, US Geological Survey and National Academy and Sciences, USGS OF-2007-1047. http://dx.doi.org/10.3133/of2007-1047.srp038

Kawasaki T, Osanai Y (2008) Empirical thermometer of TiO_2 in quartz for ultrahigh-temperature granulites of East Antarctica. In: Satish-Kumar M, Motoyoshi Y, Osanai Y (eds) Geodynamic evolution of east Antarctica: a key to the east-west Gondwana connection, vol 308. Geologocal Society Special Publication, Blackwell, London, pp 419–430

Kelsey DE (2008) On ultrahigh-temperature crustal metamorphism. Gondwana Res 13:1–29

Kleemann U, Reinhardt J (1994) Garnet-biotite thermometry revisited: The effect of Al^{VI} and Ti in biotite. Eur J Mineral 6:925–941

Kohler T, Brey GP (1990) Calcium exchange between olivine and clinopyroxene calibrated as a geothermobarometer for natural peridotites from 2 to 60 kb with applications. Geochim Cosmochim Acta 54:2375–2388

Kohn MJ, Spear FS (1989) Empirical calibration of geobarometers for the assemblage garnet + hornblende + plagioclase + quartz. Am Mineral 74:77–84

Kohn MJ, Spear FS (1991) Error propagation for barometers. Am Mineral 76:138–147

Kotov NV (1986) Termodinamicheskiye usloviya pozdnego diageneza i nachal'nogo metamorfizma (Thermodynamic conditions of late diagenesis and initial metamorphism). Nauka, Moscow, pp 90–103

Koziol AM (1989) Recalibration of the garnet-plagioclase-Al_2SiO_5-quartz (GASP) geobarometer and application to natural parageneses. EOS Trans Am Geophys Union 70:493

Koziol AM, Newton RC (1988) Redetermination of the garnet breakdown reaction and improvement of the plagiclase-garnet-Al_2SiO_5-quartz geobarometer. Am Mineral 73:216–223

Koziol AM, Bohlen SR (1992) Solution properties of almandine-pyrope garnet as determined by phase equilibrium experiments. Am Mineral 77:765–773

Kroll H, Evangelakakis C, Voll C (1993) Two-feldspar geothermometry: a review and revision for slowly cooled rocks. Contrib Mineral Petrol 114:510–518

Le Breton N, Thompson AB (1988) Fluid-absent (dehydration) melting of biotite in metapelites in the early stages of crustal anataxis. Contrib Mineral Petrol 99:226–237

Likhanov II (1988a) Chloritoid, staurolite and gedrite of the high-a lumina hornfelses of the Karatash pluton. Int Geol Rev 30(8):868–877

Likhanov II (1988b) Evolution of chemical composition of metapelite minerals during low-temperature contact metamorphism at the Karatash pluton. Int Geol Rev 30(8):878–887

Likhanov II, Reverdatto VV (2013) Mineral assemblages of the Al_2SiO_5 "triple point" in metapelites. Dokl Earth Sci 448(1):74–77

Likhanov II, Reverdatto VV (2014) P-T-t constraints on the metamorphic evolution of the Transangarian Yenisei Ridge: geodynamic and petrological implications. Russ Geol Geophys 55(3):299–322

Likhanov II, Reverdatto VV, Sheplev VS et al (2001) Contact metamorphism of Fe- and Al-rich graphitic metapelites in the Transangarian region of the Yenisei Ridge, eastern Siberia, Russia. Lithos 58:55–80

Likhanov II, Polyansky OP, Reverdatto VV et al (2004a) Evidence from Fe- and Al- rich metapelites for thrust loading in the Transangarian Region of the Yenisei Ridge, eastern Siberia. J Metamorph Geol 22:743–762

Likhanov II, Reverdatto VV, Selyatizky AY (2004b) Petrogenetic grid for ferruginous-aluminous metapelites in the K_2O–FeO–MgO–Al_2O_3–SiO_2–H_2O system. Dokl Earth Sci 394(1):46–49

Likhanov II, Reverdatto VV, Selyatizkii AY (2005) Mineral equilibria and P-T diagram for Fe- and Al-rich metapelites in the KFMASH system (K_2O–FeO–MgO–Al_2O_3–SiO_2–H_2O). Petrology 13(1):73–83

Likhanov II, Reverdatto VV, Kozlov PS et al (2015) *P-T-t* constraints on polymetamorphic complexes in the Yenisei Ridge, East Siberia: implications for Neoproterozoic paleocontinental reconstructions. J Asian Earth Sci 113:391–410

Liou JG, Maruyama S, Cho M (1987) Very low-grade metamorphism of volcanic and volcaniclastic rocks—mineral assemblages and mineral facies. In: Frey M (ed) Low temperature metamorphism. Blackie, Glasgow, pp 59–114

Lusk J, Ford CE (1978) Experimental extension of the sphalerite geobarometer at 10 kbar. Am Mineral 63:516–519

Mahar EM, Baker JM, Powell R et al (1997) The effect of Mn on mineral stability in metapelites. J Metamorphic Geol 15:223–238

Massonne HJ, Schreyer W (1987) Phengite geobarometry based on the limiting assemblage with K-feldspar, phlogopite, and quartz. Contrib Mineral Petrol 96:212–224

McCartey TC, Patino Douce AE (1998) Empirical calibration of the silica-tschermak's-anorthite (SCAn) geobarometer. J Metamorph Geol 16:675–686

Mirwald PW, Scola M, Tropper P (2008) Experimental study on the incorporation of Na in Mg-cordierite in the presence of different fluids (Na(OH), NaCl–H_2O, albite-H_2O). Geophys Res Abstr 10:EGU2008-A-04149

Moecher DP, Essene EJ, Anovitz LM (1988) Calculation of clinopyroxene-garnet-plagioclasequartz geobarometers and application to high grade metamorphic rocks. Contrib Mineral Petrol 100:92–106

Newton RC (1983) Geobarometry of high-grade metamorphic rocks. Am J Sci 283A:1–28

Nickel KG, Green DH (1985) Empirical geothermobarometry for garnet peridotites and implications for the nature of the lithosphere, kimberlites and diamonds. Earth Planet Sci Lett 73:158–170

O'Neill HSC, Wood BJ (1979) An experimental study of Fe–Mg-partitioning between garnet and olivine and its calibration as a geothermometer. Contrib Mineral Petrol 70:59–70

Pactunc AD (1998) MODAN: an interactive computer program for estimating mineral quantities based on bulk composition. Comput Geosci 24:425–431

Paria P, Bhattacharya A, Sen A (1988) The reaction garnet + clinopyroxene + quartz = 2 orthopyroxene + anorthite: a potential geobarometer for granulites. Contrib Mineral Petrol l99:126–133

Pattison DRM (1992) Stability of andalusite and sillimanite and the Al_2SiO_5 triple point: constraints from the Ballachulish aureole, Scotland. J Geol 100:423–446

Pattison DRM (2001) Instability of Al_2SiO_5 «triple point» assemblages in muscovite + biotite + quartz—bearing metapelites, with implications. Am Mineral 86:1414–1422

Pattison DRM, Newton RC (1989) Reversed experimental calibration of the garnet-clinopyroxene Fe–Mg exchange thermometer. Contrib Mineral Petrol 101:87–103

Pattison DRM, Chako T, Farquhar J et al (2003) Temperatures of granulite-facies metamorphism: constraints from experimental phase equilibria and thermo-barometry corrected from retrograde exchange. J Petrol 44:867–900

Perchuk LL (1970) Ravnovesiya porodoobrazuyushih mineralov (Equilibria of rock-forming minerals). Nauka, Moscow

Perchuk LL (1989) Vzaimosoglasovanie nekotorykh Fe–Mg geotermomotrov na osnove zakona | Nernsta: revisiya (Mutual consistence between some Fe–Mg-geothermometers based on the Nernst law: revision). Geokhimiya 27(5):611–622

Perchuk LL (1991) Derivation of a thermodynamically consistent set of geothermometers and geobarometers for metamorphic and magmatic rocks. In: Perchuk LL (ed) Progress in metamorphic and magmatic petrology. Cambridge University Press, Cambridge, pp 93–112

Perchuk LL, Gerya TV (1989) A set of internally consistent spinel-bearing geothermometers and geobarometers. In: Abstracts of the international symposium "Granulite metamorphism", University of New South Wales, Sydney

Perchuk LL, Lavrent'eva IV (1983) Experimental investigation of exchange equilibria in the system cordierite-garnet-biotite. In: Saxena SK (ed) Kinetics and equilibrium in mineral reactions. Springer, Heidelberg, pp 199–239

Perchuk LL, Ryabchikov ID (1976) Phasovye sootvetstviya v mineralnyh systemakh (Phase correspondences in mineral systems). Nedra, Moscow

Perchuk LL, Gerya TV, van Reenen TV et al (2000) Comparable petrology and metamorphic evolution of the Limpopo (South Africa) and Lapland (Fennoscandia) high-grade terrains. Mineral Petrol 69:69–107

Perkins D, Chipera SJ (1985) Garnet-orthopyroxene-plagioclase-quartz barometry: refinement and application to the English River subprovince and the Minnesota River Valley. Contrib Mineral Petrol 89:69–80

Powell R (1985) Regression diagnostics and robust regression in geothermometer/geobarometer calibration: the garnet-clinopyroxene geothermometer revisited. J Metamorph Geol 3:231–243

Powell R, Evans JA (1983) A new geobarometer for the assemblage biotite-muscovite-chlorite-quartz. J Metamorphic Geol 1:331–336

Powell R, Holland TJB (1994) Optimal geothermometry and geobarometry. Am Mineral 79:120–133

Powell R, Condliffe DM, Condliffe E (1984) Calcite-dolomite geothermometry in the system $CaCO_3$–$MgCO_3$–$FeCO_3$: an experimental study. J Metamorph Geol 2:33–41

Pownceby MI, Wall VJ, O'Neill HSC (1987) Fe–Mn partitioning between garnet and ilmenite: experimental calibration and applications. Contrib Mineral Petrol 97:116–126

Pownceby MI, Wall VJ, O'Neill HStC (1991) An experimental study of the effect of Ca upon garnet—ilmenite Fe-Mn exchange equilibria. Am Mineral 76:1580–1588

Pyle JM, Spear FS (2000) An empirical garnet (YAG)–xenotime thermometer. Contrib Mineral Petrol 138:51–58

Pyle JM, Spear FS, Rudnick RL et al (2001) Monazite-xenotine-garnet equilibrium in metapelites and a new monazite-garnet thermometer. J Petrol 42:2083–2107

Ravna EJK (2000a) Distribution of Fe^{2+} and Mg between coexisting garnet and hornblende in synthetic and natural systems: an empirical calibration of the garnet-hornblende Fe–Mg geothermometer. Lithos 53:265–277

Ravna EJK (2000b) The garnet–clinopyroxene Fe^{2+}–Mg geothermometer: an updated calibration. J Metamorphic Geol 18:211–219

Ravna EJK, Paquin J (2003) Thermobarometric methodologies applicable to eclogites and garnet ultrabasites. In: Carswell DA, Compagnoni R (eds) High pressure metamorphism. European mineralogical union notes in mineralogy, vol 5. Eötvos University Press, Budapest, pp 229–259

Ravna EJK, Terry MP (2004) Geothermobarometry of UHP and HP eclogites and schists—an evaluation of equilibria among garnet–clinopyroxene–kyanite–phengite–coesite/quartz. J Metamorph Geol 22:593–604

Rejebian VA, Harris AG, Huebner S (1987) Conodont color and textural alteration: an index to regional metamorphism, contact metamorphism, and hydrothermal alteration. Geol Soc Am Bull 99:471–479

Richardson SW, Gilbert MC, Bell PM (1969) Experimental determination of kyanite-andalusite and andalusite-sillimanite equilibria: the aluminum silicate triple point. Am J Sci 267:259–272

Sack RO (1980) Some constraints on the thermodynamic mixing properties of Fe–Mg ortho-pyroxenes and olivines. Contrib Mineral Petrol 71:257–269

Saxena SK (1976) Two-pyroxene geothermometer: a model with an approximate solution. Am Mineral 61:643–652

Schmadicke E (2000) Phase relations in perodotic and pyroxenitic rocks in the model system CMASH and NCMASH. J Petrol 41:69–86

Schmidt MW (1992) Amphibole composition in tonalite as a function of pressure: an experimental calibration of the Al-in-hornblende barometer. Contrib Mineral Petrol 110:304–310

Scott SD (1973) Experimental calibration of the sphalerite geobarometer. Econ Geol 68:466–474

Scott SD (1976) Application of the sphalerite geobarometer to regionally metamorphosed terrains. Am Mineral 61:661–670

Seitz H-M, Altherr R, Ludwig T (1999) Partitioning of transition elements between orthopyroxene and clynopyroxene in peridotitic and websteritic xenoliths: new empirical geothermometers. Geochim Cosmochim Acta 63:3967–3982

Selverstone J, Spear FS, Franz G et al (1984) High-pressure metamorphism in the SW Tauern Window, Austria: P-T paths from hornblende-kyanite-staurolite schists. J Petrol 25:501–531

Sengupta P, Dasgupta S, Bhattacharya PK et al (1990) An orthopyroxene-biotite geothermometer and its application in crustal granulites and mantle-derived rocks. J Metamorphic Geol 8:191–198

Shulters JC, Bohlen SR (1989) The stability of hercynite and hercynite-gahnite spinels in corundum- or quartz-bearing assemblages. J Petrol 30:1017–1031

Skublov S, Drugova G (2003) Patterns of trace-element distribution in calcic amphiboles as a function of metamorphic grade. Can Mineral 41:383–392

Spear FS (1986) PTPATH: a Fortran program to calculate pressure-temperature paths from zoned metamorphic garnets. Comput Geosci 12:247–266

Spear FS (1993) Metamorphic phase equilibria and pressure-temperature-time paths. Mineralogical Society of America Monograph, Washington

Spear FS, Cheney JT (1989) A petrogenetic grid for pelitic schists in the system SiO_2–Al_2O_3–FeO–MgO–K_2O–H_2O. Contrib Mineral Petrol 101:149–164

Spear FS, Selverstone J (1983) Quantitative *P-T* paths from zoned minerals: theory and tectonic application. Contrib Mineral Petrol 83:348–357

Spear FS, Peacock SM, Kohn MJ et al (1991) Computer programs for petrologic *P-T-t* path calculations. Am Mineral 76:2009–2012

Stephenson NCN (1984) Two-pyroxene thermometry of Precambrian granulites from Cape Riche, Albany-Fraser Province, Western Australia. J Metamorph Geol 2:297–314

Stormer JC JC Jr (1983) The effects of recalculation on estimates of temperature and oxygen fugacity from analyses of multicomponent iron–titanium oxides. Am Mineral 68:586–594

Taylor WR (1998) An experimental test of some geothermometer and geobarometer formulations for upper mantle peridotites with application to the thermobarometry of fertile lherzolite and garnet websterite. Neues Jahrb Mineral Abh 172:381–408

Thomas JB, Watson EB, Spear FS et al (2010) TitaniQ under pressure: the effect of pressure and temperature on the solubility of Ti in quartz. Contrib Mineral Petrol 160:743–759

Thompson AB, Frey M (1984) Illite 'crystallinity' in the Western River Formation and its significance regarding the regional metamorphism of the early Proterozoic Goulburn Group, District of Mackenzie. Current Research, Part A. Geological Survey of Canada, Paper 84-1A

Tiechmuller M (1987) Organic material and very low grade metamorphism. In: Frey M (ed) Low temperature metamorphism. Blackie, Glasgow, pp 114–161

Tomkins HS, Powell R, Ellis DJ (2007) The pressure dependence of the zirconium-in-rutile-thermometer. J Metamorph Geol 25:703–713

Triboulet C (1992) The (Na–Ca) amphibole-albite-chlorite-epidote-quartz geothermobarometer in the system S-A-F–M–C–N–H_2O. 1. An empirical calibration. J Metamorphic Geol 10:545–556

Vance D, Holland T (1993) A detailed isotopic and petrological study of a single garnet from the Gassetts schist, Vermont. Contrib Mineral Petrol 114:101–118

Vernon RH (1977) Relationships between microstructural and metamorphic assemblages. Tectonophysics 39:439–452

Vidal O, Parra T (2000) Exhumation paths of high pressure metapelites obtained from local equilibria for chlorite-phengite assemblages. Geol Mag 35:139–161

Vidal O, Goffe B, Bousquet R et al (1999) Calibration and testing of an empirical chloritoid-chlorite Mg–Fe exchange thermometer and thermodynamic data for daphnite. J Metamorph Geol 17:25–39

Vielzeuf D (1983) The spinel and quartz associations in high grade xenoliths from Tallante (S.E. Spain) and their potential use in geothermometry and barometry. Contrib Mineral Petrol 82:301–311

Wan Z, Coogan LA, Canil D (2008) Experimental calibration of aluminium pertitioning between olivine and spinel as a geothermometer. Am Mineral 93:1142–1147

Wark DA, Watson EB (2006) TitaniQ: a titanium-in-quartz geothermometer. Contrib Mineral Petrol 152:743–754

Waters DJ, Martin HN (1993) Geobarometry in phengite-bearing eclogites. Terra Abstr 5:410–411

Watson EB, Wark DA, Thomas JB (2006) Crystallization thermometers for zircon and rutile. Contrib Mineral Petrol 151:413–433

Will T, Okrush M, Schmaedicke E et al (1998) Phase relations in the greenschist-blueschist-amphibolite-eclogite facies in the system Na_2O–CaO–FeO–MgO–Al_2O_3–SiO_2–CO_2–H_2O (NCFMASH) with application to metamorphic rocks from Samos, Greece. Contrib Mineral Petrol 104:353–386

Wood BJ, Fraser DG (1976) Elementary thermodynamics for geologist. Oxford University Press, London

Wu CM (2015) Revised empirical garnet-biotite-muscovite-plagioclase geobarometer in metapelites. J Metamorphic Geol 33:167–176

Wu CM (2017) Calibration of the Al_2SiO_5-quartz geobarometer for metapelites. J Metamorph Geol 35:993–998

Wu CM, Pan YS, Wang KY (1999) Refinement of the biotite-orthopyroxene geothermometer with applications. Acta Petrol Sinica 15:463–468

Wu CM, Pan YS, Wang KY et al (2002) A report on a biotite-calcic hornblende geothermometer. Acta Geol Sinica 76:126–131

Wu CM, Zhang J, Ren LD (2004) Empirical garnet-biotite-plagioclase-quartz (GBPQ) geobarometry in medium- to high-grade metapelites. J Petrol 45:1907–1921

Wu CM, Zhao GC (2006) Recalibration of the garnet–muscovite (GM) geothermometer and the garnet–muscovite–plagioclase–quartz (GMPQ) geobarometer for metapelitic assemblages. J Petrol 47:2357–2368

Wu CM, Zhao GC (2007a) A recalibration of the garnet-olivine geothermometer and a new geobarometer for garnet-olivine-plagioclase-bearing granulites. J Metamorphic Geol 25:497–505

Wu CM, Zhao GC (2007b) The metapelitic garnet–biotite–muscovite–aluminosilicate–quartz (GBMAQ) geobarometer. Lithos 97:365–372

Zack T, Moraes R, Kronz A (2004) Temperature dependence of Zr in rutile: empirical calibration of a rutile thermometer. Contrib Mineral Petrol 148:471–488

Zenk M, Schulz B (2004) Zoned Ca-amphiboles and related *P-T* evolution in metabasites from the classical Barrovian metamorphic zones in Scotland. Miner Mag 68:769–786

Chapter 3
Causes, Geodynamic Factors and Models of Metamorphism

3.1 Types of Metamorphism

Convection currents in the Earth's mantle interact with the lithosphere, causing the transfer of mass and heat, which, as a result of deep-seated geodynamic activity, is manifested mainly in tectonics and magmatism, i.e., in the movements of blocks and plates of the Earth's crust, diapirism and injections of magma, thus destroying the original distribution of matter and temperature. The process will continue until thermal equilibrium state and/or mass balance was reached again, or will be interrupted by another tectonic or magmatic event. Changes in the *P*-*T* conditions are manifested in the replacement of early minerals and mineral assemblages by later ones, i.e., lead to metamorphism. Therefore, metamorphism can be regarded as an indicator of geodynamics.

Building upon the ideas of Miyashiro (1961, 1973) on the relationship between the manifestations of metamorphism and the magnitudes of geothermal gradients and the important conceptions of Turner (1968), Reverdatto and Sheplev (1998) identified five types of mineral transformations associated with tectonic and magmatic activities. Each type was characterized by specific *P*-*T* conditions, occupied the relevant field in *P*-*T* space and has its own geothermal gradients T/H (where H is a distance). Later, the *P*-*T* parameters were slightly changed and four types of metamorphism were recognized instead of five: I—contact ($T/H > 90$ °C/km, $T \approx 350$–1000 °C, $P \approx 0$–1.5 kbar), II—medium-pressure zonal metamorphism ($T/H \approx 25$–90 °C/km, $T \approx 300$–900 °C, $P \approx 1.5$–8 kbar), III—metamorphism with a geothermal gradient close to normal (average continental) values ($T/H \approx 18$–45 °C/km, $T \approx 100$–1000 °C, $P \approx 0.5$–13 kbar), and IV—metamorphism in collision and subduction zones ($T/H \approx 7$–25 °C/km, T not less than 200–300 °C, P not less than ~ 4 kbar; in subduction zones, T up to ~ 1000 °C and $P > 40$ kbar). The diagram illustrating the approximate location of the above types of metamorphism in *P*-*T* space is shown in Fig. 3.1. The main differences between the earlier and later schemes of Reverdatto and Sheplev (1998) are as follows:

V. V. Reverdatto et al., *The Nature and Models of Metamorphism*,
Springer Geology, https://doi.org/10.1007/978-3-030-03029-2_3

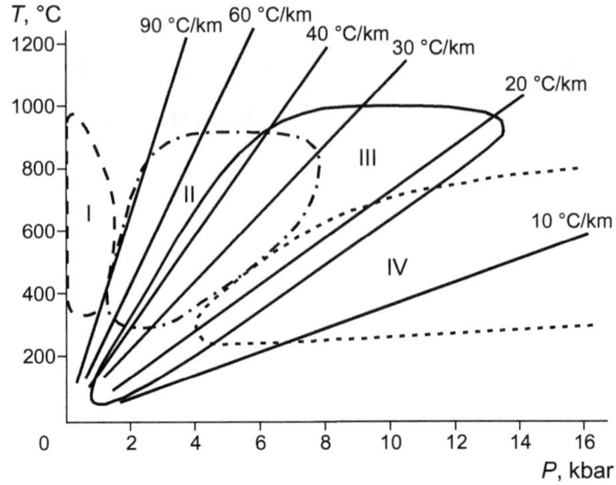

Fig. 3.1 *P-T* diagram for types of metamorphism characterized by different geothermal gradients. I–IV—Types of metamorphism: I—contact, II—zonal low-pressure/high temperature, III—burial, tectonic stacking, and Archean during formation of granulite complexes, IV—collisional including subduction complexes

pressure fields were expanded up to 8 kbar for type II and to >40 kbar for type IV metamorphism, fields for burial metamorphism, metamorphism produced by tectonic stacking and metamorphism responsible for the formation of Archean crust (type III) were combined to form a single *P-T* field.

These types are manifested in different geodynamic regimes over different time scales and can be correlated with a specific combination of metamorphic facies. According to the adopted nomenclature (Dobretsov et al. 1972), contact metamorphism produces different combinations of contact metamorphic facies (muscovite-hornfels, amphibole-hornfels, pyroxene-hornfels, and spurrite-merwinite) in thermal aureoles adjacent to shallow (<5–6 km from the surface) igneous intrusions, depending on the spatial distribution of temperatures. Medium-pressure zonal metamorphism (formerly referred to as low-*P* and high-*T* or *LP*/*HT*) accompanied by mid- and upper-crustal magmatism produces andalusite-sillimanite (in metapelites) complexes represented by different combinations of greenschist, epidote-amphibolite and amphibolite facies rocks. The appearance of kyanite marks the transition to greater depths (up to 28–30 km) in zoned (kyanite-sillimanite) complexes. Burial metamorphism during rifting in regions of continental extension produces rocks that were metamorphosed under zeolite facies and, more rarely, greenschist facies conditions. The tectonic stacking responsible for crustal thickening is accompanied by low- and medium-temperature metamorphism. Multiple stages of metamorphism accompanying the formation of Archean continental crust cannot be readily differentiated but it is evident that the Precambrian was characterized by a variety of thermodynamic regimes. Metamorphism resulted in the formation of granulite and amphibolite facies rocks that are widespread in all continental shields (the most ancient stable parts of the Earth's crust). Metamorphism in the continental collision zones results in the formation of kyanite-bearing gneiss schist complexes (in metapelites) or a combination of eclogite, kyanite schist, greenschist, and jadeite-glaucophane facies rocks.

The above types of metamorphism occur in a variety of combinations, e.g., a combination of contact and zonal metamorphism is found within medium-pressure belts, burial metamorphism and metamorphism produced by tectonic stacking are found in combination with collisional metamorphism, etc. Such combinations are not accidental; on the one hand, they indicate transitional tectonic regimes and, on the other hand, reflect the specific geological and tectonic trends in the evolution of the Earth's crust. The first two types were caused by an additional supply of heat from magmatic intrusions and diapirs into the crust, which has led to a higher than normal (average continental) geothermal gradient and a heat flow of above 50–60 mW m^{-2}. During contact metamorphism, the heat flow was between $\sim$180 mW m^{-2} and a few W m^{-2} (sometimes up to $\sim$10 W m^{-2} and higher). The third type is characterized by geothermal gradients close to average (normal) values (approximately between 20 and 40 °C/km). The heat flow values for this type of metamorphism were close to normal ($\sim$50 mW m^{-2}), while elevated heat flow values (60–100 mW m^{-2} and higher) were probably due to the influx of additional heat from magmatic intrusions. The fourth type of metamorphism is characterized by lower than normal (average continental) geothermal gradients, which can be explained by a relatively short duration of events and by the fact that thermal equilibrium between crustal blocks was not attained at corresponding depths, since temperature changes more slowly than pressure. The pressure in all types of metamorphism is believed to be controlled by the lithostatic load.

Based on the above considerations, it can be concluded that the following causes of metamorphism can be distinguished from a geodynamic perspective: (1) the influx of additional heat from magmatic intrusions and diapirism (or fluid flows); (2) rocks transformation at a near-normal geothermal gradient; (3) collision of lithospheric plates and crustal blocks due to horizontal movements, triggering deformation, convergence and subduction, over- and underthrusting. The combination of geodynamic factors of metamorphism may impose large uncertainties.

Vertical migration of magma in the Earth's crust is an effective heat transfer mechanism. The additional heat input causes the development of metamorphic aureoles around magmatic intrusions and diapirs. The key feature of this type of metamorphism is elevated horizontal thermal gradient. Conductive heat transfer tends to dominate over convective heat transfer, which is associated with fluid migration in the vicinity of magmatic bodies and is generally accompanied by metasomatism. A full review of this issue is beyond the scope of this book and can be found in, for example, Barton et al. (1991a).

Subsidence of sediments to a depth of at least $\sim$10 km causes metamorphism in depressions formed during crustal extension. Rifting (in a broad sense, including the formation of sedimentary basins through a basement detachment and, possibly, depressions along the passive continental margins, see for example, Leonov 2004) should be considered as the major and most common cause of crustal subsidence. The depths of subsidence within intra-platform syneclises and foredeeps that were formed in response to thrust sheet loading do not exceed 4–6 km. In this case, temperatures were not great enough (100–200 °C) to cause metamorphism. However, at the base of the deep depressions the rocks may have experienced metamorphism under epidote-amphibolite and amphibolite facies conditions. It seems that a rather complicated, multi-phase extension of the lithosphere accompanied by mantle upwelling played a leading role in the formation of such depressions.

It seems likely that the specific geodynamic mechanism responsible for Archaean metamorphism is still obscure. Multistage polymetamorphism in shield regions was accompanied by complex tectonic movements that resulted in crustal thickening, subsequent extension, and a strong increase in the heat flux associated with lithospheric thinning and injection of mantle magmas. One of the possible explanations for the high heat flux is the convective flow of hot mantle beneath the region with a thinned lithosphere. Although the heat conduction is the dominant heat transport mechanism through the rocks (Dobretsov et al. 1975), minor variations in temperature with depth in granulites during the same stage of metamorphism may indicate the possibility of heat transfer by a fluid flow or specific thermophysical conditions.

As a result of horizontal movements, lithospheric plates are squeezed together along plate boundaries. The resulting convergence and subduction of plates, over- and underthrusting of crustal blocks may sometimes lead to collisional metamorphism (in a broad sense). From a standpoint of mechanics, subduction as well as over- and underthrusting of smaller tectonic blocks can be treated as processes involving deformation of solid bodies, which differ mostly in the scales of interaction, duration, thermomechanical properties of materials, and P-T parameters. Collisional metamorphism during over- and underthrusting of crustal blocks takes place under conditions of a great diversity of initial mechanical, temperature, and geometrical factors governed by geological and tectonic processes. Both structural and thermal rearrangement produces different metamorphic zoning patterns, depending on P-T conditions, thermophysical properties and velocities of the interacting blocks (Ruppel and Hodges 1994). The intracontinental collision processes most likely involve large-scale overthrusting and underthrusting, lithospheric plate flexure, shearing, folding, subduction, and crustal thickening. During orogenic tectonic stacking, a very intense folding and thrusting was accompanied by metamorphism, due to the rise of isotherms associated with heat transfer to the regions of thickened crust from lower parts of the continental plates. Subduction processes involve underthrusting of oceanic crust beneath continental curst and convergence of continental plates. Subduction of crustal blocks to great depths and deformation of lithospheric plates may result in the formation of high- and ultrahigh-pressure metamorphic rocks, which can be of great geodynamic and petrological interest. During subduction, the horizontal force must be large enough to overcome the buoyancy forces in the mantle, at least, in the beginning of movement. Eclogitization of oceanic basalts tends to assist sinking of the subducting plate into the subcrustal lithosphere and further down to a significant depths. It is most likely that the exhumation of deep-seated crustal rocks is driven by detachment of the deeper portion of subducted lithosphere and ascent of the buoyant remaining portion. Intense deformation in subduction zones results in the formation of rocks associations (during subsidence and uplift), which are indicative of a complex tectonometamorphic history and inhomogeneous distribution of *P-T* parameters. Erosion usually exposes a tectonic mosaic comprising different rock units, which have been deformed and metamorphosed to a variable degree and exhumed from different depths (Reverdatto and Sheplev 1998). A reasonable model for subduction accounting for the above factors and rapid exhumation is absent. Collisional metamorphism is accompanied by other types of metamorphism during different evolutionary stages. The fragments of ancient metamorphic folded blocks are also involved in intracontinental collision. All this reflects difficulties in studying the geologic history of fold belts.

The relationship between metamorphism and geodynamic processes is very complex and far from obvious. The elucidation and analysis of these relationships, the study of the characteristics of metamorphism in different geodynamic settings are among the most topical geological problems.

3.2 Models of Metamorphism

3.2.1 Metamorphism Related to Additional Heat Supply

3.2.1.1 Contact Metamorphism

Contact metamorphism usually takes place at relatively shallow depths in the upper crust during a period of tectonic quiescence. The nature of this phenomenon is extensively studied and fairly well understood. The intrusive heating is the causal agent of contact metamorphism, which manifests itself around subvolcanic and hypabyssal igneous bodies and also inside them in the xenoliths of enclosing rocks. The emplacement of magma into relatively cold or slightly heated country rocks causes a sharp increase in temperature at contacts with intruding magma and leads to metamorphism characterized by a steep and nonlinear geothermal gradient (greater than ~90 °C/km). The *P-T* conditions of metamorphism vary widely from ~400 to ~1000 °C and from 1 bar to ~1.5 kbar.

Previous studies of contact aureoles clearly showed that in some cases the additional input of material from the intruding magma into the surrounding country rocks was insignificant, and the metamorphism was essentially isochemical (see, for example, Reverdatto 1970; Reverdatto et al. 1974). In case of isochemical mineral transformations, the conductive heat transfer prevails (Reverdatto et al. 1972, 1974; Furlong et al. 1991). Essentially isochemical metamorphism may be assumed to take place near the contacts of both mafic and felsic intrusions. Although this type of metamorphism is common near the contacts with basalt and gabbro, it is only rarely observed in contact aureoles around granitoids, and no reliable evidence has been reported at all in relation to alkaline magmatism. The fact underlying this is that it is far more likely that saturation (and oversaturation) of melts with respect to the volatile components and the release of volatiles leading to the input of the material into surrounding country rocks is much more likely to be reached in a felsic and alkaline rather than mafic magma at the early-magmatic stage. Medium- and low-temperature metamorphism in the vicinity of granitic bodies at shallow depths is essentially allochemical and is accompanied by a change in the chemical composition of rocks in contact aureoles. Because of post-magmatic hydrothermal alteration, contact metamorphic rocks of the early magmatic stage are often difficult to recognize.

The diverse geometry of intrusive bodies and the structure of contact aureoles are described by Paterson et al. (1991). Types of contact metamorphic zones (facies series) in metapelites were described using specific examples by Pattison and Tracy (1991). The classification and nomenclature of contact metamorphic rocks depending on temperature conditions and the chemical composition of protolith was proposed by Kolobov et al. (1992).

The statistics of manifestations of contact metamorphism (Reverdatto 1973a, b; Barton et al. 1991b) show that the temperatures of at least 650–700 °C were usually achieved in close proximity to mafic intrusive bodies. In most cases, these intrusions are presented by small gabbro or diorite bodies occupying less than 80–100 km^2 in plan, stocks or subvolcanic basaltic dikes and necks. Causes of high-temperature contact metamorphism can rarely be attributed to intrusions of intermediate magma and extremely rarely to intrusions of felsic magma. At the contact with granitic bodies, metamorphic temperatures may reach 550–600 °C. Gabbro or diorite intrusions may develop $\sim$1 to 1.5–2 km wide contact aureoles consisting of pyroxene-hornfels, amphibole-hornfels, and muscovite-hornfels facies rocks. The contact aureoles around granitic intrusions often display marked zoning and are composed of rocks metamorphosed under amphibole-hornfels and muscovite-hornfels facies conditions. Metamorphic rocks of the spurrite-merwinite facies (locally in association with the pyroxene-hornfels facies) are developed in xenoliths of country rocks at the contacts with dolerite dikes, necks, and sill where forced flow (in the channels) or melt convection can be assumed; much more rarely, the rocks of this facies are found in association with diorite, monzonite, etc. Muscovite-hornfels facies metamorphic rocks are observed near contacts with subvolcanic intrusive bodies such as basic, intermediate and felsic dikes, sills, laccoliths, etc., which are characterized by a small thickness and extensive magma crystallization. All these facts can be explained by the nature of an intrusive body as a source of heat for metamorphism and thus can be readily analyzed by a mathematical model, which relates the evolution of a magma body undergoing cooling and crystallization with heating and metamorphism of the surrounding country rocks. As a first approximation, and without taking into account convection in magma and heat transfer by intergranular fluids percolating through the country rocks, this problem has been discussed in many previous works (Dudarev et al. 1972; Ingersoll and Zobel 1913; Lovering 1935, 1955; Jaeger 1957, 1959, 1964, 1968; Carslaw and Jaeger 1959; Reverdatto et al. 1970, 1972; Turcotte and Schubert 1982; Buntebarth 1991; Furlong et al. 1991; Stüwe 2002 and others).

Conductive heat transfer can be expressed in general form with the equation

$$\rho C \frac{\partial T}{\partial t} = \lambda \left[\frac{\partial^2 T}{\partial x^2} + \frac{\partial^2 T}{\partial y^2} + \frac{\partial^2 T}{\partial z^2} \right] + W, \tag{3.1}$$

on the assumption of constant $a = \frac{\lambda}{\rho C}$, where a—thermal diffusivity, m^2 s^{-1}; λ—thermal conductivity, W m^{-1} K^{-1}; ρ—density, kg m^{-3}; C—specific heat capacity, J kg^{-1} K^{-1}; W—heat sources and sinks, W m^{-3}; t—time, s; T—temperature, K; x, y, z—coordinates. When applied to geological situations, the solution of this equation describing a time-varying spatial distribution of the temperature is possible under different conditions.

A one-dimensional mathematical model of contact metamorphism (for example, for a sheet-like magma intrusion, Fig. 3.2) provides a simple, effective, and very clear visualization of the general evolution of the temperature field within and near the cooling intrusion, for which the formulation of the problem should be described in detail (e.g., Dudarev et al. 1972; Reverdatto et al. 1970, 1972). The model constitutes a system of four heat conduction equations for magma (subscript 1) and rocks: magmatic (2), metamorphic (3), and country rock (4):

$$\frac{\partial T_1}{\partial t} = a_1 \frac{\partial^2 T_1}{\partial x^2}, \quad 0 < x < \xi_1(t), \tag{3.2}$$

$$\frac{\partial T_2}{\partial t} = a_2 \frac{\partial^2 T_2}{\partial x^2}, \quad \xi_1(t) < x < l, \tag{3.3}$$

$$\frac{\partial T_3}{\partial t} = a_3 \frac{\partial^2 T_3}{\partial x^2}, \quad l < x < \xi_2(t), \tag{3.4}$$

$$\frac{\partial T_4}{\partial t} = a_4 \frac{\partial^2 T_4}{\partial x^2}, \quad x > \xi_2(t), \tag{3.5}$$

under the following initial and boundary conditions:

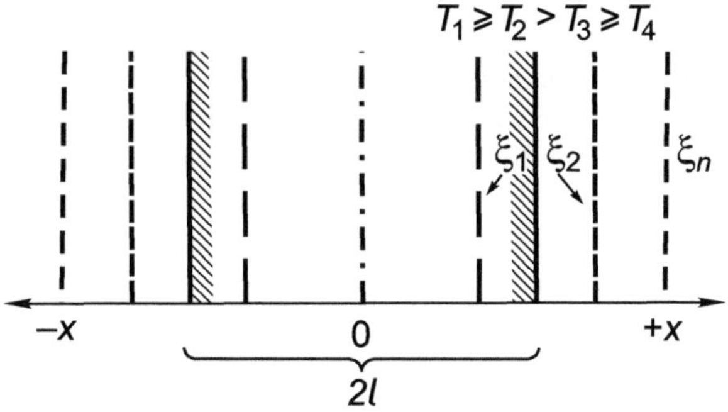

Fig. 3.2 Schematic representation of one-dimensional model for contact metamorphism around a sheet-like intrusive body (see text for explanation). $X = 0$—is the symmetry point; $2l$ —width of sheet-like intrusive bogy; ξ_1, ξ_2 and ξ_n are mobile boundaries of phase transitions (dashed lines)

$$\left.\frac{\partial T_1}{\partial x}\right|_{x=0} = 0 \text{ (symmetry condition)} \tag{3.6}$$

$$T_1(x,0) \equiv T_1 \geq T_1^*;\ \xi_1(0) = \xi_2(0) = l \tag{3.7}$$

(T_1—the temperature of the magma is taken to be homogeneous and higher or equal to the temperature of magms crystallization- T_1^*; the infilling of a magma chamber is instant in a space $2l$; at the initial moment $t = 0$ the boundaries of phase transitions $\xi_{1\text{-}2}$ and the boundaries of the magma chamber $x = \pm l$ coincide),

$$T_4(x,0) = T_4(\infty,t) = T_4 < T_3^* \tag{3.8}$$

(the initial temperature of the country rock T_4 is assumed to be homogeneous and below than the temperature of the metamorphic reaction T_3^*),

$$[T(l,t)] = 0 \tag{3.9}$$

(the equality of temperatures at the intrusive contact l; the difference in temperature is equal to zero),

$$\left[\lambda \frac{\partial T(l,t)}{\partial x}\right] = 0 \tag{3.10}$$

(the equality of heat flows at the intrusive contact l; the difference in heat flows is equal to zero),

$$\lambda_2 \left.\frac{\partial T_2}{\partial x}\right|_{\xi_1(t)-0} - \lambda_1 \left.\frac{\partial T_1}{\partial x}\right|_{\xi_1(t)+0} = \rho G_1 \xi_1'(t); \tag{3.11}$$

(the Stefan condition: at the moving boundary of magma crystallization ξ_1, the latent heat of melting $G_1 > 0$ is liberated at temperature T_1^*,), $T_1(\xi_1,t) = T_2(\xi_1,t) = T_1^*$;

$$\lambda_3 \left.\frac{\partial T_3}{\partial x}\right|_{\xi_2(t)-0} - \lambda_4 \left.\frac{\partial T_4}{\partial x}\right|_{\xi_2(t)+0} = \rho G_2 \xi_2'(t); \tag{3.12}$$

(the Stefan condition: at the moving boundary of metamorphism ξ_2 the heat of reaction G_2 is absorbed at temperature T_3^*), $T_3(\xi_2,t) = T_4(\xi_2,t) = T_3^*$, because of the irreversibility of a metamorphic reaction G_2

$$\begin{aligned} G_2 > 0, \ if\ \lambda_3 \frac{\partial T_3}{\partial x}\bigg|_{\xi_2(t)-0} &> \lambda_4 \frac{\partial T_4}{\partial x}\bigg|_{\xi_2(t)+0} \\ G_2 = 0, \ if\ \lambda_3 \frac{\partial T_3}{\partial x}\bigg|_{\xi_2(t)-0} &\leq \lambda_4 \frac{\partial T_4}{\partial x}\bigg|_{\xi_2(t)+0} \end{aligned} \tag{3.13}$$

This model assumes the instantaneous intrusion of magma. In addition, for the sake of simplicity, we assume that the phase transition in a country rock is instantaneous after reaching T_3^*. Therefore, the calculations give the maximum width of metamorphic zoning and the maximum distance between the respective isograds and the intrusive contact; the rate of metamorphism is controlled by the rate of conductive heat transfer. This, of course, is meant to be a rough approximation (see Chap. 4.2.3), but this formulation is expected to demonstrate simple relations between metamorphism and a cooling magma and identify characteristic features.

The duration of the infilling ranges from a few days for a basaltic magma chamber with a volume of about 1 km^3 (as estimated by Sharapov et al. 2000) to a few tens of days or a few months for large intrusions emplaced in a single pulse. This duration is too small compared to the duration of cooling and solidification of magmatic bodies. Therefore, the assumption on instantaneous intrusion could be feasible in thermophysical modeling, particularly, for shallow mafic intrusions. The assumption on a short duration of emplacement (instantaneous) is supported by the presence of chilled (glassy or micrograined) margins at the contact of an intrusion and relatively cold country rocks (Jaeger 1964, 1968).

Under otherwise equal conditions, the width of the contact aureole appears to correlate linearly with the thickness width of the intrusion while the duration of contact metamorphism is proportional to the square of the half-width of contact. The difference between these model relationships and the actual geological situation can be explained by the kinetics of metamorphic reactions (see Chap. 4.2.3), which is especially obvious for thin shallow dikes or crystallized melts. Geological observations show that large (>8–10 km across) intrusions are rare in the upper-crustal levels at depths less than ~5 km. Dikes and sills (single-pulse emplacement of magma) are not more than a few hundreds of meters thick. Some laterally extensive dikes related to rifting can rarely measure up to a few kilometers in thickness; the Great Dike of Zimbabwe is a 10–12 km wide and ~560 km long vertical stratified mafic and ultramafic intrusive body. Contact aureoles at subvolcanic and hypabyssal depths can rarely attain ~1.5 km in width. However, this is not the case for much greater depths: an increase in temperature of the adjacent country rocks with depth increases the thickness of the metamorphic contact aureole.

Numerical calculations are carried out using thermophysical parameters typical of rocks and magmas of different composition (see, for example, Jaeger 1964, 1968; Clark 1966; Turcotte and Schubert 1982; Buntebarth 1991; Furlong et al. 1991 and others). Thermal diffusivity (a) and thermal conductivity (λ) of the rocks decrease with increasing temperature. Over the temperature range 100–1000 °C, the parameter a varies from $\sim 1.4 \times 10^{-6}$ to $\sim 0.7 \times 10^{-6}$ m^2 s^{-1} for granite and from (0.9–1.0) $\times 10^{-6}$ to (0.3–0.5) $\times 10^{-6}$ m^2 s^{-1} for basalt, limestone and sandstone. Approximately in the same temperature range, λ decreases from 2–4 to 0.7–2.0 W m^{-1} $°C^{-1}$ for granite, sandstone, limestone, and basalt (Delaney 1988). The thermal diffusivity of magmas is lower than that of rocks of the same composition. The thermal diffusivity of mafic to felsic aluminosilicate melts is generally much lower than 1×10^{-6} m^2 s^{-1}; the thermal conductivity of aluminosilicate melts is approximately 1.5–2 times lower than that of the rocks. The notation and parameters used for the calculations are summarized in Tables 3.1 and 3.2.

The ranges of magmatic temperatures are estimated to be 650–750 °C for felsic melts, 1100–1200 °C for mafic melts and above $\sim$1300 °C for ultramafic melts. The final solidification temperatures are placed at 650–670 °C for granites and

Table 3.1 Parameter designation for the models of the contact metamorphism and the zoning metamorphism of moderate pressure

Parameters	Dimension
T	Temperature °C
ρ	Density kg m^{-3}
t	Time c
a	Thermal diffusivity m^2 c^{-1}
C	Heat capacity J kg^{-1} $°C^{-1}$
λ	Thermal conductivity W m^{-1} $°C^{-1}$
G	Latent heat of phase transition J kg^{-1}
x	Horizontal coordinate m
y	Vertical coordinate m
H	Position of the Earth's surface
R	Position of upper intrusive contact
P	Position of lower intrusive contact
K	Position of the right vertical intrusive contact
α	The angle between the Earth's surface and intrusive contact
ε	Position of the moving boundary of phase transition
Indices	
o	Initial value
*	Solidus
m	Magma
mr	Magmatic rock
ax	Anatectic melt
er	Country rock
tr	Metamorphic rock
m-mr	Phase transition: magma/magmatic rock
ax-tr	Phase transition: anatexis/metamorphic rock
tr-er	Phase transition: metamorphic rock/country rock

Table 3.2 The values of parameters used for calculations in the models of contact and medium-pressure zonal metamorphism

Parameter	Value
T_m°	1180–1300 °C
T_{m-mr}^*	1100 °C
T_{ax-tr}°	700 °C
T_{tr-er}°	550–650 °C
ρ_m	3000 kg m^{-3}
ρ_{mr}	3300 kg m^{-3}
ρ_{ax}	2600 kg m^{-3}
ρ_{er}	2800 kg m^{-3}
C_m	1200 J kg^{-1} $°C^{-1}$
C_{mr}	1200 J kg^{-1} $°C^{-1}$
C_{ax}	900 J kg^{-1} $°C^{-1}$
C_{er}	900 J kg^{-1} $°C^{-1}$
G_{mr}	430,000 J kg^{-1}
G_{ax}	250,000 J kg^{-1}
G_{tr}	160,000 J kg^{-1}
λ_m	1.4 W m^{-1} $°C^{-1}$
λ_{mr}	2.6 W m^{-1} $°C^{-1}$
λ_{ax}	1.1 W m^{-1} $°C^{-1}$
λ_{er}	2.0 W m^{-1} $°C^{-1}$

~1050 °C for basalts (felsic derivatives of basalts formed by fractionation of basaltic magmas crystallize at 650–750 °C).

The minimum temperatures reached during contact metamorphism usually coincide with the so-called "kinetic threshold" of 400–500 °C for low pressures (Dobretsov et al. 1972; Reverdatto 1973a, b; Reverdatto et al. 1970), which defines the outer boundary of metamorphic zoning (in the absence of metasomatism). However, mineral transformations are also possible at even lower temperatures (at 200–300 °C) but are then not complete. The maximum temperature for metamorphism at the intrusive contact is typically ~1000 °C, which is only possible in the case of a pressure flow of mafic magma.

The latent heat of melting ranges from ~420 J g^{-1} for basalt to ~300 J g^{-1} and less for diorite and granite. The latent heat may account for ~25% and more of the total heat carried by the magma (Buntebarth 1984). The heat of phase transition during metamorphism (per gram of fluid produced during reaction) is generally 3–4 times lower than the latent heat of melting of magmatic rocks.

The calculations provide different models to approach cooling of intrusive bodies of simple shape (without taking into account convection) and heating of the country rocks. Some examples are given in Fig. 3.3 to demonstrate such calculations for contact zones surrounding several 100 m thick dikes, illustrating the development of a thermal aureole and a zone of contact metamorphism in time at a fixed model temperature of metamorphic changes (200 and 500 °C).

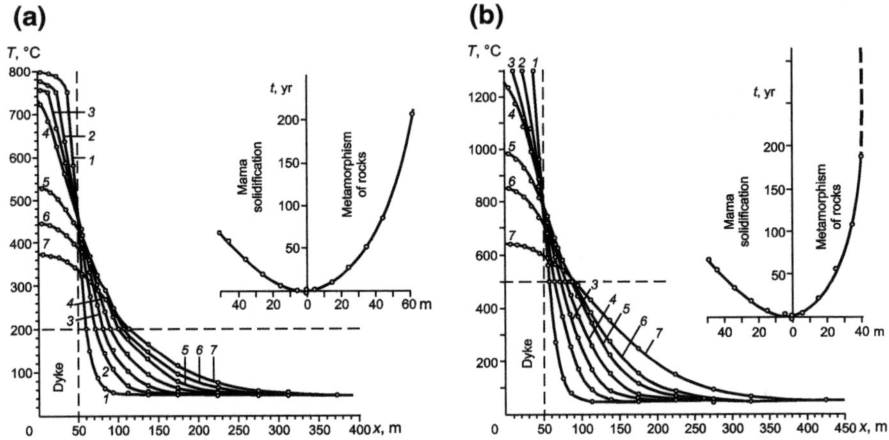

Fig. 3.3 The estimated changes in temperature inside and around a sheet-like intrusive body (the x-axis gives the distance from the center of symmetry of the body where $x = 0$). **a** For the granitic dike: thickness $2l$ = 100м, T_1 = 800 °C, $T_1^* = 750$ °C, $T_3^* = 200$ °C, $T_4 = 50$ °C; the numbers on isochrons denote temperatures after a period of: (1) 5 years, (2) 20 years, (3) 40 years, (4) 70 years, (5) 110 years, (6) 150 years, (7) 210 years. **b** For the ultramafic dike with thickness $2l$ = 100м, T_1 = $T_1^* = 1300$ °C, $T_3^* = 500$ °C, $T_4 = 50$ °C; the numbers on isochrons denote temperatures after a period of: (1) 5 years, (2) 20 years, (3) 40 years, (4) 70 years, (5) 110 years, (6) 150 years, (7) 290 years. On the side panels for each case are shown: on the left—the dynamics of solidification of magma from the intrusive contact, where $x = 0$, toward the center; on the right—migration of the metamorphic front in the country rocks (Reverdatto et al. 1970; Dudarev et al. 1972)

The temperature evolution within and around basaltic and granitic dikes of any thickness is summarized in Fig. 3.4. To further study the thermal evolution of contact metamorphism around a sheet-like intrusive bodies of simple shape, it would be useful to construct a plot in T and $\frac{x}{\sqrt{t}}$ coordinates, as shown in Fig. 3.5. The initial temperature at the direct contact immediately after magma intrusion can be expressed as $T_c = \sigma T_1/(1+\sigma)$, where $\sigma = \lambda_1\sqrt{a_4}/\lambda_4\sqrt{a_1}$ (Lovering 1936). The value of parameter σ varies between 0.6 and 2, generally approaching 1, therefore in the first approximation $T_c \approx 0.5T_1$. If we take into account the latent heat of crystallization, T_c would exceed $0.5T_1$ and can be as high as $\sim 0.6T_1$ or slightly higher. The calculations (assuming that $T_4 \approx 0\,°\mathrm{C}$ and $T_1 = T_1^*$) indicate a general decrease in the effect of latent heat of crystallization on T_c at $\sigma < 1$ and an increase at $\sigma > 1$ (Jaeger 1964).

There are intrusions in which the magma has been partially or almost totally solidified. In this case, the contact metamorphic zone can be very restricted or even absent due to a low temperature of the intruding material and a small amount (or even absence) of heat of crystallization, W. Some examples of the limited thermal effect of the intrusion on country rocks have been reported in the literature. These include the Alpine-type peridotite massifs (Turner and Verhoogen 1951), Ayu-Dag

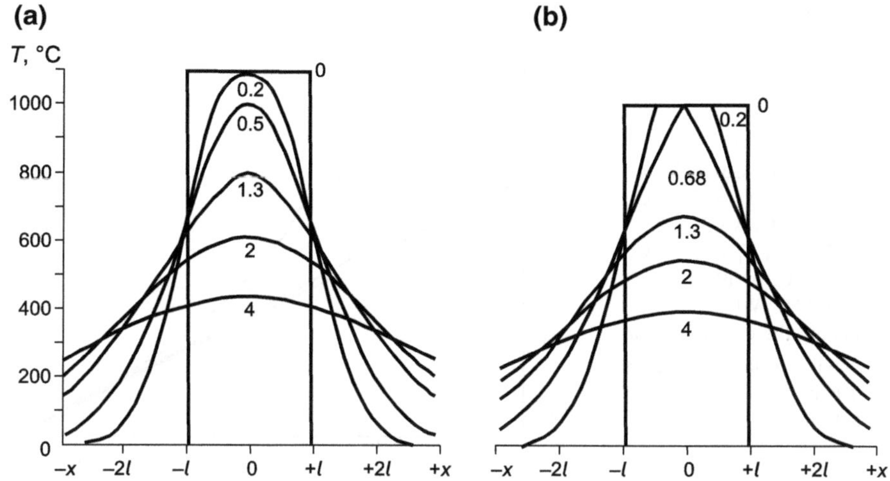

Fig. 3.4 The model evolution of temperature inside and around the cooling sheet-like intrusive body with a thickness of $2l$; the numbers on isochrons indicate a dimensionless time $\tau = \frac{at}{l^2}$. **a** Basaltic dike, T_l = 1100 °C, magma solidification occurs in the temperature range ΔT = 800–1100 °C. **b** Granite dike, $T_l = T_1^* = 1000$ °C (Jaeger 1968)

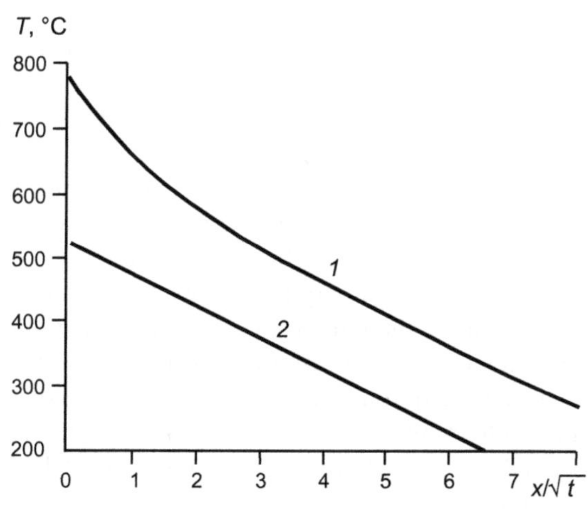

Fig. 3.5 The calculated dependence T on $\frac{x}{\sqrt{t}}$ (where x is the width in meters, and t is time in years) for the formation of contact metamorphic rocks. A metamorphic reaction occurs at a fixed $T_3^* = 500$ °C, $T_4 = 50$ °C. Curve 1 for the case of cooling of sheet-like ultramafic intrusion, when $T_l = T_1^* = 1300$ °C. Curve 2 for the case of cooling of felsic intrusion, when $T_l = T_1^* = 750$ °C (Reverdatto et al. 1970)

gabbro-diorite laccolith of Crimea (Anan'ev 1999), andesite necks around Lake Badkhyz, Turkmenia (Likhanov et al. 1996), etc.

If the temperature of the magma is well above the solidus, i.e., overheated, then, all other conditions being equal, it will affect the character of contact metamorphism. For example, overheating of the felsic magma by 450 °C as compared to $T_1^* = 750$ °C should increase the temperature at the dike contact by ~40 °C and increase the size of the metamorphic zone (Fig. 3.6). However, a high degree of

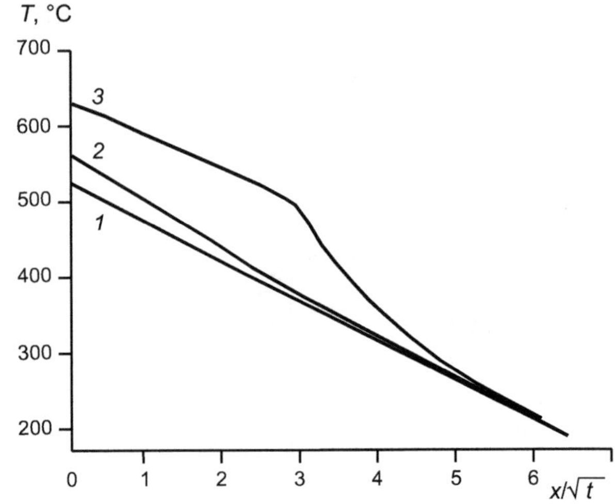

Fig. 3.6 Change in the width of a contact metamorphic zone as a function of the initial temperature of magma and surrounding rocks. Curve 1: T_I = 800 °C, T_1^* = 750 °C, T_4 = 50 °C. Curve 2: T_I = 1200 °C, T_1^* = 750 °C, T_4 = 50 °C. Curve 3: T_I = 1200 °C, T_1^* = 750 °C, T_4 = 120 °C. In all cases T_3^* = 500 °C. When calculating the criterion $\frac{x}{\sqrt{t}}$ on the x-axis, x is taken in meters and t is in years (Reverdatto et al. 1970)

overheating of the intruded magmas is very unlikely: magma temperatures at near-surface conditions approach the liquidus value or lie in the range between the liquidus and solidus (Turner and Verhoogen 1951). This is indicated, for example, by the presence of porphyritic segregations of minerals in the chilled marginal zones of shallow magmatic bodies. As shown in Fig. 3.5 the crystallization temperature of the intruded magma T_1^* and the latent heat of crystallization W that depend on melt composition have much more pronounced effect on contact metamorphism. Broadly speaking, an increase in T_1^* of ~100 °C will increase the temperature at the contact of ~50 °C.

The way latent heat is released during magma solidification has also a large effect on contact metamorphism. The release of latent heat during cotectic crystallization occurs during the interval $\Delta T = T_1^{**} - T_1^*$, where T_1^* is solidus temperature, and T_1^{**} is the liquidus temperature characterized by a slight increase in the temperature at the intrusive contact in the initial period after magma intrusion. The way latent heat is released during magma solidification has also a large effect on contact metamorphism. However, the value of the interval $\Delta T = T_1^{**} - T_1^*$ has a minor effect on T_c (Jaeger 1964). Such an increase in T_c is not typical of eutectic crystallization at the fixed T_1^*. As shown in Fig. 3.7, the duration of temperature maximum at the intrusive contact during cotectic crystallization is insignificant compared to the overall duration of contact metamorphism (near relatively thick intrusions). The duration of a cotectic type of magma crystallization is greater only by ~10% than that of a eutectic type (Reverdatto et al. 1972). Therefore, the duration of contact metamorphism is essentially independent of whether (under otherwise equal conditions) magma crystallization would occur at the fixed solidus temperature or over a range of temperatures (would be of the cotectic type).

Figure 3.6 shows changes in the width of the contact metamorphic zone at an elevated initial temperature of the country rocks. Under otherwise equal conditions,

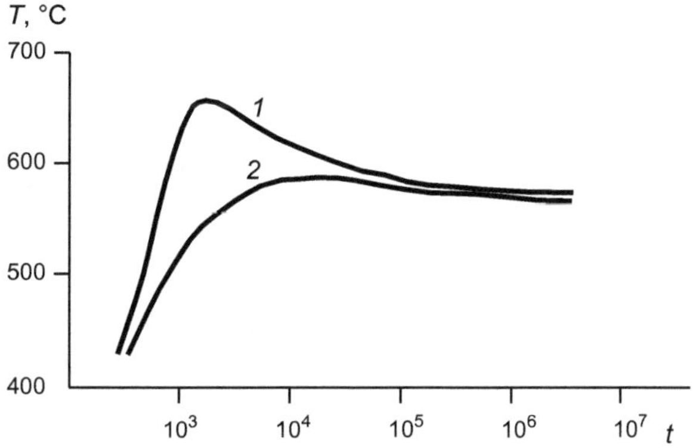

Fig. 3.7 Temperature change at the intrusive contact after magma emplacement depending on the type of its crystallization. T_1 = 900 °C, $T_1^* = 750$ °C, $T_1^{**} = 850$ °C, $T_4 = 150$ °C. Curve 1 is the cotectic type of magma crystallization in the interval $\Delta T = T_1^{**} - T_1^*$, $T_1^* = 750$ °C. Curve 2 is the eutectic type of magma crystallization at $T_1^* = 750$ °C. Time *t*—in hours (Reverdatto et al. 1972)

the increase of *T*4 is accompanied by the increase of temperature at the direct intrusive contact. These data indicate that the effect of the initial temperature of the country rocks on the temperature and size of a metamorphic aureole begins to appear at a depth of ~5 km given an average thermal gradient. Therefore, contact metamorphism should be restricted to relatively shallow depths, not exceeding 5–6 km.

Modeling the thermal effects of emplacement of a sheet-like magma intrusion on the country rocks helps to understand the behavior of the metamorphic reaction when a phase transition is not fixed at constant temperature T_3^* but occurs over an extended temperature interval $\Delta T = T_3^{**} - T_3^*$; the heat of phase transition is also distributed over the same interval. Assuming that the endothermic effect takes place linearly over the range of temperatures, the outer (low-temperature) boundary of the reaction will move further away from the intrusive contact than in the case of the phase transition at a fixed T_3^*. The outer boundary of phase transition will propagate a maximum distance from the intrusion, under otherwise equal conditions, in the case of slight heat absorption in the beginning of the reaction associated with a broad, nonlinearly extended endothermic effect. In contrast, if the heat absorption is relatively large in the beginning of the reaction, the outer boundary of metamorphic zone will retreat because the larger the endothermic effect at the reaction front, the lower the heating of country rocks should be ahead of the reaction front. As shown in Fig. 3.8, taking the above consideration into account, the following conclusion can be drawn: the number and character of the reactions and the distribution of the endothermic effect during protracted metamorphism around a large intrusive body have insignificant effect on the distribution of internal isograds within a metamorphic zoning (Reverdatto et al. 1970). However, since the position of the most external isograd depends on the character of the endothermic effect and reaction kinetics at low temperatures, this should have an effect on the total width of contact metamorphic zones, especially for small intrusions and a short period of heating.

The presence or absence of metamorphic reactions in the contact aureole has an effect on the intensity of heating of country rocks and cooling of a magma body. Model calculations show that the heat sink induced by metamorphic reactions may

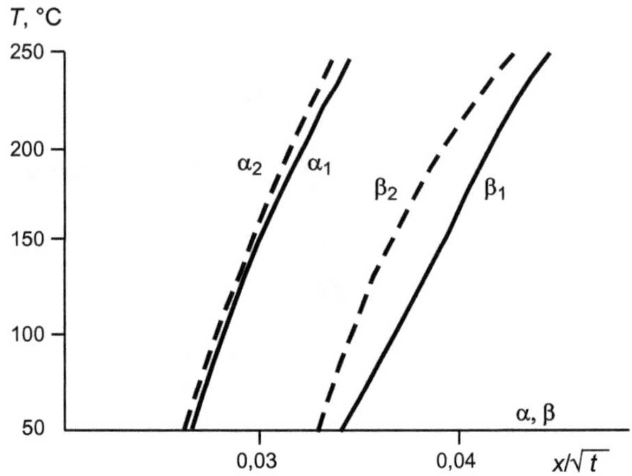

Fig. 3.8 The behavior of a complex metamorphic front depending on the distribution of an endothermic effect distribution in phase transition. α is the reaction rate constant at $T_3^{**} = 450$ °C, β is the reaction rate constant at $T_3^* = 400$ °C. Index 1 corresponds to the case when 14% of heat of phase transition is absorbed at T_3^*, and the remaining heat is absorbed in the temperature interval $\Delta T = T_3^{**} - T_3^*$; index 2 corresponds to another case, when 86% of reaction heat is absorbed at T_3^*, and the remaining heat is evenly distributed in the interval $\Delta T = T_3^{**} - T_3^*$. For greater clarity, the endothermic effect of the reaction is increased to 300 J g^{-1}. Initial temperatures of the country rocks are shown on the y-axis; the rate constants α and β in the form $\frac{x}{\sqrt{t}}$; *x*—in meters, *t*—in hours are shown on the x-axis (Reverdatto et al. 1970)

cause a decrease in temperature in the contact aureole. For example, around large sheet-like magma intrusions with a thickness of up to 3–4 km at $T_1 > 800$ °C, at a depth where $T_4 > 100$ °C, the endothermic effect of the reaction will decrease the temperature by 60–70 °C relative to a corresponding isotherm in the country rocks where metamorphic reactions are absent (and all other things being equal). The above-mentioned heat sink also tends to increase the rate of cooling of a magma intrusion. Figure 3.9 illustrates the dependence of the parameter $\frac{x}{\sqrt{t}}$ describing the rate of a moving crystallization front on the presence or absence of a metamorphic reaction in the country rocks. The rate of magma solidification slows down as the initial temperature of the country rocks increases (Reverdatto et al. 1972).

The above-considered one-dimensional models of contact metamorphism are applicable to sheet magmatic intrusions like an infinitely long dike or sill. The thermal history of such intrusions with simple geometries, as in the case of an infinite cylinder or sphere, can be readily studied using analytical formulations. Comparing solutions for different heat sources with simple geometries suggests that the temperature near convex contacts will be lower than that near planar contacts due to greater heat losses to the surrounding rocks (Carslaw and Jaeger 1959; Jaeger 1968). This directly follows from the solution of the heat conduction equation in a cylinder or sphere. If for an infinite magmatic sheet, the temperature at the contact is equal to 1/2 of the sum of T_1 (without taking into account the latent heat of

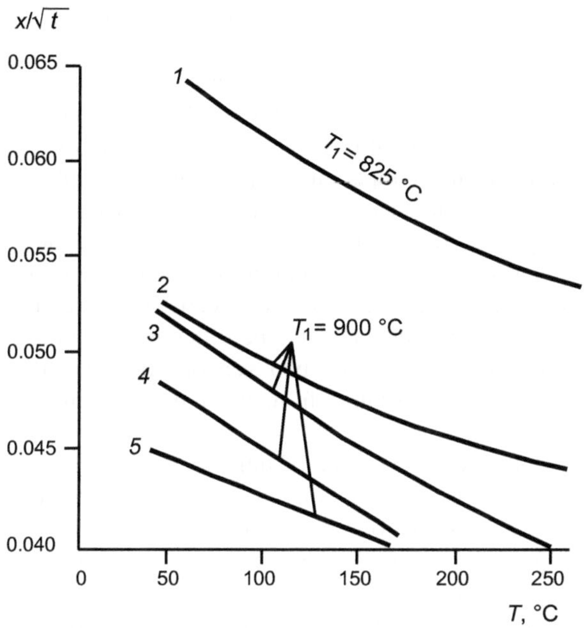

Fig. 3.9 Variation in the rate constant of magmatic crystallization $\frac{x}{\sqrt{t}}$ in a sheet-like intrusive body as a function of the initial temperature of country rocks T_4. Two types of magma solidification are modeled: eutectic type at a fixed crystallization temperature T_1^* (curves 1–3) and cotectic type when $T_1^{**} - T_1^* = \Delta T$ (curves 4 and 5). Curves 1, 3 and 5 correspond to the case when magma crystallizes in the absence of metamorphism in country rocks; curves 2 and 4—in the presence of metamorphism. T_1 = 825 or 900 °C, T_1^{**} = 850 °C (for curves 4 and 5), T_1^* = 750 °C (for all curves); *x*—in meters, *t*—in hours (Reverdatto et al. 1972)

crystallization) and T_4, then for a cylinder or sphere, this temperature T_c will be equal only to ~2/5 of the sum. Some quantitative estimates of this phenomenon were presented in Lovering (1955) and Jaeger (1964, 1968). The temperature near concave contacts will be higher than near planar contacts due to smaller heat losses. Since xenoliths of the country rocks are heated from all sides and do not act as heat sinks, the temperature of small fragments enclosed into large magmatic bodies may approach that of the magma. Simultaneous emplacement of multiple infinitely long sills or dikes with parallel contacts causes strong overheating of the layers of country rocks (interfaces) between intruded magma bodies as a result of superposition of temperature fields; for basaltic intrusions, the metamorphism with a temperature T_3 > 400 °C occurs if the widths of the interfaces are comparable or even >1.5–2 larger than the thickness of sills (Reverdatto et al. 1982; Reverdatto and Melenevskij 1983; Jaeger 1964).

Thermophysical modeling using the constant values of *a*, λ, and *G* helps to understand the most characteristic features of conductive cooling of magmatic bodies, heating and metamorphism of country rocks. These features of thermal evolution have been analytically expressed in many previous papers (Ingersoll and Zobel 1913; Lovering 1935, 1936, 1955; Jaeger 1957, 1959, 1964 and others). However, the parameters such as thermal conductivity, λ, thermal diffusivity, a, and latent heat of phase transitions, G, depend on the temperature, thus providing inexact analytical solutions. In this case, a better choice would be to perform numerical modeling using temperature-dependent thermophysical parameters. Modeling of the thermal aureoles indicates that the decrease in the thermal diffusivity of the country rocks with increasing temperature would increase the duration of metamorphism as compared to model results at constant a. The maximum temperatures reached in the contact aureole at constant a are slightly lower than those where a is temperature-dependent. The difference in the modeled temperatures of metamorphism (between the two versions of the parameter a used in calculations) decreases with distance from the intrusive contact (Nabelek et al. 2012).

As noted above, the complex shape of intrusive bodies can cause large spatial variations in the temperature distribution within the country rocks and thicknesses of contact aureoles. The evaluation of thermal histories of real magmatic intrusive bodies with complex geometries, as in the case when temperature-dependent thermophysical parameters are applied, requires the use of two- or three-dimensional models with a finite difference approach. Modeling is performed on finite-difference grids and allows accounting for the shape of intrusive bodies, varying thermal properties of rocks and melts, thermal effects of magma crystallization and metamorphic reactions. Map projections of the intrusions can be used to study the thermal evolution in a two-dimensional approximation. The comparison of modeled isotherms reflecting maximum temperatures with actually observed isograds in the vicinity of mapped magmatic intrusions was used to verify the models and prove the correctness of the conclusion that conduction is the dominant heat transfer mechanism during contact metamorphism (see above). For example, such a comparison was made for many contact metamorphic aureoles worldwide (Furlong et al. 1991).

The above considerations can be illustrated by modeling thermal metamorphism near the Santa Rosa granodiorite stock of Nevada, USA. According to Compton (1960), this stock was intruded into greenschist facies metapelitic rocks at a depth of 3.5–4 km, where $P_{\text{lith}} \approx 1$ kbar. A 1.5–2 km thick contact aureole around this granitic stock is composed of andalusite- cordierite hornfelsed schists (in the outer part of the contact aureole) and massive hornfelses (inner part), the critical mineral assemblage of which is Crd + Bt + Kfs + And/Sil + Pl + Qz. As can be seen (Fig. 3.10), there is a close match between the calculated isotherms and metamorphic facies boundaries (which can be taken as paleoisotherms corresponding to the respective isograds) in the contact aureole.

The availability of reliable information about the geometries and thicknesses of intrusive bodies makes 3D modeling possible. In the absence of such data, one tends to rely only on 2D models, which provide inexact estimates. This is illustrated in Fig. 3.10, which presents a comparison of the calculated isotherms near the Santa Rosa granodiorite stock for both a 2D planar mode and a 3D model. In the 2D

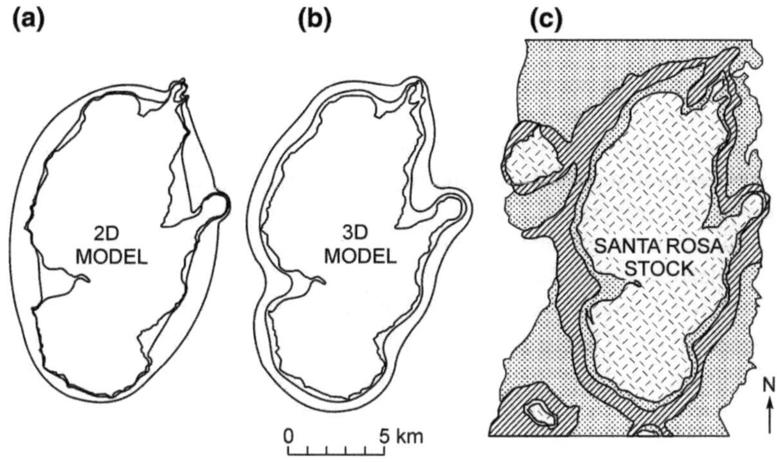

Fig. 3.10 Comparison between configurations of the calculated isotherms and the boundaries of metamorphic facies (isograds) taken as paleoisotherms in the contact aureole near the Santa Rosa granodiorite stock, Nevada, USA. The position of the calculated isotherms is from Furlong et al. (1991) for the case of: **a** two-dimensional and **b** three-dimensional (for a 2 km thick discoid, irregularly shaped magmatic intrusion) models; **c** metamorphic zones around the granodiorite body after Compton (1960)

planar model, the internal isotherm is located closer to intrusive contacts than in the 3D case, in which the intrusion is modeled as a disc-shaped body with an irregular outline and a thickness of up to 2 km. With a decrease in intrusion thickness in 3D models (under otherwise equal conditions), the isotherms tend to move closer to the intrusive contacts (Furlong et al. 1991).

An important problem for contact metamorphism is convection in magma chambers. In the case of shallow sheet-like magma bodies, having a thickness of less than 100 m, the problems of cooling can be solved without taking into account convection, since the error is not very large (Jaeger 1964, 1968; Reverdatto et al. 1970, 1972; Furlong et al. 1991). However, for large intrusions that may reach hundreds of meters or more across, the exact solution for the problem must account for the presence of convection; here, it is predicted by the value of the Rayleigh number: $Ra = \frac{L^3 \rho \alpha g \Delta T}{\eta a}$, where L is the characteristic length scale for convection, α is the coefficient of thermal expansion, g is the acceleration due to gravity, ΔT is the temperature difference, η is the dynamic viscosity, ρ is the density, a is the thermal diffusivity (Landau and Lifshitz 1986). If Ra exceeds a certain critical value (Turcotte and Schubert 1982), thermal convection is predicted to occur in a magma chamber (e.g., Jellinek and Kerr 1999). The viscosity of more mobile mafic magmas ranges from $\sim$0.1 to 10^2–10^3 Pa s and that of intermediate and felsic magmas is between 10^1–10^2 and 10^7–10^8 Pa s; in this case, the viscosity decreases with increasing temperature and pressure (Persikov 1984). Convection will be more rapid, under otherwise equal conditions, in a low-viscosity melt filling the volume of a magma chamber, where the temperature difference will develop between the floor and the top.

The magnitude of convection is smaller in the high-viscosity fluid; however, thermal convection can be expected in larger granitic magma bodies (e.g., McCarthy and Fripp 1980; Shimizu 1986). Convection increases the rate of cooling and promotes heat supply flux to the upper intrusive contact, thus causing a slight increase in its temperature, *Tc*, as compared to the absence of convection and mixing during cooling (Jaeger 1964, 1968; Hodge 1974). The efficiency of heat transfer is determined by the Nusselt number (Nu), which can be defined as the ratio of convective to conductive heat transfer. Within the convective regime, the low thermal diffusivity of the magma will increase Nu and Ra, which are defined by the relationship $Nu = c(Ra)^{1/3}$, where *c* is the constant (Turner 1973). This illustrates that an increase in thermal diffusivity of magma with progressive cooling will decrease the convective heat transfer (Nabelek et al. 2012).

The development of chilled margins at the intrusive contact immediately after magma injection will introduce additional "thermal resistance" (Pekhovich and Zhidkikh 1976), which results in slightly decreasing T_c. Upon subsequent cooling and solidification of the magma from contacts toward the center of the intrusion, the convective heat transfer slows down and finally stops (when Ra does not reach a critical value), leading to a considerable decrease in T_c. The slow solidification of a cooling magma in the gravity field is usually accompanied by its differentiation. As the melt crystallized, heavy minerals precipitate and sink to the lower part of the magma chamber, where they accumulate and lead to the formation of a heterogeneous mass which behaves as a Bingham body. The residual less dense liquid rises upward and participates in convection, which, nevertheless, gradually slows down due to the increasing viscosity. In mafic magmas, convection ceases when the magmatic temperature drops below 900 °C (Hodge 1974).

The discrepancy between the calculated maximum temperatures of metamorphism at the contacts of igneous dikes and the temperature estimates based on the *P-T* stability of mineral assemblages, may indicate, in some cases, forced magma flow through fractures. A good example is a gently-dipping dike crosscutting the marly limestone strata on the Kochumdek River, a tributary of the Podkamennaya Tunguska River (Pertsev 1977). The dike is composed of gabbro-dolerite and has a true thickness of about 20 m (apparent thickness of 70–90 m). The contact of the dike shows a distinct metamorphic zoning. Zone I (inner) located about 1 m (true thickness) from the contact is characterized by the stable assemblage Mw + Spu + Cal и Mw + Mll (with 20% åkermanite) + Cal. Zone II (2–4 m away from the contact) is characterize by the stable assemblage Mtc + Wo + Grt + Cal, whereas Zone III located some 4–5 m and more from the contact, pyroxene (diopside?) is found in association with wollastonite, garnet, vesuvianite, and calcite. Contact metamorphism took place at $P_{CO2} \approx$ 20–30 kbar. The analysis of mineral parageneses in the system $CaO–MgO–SiO_2–CO_2–H_2O$ showed (Pertsev 1977) that peak temperatures during the contact metamorphism ranged from 820–865 °C in Zone I to 645–690 °C in Zone II and to 615 °C and below in Zone III.

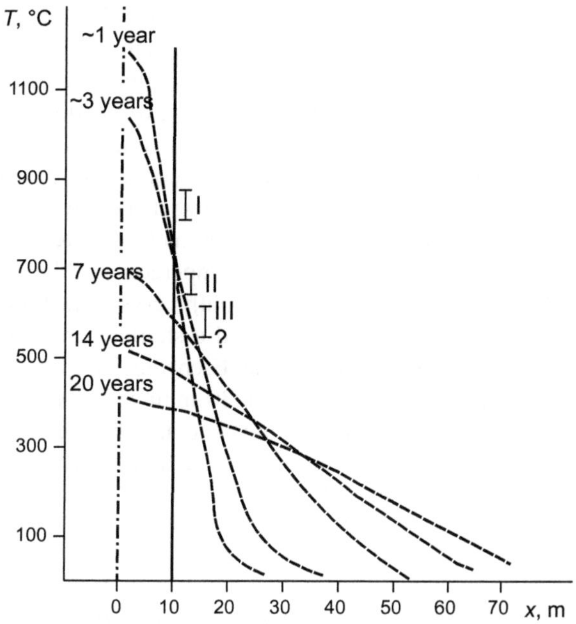

Fig. 3.11 Temperature evolution around a dike of basaltic magma (T_1 = 1200 °C) emplaced in limestones (see in the text). The dike is 20 m thick, x is the distance from the center of the dike (dash-dotted line) in meters. The calculations were performed by solving the one-dimensional heat transfer equation without taking into account the magma flow through the crack. The heat of crystallization of melt and changes in the thermal conductivity during phase transition are taken into account. The dashed curves denote the calculated temperature distributions in the country rocks after magma intrusion in years. The temperature estimates and the boundaries of metamorphic zones I, II and III according to Pertsev (1977) at the appropriate distance from the intrusive contact are also shown in the figure

These temperature estimates and results of thermophysical calculations on the evolution of the temperature in limestones near a sheet-like basaltic intrusion are shown in Fig. 3.11. It can be seen that the temperatures determined from mineral equilibria are higher than the calculated temperatures (especially, in Zone I). This can be explained by invoking magma flow through fractures during metamorphism of the country rocks. The duration of heating was long enough to allow for metamorphic reactions to proceed (Reverdatto et al. 1970).

The rates of basaltic lava (magma) flow along volcanic conduits (at Kilauea volcano) were estimated to vary from 0.02–0.5 to 10–20 m/s (Sharapov et al. 2000). As low-viscosity lava moves rapidly through a feeder conduit (vent or fracture in the crust), the temperature at the contact T_c may approach T_1 (Jaeger 1964). At a distance x from the planar contact at the time t after magma begins to flow through such a fracture, the temperature of the rocks is determined by $T = T_1\left[1 - erf\, x/2(at)^{1/2}\right]$ (Carslaw and Jaeger 1959; Jaeger 1964).

A good example of high-temperature contact metamorphism is melting of arkose sandstone near the peridotite (~200 m in diameter) and basalt (~50 m in diameter) necks on the Isle of Rum, Scotland (Holness 1999). A model of temperature distribution around necks was calculated by solving a one-dimensional heat conduction equation in a vertical cylinder (Fig. 3.12). Melting reached 75–85 vol.% for the anatectic melt and occurred up to 6 m from the contact for peridotites and up to 15 m for basalts. In both cases, it was assumed that the temperature at the neck contacts remained constant due to the intense magma flow lasted for 3–10 years for peridotites and 35–40 years for basalts. The temperature of the magma was estimated at 1000–1200 °C for peridotites (partially crystallized melt) and 1200–1250 °C for basalts. Koritnig (1955) suggested a shorter period of heating of the host sandstones around a dolerite body near Eschwege, Germany. He calculated that the magma flow through a fracture (about 30 m in width) with a temperature of 1200 °C ensures heating and melting of sandstone at a distance of 10 cm from the contact over a period of 1 day, ~50 cm from the contact over a period of 20 days, ~2 m from the contact over a period of 180 days. The melting was supported by the abundance of glass present in the contact rocks. Another example of the generation of very high temperatures at the intrusive contact due to relatively high rates of magma flow through the feeder conduit is given by melting of sandstone near the basalt neck (Tasmania) at $T \approx 1000$ °C and at 50 bars P_{H_2O} (Spry and Solomon 1964). Low flow rates typically result in a relatively low contact temperature because of the relatively low rates of heat loss replacement and the formation of chilled margins, which may increase in thickness with time and lead to the blockage of the conduit.

In multiphase intrusions, late magma batches are usually emplaced into relatively cold country rocks, as indicated by the development of chilled margins at boundary zones. However, if the time elapsed between batches is short enough, the magma may intrude relatively hot rocks, thus leading to an increase in T_c and extended duration of crystallization compared to the case when country rocks were cold. The injection of mafic magmas into felsic warmer rocks may result in their melting. "Cold" granites at the contact with a mafic magma may begin to melt in case of intense convection or rapid magma flow along a fracture.

To keep within the framework of conductive heat transfer and isochemical metamorphism, some special and complex issues such as hydrothermal-metasomatic processes and models of convective heat transfer by a fluid are not considered further because it is beyond the scope of this book. These questions were discussed in a number of previous works (e.g., Golubev and Sharapov 1974; Norton and Knight 1977; Sharapov and Averkin 1990; Polyansky and Reverdatto 2002; Baumgartner and Ferry 1991; Furlong et al. 1991; Polyanskii et al. 2002, 2003).

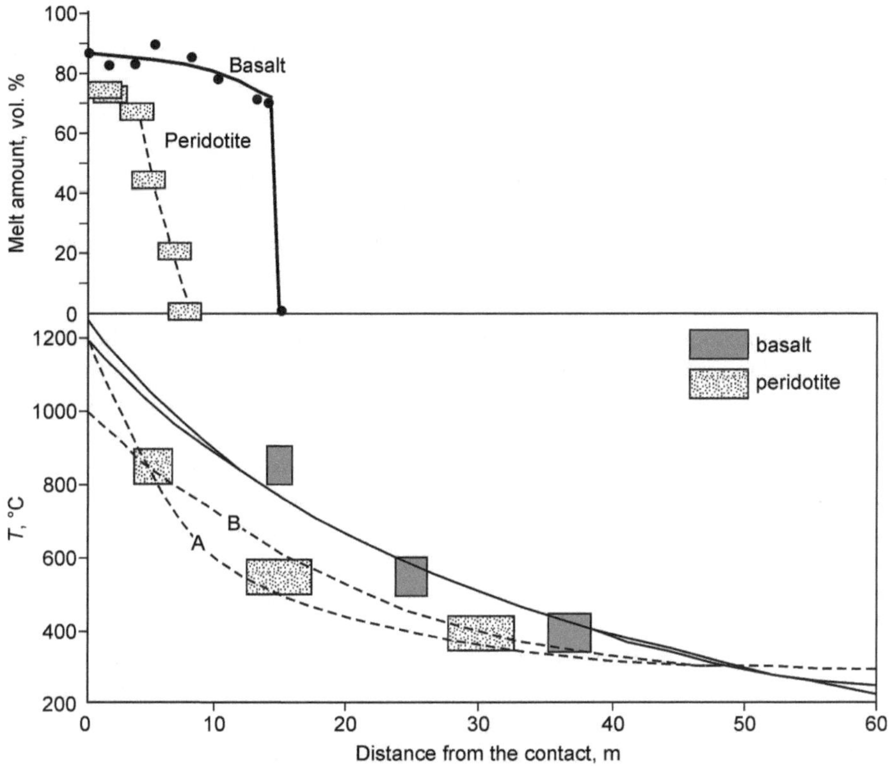

Fig. 3.12 Quantity (vol. %) of melt during melting of sandstone (upper panel) and calculated temperature distributions (lower panel) at contacts of gabbro and peridotite necks on the island of Rum, Scotland. Curve dotted line A refers to peridotite magma with temperature of 1200 °C; this temperature is maintained at the contact for 3 years. Curve B refers to peridotite magma with temperature of 1000 °C, flowing along the supply channel for 10 years. Two solid curves characterize the temperature distributions near the basaltic neck with initial T = 1200–1250 °C, flowing through the channel for 35–40 years. Rectangles show uncertainties and variations in temperature estimates (according to melting experiments) and distances (Holness 1999)

3.2.1.2 Medium-Pressure Zonal Metamorphism

Contact metamorphism and medium-pressure zonal metamorphism have the same physical nature, being genetically and spatially associated with intrusive magmatism. However, contact metamorphism is usually restricted to shallow depths (<5–6 km) where the initial temperature of the country rocks, T_4, rarely exceeds 100 °C. In a first approximation, this might not be accounted for by model calculations of the width of contact aureoles. Medium-pressure zonal metamorphism occurs at a depth greater than 5–6 km. Therefore, under otherwise equal conditions, metamorphic temperatures at such depths around magmatic intrusive bodies are greater than those during contact metamorphism, heating times are longer and the width of zones of thermal transformations is greater. The latter factor also results in much steeper geothermal gradients (25–90 °C/km) that those produced during contact metamorphism. The thermal effects during solidification of the magma or metamorphic reactions taking place in contact aureoles (see Sect. 3.2.1.1) may be limited or extended in time. Because redistribution of elements in the country rocks surrounding magmatic (mostly mafic) bodies is not significant, contact metamorphism and medium-pressure zonal metamorphism are largely isochemical, as shown in previous studies (Pitcher and Sinha 1958; Reverdatto et al. 1974). However, at greater depths, where anatexis takes place, the mineral changes in country rocks around an igneous intrusion are usually accompanied by metasomatism.

The compositions of coexisting minerals in metapelites and metabasites formed during medium-pressure zonal metamorphism should be different from those of contact metamorphic mineral assemblages, although the criteria for such difference are poorly understood (Spear 1993). This is true, in particular, for pressure differences between pyroxene-hornfels (A1) and granulite (two-pyroxene gneisses) (B1) facies and between amphibole-hornfels (A2) and amphibolite (sillimanite-biotite gneisses) (B2) facies (Fig. 1.13 in Chap. 1, see also Dobretsov et al. 1972). It seems that conditionally univariant mineral equilibria with fixed X_{Fe} and/or X_{Al} must be used as facies boundaries in metapelites, although specific estimates have yet to be developed.

Since the temperature of medium-pressure zonal metamorphism does not generally exceed ~900 °C, it should be restricted to crustal depths of 25–28 km, as indicated by estimated temperature conditions at intrusive contacts, given an initial temperature of the country rocks $T_4 \approx 600$–700 °C, an initial geothermal gradient 25 °C/km, and the mafic magma temperature $T_1 \approx 1100$ °C. At a geothermal gradient of ~30 °C/km, zonal metamorphism should be restricted to depths of 20–23 km. At the same time, the depth at which the temperature difference between the intrusive contact and the surrounding country rocks is small we took as maximum depth of metamorphism caused by the intrusion of a granitic magma. At the initial geothermal gradient of ~25 °C/km, the depth of the intrusion of granitic magmas will be 26–28 km. The outer boundary of metamorphic zoning defined on the basis of mineral transformations can be only vaguely determined at such a great depth (at mid-crustal levels). Accordingly, the depth of 25–28 km can be taken as a maximum depth of medium-pressure zonal metamorphism. The maximum pressure at this depth is limited to 7–8 kbar. Under such conditions, regionally elevated temperatures lead to widespread melting of felsic rocks and migmatite formation; however, metamorphism caused by intrusions of magma may occur as well. This is supported by a few, but characteristic examples of metamorphism at 5–7 kbar (Zen et al. 1968; Hermes and Murray 1988; De Yoreo et al. 1989a, b; Schumacher et al. 1989) or at 8 kbar (Elan 1985; Ross 1985; Sams and Saleeby 1988). The minimum pressure of medium-pressure zonal metamorphism corresponds to the maximum pressure of contact metamorphism (~1.5 kbar). The average pressure of medium-pressure zonal metamorphism is about 3–4 kbar (De Yoreo et al. 1991). The maximum and average temperatures of zonal metamorphism are approximately estimated at 800–900 °C and 500–750 °C, respectively (De Yoreo et al. 1991). As in the case of contact metamorphism, temperature in the country rocks should correlate with the temperature and composition of the intruding magma: higher temperature with relatively hot mafic magma and lower temperature with relatively cold felsic magma. However, this can be reliably traced only for conditions observed at the same depth and at the same initial temperature of the country rocks, T_4. Such observations are lacking at present, while the available temperature estimates for zonal metamorphism depend both on T_1 and T_4. The minimum temperature of zonal metamorphism is 300–350 °C, which is consistent with the "kinetic threshold" for these pressures (see above and also Dobretsov et al. 1972, 1973, 1975; Reverdatto 1973a, b).

Magmatic intrusions that cause medium-pressure zonal metamorphism are often represented by large felsic, intermediate, and mafic plutons, ranging in size from 500 to 1000 km^2 and more. Typical examples of such plutons include the Donegal granite and granodiorite, northwestern Ireland (Pitcher and Berger 1972; Siegesmund and Becker 2000), Liberty Hill granite and diorite, southern Appalachians, USA (Fullagar and Butler 1979; Gibson and Speer 1986), as well as Conway, Sebago, and Chelmsford granite, granodiorite, and monzonite, northern Appalachians, USA (New Hampshire and Main), CIIIA (Zen et al. 1968; Hermes and Murray 1988; de Yoreo et al. 1989a, b; Schumacher et al. 1989). Of considerable interest are high-temperature ultramafic intrusions with wide contact metamorphic aureoles, such as the Beni Buchera massif of Morocco (Milliard 1959; Pearson et al. 1995; Montel et al. 2000), Ronda, Spain (Obata 1980; van der Wal and Vissers 1993), Tinaquillo, Venezuela (MacKenzie 1960; Green 1967), etc. In some cases, the metamorphic aureoles (Beni Buchera, Tinaquillo) may reach 4–5 km in width. Metamorphic rocks may have formed around layered mafic-ultramafic intrusions such as the Bushveld igneous complex of South Africa (Nell 1985; Englebrecht 1990) or around massif-type anorthosite plutons, such as the Nain complex, Labrador, Canada (Berg 1977a, b). Alkaline plutons responsible for isochemical metamorphism at mid-crustal levels are not known.

The width of zoned metamorphic aureoles around large deep intrusive bodies can be much greater than that of contact metamorphic aureoles. Zoned metamorphic aureoles are commonly represented by granulite amphibolite, and epidote-amphibolite facies rocks; migmatites are widespread in metapelitic rocks around intrusions; in some cases, the outer part of the metamorphic zoning is composed of greenschist and zeolite facies rocks (Barton et al. 1991b). The latter is formed as a result of magma intruding into rocks, which have been only weakly metamorphosed.

Based on the average geothermal gradient of about 50 °C/km, the more or less reliable manifestations of medium-pressure zonal metamorphism (with additional heat supply from magmatic intrusions) include the Lepontine Alps, the northeastern part of the Grampian Highlands of Scotland, central parts of the South Chuya Range and the Chulyshman highlands of Gorny Altai, the axial part of the Turkestan Range in Central Asia, the Muzkol complex in the Eastern Pamirs, the Chota-Nagpur in eastern India, the Cariboo Mountains in the Canadian Cordillera, and a number of areas in New England (USA) (Reverdatto 1973a, b).

Here we present the formulation of a mathematical model for medium-pressure zonal metamorphism restricted to the mid-crustal level. As is the case with contact metamorphism, for simplicity, this model does not account for convection in igneous magma and heat transfer via an intergranular fluid in the rocks undergoing metamorphism. The main difference between the models for medium-pressure zonal metamorphism and contact metamorphism (which was considered in the one-dimensional formulation) is the increased initial temperature of the country rocks. To cover a wider range of simulation capabilities, the mathematical model of medium-pressure zonal metamorphism will be presented in a two-dimensional formulation. Equations (3.2)–(3.13) can be easily transformed into two-dimensional equations and thus are not duplicated here.

If the country rocks are metapelites, melting would take place at a great depth near the intrusive contact. The moving boundary of phase transition between the zone of anatexis and the metamorphic rock is designated as $\varepsilon_{ax\text{-}tr}$, the boundary between the metamorphic and country rock is designated as $\varepsilon_{tr\text{-}er}$, and the boundary between the magma and the magmatic rock is designated as $\varepsilon_{m\text{-}mr}$ (Fig. 3.13).

As a first approximation, the initial temperature distribution in a vertical cross-section of the crust is assumed to be linear. However, taking into account radiogenic heat generation, and also the average continental estimates of geothermal regime during orogeny (Watson 1978), it would be more appropriate to assume the initial temperature distribution to be nonlinear (IT curve in Fig. 3.14).

The system of equations was solved using a finite difference method and typical thermophysical parameters for the rocks and magmas of different compositions (Table 3.2). Modeling with the use of these parameters allows us to characterize in general the different cases of medium-pressure zonal metamorphism. In model versions, we used different initial magma temperatures, thermal effects at the moving boundaries of phase transitions, diverse shapes and thicknesses of magmatic bodies, inclinations of intrusive contacts, etc. Taking into account the

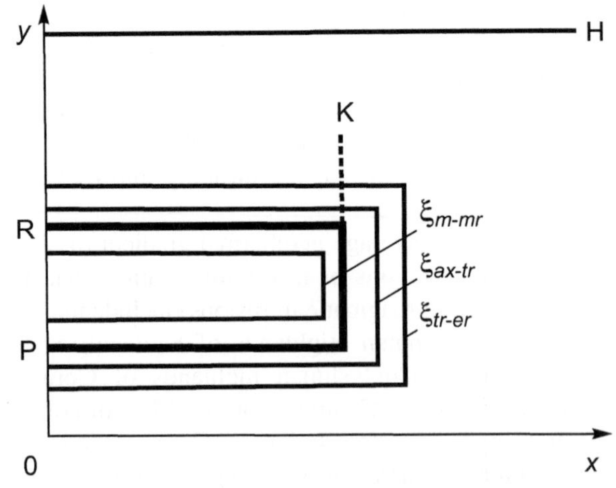

Fig. 3.13 Approximation of a sheet-like magmatic intrusive body in the form of a rectangle in the X–Y plane (only the right half of the intrusive, symmetric to the left) is shown. H is the position of the earth's surface. Bold lines indicate the boundaries of intrusive body; distance P–R is the thickness of a sheet; distance R–K is the half-width of a sheet; K is the lateral contact. Thin lines show the mobile boundaries of phase transitions $\xi_{m\text{-}mr}$, $\xi_{ax\text{-}tr}$ and $\xi_{tr\text{-}er}$

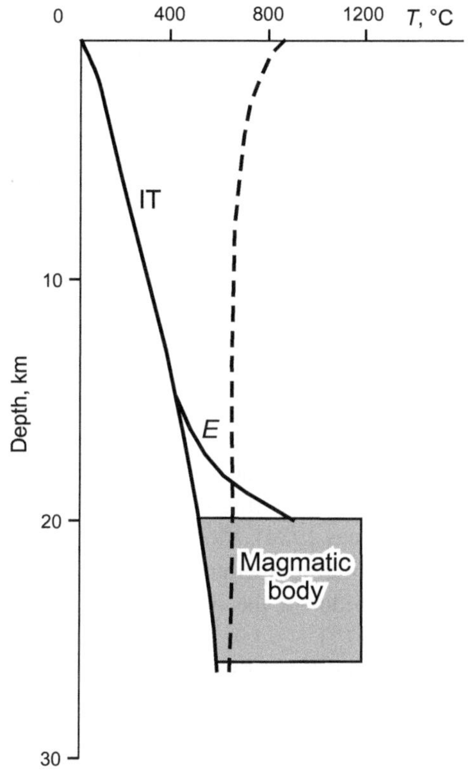

Fig. 3.14 Distribution of temperatures in the vertical section of the Earth's crust: IT is the initial distribution, E is the calculated envelope of maximum temperatures (for all times) above the upper contact of a 6 km thick horizontal sheet-like magmatic intrusive body (Tm = 1180 °C) with a roof at a depth of 20 km. The dashed line shows the onset of granite melting in the presence of H_2O (Johannes 1984)

approximations conditioned by the formulation of the model, numerical solutions and so on, the accuracy of temperature estimation and duration of metamorphism can be taken to be 10–15% (Jaeger 1968; Reverdatto et al. 1970, 1972; Reverdatto and Kalinin 1989a; Furlong et al. 1991). As regards magma convection within an intrusive body, it should be noted that the transfer of heat by moving magma in this case leads to a slight increase in temperature at the roof and a faster cooling rate of the intrusion (Jaeger 1968; Hodge 1974). Given the approximations used in this model, the refinements related to magma convection are hardly necessary.

The main characteristic features of medium-pressure zonal metamorphism can be reliably traced in a simple planar model (Figs. 3.13 and 3.14). Let us assume that a horizontal mid-crustal magma sheet is instantaneously emplaced at depths up to 30 km (where the lithostatic pressure reaches 8–9 kbar). Numerical calculations (using parameters given in Table 3.2) can be used to obtain characteristic temperatures and to study temperature evolution around an intrusive body, taking into account the distribution of initial temperature of the enclosing rocks T_{er}^0 (IT curve in Fig. 3.14). The calculations show that for a magma sheet with a thickness of 6 km and a half-width of 20 km, the maximum temperatures will be equal to ~940 °C at the roof of the intrusion at a depth of 20 km, at T_m^0 = 1300 °C (ultramafic magma), and to ~1050 °C at the floor of the intrusion; and the same parameters at T_m^0= 1180 °C (mafic magma) will be equal to 870–880 and 900–920 °C, respectively. Such temperatures are achieved at the contact of the sheet immediately after magma intrusion and decrease with further cooling of the intrusion and the development of a thermal aureole. With time, the maximum temperature in the cooling magma body after crystallization is shifted from the interior toward the floor. The distance over which the thermal effect of the intrusion would be significant at the upper-crustal level, i.e., where temperatures would be different from the initial ones at a respective depth, will be equal to 4–5 km for T_m^0 = 1180 °C and 7–8 km for T_m^0 = 1300 °C at the roof and 4 km at the floor of the intrusion. Depending on T_m^0 the thickness of anatexis zone could exceed ~1 km at the roof or approach ~2.5 km at the floor of the mafic intrusion. It is worth noting that the temperature of anatexis is taken to be about 700 °C, which is consistent with the experimentally derived temperatures of complete melting of granitic rocks at $P_{H_2O} \approx$ 5–6 kbar (at a depth of 20 km); at the same water pressure, the temperature at the onset of melting in the system Qz–Or–Ab–An–H_2O is estimated at 640–650 ° C (Johannes 1984). More intense anatexis at the floor of the intrusion is caused by a higher initial temperature of the country rocks in accordance with $T_{er}^0 = f(y)$.

The width of the metamorphic aureole at the roof may reach on average (for different T_m^0) ~6 km. For enclosing metapelites containing mafic minerals, partial melting and heat absorption will occur only in the quartzo-feldsparthic-micaceous domains of the rocks, and the amount of felsic melts will directly depend on the original composition of the metamorphic rock (Hodge 1974). The total lifetime of thermal anomaly in the crust associated with magma intrusion (T_m^0 = 1180–1300 ° C) is greater than ~10 Ma. At the end of this period, the temperature almost in all parts of the section returns to its initial values due to $T_{er}^0 = f(y)$ at the corresponding depth.

The model results on the temperature distribution and evolution around a 5 km thick magma sheet (under otherwise equal conditions) are almost identical to those calculated (within the error limits) for a 6 km thick magma sheet. A decrease in the thickness of the magma sheet to 3 km will reduce the duration of the thermal anomaly induced by intrusions of magma to 4–6 Ma, as well as the temperature and distance between isotherms and intrusive contacts. An increase in the thickness of the magma sheet to 8–10 km does not change the temperature (isotherms) distribution around intrusion but permits extended (up to 2 times) duration of thermal anomaly in the crust as compared to the 5–6 km thick magma sheet.

Let us consider the temperature variations close to the vertical contact of a horizontal sheet-like intrusive body. At a depth where T_{er}^0 = 500 °C, the lateral thickness of anatexis zone is >1 km at T_m^0 = 1300 °C and <1 km at T_m^0 = 1180 °C. The maximum width of the lateral zone of heating (to initial temperatures at this depth) is about 11 km at T_m^0 = 1180 °C. At the same T_m^0 and with increasing thickness of the magma sheet up to 8–10 km, the lateral distance tends to increase to ~12 km.

On the basis of the above estimates, it could be concluded that anatectic liquids in deep-seated metapelites in the vicinity of a mafic intrusion, due to a lower melting point, can remain mobile even after solidification of the basaltic magma bodies. The felsic melts, lying beneath the dense intrusive rocks, are gravitationally unstable. In this case, we should expect a redistribution of the material in the crust due to subsidence of the heavy overlying rocks, rise of buoyant felsic magma and formation of granite-gneiss diapirs and granitic intrusions. This question is discussed in detail below (Sect. 3.2.1.3).

It should be also noted that the presence of hot (with a temperature above ~550 °C) fluid fluxes of H_2O- and CO_2 percolating through the permeable rocks at a rate of over ~10^{-9} g cm^{-2} s^{-1} will permit increased distance between intrusive contacts and corresponding isograds and extended duration of metamorphism and anatexis (Reverdatto and Kalinin 1990). However, the presence of low-permeability rocks (Shmonov et al. 2002) provides a strong argument that metamorphism and anatexis in zoned complexes were primarily a result of the thermal effect of mafic-ultramafic intrusions without any significant heat transfer by infiltrating fluid. This is consistent with the concept of isochemical metamorphism.

The results of 3D numerical modeling show that the width of the heated zone increases with depth close to the magma intrusion (Fig. 3.15). Tables 3.3 and 3.4 present the calculated distances from the intrusive contacts to the respective

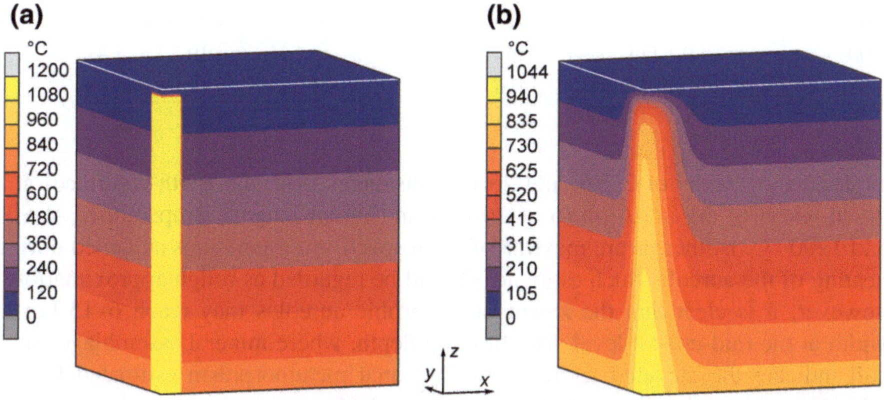

Fig. 3.15 Cuttings from a three-dimensional model space as a rectangular parallelepiped with a square base of 52 × 52 km and height of 32 km (1/4 of the model area is shown). In the center of space there is a magmatic intrusive body in the form of rectangular parallelepiped with height of 31 km. The upper and lower faces of parallelepiped are squares. In different versions of calculations, the side of smaller (inner) square (the width of intrusive body) is equal to 6 or 20 km. The initial depth distribution of temperatures in the Earth' crust is given in accordance with a linear geothermal gradient of 25 km/°C (without taking into account radioactive sources of heat generation). The figure shows the position of isotherms near the vertical intrusive body in the depth of the earth's crust. **a** is the temperature distribution at the initial moment; **b** is the temperature distribution 1 million years after magmatic intrusion. On the left are temperature scales in gradation of color for different times

Table 3.3 The distance from the intrusive contact to the corresponding isotherms at different depths at an initial magma temperature of 800 °C (granodiorite)

Near-intrusive isotherms, °C	Half-width of the intrusive body (parallelepiped), km					
	3			10		
	Depth, km			Depth, km		
	2	16	25	2	16	25
700	–	–	>0.1 km	–	–	~1.5 km
650	–	–	~5.5 km	–	–	~11.5 km
600	–	>0.1 km	–	–	>0.1 km	–
500	–	~2.5 km	–	–	~4.0 km	–
450	>0.1 km	~6.0 km	–	>0.1 km	~10.0 km	–

Table 3.4 The distance from the intrusive contact to the corresponding isotherms at different depths at an initial magma temperature of 1200 °C (gabbro)

Near-intrusive isotherms, °C	Half-width of the intrusive body (parallelepiped), km					
	3			10		
	Depth, km			Depth, km		
	2	16	25	2	16	25
800	–	–	>1.0 km	–	>0.1 km	~2.5 km
700	–	~0.5 km	~6.0 km	–	~2.0 km	~8.5 km
650	–	~1.5 km	~13.0 km	–	~3.0 km	~16.5 km
600	–	~2.5 km	–	>0.1 km	~5.0 km	–
500	>0.1 km	~6.0 km	–	~0.2 km	~10.5 km	–
450	~0.2 km	~12.5 km	–	~0.3 km	~16.0 km	–

isotherms at depths of 2, 16, and 25 km (distances from the Earth's surface) at a given width of the intrusion (6 or 20 km) and initial magma temperatures of 800 and 1200 °C. Isotherms are maxima of the respective temperatures achieved during heating of the aureole. Such estimates should be regarded as rough approximations; however, it is clear that the zoned metamorphic aureoles may reach 6–15 km in width at the mid-crustal level. The limiting depth, where mineral assemblages may still indicate the elevated temperatures of zonal metamorphism estimated by the 650 °C isotherm, hardly exceeds 25 km (Fig. 3.16).

3D modeling can be used for the approximate estimation of the emplacement depth of magmatic intrusions based on data for the distribution of isotherms in the near-intrusive area. As an example, we calculate the possible depth of emplacement of the Devonian Kharlovo gabbro-massif, which is located in the northwestern foothill belts of Altai, in the middle course of the Charysh River (Reverdatto et al. 1974; Likhanov et al. 1994) (Fig. 3.17). The massif represents an intrusive body that has been exposed on the surface by erosion and has a shape of an irregular circle with a diameter of about 4.5 km. The body is composed of olivine-titanomagnetite melanogabbro and olivine-free leucogabbro; a small

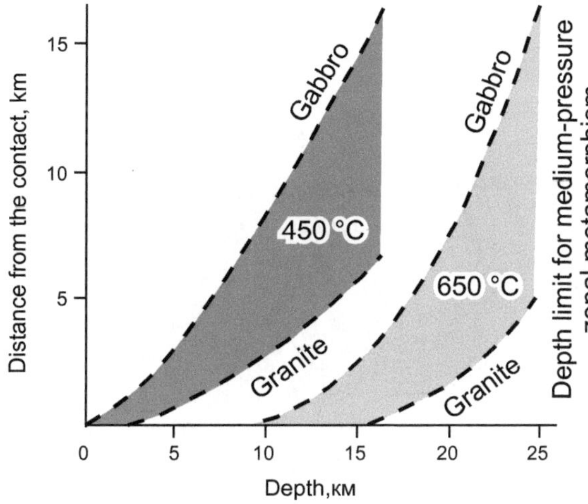

Fig. 3.16 Distance from intrusive contacts to 450 or 650 °C isotherms at different depths around mafic and felsic magmatic bodies. The minimal metamorphism caused by intrusion of magma bodies takes place at a depth where only little temperature exists between the intrusive contact and surrounding rocks. At this depth the outer limit of the metamorphic zoning defined by mineral transformations in the mid-crustal levels becomes uncertain

granodiorite stock represents a later intrusive phase. At depth, the gabbro massif has the shape of a vertical cylinder. The country rocks hosting this intrusion, Ordovician chlorite-sericite schists experienced metamorphism with the formation of biotite-cordierite hornfels. Biotite appears first; cordierite and biotite may have formed simultaneously (in spotted schists) by the reaction Ms + Chl + Qz → Crd + Bt + H_2O. At the intrusive contact, the formation of orthoclase-cordierite-biotite hornfelses takes place by the reaction Ms + Bt + Qz → Crd + Kfs + H_2O. Taking into account the estimates of Pattison (1991), the first reaction (the formation of biotite + cordierite) occurs at the temperature of ~560 °C at 3 kbar, ~530 °C at 2 kbar, and ~480 °C at 1 kbar (extrapolation). The second reaction (the formation of cordierite +K-feldspar) takes place at the temperature of 630–640 °C at 3 kbar, 580–600 °C at 2 kbar, and 530–570 °C at 1 kbar (extrapolation) (Ibid). To determine the depth of emplacement of the Kharlovo massif by solving the problem of cooling an infinite vertical cylinder of a gabbro magma (with an initial temperature of 1100 °C), we should first calculate the distance from the intrusive contact to the respective isograds/isotherms. The calculations show that the distance between the ~480 °C isotherm at a depth of ~4 km (lithostatic pressure of ~1 kbar) and the intrusion is ~0.5 km (given a normal continental geothermal gradient of ~25 °C/km); the distance between the 530 °C isotherm and the intrusion is 0.9–1 km at a depth of 7–8 km (lithostatic pressure of ~2 kbar); the distance between the 560 °C isotherm and the intrusion is 1.5–2 km at a depth of 10–11 km (lithostatic pressure of ~3 kbar). It can be seen that the possible depth of emplacement of the Kharlovo massif can be taken to be 7–8 km, while the estimated depths of ~4 km and 10–11 km appear to be less reliable. However, such estimates may be different for a lower crust geothermal gradient. At a depth of 7–8 km, the formation of cordierite + K-feldspar (in the K-feldspar zone) at the expense of muscovite, biotite, and quartz, takes place at 580–600 °C (Pattison 1991). The calculations show that this temperature can be

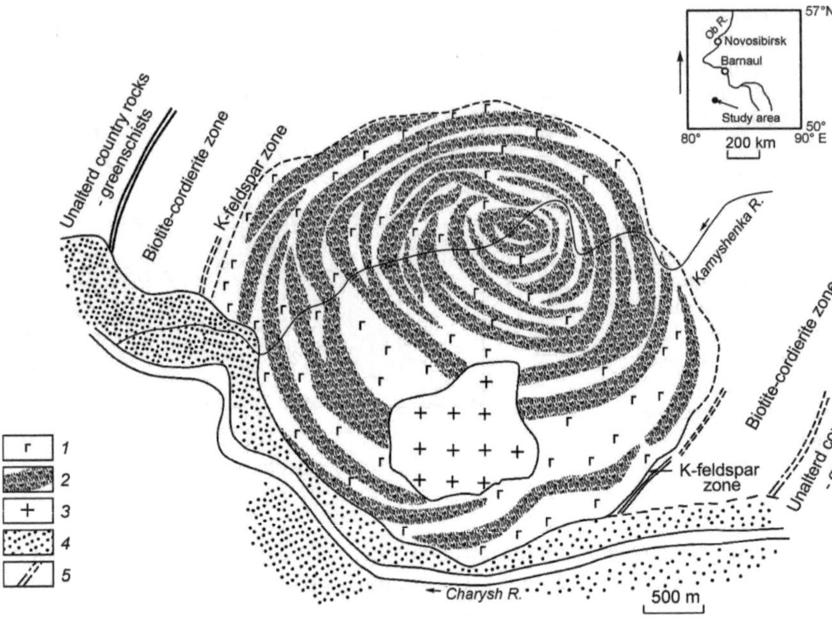

Fig. 3.17 Geological sketch map of the Kharlovo gabbro massif and contact aureole at the northwest foothills of the Altai in Siberia (Reverdatto et al. 1974; Likhanov et al. 1994). 1—leucogabbro, 2—titanomagnetite melanogabbro, 3—granodiorite, 4—alluvium deposits, 5—metamorphic zone boundaries—isograds (mapped and inferred)

achieved at a distance of a few tens of meters away from the intrusive contact, which is consistent with the geological observations (Reverdatto et al. 1974; Likhanov et al. 1994).

Some additional examples of medium-pressure zonal metamorphism will be discussed below. Then excellent agreement between the observed structure of a metamorphic zoning and the model temperature distribution obtained by solving the heat conduction equation is illustrated by the study of thermal transformations in rocks surrounding the quartz-monzonite Cupsuptic pluton, Maine, USA (Harwood 1973; Bowers et al. 1990; Furlong et al. 1991). The pluton, some 6×8 km in diameter, was emplaced at a depth of 11–12 km, i.e., at a lithostatic pressure of about 2.5 kbar. The initial temperature of the country rocks at a geothermal gradient of ~ 30 °C/km is taken to be ~ 350 °C; the initial magma temperature is estimated at 950 °C. The metamorphic aureole (varying from ~ 0.3 to ~ 1.3 km in width) which comprises the biotite, andalusite, and sillimanite zones was formed at the expense of the greenschist facies metapelitic rocks. The biotite zone is represented by spotted dense schists, in which biotite, garnet, chlorite, and cordierite porphyroblasts are embedded in a fine-grained quartz-muscovite-albite matrix. Large porphyroblasts of almandine-spessartine garnet appear at the external boundary of the biotite zone; smaller garnet grains are found close to the internal boundary. The andalusite zone is composed of granoblastic andalusite-biotite-chlorite-cordierite-plagioclase-muscovite-quartz hornfelses. The sillimanite zone is composed of hornfelses characterized by the presence of sillimanite, biotite, cordierite, plagioclase, muscovite, and quartz, sometimes with garnet. Toward the intrusive contact and in roof pendants, muscovite is locally replaced by sillimanite and K-Na-feldspar and biotite by sillimanite and magnetite. The temperatures of the sillimanite, andalusite, and biotite zones were determined with the garnet-biotite geothermometer (Ferry and Spear 1978) at 580, 510, and 460 °C, respectively. Numerical simulation of heating and metamorphism of the country rocks around a complexly-shaped magmatic body was performed for two- and three-dimensional versions (Fig. 3.18). The calculations allow taking into account release of latent heat of magma crystallization (~ 150 J g^{-1}) and thermal effect of a low-temperature metamorphic reaction, i.e., the formation of biotite, cordierite, and water at the expense of muscovite, chlorite, and quartz (~ 85 J g^{-1}). These thermal effects, although severely underestimated compared to the values presented in Dudarev et al. (1972), Birth et al. (1942) and Buntebarth (1984), are not regarded to play an important role, and the temperature distribution around a heat source was dependent primarily on T_m^0 and intrusion geometry. Nevertheless, fairly good agreement between the calculated spatial temperature distribution (isotherms) and

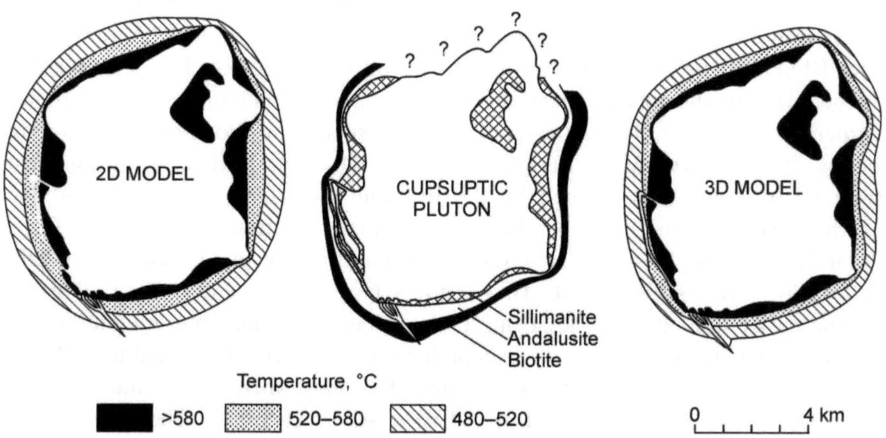

Fig. 3.18 Results of mathematical modeling of metamorphic zoning around the Cupsuptic pluton, USA (Bowers et al. 1990; Furlong et al. 1991). Picture in the center shows the shape of the pluton and the configuration of metamorphic zones; their boundaries are mapped by the first appearance of biotite, andalusite and sillimanite (Harwood 1973). Pictures on the left and right show, respectively, the results of two- and three-dimensional numerical modeling of temperature distribution of metamorphic zones near the pluton (Bowers et al. 1990; Furlong et al. 1991). Picture on the left shows the contacts of the pluton are assumed to be vertical and of infinite length (perpendicular to the plane of the figure); heat from contacts spreads only horizontally. In the three-dimensional model (picture on the right), it is assumed that the heat source, i.e. a 4 km thick magmatic intrusive, is discoid in shape (or "washer" with beveled edges). The temperature distribution is shown in the horizontal section across the center of a discoid

metamorphic zone configuration (isograds) confirms the correctness of the model and suggest prevalence of conductive over convective heat transfer from a magma intrusion to the country rocks.

The Ballachulish igneous pluton and the surrounding metamorphic aureole is one of the world's most extensively studied metamorphic complexes. It is located in the west of Scotland, on the southern side of Loch Linnhe (Voll et al. 1991). Although the Ballachulish zoned aureole was described as an example of contact metamorphism, we consider it in the chapter on medium-pressure zonal metamorphism, because this two-phase intrusion was emplaced at a depth of ~10 km at a pressure of 3 ± 0.3 kbar. At this depth, the initial temperature of the country rocks was ~250 °C (at initial geothermal gradient of ~25 °C/km). The intrusion measures in plan 7.5×4.5 km and consists of two magmatic phases: diorites (quartze diorite and monzonite-diorite) and granites. The former comprise an outer envelope and the latter occupy an inner core of the pluton. The age of emplacement of this two-phase intrusion is constrained at 412 ± 28 Ma. The metamorphic aureole varies in width from ~0.4 to ~1.7 km. The pluton was intruded into Dalradian (Neoproterozoic–Lower Paleozoic) metasediments represented by schists and quartzites, with thin interlayers of carbonate rocks (Fig. 3.19). The metapelitic rocks (Zone I) typically contain mineral assemblages consisting of Ms + Qz ± Chl ± Bt и Ms + Qz + Grt + Bt ± Chl. The outer part of the metamorphic aureole (Zone II) is characterized by the appearance of porphyroblastic cordierite in association with muscovite, chlorite, biotite, albite, and quartz (matrix). Where present, garnet is partially replaced by chlorite and biotite. The transition to Zone III in these pelitic rocks is defined by the disappearance of chlorite. Zone IV is characterized by the appearance of andalusite in association with muscovite, biotite, cordierite, and quartz. The appearance of K-feldspar in this zone can be accounted for by the reaction Ms + Bt + Qz → Crd + Kfs + H_2O. The transition from Zone IV to Zone V is defined by the reaction Ms + Qz → And + Kfs + H_2O, which may occur in parallel with the reaction And + Bt + Qz → Crd + Kfs + H_2O. In quartz-free rocks, corundum may appear by the reaction Ms → Crn + Kfs + H_2O and Bt + And → Crn + Kfs + Crd + H_2O. Partially molten rocks immediately adjacent to the igneous contact are mineralogically similar to the rocks in Zone V, except for the presence of migmatites. Mineral transformations during anatexis can be described by the reaction Kfs + Qz + H_2O ± Bt ± Pl → Liq ± Crd ± Als (Al_2SiO_5). The presence of hypersthene and garnet is documented in the quartz-rich molten products, while quartz-free products may contain spinel formed by the reactions Bt + Grt + Qz → Crd + Hyp + Kfs + Liq, Bt + Als (Al_2SiO_5) → Spl + Crn + Crd + Kfs + Liq and/or H_2O, Bt + Crn + Crd → Spl + Kfs + Liq and/or H_2O, etc. In the siliceous dolomites, the low-temperature metamorphic grade is manifested by the reaction Dol + Qz + H_2O → Tlc + Cal + CO_2; the appearance of tremolite, forsterite + calcite, diopside + forsterite, diopside + dolomite etc. is noted close to the contact aureole. At higher temperatures, periclase + calcite are formed from dolomite. At a distance of 200 m from the contact, the siliceous limestones were subjected to metamorphism with the formation of wollastonite.

Temperatures marking the boundaries between metamorphic zones (isograds) were estimated using a petrogenetic grid for metapelites and garnet-cordierite and garnet-orthopyroxene geothermometers at $P = 3 \pm 0.3$ kbar: 550–560 °C between Zones I and II, 600–620 °C between Zones III and IV, 625–640 °C between Zones IV and V; muscovite disappears from quartz-free ricks at 655–670 °C; melting of quartzo-K-feldsparthic rocks takes place at 670–775 °C; periclase appears in dolomites at 750–800 °C. Thermal modeling of Buntebarth (1991) assumed a cylinder-type intrusion, 5800 m in diameter, with an inner core of granite, 3400 m in diameter, and an outer envelope of diorite. It was taken that for diorite $T_m^0 = 1050$ °C and $T_m^0 = 750$ °C and for granite $T_m^0 = 850$ °C and $T_m^0 = 680$ °C; the latent heat of magma crystallization $G_{m\text{-}mr} = 310$ J g^{-1}. Figure 3.20 shows variation of the maximum temperatures (shown by an envelope curve by isochrons for different times of heating) in country rocks of various compositions in four profiles. These temperatures are in good agreement with temperature estimates from mineral thermometers.

The well-studied extensive metamorphic zoning in the Connemara region of western Ireland (Barber and Yardley 1985; Yardley et al. 1987; Friedrich et al. 1999; Reverdatto and Polyansky 2004) is regarded as a fragment of the Caledonian orogen of the Great Britain (Fig. 3.21). The metamorphic zones developed in metapelites which are correlated with the Dalradian rocks of Scotland. These metapelites contain kyanite and staurolite and were subjected to the Barrovian-type regional metamorphism. Four subparallel metamorphic zones are recognized, which contain the following mineral assemblages: (1) Ms + Bt + St + Pl + Qz + Ilm ± Grt ± Chl ± Crd ± Ky or And (some minerals in this zone, e.g., kyanite, were inherited as relics from the preceding regional metamorphism), (2) Ms + Bt + Pl + Qz ± Sil ± And ± St ± Grt (the amount of staurolite and andalusite decreases approaching Zone III), 3) Sil + Kfs + Bt + Pl + Qz ± Grt ± Ms ± Crd, 4) Sil + Kfs + Bt + Pl + Qz ± Grt together with migmatites. All four zones vary in width from 6–8 km (Zone I) and 4–5 km (Zone II) to 2.5–4 km (Zone III) and 1–3 km (Zone IV). In the south, the latter zone is in contact with gabbro and diorite intrusive bodies. The total width of metamorphic zoning is 16–17 km. The metamorphic pressure is estimated to be 5.5 ± 1 kbar and remains almost unchanged across the entire metamorphic zoning (Yardley et al. 1987). Using mineral equilibria it was established that Zone I was formed at a temperature of 520–580 °C; sillimanite at the boundary between Zone I and II appears at ~650 °C; the migmatites were developed (melting of metapelites) in Zone IV at 700–750 °C (Barber and Yardley 1985; Yardley et al. 1987). These estimates suggest that the horizontal thermal gradient across the metamorphic zoning (indicated by the arrow in Fig. 3.21) was close to 14 °C/km. Both geological and geochronological data indicate a distinct spatial and temporal link between the zonal metamorphism and magmatic activity. Gabbros and dolerites, forming a single igneous complex, have the age of 475–470 and 467 Ma, respectively; metamorphism at temperatures between 500 and 750 °C is bracketed to between 468 and 461 Ma; subsequent cooling of the metamorphic rocks to a

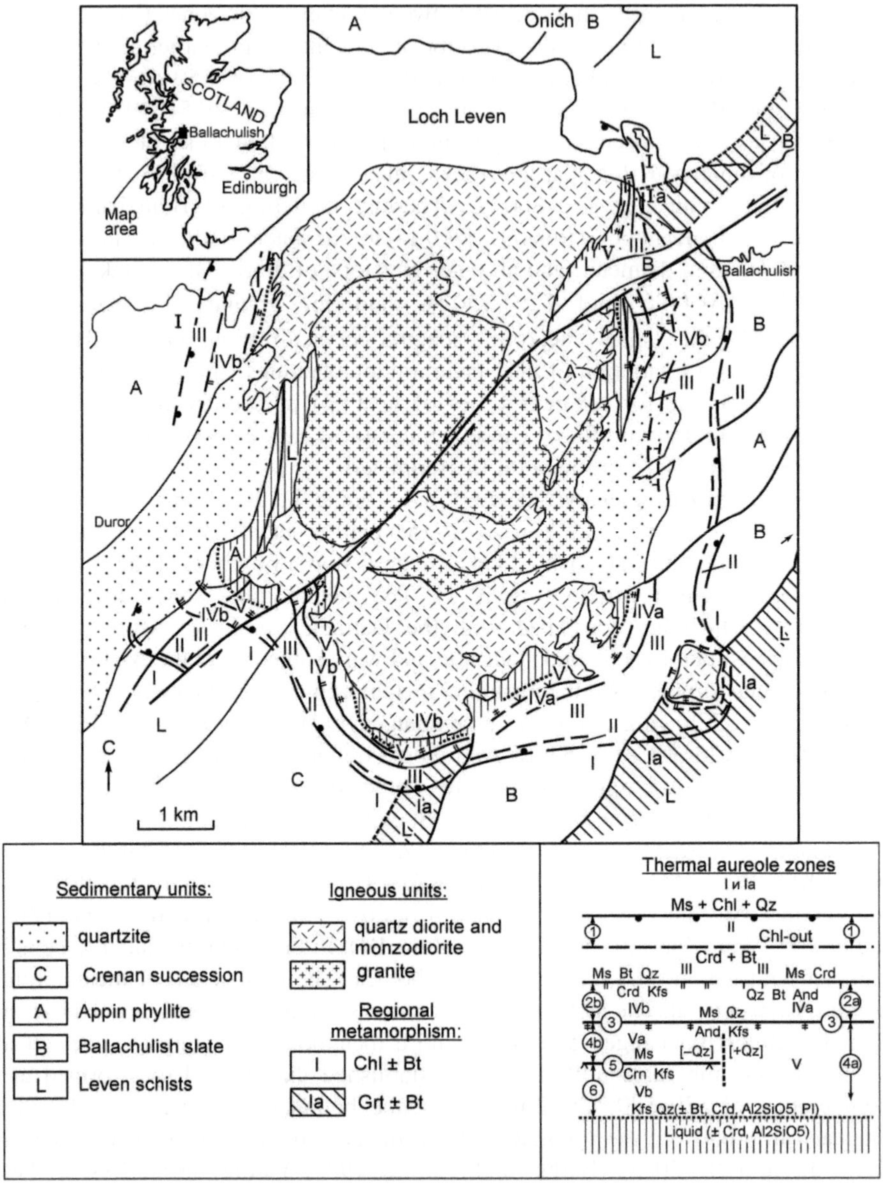

Fig. 3.19 The geological scheme of Ballachulish intrusive and surrounding metamorphic zoning. The boundaries of metamorphic zones in the rocks of different composition (I–V) are represented by isograds which are shown by symbols

temperature of ~350 °C lasted ~10 m.y. (based on 40Ar/39Ar ages of muscovite). The shape of a heat source was modeled based on the best fit of the observed temperature distribution within the zones, which was determined from

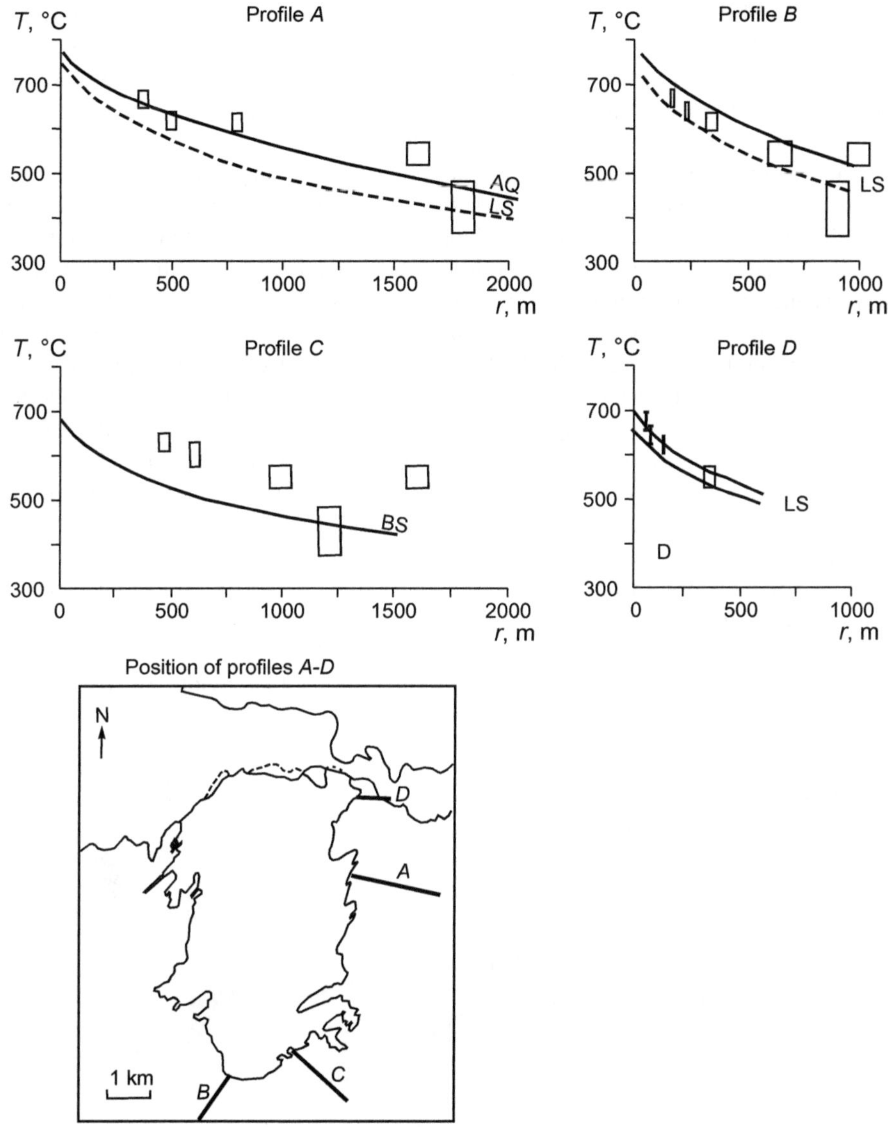

Fig. 3.20 The calculated distribution of the maximum temperatures near the Ballachulish intrusive along the profiles *A*, *B*, *C*, *D* in comparison with the temperature estimates from mineral geothermometers (rectangles and segments of vertical straight lines). The intrusive is modeled in the form of a cylinder made up of diorite and granite magma with a diameter of 5800 m; *r* is the distance from the contact along the external radius in meters. The initial temperature of the surrounding rocks is 250 °C. Profile *A* is calculated in quartzites (AQ) and shales (LS), the remaining profiles in schists (BS and LS); the dashed lines show the average value (Voll et al. 1991)

mineral equilibria, and the calculated spatial distribution of maximum temperatures, which was derived by solving the heat conduction equation for different times (shown as an envelope curve by isochrones for different times of heating around the intrusion) at different initial conditions. The mathematical modeling shows that the most plausible heat source for the Connemara zoning is best explained by modeling (the magmatic event was treated, for simplicity, as a single-stage instantaneous intrusion of mafic magma) a 6 km thick sheet-like magmatic body tilted at an angle $\alpha \approx 20$–$40°$ that was emplaced at a depth of ~20 km. The different versions of calculated temperature distributions (along a horizontal section of the Earth's crust at a distance from the tilted intrusive contact) are shown in Fig. 3.22. The model results obtained for a 5 km thick sheet-like body are almost identical to those for a 6 km thick body. A decrease in thickness to 3 km reduced the duration of metamorphism to ~4 Ma; an increase in thickness of the body to 8–10 km does not result in any change in the spatial temperature structure and total width of the metamorphic zoning, but extends the estimated duration of metamorphism to 10–15 Ma. These two estimates are inconsistent with radiometric dates for the timing of formation of the metamorphic zoning (~7 m.y.) (Friedrich et al. 1999). The assumption on the development of the metamorphic zoning above a horizontal sheet-like magma body which was subsequently tilted as a result of tectonic activity lacks support from the geological data and non-isobaric conditions of metamorphism (Yardley et al. 1987; Friedrich et al. 1999). In addition, the model spatial temperature structure of this zoning version (with a total width of 16–17 km)

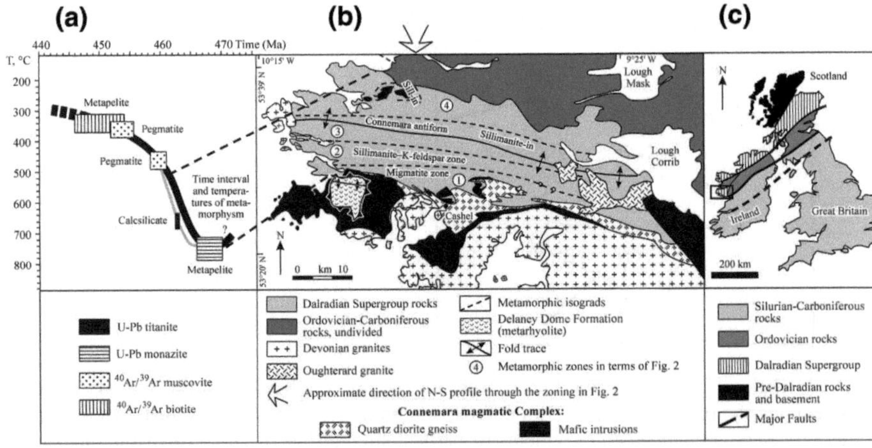

Fig. 3.21 Metamorphic zoning of Connemara. **a** The relationship between age and temperature of metamorphism for the rocks in the aureole (Friedrich et al. 1999); the temperature boundaries of metamorphism are established according to Barber and Yardley (1985) and Yardley et al. (1987). Dimensions of rectangles reflect uncertainties in the estimates of temperature and age, obtained by different methods and for different minerals in the metamorphic rocks. **b** Schematic map of the metamorphic zoning and magmatic rocks of the Connemara region. **c** Connemara region (rectangle) in the Caledonides of Ireland and the United Kingdom

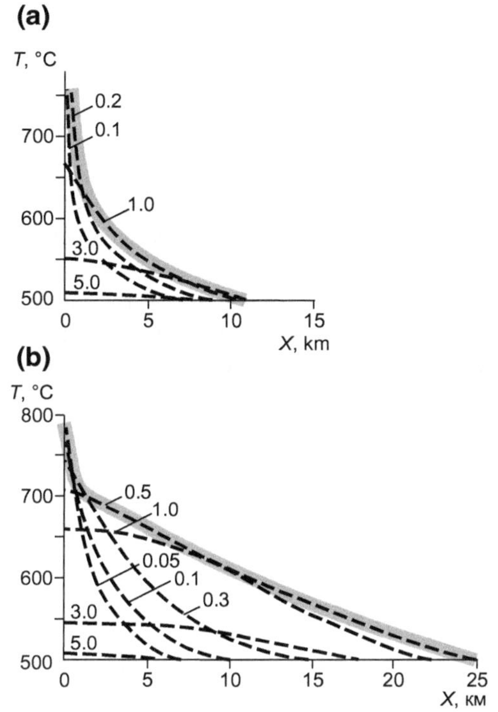

Fig. 3.22 The calculated thermal evolution along the horizontal section of the earth's crust at a depth of 20 km where T^0_{er} = 500 °C for the model version with T^0_m = 1180 °C: ***a*** near the lateral vertical contact (K in Fig. 22) of a 6 km thick sheet-like magmatic intrusive body; **b–d** above the upper contact of a 6 km thick sheet-like magmatic intrusion (distance P–R in Fig. 22); the sheet is inclined to the horizon at an angle: **b** α = 20°, **c** α = 35°, **d** α = 40°. The numbers on dotted curves indicate the timing of intrusion cooling in million years after the emplacement of the intrusion. A thick grey line is the envelope curves for maximum temperatures at different times

proved to be incompatible with its observed internal structure (Reverdatto and Polyansky 2004).

Another example of the metamorphic zoning is the Tongulak complex of the Kurai mountain range, southern Gorny Altai (Lepezin 1972; Anan'ev et al. 2003). Four prograde metamorphic zones are recognized: Chl + Ms + Qz (chlorite zone) → Bt + Crd + Grt + H_2O (cordierite zone); Ms + Crd + Grt → St + Bt + Qz + H_2O (the appearance of staurolite + quartz within the cordierite zone); Ms + Crd + St → Bt + Sil + Qz + H_2O (sillimanite zone); Ms + St + Qz → Bt + Sil + Grt + H_2O (the appearance of garnet within the sillimanite zone); St + Qz → Crd + Grt + Sil + H_2O (staurolite-free zone). The first reaction is responsible for the first appearance of cordierite; the second reaction produces staurolite + quartz; the third reaction produces sillimanite, while the fourth and fifth reactions produce garnet by breakdown of staurolite + quartz. The transition from the beginning of the cordierite zone to the central part of the staurolite-free zone is 7–8 km wide. The metamorphic zoning is symmetric, forming an apparent "thermal anticline" (Fig. 3.23). Temperature conditions of metamorphism are estimated from mineral equilibria: the minimum temperature corresponds to the chlorite zone, $T < 500$ °C, the peak temperature corresponds to the staurolite-free zone, $T \approx 650$–670 °C (Anan'ev et al. 2003). Pressures are estimated to vary from ~3.5 kbar in the chlorite zone, where andalusite is present (Bohlen et al. 1991), to ~5 kbar estimated from the X_{Fe} of garnet (Fe/(Fe + Mg) ≈ 0.8) in association with cordierite, sillimanite, and quartz (Aranovich and Podlesskii 1983).

The following alternative mechanisms are possible to explain the formation of the Tongulak metamorphic zoning: (1) tectonic overthrusting with subsequent uplift and thermal restructuring, (2) supply of additional heat by the intruding magma or

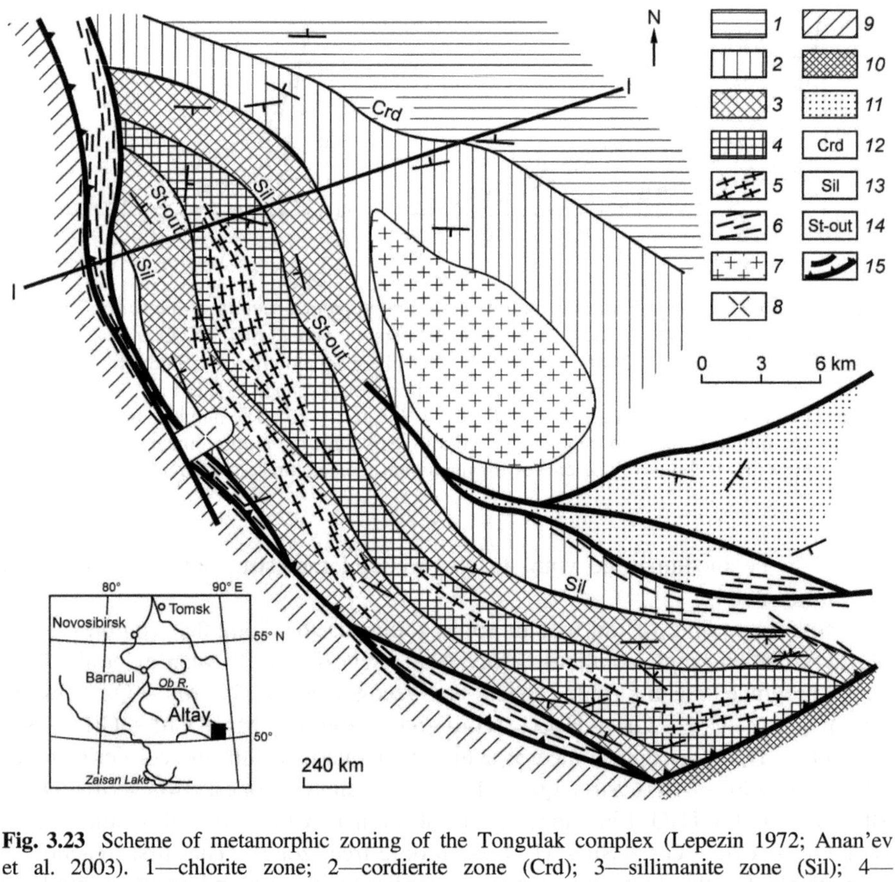

Fig. 3.23 Scheme of metamorphic zoning of the Tongulak complex (Lepezin 1972; Anan'ev et al. 2003). 1—chlorite zone; 2—cordierite zone (Crd); 3—sillimanite zone (Sil); 4—staurolite-out zone (St-out); 5—oligoclase-quartz gneiss; 6—diaphthorite; 7—granite intrusion; 8—gabbro-monzonite-plagiogranite stock; 9—Lower Cambrian deposits; 10—Devonian deposits; 11—Cambrian-Ordovician siltstones and shales; 12–14—boundaries of zones defined by corresponding isograds; 15—faults; I–I—sampling profile

hot fluid flow percolating through a permeable zone. The first mechanism is not realistic for geological reasons: field observations do not indicate overlapping of strata, tectonic rearrangement and thermal restructuring. Alternatively, such conditions should lead to an asymmetrical, flexure-like inverse metamorphic zoning, in which the lower "hot" part of the section is thrust over the upper "cold" part producing the intense folding, deformation and shearing in the vicinity of the contact. The second mechanism provides a more plausible explanation for the development of the metamorphic zoning. The main evidence for this is the andalusite-sillimanite transition from the periphery toward the center of the zoning, which can be caused at medium pressures only by an additional heat input to the crust. The isochemical character of metamorphism contradicts the assumption on intensive heat transfer by fluid flow, therefore, the additional heat can be only introduced by igneous intrusions (Reverdatto and Kalinin 1989a, 1990; Reverdatto et al. 1972, 1974; Furlong et al. 1991).

As in the previous case (Connemara), the modeling of the heat source shape was performed by the identification of the best fit between the calculated (by solving the heat conduction equation) spatial distribution of maximum temperatures around a hidden deep-seated igneous intrusion and temperatures of the metamorphic zoning estimated from mineral thermometry. The calculations were performed for a two-dimensional domain with an area of 30 × 40 km, located along a vertical section of the crust, transverse to the zoning (along the profile shown in Fig. 3.23). Two versions were considered: (1) an igneous intrusive body conceived as a rectangle in the plane section and (2) emplacement of an upward-projected magma "wedge".

The model assumed an instantaneously emplaced intrusion (mafic magma) with T_m^0 = 1100 °C, a temperature distribution in the crust corresponding to a regional thermal gradient of 20 °C/km, the initial temperature of the country rocks of 400 °C at a depth of 20 km at the onset of emplacement of a magmatic intrusion, and the surface temperature of 0 °C. The model results showed that the vertical intrusive body having a flat roof of any width cannot generate a metamorphic zoning similar to that shown in Fig. 3.23. For the second version, we run a series of calculations with different angles of incidence of the sides of the wedge-shaped roof: from 73° to 33°. The results are shown in Fig. 3.24. The calculated temperatures were compared with temperature estimates using the biotite-garnet and garnet-chlorite geothermometers. The best fit was obtained for the version B. This suggests that the roof of the intrusion is wedge-like, with the both sides dipping at about ~45°. The calculations show that the metamorphic zoning in the form shown in Fig. 3.23, i.e., at the level of the present-day erosion surface, is located only a few hundreds of meters away from the apical part of the intrusion. For the version B, this distance is equal to ~650 m (Anan'ev et al. 2003).

In the above example, the conclusion about the shape and depth of the heat source is made only based on data on the structure of the metamorphic zoning and calculated temperature distribution. Information on the shape of the intrusion and its vertical extent can also be obtained from geophysical studies and drilling data. One

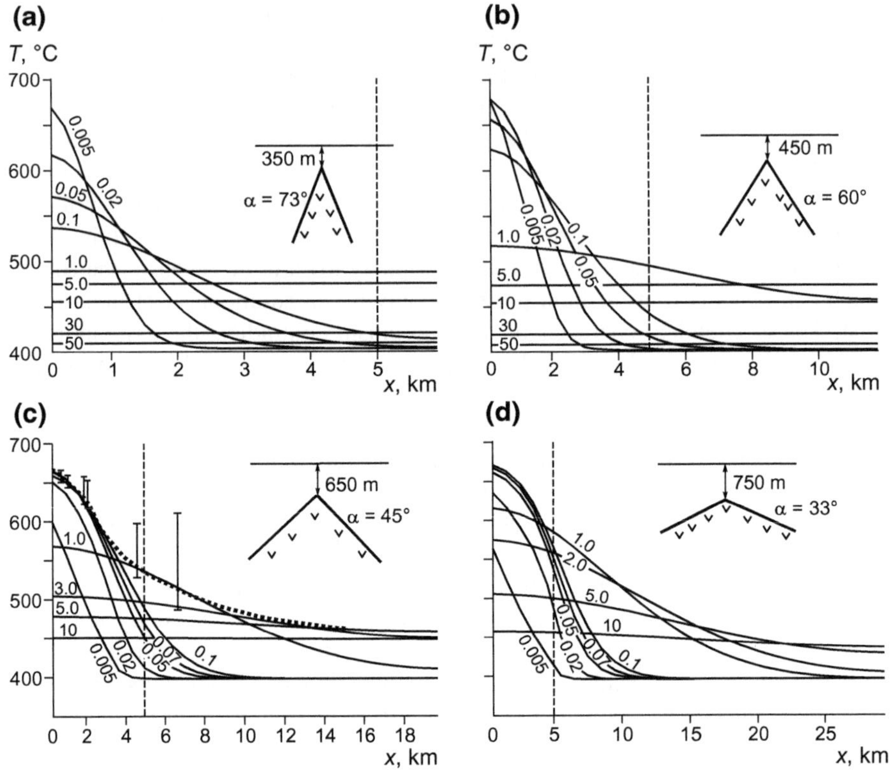

Fig. 3.24 Versions of calculations (A-D) of temperature spatial distribution in the horizontal section above the right wall of the wedge-shaped roof of the intrusion (Anan'ev et al. 2003). The results are presented for different dips of the roof walls. The numbers in the curves indicate the time after the emplacement of magma intrusion in millions of years. The best variant of model calculations and temperature estimates from mineral geothermometers (shown in vertical segments) corresponds to the case "C"; the envelop curve for maximum temperatures is given by a dotted line. The shapes of the intrusion roof with different dip angles and different distances of the top wedge from the modern erosion surface are depicted in the upper right corner. X is the horizontal distance from the symmetry axis of the metamorphic zoning; the section at a distance of 5 km (from the axis of symmetry) is shown by a vertical dashed line

example is the Devonian Skiddaw granite massif (Cumbria, NW England), which has been exposed only at its top. The metamorphic zoning was mapped by Mason (1978) and is shown in Fig. 3.25. A regional gravity survey of the area has formed the basis of a 3D interpretation of this granite massif (Lee 1986). Figure 3.26 shows the shape of the massif and the calculated depth temperature distribution (Furlong et al. 1991).

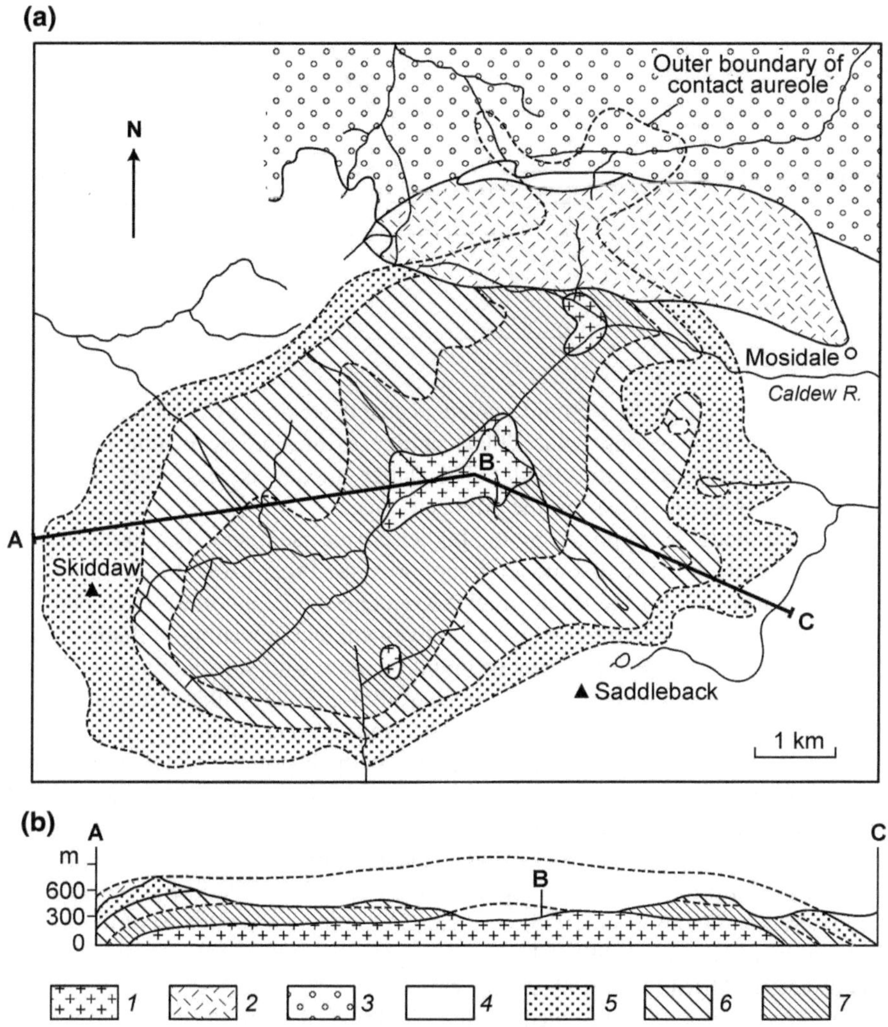

Fig. 3.25 Sketch geological map (**a**) and cross-section (**b**) of the Skiddaw granite (Cambria, NW England) and its contact aureole (Mason 1978). 1—Devonian granite, 2—Carrock Fell gabbro pre-granite complex, 3—Borrowdale volcanic series, 4—Ordovician slate, 5—Outer spotted hornfels zone, 6—Inner granoblastic biotite-cordierite- andalusite hornfels zone

3.2.1.3 Metamorphism Related to Magmatic Diapirism

Magmatic diapirs are thought to form in the lower crust and mantle and represent deep-seated heat sources. Diapirs can be divided into two types based on the age and tectonic setting. The first type includes the diapiric structures beneath the continental crust of ancient cratons that are widespread within the Precambrian granite-greenstone belts (Eskola 1948). Two models are invoked to explain their

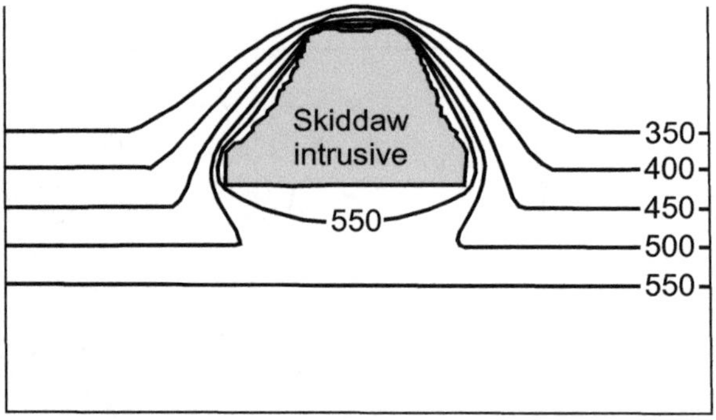

Fig. 3.26 Distribution of isotherms in the country rocks round the Skiddaw granite; its shape in a three-dimensional model is derived from gravimetric survey (Lee 1986; Furlong et al. 1991)

formation: either due to tectonic subsidence and heating by internal radiogenic sources (Van Kranendonk et al. 2004) or due to heating by mantle convection (Artyushkov and Batsanin 1984). The latter model is discussed below in Sect. 3.2.2.3 Metamorphism in the Archean crust.

The first model is supported by the higher crustal radiogenic heat production in the Archaean than in The first model is supported the Proterozoic; for example, 1.47 versus 0.4 μW m^{-3} for the Baltic shield (Kremenetsky et al. 1989). High plasticity of the ancient crust promotes an upward movement and upwelling of diapirs (Arndt 2013). Classic examples of groups of granite diapirs with surrounding metabasites and schists were reported from the Archean (3.4–3.3 Ga) Pilbara craton, western Australia (Van Kranendonk et al. 2004) and 2.5 Ga Dharwar craton, India (Choukroune et al. 1997). They are interpreted in the literature as "dome-and-basin" structures representing the buoyant upwelling of a low-density granitic substrate and simultaneous downwelling of supracrustal rocks, mostly high-density metabasites and metasedimentary rocks. Other ancient structures formed in both collisional and platform settings also include the 1.84–1.77 granite-gneiss domes of the Ladoga region (Mints et al. 1996), 1.2–1.0 Ga granite-gneiss domes of the Yenisei Ridge (Nozhkin et al. 1999), 542 ± 6 Ma granite diapirs of the Damara orogen, southwestern Africa (Toé et al. 2013), etc.

The second type is represented by diapiric domes that are developed in younger structures of collision and suprasubduction zones. In these cases, diapirism is caused by melting of thickened crust in the orogenic areas with elevated radiogenic heat production, or decompression melting of a volatile-rich substrate. The natural examples include the ~56 Ma Thor-Odin migmatite domes of the Cordilleran orogen (Norlander et al. 2002), 12–17 Ma migmatite dome of Naxos Island in Attic-Cycladic massif (Greece) (Kruckenberg et al. 2011), Late Quaternary gneiss domes of D'Entrecasteaux Islands, Papua New Guinea (Little et al. 2011), etc.

In addition to the two main types of diapirs described above, the diapiric rise of magma that has been documented in spreading (Katz and Weatherley 2012) and suprasubduction zones (Hasenclever et al. 2011), is not considered here.

The mechanism of ascent of granite magmas through the crust remains controversial. According to the review of Brown (2013), transport through fractures (diking) and magmatic intrusions in shear zones are the most commonly postulated mechanisms. Although discredited in the previous publications (e.g., Petford 1996), the viability of the mechanism of diapiric ascent of granite magmas through the crust has been demonstrated in a number of previous studies (Weinberg and Podladchikov 1994; Bittner and Schmeling 1995; Burov et al. 2003). The characteristic features of granitic diapirism have been illustrated using natural examples (Norlander et al. 2002; Vanderhaeghe 2004; He et al. 2009; Little et al. 2011; Toé et al. 2013). In these papers, geologic, structural, microstructural, thermochronologic, and thermobarometric data were combined to justify diapirism as a viable mechanism of the dome development.

Crustal diapirs have been studied in much more detail than mantle diapirs. This difference in the knowledge of crustal and mantle diapirs can be explained by insufficient erosion of the continental crust and the higher density of mafic-ultramafic magmas compared to felsic magmas, which does not allow buoyant ascent through the crust. For these reasons, only rare examples of buoyant mantle diapiric structures have been reported to date from axial parts of rifting zones. They are usually composed of a core of lherzolite surrounded by lower crustal rocks, as is the case with the metamorphic complex of Zabargad Island, Red Sea rift (Sklyarov et al. 2001). There are some examples showing that the the ultramafic magmas may represent the melting products of a rapidly ascending diapir of previously depleted subducted oceanic lithosphere. The mafic rocks of the Seiland Complex, Scandinavian Caledonides may have been extracted from this diapir (Griffin et al. 2013).

The mechanism of formation of crustal granite-gneiss diapirs for thermoelastic-viscoplastic rheologies was described in previous works (Polyansky et al. 2009, 2010a). The same approach was used to study diapiric ascent of ultramafic-mafic magmas through the mantle lithosphere (Polyansky et al. 2012, 2014). In this chapter, we attempted to explore further the possibility of modeling mantle diapirism—magmatic underplating—crustal diapirism as sequential processes. A similar model for sequential stages of diapir ascent was first considered by Weinberg and Podladchikov (1994) using the Hadamard–Rybczynski equation for the velocity of viscous spherical drops rising through Newtonian ambient fluids. However, the analytical approach alone is insufficient to calculate the ascent time, depth and shapes of the rising magmatic bodies. Therefore, we employed 2D numerical simulation, which allows implementation the temperature-dependent nonlinear rheology of rocks and an arbitrary (unknown) shape of the rising bodies (Polyansky et al. 2016). This numerical method was used to answer the following questions.

1. What is the mechanism that allows the diapiric ascent of partially molten material through a superviscous but deformable lithosphere?
2. What will be the duration of action of a sublithospheric heat source to ensure thermal softening and ascent of magma through the cratonic lithosphere and what will be the time interval between the occurrence of bimodal mantle (mafic) and crustal (felsic) magmatism?

3. What degree of melting of the mantle lithosphere and crust would be responsible for generating partially molten masses that can effectively rise as diapirs through mantle and crust?
4. What are the physical processes that govern interaction between rising mantle diapirs and the crust: thermomechanical erosion and thinning or magma underplating and subsequent crustal melting? In the case of crustal melting, under what conditions secondary (crustal) diapirism is likely?

Each particular dome has been attributed to various mechanisms because geological observations do not provide an answer to the question of what was the governing process. Possible mechanisms responsible for the development of dome-shaped structures include:

1. diapirism first proposed by Eskola (1948) was thought to be driven by inversion of the rocks densities with depth due to partial melting or reactions involving an increase in the specific volume;
2. the mechanism that combines ballooning and intrusion of granitic magmas into country rocks, i.e., what Pitcher and Berger (1972) called piercement diapir with respect to the Ardara pluton in Ireland;
3. isostatic unloading due to tectonic extension and unroofing leading to the formation of metamorphic core complexes (Sklyarov 2006; Buck 1991; Rey et al. 2009).

As noted above, because mantle diapirs are inaccessible to direct observation, their effect is determined from indirect evidence. The Vilyui Igneous Province in the eastern part of the Siberian platform is an example of a mantle diapir (superplume) (Kuzmin et al. 2010). It is assumed that a superplume rising under the Vilyui rift in the Middle Paleozoic may have transported significant amounts of molten material to the base of the lithosphere, which was in part ($\sim$320,000 km^3) erupted to the surface or intruded into the sedimentary successions (Kiselev et al. 2014). Indirect evidence of the existence of a mantle diapir beneath the Vilyui province comes from plutons of alkaline ultramafic rocks, flood basalt eruptions at 380–350 Ma, formation of rift zones and mafic dike belts (Kuzmin et al. 2010; Polyansky et al. 2013, 2017).

The formation of ultramafic plutons making up the Caledonian Seiland complex of Norway was explained by Griffin et al. (2013) using a model of multistage melting and ascent of a lherzolite thermochemical plume similar to that proposed by Dobretsov (2010) and Kirdyashkin and Kirdyashkin (2013). It was suggested that when the lherzolite diapir rising from a transition zone 200–250 °C hotter than the mantle adiabat impinges on the base of a 65 km thick lithosphere, it is expected to spread out horizontally. The Seiland Igneous Province was emplaced in an extensional setting, which would have facilitated the rapid rise of such a diapir and the generation of high-T magmas represented by late-stage intrusions of the picrite, mafic and ultramafic dikes into the igneous complex.

Granite-gneiss domes are widespread in the Murmansk, Kola–Belomor, and Savo–Ladoga granite-migmatite belts of the Baltic Shield (Grigor'eva and

Shinkarev 1981; Mints et al. 1996). These rising structures have a drop-like shape and typically comprise a migmatite-granite core 5–10 km (rarely up to 50 km) across surrounded by an envelope of amphibolite or two-pyroxene-plagioclase schist (Fig. 3.27a). In some cases, they consist of a tail and a large mushroom-shaped head spreading at the base of a more rigid suprastructure. The largest Northern Ladoga—Kokkosel'kya granite-migmatite-gneiss dome (Morozov and Gaft 1985) reaches ~ 10 km across (Fig. 3.27a). It is assumed that the masses of partially molten material of migmatites and felsic magmas that rose from a zone of ultrametamorphism underwent plastic deformation simultaneously with the overlying amphibolites (Grigor'eva and Shinkarev 1981). The estimated *P-T* conditions for amphibolites conformably overlying the dome core are 6–7 kbar and 525–550 °C and those for granite-gneiss in the core are 4.2–4.8 kbar and 500–520 °C (Polyansky and Efremov 1989). Based on a 1.5–2.5 kbar difference in lithostatic pressure between the core and envelope, the vertical ascent distance of the dome can be constrained to be at least 5.0–8.5 km.

It is quite difficult to establish the cause of the Rayleigh-Taylor gravitational instability, but in some cases it can be determined from geophysical data in areas with the widespread development of granite-gneiss domes. For example,

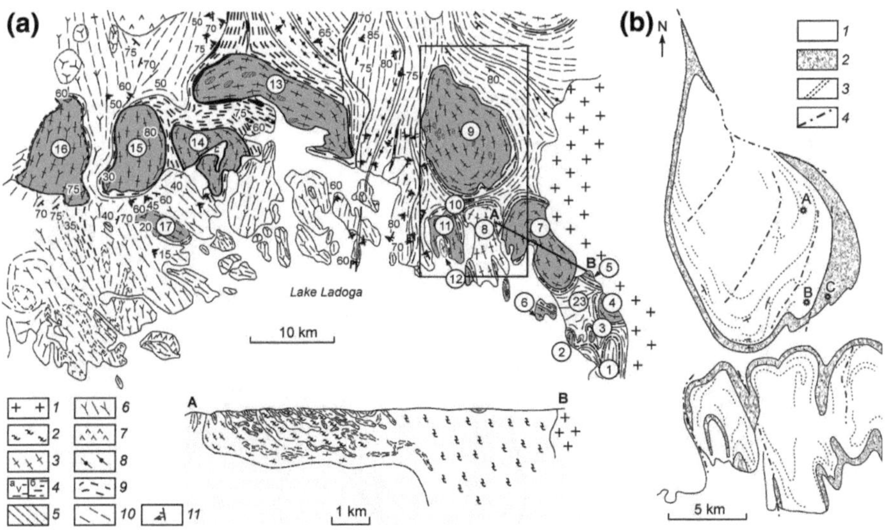

Fig. 3.27 **a** Occurrence of granite-gneiss domes of the Northern Ladoga region, modified after Grigor'eva and Shinkarev (1981). Numbers in circles 1–17 indicate domes: 7, Pitkyarant, 9, Kokkosel'kya. The A–B line denotes the cross-section in the bottom part of Fig. 1, rapakivi granite; 2, granite-gneiss; 3, undivided gneiss-granite and migmatite; 4, Sortaval Group [limestone, skarnoid, schist (**a**), amphibolite (**b**)]; 5, gneiss; 6, gabbro; 7, mafic dike; 9, 10, Ladoga Group (quartz-biotite schist); 11, strike, dip, and direction of foliation. *b*, Structure of the Kokkosel'kya dome of the Northern Ladoga region. 1, granite-gneiss; 2, amphibolite; 3, mafic dike; 4, faults. Metamorphic *P-T* estimates at locations: A, *T* = 450–500 °C; *P* = 6.2–8.0 kbar; B, *T* = 500–520 °C; *P* = 4.2–4.8 kbar; C, *T* = 525–550 °C; *P* = 6.0–7.2 kbar

(a)

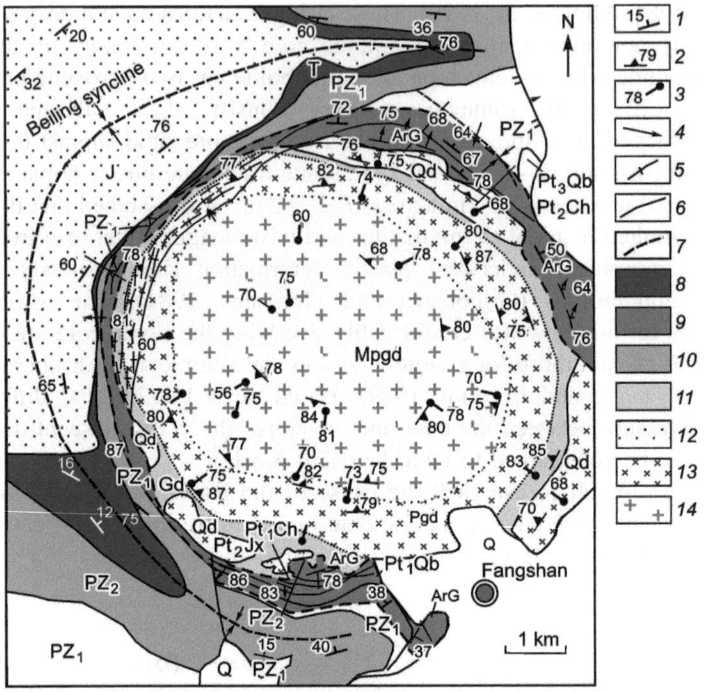
Beiling syncline
N
Mpgd
Pgd
Qd
Gd
ArG
J
T
PZ_1
PZ_2
Pt_3Qb
Pt_2Ch
Pt_1Ch
Pt_2Jx
Pt_1Qb
Q
Fangshan
1 km
1
2
3
4
5
6
7
8
9
10
11
12
13
14

(b)

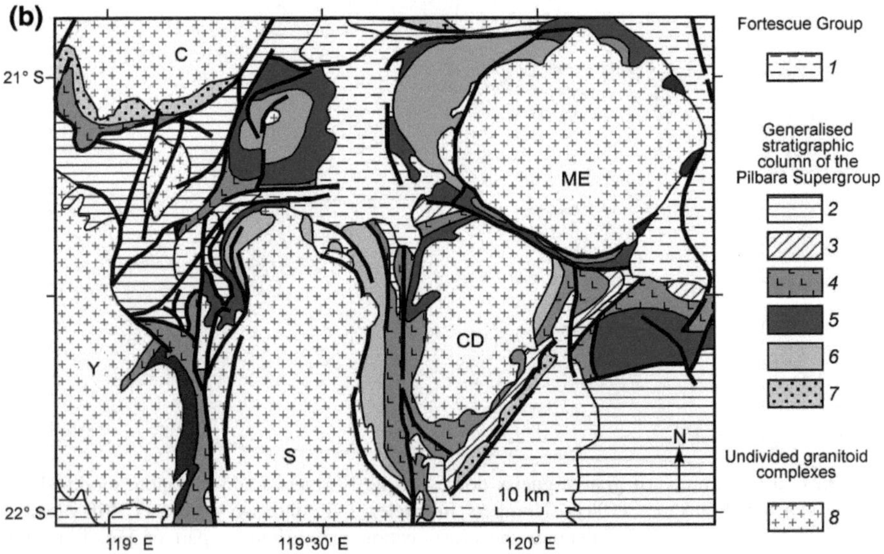
C
ME
CD
Y
S
21° S
22° S
119° E
119°30' E
120° E
N
10 km
Fortescue Group
1
Generalised stratigraphic column of the Pilbara Supergroup
2
3
4
5
6
7
Undivided granitoid complexes
8

◂**Fig. 3.28** **a** Structural and geological map of the Fangshan pluton and its wall rocks (He et al., 2009). (1) foliation, (2) magmatic foliation, (3) magmatic lineation with plunge, (4) stretching lineation, (5) schistosity, (6) faults, (7) dashed line roughly marks the exterior boundary of the high-temperature shear aureole of the Fangshan pluton. Stratigraphic units and their symbols (8–12): (8) Triassic (T), (9) Archean Guandi complex (ArG), Mesoproterozoic Changcheng group (Pt_2Ch), Jixian group (Pt_2Jx), Neoproterozoic Qingbaikou System (Pt_3Qb); (10) Lower Paleozoic (Pz_1), (11) Upper Paleozoic (Pz_2), (12) Jurassic (J). The pluton has four units: fine-grained quartz diorites (Qd), medium-grained granodiorites (Gd), (13) porphyritic granodiorites (Pgd) and (14) megaporphyritic granodiorites (Mpgd). Dotted line defines the internal boundary within the pluton of subsolidus deformation. **b** Simplified geological map of the East Pilbara, showing distribution of main groups around the granitoid complexes. Stratigraphic units and their symbols: (1) Fortescue Group ca. <2770 Ma; (2) Gorge Creek Group <3230 Ma, (3) Budjan Creek Formation ca. 3308 Ma, (4–6) Warrawoona Group, 3320–3490 Ma, (7) Coonterunah Group, 3515–3498 Ma. (8) undivided granitoid complexes, letters denote names: C = Carlindi, CD = Corunna Downs, ME = Mount Edgar, S = Shaw, Y = Yule

dome-shaped structures of the Northern Ladoga region may have been formed by emplacement of mantle-derived magmas, which initiated partial melting of the lower curst and upwelling of partially molten masses to higher levels in the crust (Baltybaev et al. 1996). The modeling results and geophysical data (Baltybaev et al. 2009) indicate the presence of a large magmatic body with density of up to 2.85 g cm^{-3} at deeper crustal levels, which can be interpreted as a mafic–ultramafic intrusion and a heat source triggering the formation of Ladoga "thermal domes". The thermal aureoles around Lagoda domes may reach 50–60 km in size with thermal anomalies of up to 680–780 °C compared to the background premetamorphic temperature of 500–600 °C (Korsman et al. 1999). In addition, the effects of granitoids on their country rocks are manifested as the development of porhyroblastic textures and skarnoids at contact aureoles, complex interlaying of migmatites and amphibolites (Grigor'eva and Shinkarev 1981).

Diapiric domes of the Archean Pilbara craton (Western Australia) are interpreted as isolated steep-sided granitoid-cored complexes flanked by greenstones (Fig. 3.28b). These tonalite-trondhjemite-granodiorite (TTG) diapirs are extrapolated as 35–120 km diameter subvertical cylinders to depths of 14 km (van Kranendonk et al. 2004). Diapiric emplacement of the granitoid domes is ascribed to vigorous heating of the lower crust during plume-generated magmatism at 3.3 Ga. The phase of granitoids diapirism occurred at ~3.2 Ga. The granitoid-cored domes rose from the lower crust level to depths of 10 km over a period of 17–20 Myr (Sandiford et al. 2004).

Some of the granite domes are interpreted to have developed through the concurrent or sequential action of two different mechanisms, diapirism and intrusion. An example is the Fangshan pluton of the North China craton (Polyansky et al. 2010a; He et al. 2009). It represents an approximately spherical body composed of a granodiorite core, some 8–10 km in diameter, surrounded by a contact metamorphic aureole, 0.3–2 km in width (Fig. 3.28a). This pluton is interpreted as a magmatic diapir that formed by the upward movement of magma from ductile middle crust through the uppermost brittle crust. It was shown that the emplacement of this diapiric granodiorite pluton occurred at a pressure of about 2 kbar corresponding to a depth of about 6–7 km.

The Thor-Odin granitic-migmatitic dome is a typical example in a series of domes located in the plutonic hinterland of the Rocky Mountain Foreland Belt, Canadian Cordillera (Norlander et al. 2002). The dome has an oval shape in plan and is (5–7) × 20 km in size. Its core is composed of migmatitic gneiss and leucogranite surrounded by greenschist facies metabasites. Thermochronological reconstructions reveal the following stages of the dome formation: (1) subsidence of the rocks during collision and orogeny, mid-crustal heating and the beginning of anatexis as a consequence of deep heat transfer, (2) accumulation of melt at the mid- and lower crustal levels leading to magma upwelling of at least 20 km isothermal decompression from 10 to 4–5 kbar, and (3) crustal extension and exhumation of diapir cores and surrounding metamorphic rocks leading to the formation of metamorphic core complexes. The duration of diapiric upwelling is estimated at ~4 Myr, as recorded by U-Pb SHRIMP ages of zircon cores and rims of 60–56 Ma (Norlander et al. 2002).

The Naxos dome is located in the back-arc region of the Hellenic island-arc of the Aegean Sea; its emplacement is thought to take place under conditions of back-arc extension (Vanderhaeghe 2004). The dome is composed of the migmatitic core and the mantling metamorphic rocks, including marbles, schists, amphibolites, and metavolcanics (Fig. 3.29). The migmatitic core, some 5 × 12 km in size, exhibits a composite syn-migmatitic foliation. A petrographic and structural analysis showed that the migmatite core of the Naxos dome was a mixture of partially molten material and felsic magma. Extrusion of the core was accompanied by metamorphism of the surrounding host rocks leading to the appearance of corundum and biotite and disappearance of chloritoid; the kyanite/sillimanite transition took place at a pressure of 6 kbar and a temperature of 750 °C (Jansen and Schuiling 1976). The width of the heating zone defined by the 420–480 °C isotherm is not less than 10–15 km. The onset of melting is taken to occur at 25 Ma; isothermal exhumation of migmatites from a depth corresponding to 10 kbar to a depth corresponding to 6 kbar was completed by 17 Ma and was accompanied by crustal extension (Vanderhaeghe 2004). Dome formation probably resulted from the simultaneous action of upper crustal extension, ascent of partially molten material and density-driven crustal convection.

The above examples demonstrate that diapirism can be regarded as the main transport mechanism for both felsic and ultramafic magmas through the lithosphere.

The process of diapiric emplacement can be studied by numerical simulations based on a realistic rheological-petrological model of crust and mantle. Unlike conceptual models which are largely based on geological knowledge of a process, numerical model provide rigorous quantitative estimates. The equations of thermomechanics of deformable solids were solved in a coupled formulation: the mechanical equilibrium equations and equations of heat transfer were coupled with a constitutive equation describing the rheological properties of material. The mechanical setting of the problem uses a plain strain formulation. These equation formulations and numerical method are described in detail in the Appendix.

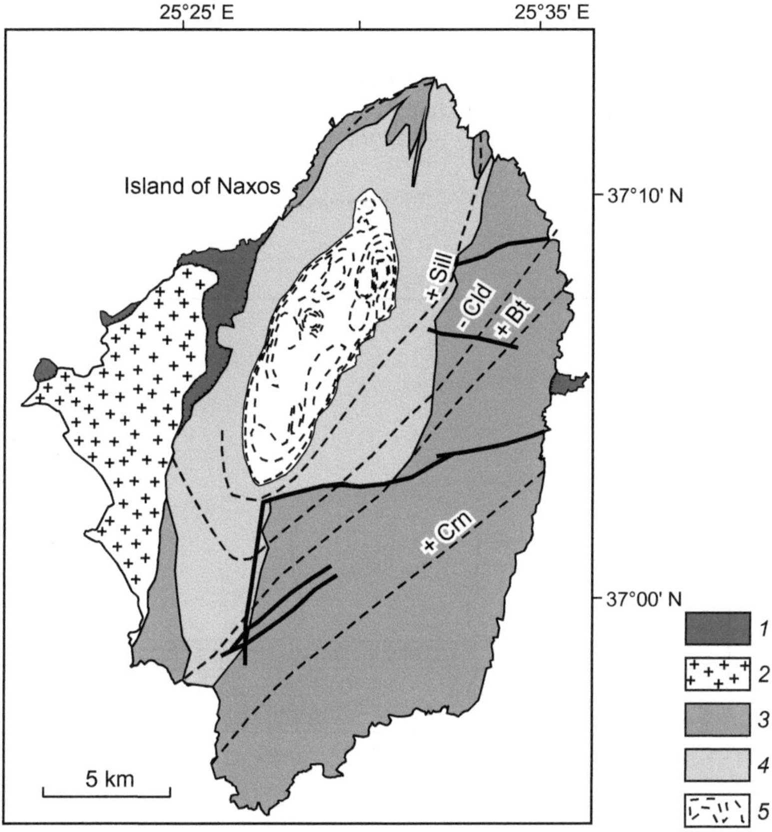

Fig. 3.29 Geologic map of the island of Naxos, Greece, after Vanderhaeghe (2004). Isograds are modified from Jansen and Schuiling (1976). Upper Unit: 1—low-grade metamorphics and Tertiary sediments; Middle Unit: 2—granodiorite, 3—low grade schists and marbles, 4—high grade schists and marbles; Lower Unit: 5—migmatite and leucogneiss

Let us first consider the model for the emplacement of granite-gneiss diapirs into the crust and then present a generalized model of diapiric magma ascent through the cratonic lithosphere.

The model setup, geometry, boundary and initial conditions are shown in Fig. 3.30a. The results of the numerical experiments are presented in Fig. 3.31 as temperature fields reflecting successive stages of diapiric ascent. At the initial stage (duration of 2 Ma), the front of partial melting is spread over a heat source. Conductive heating leads to the formation of partial melting area with immobile melt. The period of about 2 Ma involves the formation of the area with dome-like surface 6–7 km in height and width determined by given size of sheet-like mafic intrusion (about 20 km) that caused heating. Then the partially molten material begins to rise. Convective motion converts the surface of the dome into mushroom-shaped one (Fig. 3.31), thereafter the diapiric feeder channel from below is formed; it pinches out in the course of upwelling. The width of the channel is about 3–3.5 km, while the width of the diapir head is about 13 km; the height of diapiric body together with a feeding channel accounts for 23.7 km, given the 30 km thick crust. This means that diapir can rises to a depth of 6–7 km, as was the case with the Fangshan dome (He et al. 2009). The upper free surface moves for

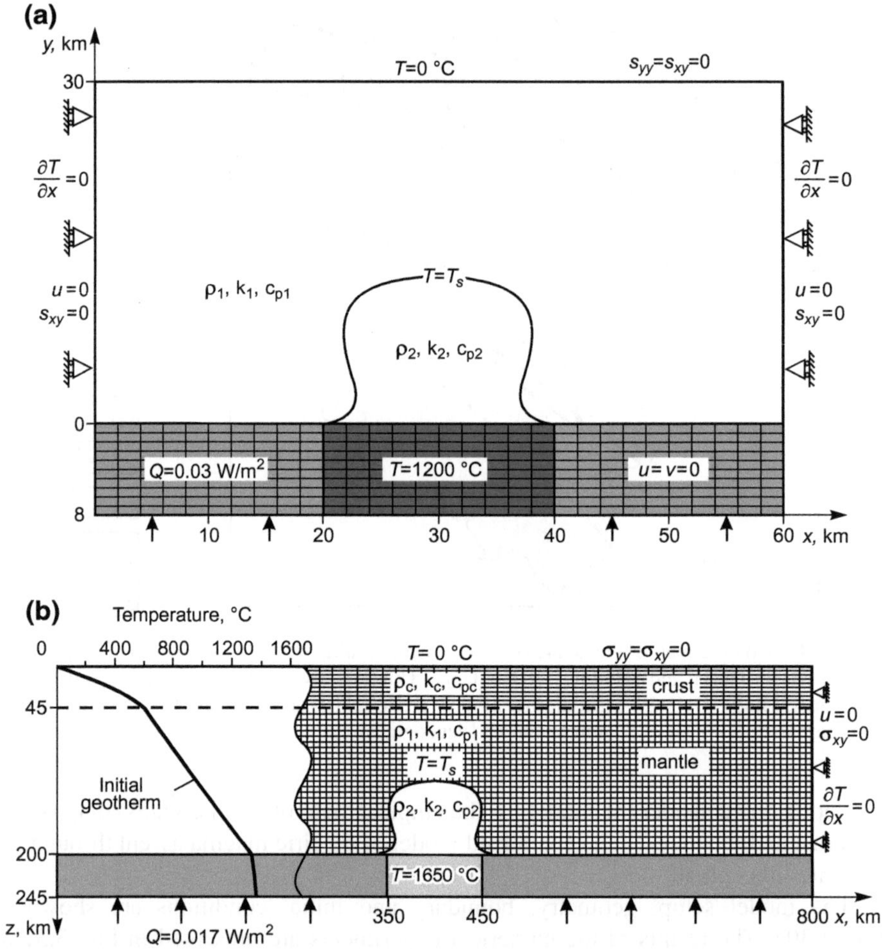

Fig. 3.30 **a** Setting up the problem of melting and diapirism in the lower crust. Shown are the modeling areas, parameters, boundary and initial conditions, initial numerical grid of finite elements. Mafic magma (dark gray) with initial temperature of 1200 °C intruded immobile medium of the lower crust (gray). Thin solid line denotes the melting boundary, which separates the partially crystallized melt and host rocks with different properties. **b** Model setup, boundary and initial conditions for 2D modeling of diapir ascent during melting of the lithosphere. Boundary conditions shown on the right are assigned on sides of the box. The steady-state geotherm taken as an initial temperature is shown on the left. Finite element grid updated during calculation is shown. The T = Ts line denotes the melting front of mantle peridotite in (**b**) or felsic crust in (**a**)

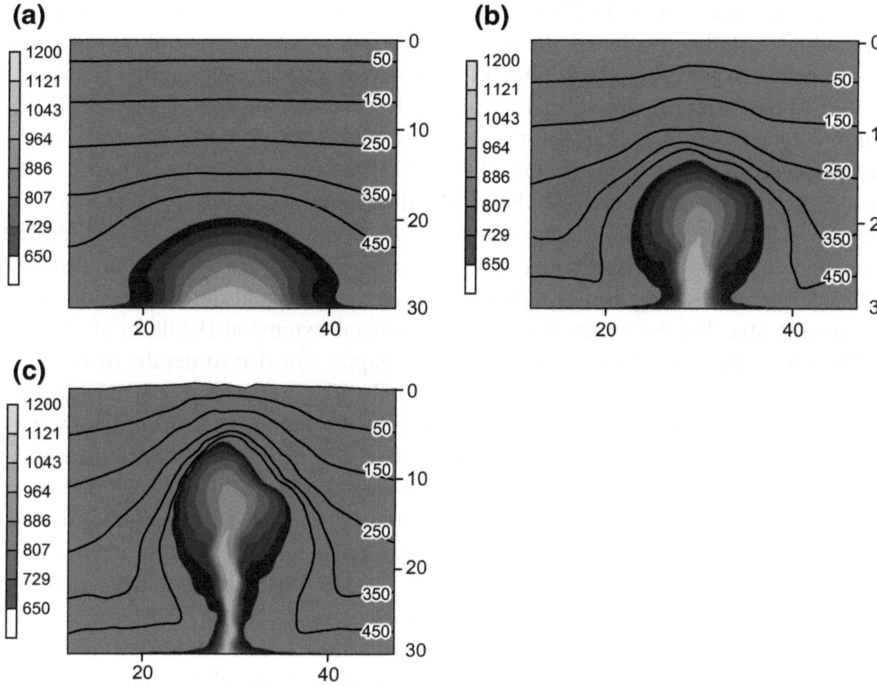

Fig. 3.31 Modeling results of diapir upwelling assuming constant yield stress of crustal material. Shown is the middle part of model area (sizes are in km), with pattern of temperature field in the diapir body (in gray gradations) and beyond it (in isolines), from onset of upwelling at 2 Myr (**a**), at 2.08 Myr (**b**), to finish of upwelling at 2.16 Myr (**c**). Scale in range of 650–1200 °C is shown on the left. Grey field indicate the area with temperature below solidus (650 °C), the values of isotherms are shown in °C

about hundreds of meters with maximal upwelling above diapiric head. After ~2.16 Ma, the diapiric ascent ceased and followed by a stage of cooling (solidification). This moment is defined by pinching out of the channel and detachment of the diapir head. Based on simulation results, we estimated the rates of melting in the lower crust during a conductive (initial) stage of heat transfer and the rates of diapiric rise during a convective (subsequent) stage. The melting front is propagated at a rate of ~4.7 mm/year, and then rapidly rises up at a rate of a few to tens of meters per year. The relatively high ascent rate and low temperatures of ambient rocks prevent melting of the upper crust, and diapir is exclusively composed of the material derived by melting of the lower crust.

The numerical modeling of granite-gneiss diapirism in the continental crust enables a quantification of the parameters of magma ascent and thus can be used to conclusively prove or disprove the possibility of such mechanism for particular natural objects. One of the key parameters is the maximum level of diapiric ascent of granitic bodies in the crust, which was estimated to range from 15–12 km

(Bittner and Schmeling 1995) to 1–5 km (Burov et al. 2003). To check the applicability of the results of the numerical experiments to natural processes, it should be taken into account to which levels may diapirs rise in nature. These data can be obtained from the estimates of the lowest metamorphic pressure in the country rocks: 4.2–4.8 kbar for the Kokkosel'kya dome, corresponding to the ascent level of 15–17 km (Polyansky and Efremov 1989), 4–5 kbar for the Thor-Odin dome, i.e., about 14–18 km (Norlander et al., 2002), ~3 kbar for the Naxos dome, i.e., about 11 km (Jansen and Schuiling 1976), 2 kbar for the Fangshan magmatic diapir, i.e., 6–7 km (He et al. 2009), 2 kbar for the D'Entrecasteaux Islands domes rising during near-isothermal decompression and a less than 2 kbar for the ascent during upper crustal extension (Little et al. 2011).

Therefore, the modeling confirms that the diapirs can rise to depths of 6–7 km, as indicated by the lithostatic pressure recorded using mineral geobarometers and structural field observations from natural objects. Probably, in nature, diapirs may rise to the upper crustal levels only in the regional extension setting or in the course of tectonic erosion (unroofing) of the dome cover. These results allow an important conclusion to be drawn: diapirism is responsible for transport of a large volume of molten material through the mantle and lower crust with viscoplastic properties. The upward movement of diapir trough the upper crust causes partial melting, rupture and solid-state deformation of wall rocks (Arndt 2013).

The model setup, geometry, boundary and initial conditions for modeling mantle-crustal diapirs are shown in Fig. 3.30b. It is assumed that the hot plume material (as a jet) with constant temperature rises below the craton as a jet from the upper-mid mantle (Burov and Cloetingh 2009) or core–mantle (Dobretsov 2010) boundary. The sublithospheric hot plume causes melting of a peridotite mantle lithosphere and diapiric upwelling of partially molten material. The lateral dimension of the mantle plume is chosen on the basis of the diameter of volcanic channels of hotspots that were estimated from seismic tomography data (Montelli et al. 2004). We accepted that the lowest of available estimates is equal to 100 km. For the sake of simplicity, we considered a sublithospheric plume to be stable, with a fixed size and constant temperature of 1400, 1500 or 1650 °C. The temperature of a heat source was inferred to be about 50–300 °C higher than the adiabatic temperature at the same depth at the base of the lithosphere. This temperature estimate was substantiated in the works of Sobolev et al. (2009) and Griffin et al. (2013), based on the compositions of ultramafic xenoliths from the Maimecha–Kotui and Seiland igneous provinces.

At the lateral boundaries, the horizontal velocity and heat flow were set to zero, i.e., the effect of the lateral boundaries was of minor importance. The base of the lithosphere was set as an undeformable boundary with a constant mantle heat flow $Q = 17$ mW m^{-2}, which is assumed to be the average value for the Siberian platform based on the earlier data (Duchkov et al. 1987).

We have conducted three series of numerical experiments with different crust and mantle rheologies: (1) diabase/wet dunite (Chopra and Patterson 1984), (2) wet quartzite/olivine basalt with the melt under hydrous conditions (Mei et al. 2002), (3) dry diabase (Carter and Tsenn 1987). First, we considered the problem of

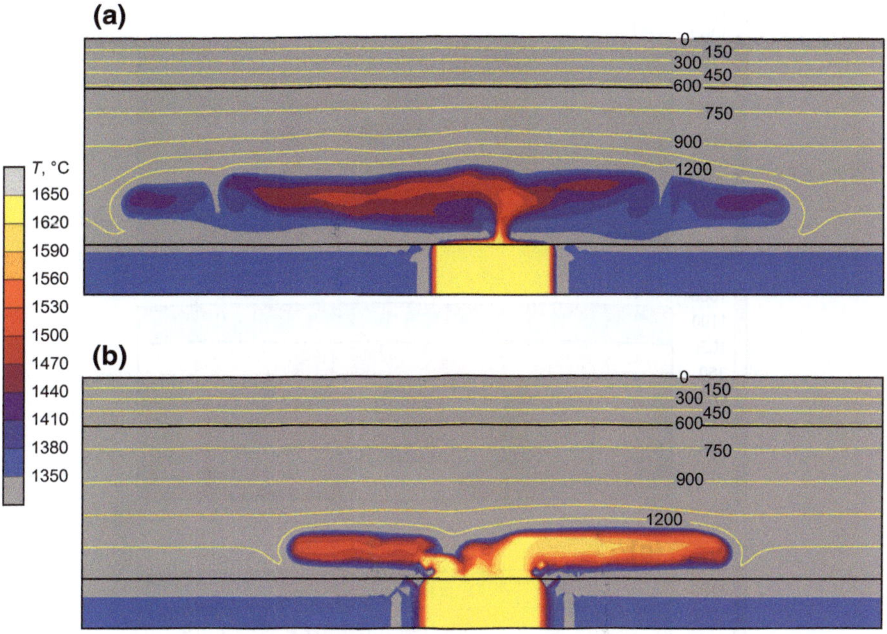

Fig. 3.32 Experiments with horizontal spreading and emplacement of sills according to model 1 for a viscous rheology of the crust/mantle corresponding to dry diabase/wet dunite (Chopra and Patterson 1984). Two variants of the model at different density contrasts between a diapir and a mantle: **a** 430 kg/m^3, **b** 215 kg/m^3. Experiments are shown for periods of time 0.39 Myr (**a**) and 2.63 Myr (**b**) after the initiation of heating

magma upwelling through the lithosphere above the mantle plume (cases 1 and 2) and then built a composite model for mantle-crustal diapirism accounting for sequential melting of the mantle lithosphere and crust. In the last case (3), we assumed that propagation of the mantle diapir of mafic or ultramafic composition to Moho boundary may cause melting of the felsic crust.

Depending on the selected rheology and melting parameters, the numerical experiments yielded different ascent regimes. Figures 3.32, 3.33 and 3.34 show temperature fields that characterize the ascent regime of the viscous-plastic diapir and deformation of its surrounding material inferred from each model.

For the case assuming the production of low-density and low-viscosity melts, the modeling revealed negligible magma upwelling. The magmatic intrusion was formed as extended sheet-like bodies with horizontally flattened upper boundaries ("lithospheric sill"). Figure 3.32 shows the results for two models with the density contrasts between the partially molten diapir and its surrounding mantle material: 430 kg/m^3 (Fig. 3.32a) and 215 kg/m^3 (Fig. 3.32b), respectively. In both cases, the viscosity contrast between the mantle and the diapir was accepted to be maximum and equal to 3×10^4. Modeling yielded narrow melting regions, up to 600 km long and 60 km thick in the first case and 420 km long and 40 km thick in the second

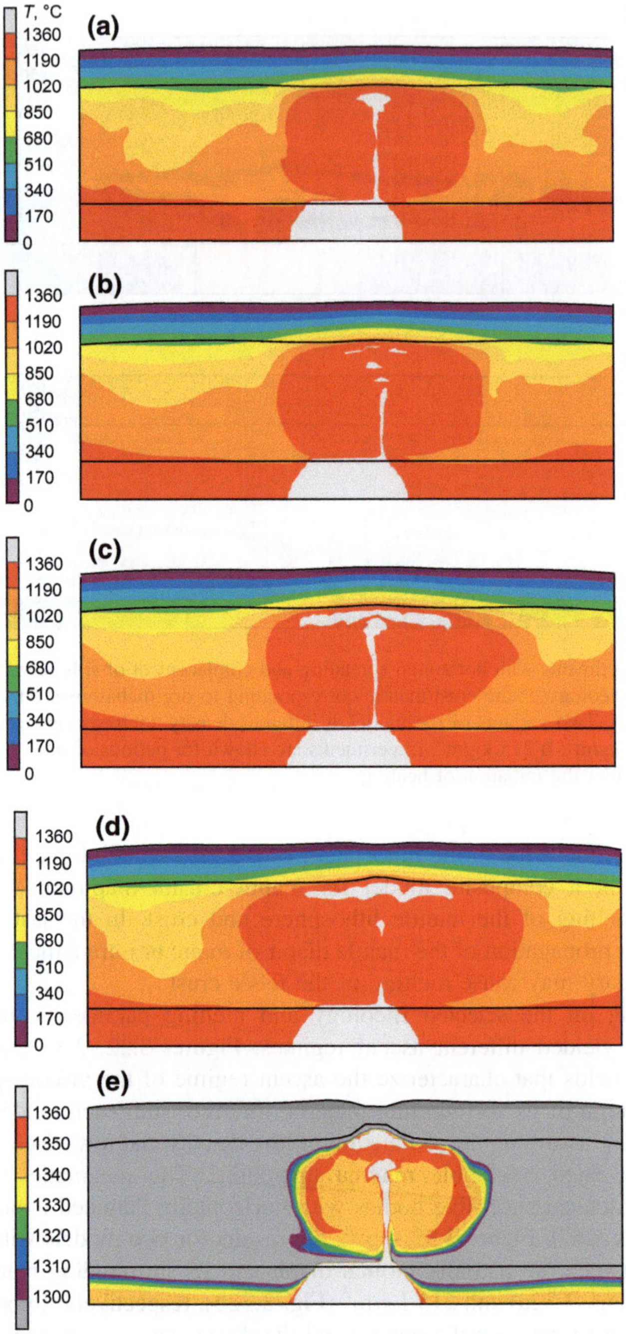
T, °C
(a)
1360
1190
1020
850
680
510
340
170
0
(b)
(c)
(d)
(e)
1350
1340
1330
1320
1310
1300

◂**Fig. 3.33** Experiments with the ascending mantle diapir in an underplating regime (model 2) for a viscoelastoplastic rheology of the crust/mantle corresponding to wet quartzite (Kronenberg and Tullis 1984) and olivine basalt with the presence of melt under hydrous conditions (Mei et al. 2002). Stages of ultramafic magma ascent through the mantle and its impingement on the crustal base are shown at 5.2, 7.2, 8.7, 10.5, 10.8 Myr

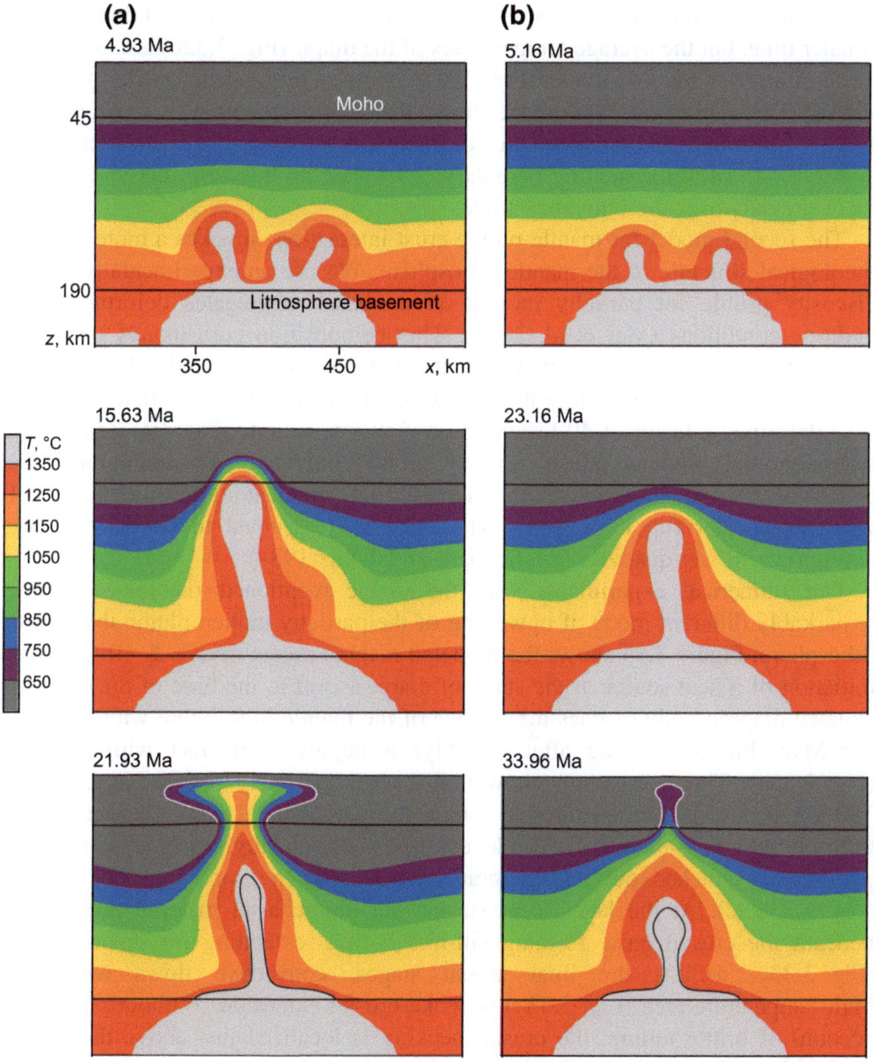

Fig. 3.34 Experiments with a successive emplacement of mantle/crustal diapirs (model 3). The entire lithosphere is described by dry diabase rheology (Carter and Tsenn 1987). The variants of models with temperatures in a sublithospheric heat source of 1650 °C (**a**) and 1500 °C (**b**). Boundaries of granite and ultramafic diapirs are shown by white and black solid lines

case. The melt did not propagate above a certain lithospheric level due to the interplay of buoyancy of the lighter magma and viscous drag of surrounding mantle with the temperature-dependent viscosity. As magma rises to a certain height, the equilibrium is attained and the ascending body is blocked by a rheological barrier above which deformations are negligible due to the high viscosity of surrounding material. A comparison of these two models indicates that in the case of a greater degree of partial melting, the diapir spreads out over larger horizontal distances in a smaller time, but the average temperatures of the diapir (Fig. 3.32b) are higher than those inferred in the first model (Fig. 3.32a) due to a prolonged activity of the heat source. This numerical experiment shows that mantle rheology is the main controlling factor and the ascent distance of the diapir is only weakly dependent on the density contrast at the melting boundary. Therefore, in this case, the viscous drag force prevails over the buoyancy force.

The second model for mantle plume-crust interactions invokes a more complex rheological structure of the mantle lithosphere. We use an experimentally-derived viscosity profile for partially molten olivine-basalt aggregates deformed under hydrous conditions (Mei et al. 2002). This composition corresponds to a more enriched (fertile) mantle as compared to the first variant of the model. Experimental results (Mei et al. 2002) show that an Ol-basalt sample with the melt fraction of 0.12 deformed a factor of 40 faster than a melt-free sample. The data demonstrate that due to the combined effects of water and melt the rate of deformation increased by 3–4 orders of magnitude with the increase in the concentrations of melt (up to 15 wt%) and water (up to 5 wt%). The crust was assumed to have a rheology appropriate to wet quartzite (Kronenberg and Tullis 1984).

Our numerical experiments with the above mentioned rheologies show a remarkably different mode of upwelling of the partially molten plume through the lithosphere. Figure 3.33 shows the modeled magma ascent over ~5–10 Myr since initiation of a heat source at the stage of diapir ascent to the base of the crust. An isolated diapiric body reaches the surface of the mantle lithosphere within the first ~5 Myr (Fig. 3.33a) and after 7–8 Myr it begins to interact with the crust (Fig. 3.33b). The calculations show that a weak/thermally softened region 200–350 km in size is formed through which the partial melt may penetrated in the lithosphere via a ~5–10 km wide conduit. Occasionally, such cannel can be periodically pinched out and generates isolated batches of ascending magmas (Fig. 3.33b, c). At the late ascent stages, the plume head material accumulates beneath the crustal base, forming sill-like bodies, extending for up to 200 km (Fig. 3.33c, d). At the final ascent stages, convective flows within the high-temperature area (Fig. 3.33e) provoke crustal extension. Without taking into account of brittle failure, the crustal necking is localized just above the vertical feeder channel of a diapir. The crust is thinned to 22 km, which corresponds to a twofold local extension in this zone. This extension zone has a width of 90–100 km and subsidence of the crustal surface is about 7–7.5 km. The calculated parameters of crustal extension/subsidence are typical for rift zones above crustal underplating by mantle-derived basaltic magma (Polyansky et al. 2013, 2014; Thybo and Artemieva 2013).

In the next case, we consider emplacement conditions of paired diapirs: mafic–ultramafic diapirs in the mantle lithosphere and granit-gneiss diapirs in the crust. The model accounts for the melting effects in the lithospheric mantle and crust during heating above the wet peridotite and wet granite solidus, respectively. For the sake of simplicity, the modeling of two-stage melting assumes that the compositions of the mantle and crustal material remain unchanged. Therefore, the melting temperatures of peridotite and granite systems are taken to be constant. The model assumes that melting temperature of the mantle is 1360–1370 °C and that of crust is 650–655 °C. Variations in the melting temperature influence the time of diapir initiation, e.g., the increase of the melting temperature of the granite system to 700–705 °C tends to give a twofold slower rate of the diapir formation.

The calculations were made using different model parameters shown in Table 3.5. It appeared that the key parameters that control diapir emplacement are the temperature in a sublithospheric plume T_0 (°C), the degree of melting of mantle material φ_m and the density contrast between a diapir and its surrounding rocks $\Delta\rho$ (kg/m^3). Summarizing all the numerical experiments, we can distinguish three regimes: (1) diapiric ascent of magma through the mantle and crust; (2) diapiric ascent of magma through the mantle only; (3) no diapiric ascent. Table 3.5 shows that the efficiency of diapirism decreases with the lowering of the source temperature and a degree of melting. The timescales in Table 3.5 designate the duration of diapir upwelling through the mantle to the base of the crust and the duration of granite diapir ascent to the maximum crustal level. The interval between these two ascents can be considered as a time lag between mafic–ultramafic and felsic magmatism in polychronous complexes. In our cases, this interval covers a time of 5 Myr. Figure 3.34a shows the results of numerical experiments for this case with parameters indicated in the top left cell of Table 3.5. Figure 3.34a shows three stages of diapir formation: (1) onset of diapir ascent at the base of the lithosphere ($\sim$ 5 Myr), (2) ascent of an ultramafic diapir through the lithosphere to the base of crust and initiation of crustal heating ($\sim$ 15 Myr), (3) ascent of felsic magma through the crust and development of a granite-gneiss dome ($\sim$ 22 Myr). The mantle diapir comprises a 10–20 km wide feeder channel, while the crustal diapir is characterized by a mushroom shape and head flattening, and has a diameter of up to 100 km.

Figure 3.34b shows modeling results at a lower sublithospheric plume temperature of 1500 °C and a melt fraction of 40% in comparison with the previous model. Such changes have an adverse effect on the efficiency of the heat transport and thus on the volume and shape of a granite diapir within the crust. In this case, the width of the modeled diapir would be no greater than 20 km and the calculated temperature would not exceed $\sim$1000 °C. A comparison of these two cases (Fig. 3.34) shows that higher magma source temperatures and melting degrees result in a faster rate of the diapir ascent either through the mantle or crust. A similar conclusion can be made for other cases shown in Table 3.5.

The strain rate field and a generalized model of mantle-crustal diapirs are shown in Fig. 3.35a, b. Figure 3.35a shows the temperature distribution in close vicinity of a granite-gneiss diapir (50–750 °C isotherms) and superimposed strain rate field. The maximum strain rate is observed on top of the granite diapir and within the

Table 3.5 Timescales for emplacement of mantle-crustal diapirs based on the results of numerical modeling

	Melting degree							
	$\varphi_m = 0.5$		$\varphi_m = 0.4$		$\varphi_m = 0.3$		$\varphi_m = 0.25$	
	Density contrast between mantle and diapir (kg/m^3):							
	$\Delta\rho = 215$		$\Delta\rho = 172$		$\Delta\rho = 129$		$\Delta\rho = 115$	
	Mantle diapir	Crustal diapir	Mantle diapir	Crustal diapir	Mantle diapir	Crustal diapir	Mantle diapir	Crustal diapir
	Myr							
$T_0 = 1650$ °C $\Delta T = 300$ °C	15.6	20.9	25.6	31.1	27.2	32.1	31.4	36.5
$T_0 = 1500$ °C $\Delta T = 150$ °C	23.2	27.5	28.8	34.6	32.4	32.4	33.7	Not emplaced
$T_0 = 1400$ °C $\Delta T = 50$ °C	Not emplaced	Not emplaced	Not emplaced	Not emplaced	Not emplaced	Not emplaced	Not emplaced	Not emplaced

hottest region of the lithospheric channel. The analysis shows that the increased temperature in a zone ahead of the melting front would decrease the viscosity and soften crustal rocks to a certain extent. The roof of the diapir may reach depths of 6–8 km (Fig. 3.35b). At such a depth and ascent rate of *ca.* ~19 mm/year, the diapir material rising through the crust remains viscoplastic in a subsolidus state and spreads laterally acquiring a dome shape due to driving buoyancy forces of the underlying partially molten core. These results are consistent with the mechanism of diapiric ascent proposed by Weinberg and Podladchikov (1994). These authors

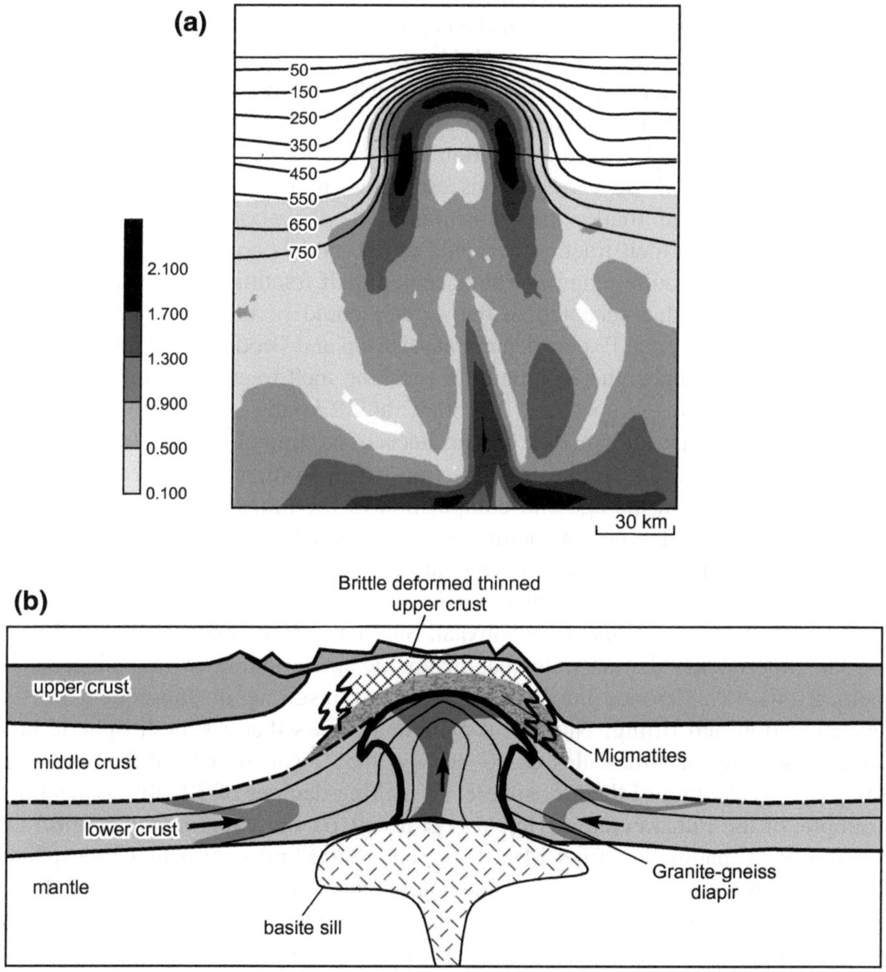

Fig. 3.35 **a** Temperature and strain rate fields in close vicinity of a diapir at 16 Myr for the model shown in Fig. 3.43a. The diapir boundary corresponds to the isotherm of 650 °C. Isolines show the temperature in a range of 0–750 °C, strain rate (dimensionless) is shown in gray gradations (left scale bar); **b** schematic representation of mantle-crustal diapirs generalized from model results and geological observations (see text for explanation)

suggest that strain rate softening may be an effective substitute for thermal softening and allows a magmatic diapir to ascent to such depths due to a decrease in the effective viscosity of the ambient rocks with a non-Newtonian rheology.

The application of the results of the numerical experiments to natural processes may give insights into the mechanism of magma ascent through the mantle and crust. The conditions under which the ascent of mantle diapirs is realized are determined by the melting parameters and properties of magma and the ambient rocks. Table 3.5 presents the parameters at which mantle diapirs are able to reach Moho depths, thus leading to melting and secondary diapirism in the crust. One of the key parameter controlling diapirism is the degree of melting or melt fraction in the rising magma mass. Estimates of the degree of melting vary widely. As regards melting of the crust, Rosenberg and Handy (2005) identified two rheological thresholds with characteristic melt fractions, which correspond to a dramatic change in the strength of partially molten rocks. The first threshold enables the attainment of melt interconnectivity at a melt fraction of 6–7% and the second threshold value of a melt fraction (35–45% by various estimates) induces the breakdown of the solid framework and marks the transition to a viscoplastic state. The second threshold value of a melt fraction accepted in the models could be attained during diapiric ascent of granite magma. The accepted melt fraction corresponding to the second threshold value accepted in the models could be attained during diapiric ascent of granite magma. Experimental data (Droop and Brodie 2012) on melting of H_2O-rich pelites support the argument for reaching melt fractions of up to 50–60% above the wet solidus temperature of 680 °C at 2.2 kbar.

The available experimental data on the effects of melting of the mantle lithosphere on rock rheology are uncertain; melt fraction varies from a few percent to tens of percent depending on the age and composition of the lithosphere (Herzberg et al. 2010). The highest degrees of melting were estimated from geochemical data on xenolith compositions: 5–20 vol.% for relatively young (<250 Ma) oceanic mantle, 25–30 vol.% for Proterozoic–Phanerozoic off-craton subcontinental lithosphere, ~45 vol.% for cratonic mantle (Kaapvaal, Siberian, Slave cratons) (Walter 2003).

Our experiments demonstrate that magmatic underplating at high degrees of melting (40–45%) erodes the base of the crust resulting in subsequent crustal destabilization and rifting. Numerical modeling shows that a sublithospheric heat source operating on timescales of 20–30 Myr may result in a local heating, softening and weakening of the lithosphere. At melting degrees of 30–40% typical, for example, of the Pilbara craton (Herzberg et al. 2010), our experiments confirm the possibility of magmatic rifting of a thick (up to 200 km) continental lithosphere. This case involves the presence of an intermediate magma chamber beneath the base of the crust from which batches of mafic magmas could escape to the surface and intrude the sedimentary cover in the form of dikes and sills. Such lenses and sheet-like bodies of mafic intrusions more than 10 km thick have been observed by seismic data below the rift structures of the Donetsk basin, Oslo graben, etc. (Thybo and Artemieva 2013).

Moreover, our models confirm the emplacement regime involving horizontal spreading of magma and formation of sill intrusions under the crustal basement (underplating) or to deeper mantle levels at the same source temperature in both cases. The results of Masaitis (2007) reveal the presence of intermediate mafic magma chambers at variuos depths of 60 and over 90 km in the Vilyui province. This suggests that plume ascent is controlled not only by its thermal power (Kirdyashkin and Kirdyashkin 2013) but also by contrasting rheological properties of plume material and surrounding mantle.

The lifetime of the heat source (mantle plume) is also an important parameter controlling diapirism, but the reliable estimates are scarce. Geochronological data for sequential mafic and felsic magmatic events within a single polychromous magmatic complex may provide tentative constraints for the lifetime of the heat source. For example, the ages of alkaline granites and synplutonic mafic intrusions constrain the duration of the emplacement of the Hangayn batholith of Central Mongolia within 268–241 Ma (i.e., 27 Myr) (Yarmolyuk et al. 2016). Our numerical estimates of the sequential diapiric ascent of magmatic bodies through the mantle and crust show that the bimodal magmatic processes occur on timescales of 20–35 Ma (Table 3.5). Therefore, a comparison of the timing of magmatic activity within the Hangayn batholith with the numerical modeling results suggests that the mantle source will have a lifespan of a few tens of millions of years.

To summarize we can conclude that three different regimes of diapiric ascent were established depending on the chosen model rheology:

(1) Single-stage diapir ascent occurs at a low degree of melting with a melt fraction of 5 vol.% over a lifespan of 30 Myr;
(2) Pulsating ascent of successive batches of mantle-derived magma to the base of the crust with a periodicity of 2–3 Myr;
(3) Emplacement of extensive magma bodies in the form of sills either beneath the base of the crust (underplating) or to deeper mantle levels.

The lifespan of 30 Myr for a heat source at the base of the lithosphere is sufficient to initiate the ascent of a diapir through the mantle and crust. The formation of some critical volume of buoyant partially molten material with a thickness of 7–18 km over a period of thermal incubation of 4–5 Myr is required to trigger the ascent and development of a "channel-diapir head" structure.

The ascending mantle diapir can reach the crust–mantle boundary over about 15–35 Myr, and a time lag between mafic and felsic magmatic events is about 5–6 Myr. Certain rheological parameters of the lithosphere and partially molten material are required for diapirism to occur. In the case of a large viscosity contrast between the melt and the lithosphere, the magma reservoir spreads out as a sill, whereas at small viscosity contrast the magma rises as a diapir. At mantle viscosity greater than 5×10^{17} Pa s and crustal viscosity greater than 10^{19} Pa s, the diapiric ascent requires an unrealistically long time, which is inconsistent with geological observations. If these conditions are not attained, the deformation of the lithosphere induces extensive fracturing, which results in fracture—derived magma ascent and diking (Babichev et al. 2014).

3.2.2 Metamorphism at a Geothermal Gradient Close to Average Continental Values

In this section, we examine examples of (1) metamorphism in depressions during continental rifting, (2) metamorphism accompanied by the processes of tectonic stacking in orogens, and (3) Archean metamorphism. Heat flow values for these types of metamorphism approach the continental average ("normal") level of ~50 mW m^{-2} and the geothermal gradient varies from ~20 to ~40 °C/km.

3.2.2.1 Metamorphism Associated with Crustal Subsidence, in Depressions During Continental Rifting

Lithospheric extension is one of the most widespread geodynamic regimes. The processes of lithosphere extension perturb the thermomechanical equilibrium of the lithosphere and create additional heat and mass flows tending to restore thermal equilibrium. Rift-related sedimentary basins are a fundamental manifestation of continental extension. Volcano-sedimentary successions accumulated within the depressions that were formed during isostatic relaxation of the system "crust–mantle lithosphere–asthenosphere" undergo metamorphism. Most of the rocks in such sedimentary basins were subject to low-grade metamorphism, up to greenschist facies conditions. The rocks in the lower levels of the crust and mantle which are undergoing extension were subject to higher-grade metamorphism.

Rift-related metamorphism is described within the framework of two extreme models: active rifting and passive rifting, differing in the character of the driving forces leading to lithospheric extension. These forces can be divided into (1) intraplate forces, acting at plate boundaries and resulting in a passive rifting process and (2) sublithospheric or mantle forces giving rise to active rifting (Sengor and Burke 1978; Turcotte and Emerman 1983; Christensen 1992). Accordingly, metamorphism associated with one or another type of rifting occurred in response to lithospheric extension, but the possible causes that give rise to thermal effects are different.

The difference in the interpretation of active and passive rifting is often obscured due to superposition of processes operating beneath the rift zones at the deeper levels of the lithosphere and crust. For example, lithospheric thinning during extension can produce passive upwelling of the asthenosphere, decompression melting, volcanism, and anatexis at lower-crustal levels. However, similar geological and petrological processes may occur in the case of active rifting, caused by mantle heat sources. For example, based on the same data different author may interpret the East African, Rio Grande and Baikal rifts either as active rifts (Logatchev et al. 1978; Sengor and Burke 1978; Zorin 1981; Parker et al. 1984) or passive rifts (Oxburgh and Turcotte Oxburgh and Turcotte 1974; Tapponnier and Molnar 1979). The characteristics of both types of rifting have much in common with each other, so it is often difficult to distinguish between them; apparently, the difference lies mostly in the sequence of events. Nevertheless, lithospheric extension is a common mechanism behind these two types of rifting.

Passive rifting begins with extensional thinning of the crust and mantle lithosphere caused by the action of forces due to convective flows in the underlying mantle. The process of passive rifting is schematically depicted in Fig. 3.36. Extensional thinning of the crust and mantle lithosphere (Fig. 3.36a) is driven by remote forces acting on plate boundaries. In this model, an upward flow of hot, buoyant asthenospheric material rises to fill the space created by the thinning of the lithosphere (Fig. 3.36b). The possible results of the passive upwelling of the asthenospheric material include decompression melting, mafic magmatism, eruption of continental flood basalts, and formation of zones of high heat flow (Fig. 3.36c). An upward flow in response to heating of the lithosphere from below may lead to the uplift of the rift flanks. The rise of hot mantle material at the base of the crust causes heating and partial melting of the overlying crust and associated metamorphism.

In such a setting, metamorphism is expected to occur either synchronously with thinning or following thinning of the lithosphere. Passive rifting is generally not associated with topographic swells at scales much wider than the rift itself, but uplift occurs, swells would develop only upon heating of the lithosphere. The uplift of the rift flanks results from mechanical unloading (Weissel and Karner 1989), either due to induced small-scale convection (Buck 1986) or due to a thermal perturbation associated with large-scale lithospheric extension.

The upwelling hot asthenosphere below a thinned continental lithosphere acts as a heat source that leads to metamorphism during passive rifting. If active diapirism is associated with mantle convection and a hot plume, in the passive upwelling processes, an upward flow of asthenospheric material rises to fill the space created

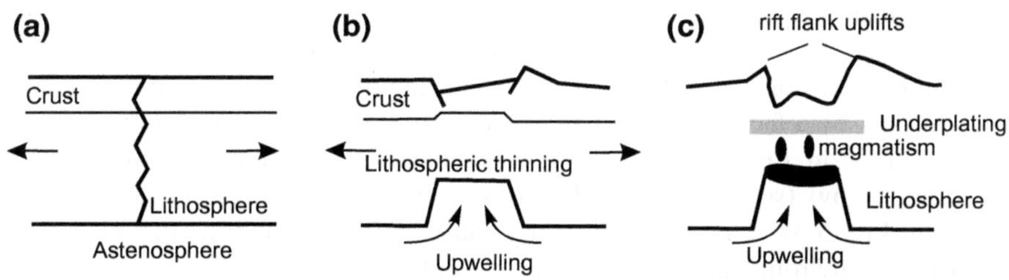

Fig. 3.36 Passive rifting processes after Ruppel (1995). **a** The prerifting configuration. Far-field stresses acting on the lithosphere cause rifting to begin at the site of a favorably oriented preexisting weakness or other lateral inhomogeneity. **b** Continued far-field extensional stresses drive thinning of the crust and mantle lithosphere. In the passive upwelling model of McKenzie and Bickle (1988), hot asthenosphere flows into the void created by the thinning lithosphere. **c** Continued rifting may produce a wide range of consequences. Melts generated deep in the lithosphere or at the top of the asthenosphere rise and produce magmatically underplated layers, which have been widely observed on volcanic margins and in some rift zones. Thermal uplift may occur in response to heating of the lithosphere from below, but large-scale, long-wavelength is not considered a defining characteristic of passive rift zones

by the thinning of the lithosphere. Passive upwelling may give rise to decompression melting (McKenzie and Bickle 1988), partial fusion of the crust and lithosphere (White and McKenzie 1989; Lister et al. 1991), generation of the large horizontal temperature gradients between deformed and undeformed domains (Moretti and Froidevaux 1986), vast outpourings of flood basalts (White and McKenzie 1989) and secondary convection (Buck 1986).

Asthenospheric upwelling and decompression melting can be understood from the study of the thermal structure of the lithosphere and sublithospheric mantle before and after thinning. The thermal state of the lithosphere at the onset of extension is characterized by the conductive geotherm in the lithosphere and adiabatic geotherm in the asthenosphere. Instantaneous homogeneous thinning of the lithosphere increases the temperature gradient and decreases the depth of the top of the asthenosphere. During ascent of hot asthenospheric material, the geotherm intersects the basalt solidus, thus defining the possible volume of melting. McKenzie and Bickle 1988 showed that the melt region can reach a thickness of 15–20, 5–10 and 1–3 km resulting from a five-fold maximum stretching of the lithosphere, given a thickness of the mechanical boundary layer of 70–130 km and mantle potential temperature of 1480, 1380 and 1280 °C, respectively.

Passive upwelling and sublithospheric magmatism have important consequences for lithospheric thermal structure, isostasy and rift zone subsidence. Intrusions of hot diapirs the base of the crust warm the overlying rocks (Lachenbruch and Sass 1978), reduce the viscosity of the lower crust (Nakada 1994), give rise to thermally-induced buoyancy and high-temperature metamorphism (Ruppel 1995). Conductive heating of the overlying lithosphere or sidewalls of the asthenospheric buldge can contribute to thermal expansion and uplift.

To estimate the degree of crustal extension and mantle ascent during passive rifting, a computer model was developed within the framework of the deformable solid mechanics, accounting for the elastoplastic rheology (Polyansky et al. 2013). Details of the equation formulation and numerical method are given in the Appendix. In the example studied here using the data on the Vilyui trough, the maximum extension of 60 km was set at the initial width of the model area of 300 km (i.e., by a factor of 1.2). This value corresponds to the crustal extension in the Vilyui basin (average $\beta = 1.17$), as obtained by the backstripping method described below (see Fig. 3.42). The model for symmetric crustal extension is shown in Fig. 3.37. The crustal thinning can be accomplished in such symmetric extensional environments by a system of individual depressions and localized deformation bands, which can be interpreted as extension faults. This interpretation may explain the concentration of mafic dikes both in the axial part of the trough and on its periphery. The reconstruction of the thermal history of the Vilyui basin shows that the temperatures marking the onset of metamorphism ($\sim$300 °C) were not attained in the local depressions at the maximum depth of 6.5 km. Apparently, such temperatures can be expected only at the base of the deepest depressions at depths greater than 8–10 km. The metamorphic rocks found locally in the Vilyui basin may be a product of contact metamorphism around basalt and monzonite intrusions (Masaitis et al. 1975).

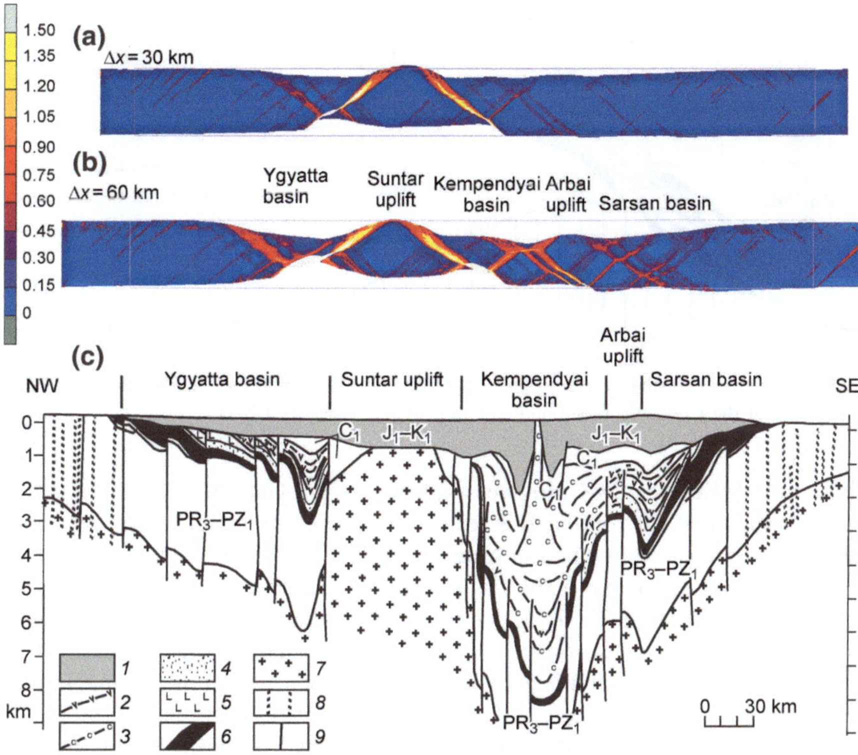

Fig. 3.37 Model for symmetric extension and subsidence in the Vilyui basin under passive rifting. Plate configuration and the equivalent plastic strain at the stages of extension by 30 (**a**) and 60 km (**b**). **c** Geologic section of the western Vilyui basin, after Gaiduk (1988). 1, Jurassic–Cretaceous sediments; 2, 3, Upper Frasnian–Famennian sediments; 4, 5, Frasnian sediments; 6, Upper Devonian basement complex, basalts; 7, crystalline basement; 8, mafic dikes; 9, faults. The vertical dimensions have been magnified by 20 times as compared with the horizontal ones

The formation of metamorphic core complexes can be considered as a special case of passive extension (Sklyarov 2006; Wernicke and Burchfiel 1982). The exhumation of metamorphic core complexes is thought to occur as a result of upper crustal brittle deformation, lower crustal ductile extension, and mantle upwelling (Fig. 3.38). Extension is initiated above a gently dipping section of the detachment in the upper crust or takes form of ductile deformation in the mantle lithosphere.

Recent geological, geophysical, petrological, and geochronological data from many rift zones reveal a certain sequence of events: from crustal uplift and volcanism to crustal extension and metamorphism (Burke and Dewey 1973; Sengor and Burke 1978). This sequence is inconsistent with passive rifting scenarios. In addition, present-day plate boundary forces cannot fully explain regions of extensional stress within plate interiors. The close spatial association between narrow belts of crustal faults, extensive uplift (e.g., East African Rift), and flood basalts

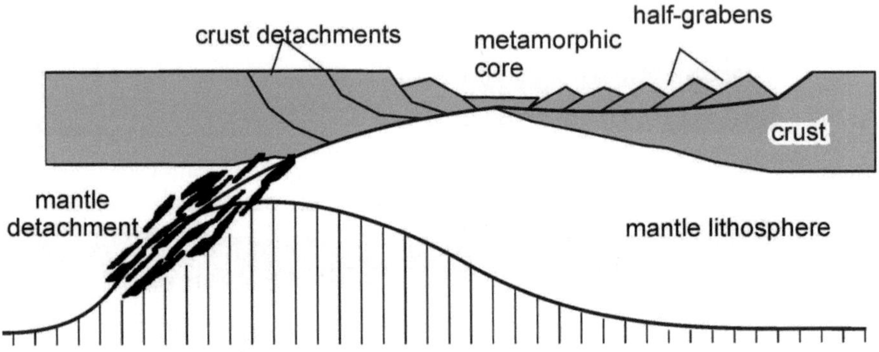

Fig. 3.38 The mechanism of metamorphic core formation after Wernicke and Burchfiel (1982) on the example of the Cenozoic Basin and Range Province (western USA)

(White and McKenzie 1989) provides evidence for a relation between heat sources beneath continental rift zones and deformation. Therefore, a mechanism other than passive rifting must be invoked in order to explain rift-related metamorphism.

The impingement of hot diapirs as a result of mantle dynamics will define the principal driving mechanism for active rifting. Such diapirs or plumes are formed as a result of thermal convection in the mantle due to the buoyancy forces of the heated material. Deformation is associated with the penetration of a plume into the mantle lithosphere (Dobretsov and Kirdyashkin 1994a; Christensen 1992). The typical sequence for active rifting is uplift followed by volcanism, which in turn is succeeded by extension and faulting. The first stage is marked by the rise of a mantle plume beneath the base of the lithosphere, which causes heating of the lithosphere and uplift of the Earth's surface. At the second stage, as the plume head reaches the lithosphere it is expected to erode its base. At the final stage, deformation of the crust gives rise to extensional structures such as rifts.

The study of the driving mechanisms for active rifting is focused on three key problems: (1) whether or not the predicted progression of events uplift—volcanism—rifting is reproducible in the models, (2) whether or not surface uplift causes extensional deformation in the upper crust, (3) whether or not the upwelling mantle plume can lead to thinning of the lithosphere upon impingement on its base.

Figure 3.39 shows a model of active rifting after Artemieva (2003). In this model, mantle upwelling initiates rifting of the crust and underlying mantle and leads to widespread magmatism and metasomatism within the lithosphere (phase 1). Post-rifting thermal cooling (phase 2) leads to thermal contraction, subsidence, and basin formation. Magmas accumulated near the base of the crust and at the mid-crustal depths favor thickening of the upper crust. Plastic material of the middle crust can be squeezed out from the sides. The lower part of the mantle lithosphere (from the base to depths of 100–150 km) undergoes metasomatic alteration and may be removed by thermal erosion and/or by delamination of a dense metasomatised lowermost lithosphere. Further subsidence (phase 3) is a result of fertilization of cratonic lithosphere during mantle-plume interaction and

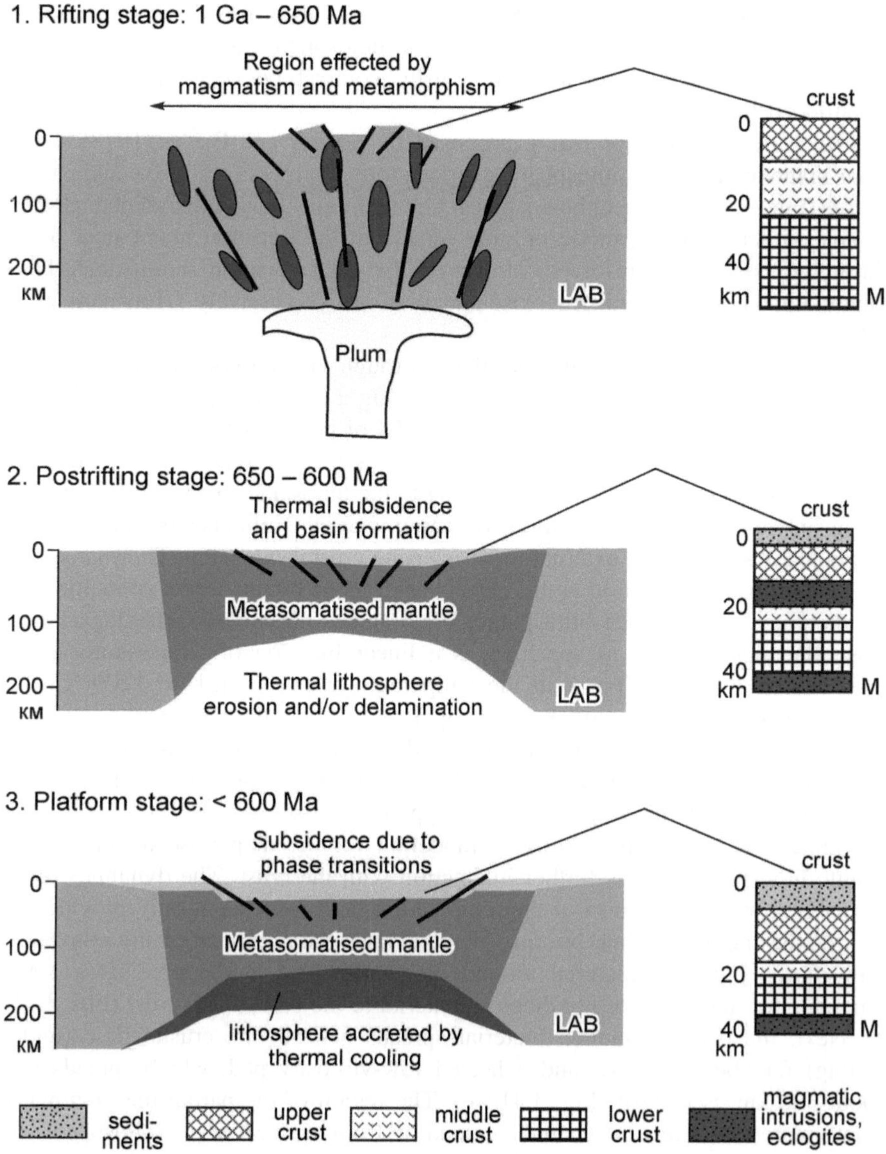

Fig. 3.39 Scheme of plume-related rifting of the cratonic crust and the subcrustal lithosphere of the Russian platform during Proterozoic after Artemieva (2003). Thermal cooling at the post-rifting stage (2) leads to thermal subsidence and basin formation. Further subsidence (3) is a result of fertilization of cratonic lithosphere during mantle–plume interaction and compositional change (Artyushkov 1993). LAB = lithosphere–asthenosphere boundary; M = Moho

involves a much larger area than affected by rifting. This process is accompanied by an accretion of a new basal part of the fertile lithosphere. In this case, a depth of 100–150 km is marked by a sharp density contrast between Fe-depleted and Fe-enriched lithosphere.

Two extreme modes of rifting proposed by Sengor and Burke (1978) should differ in the dynamics of lithosphere's surface topography during active and passive phases (Fig. 3.40). Some authors suggest that there may be a temporal change from passive to active rifting mode or vice versa, but the question about their actual chronology remains unresolved. However, the data on some sedimentary basin suggest that both mechanisms exist and may act successively (Huismans et al. 2001).

The effects of rising of a mantle diapir through the cratonic lithosphere on the deformation of the crust were studied by solving a combined problem (Polyansky et al. 2014, 2017). For this purpose, the model of passive rifting was modified as follows. On the segments of the side boundaries that are related to the crust, we specified the normal stresses corresponding to the lithostatic stress at this depth. By doing so, we modeled the conditions of rifting under the influence of mantle forces, i.e., in the absence of the external forces causing crustal extension. The character of the lithospheric deformation in this case is determined by the degree of melting and the rheology of the mantle lithosphere. We assumed elastoplastic rheology for the crust; the yield stress was specified as a linear function of temperature, which ranges from 1 to 0.1 MPa, with the temperature rising from 0 to 1200 °C. The mantle was assumed to be viscoplastic, with the properties of the olivine + basalt + water + melt mixture (Mei et al. 2002). At the crust-mantle boundary, we specified the no-slip conditions. Details of the equation formulation and numerical method are given in the Appendix.

Figure 3.41 shows the results of modeling for the diapir ascent through the mantle lithosphere and its further interaction with the crust. The dynamics of the ascent is shown in the form of the temperature field, which not only provides the idea of the thermal regime but also demonstrates the character of the convective flow of the light hot material around the diapir. The model predicts a 4 Myr timescale for the ascent of hot diapiric material to the base of the crust (Fig. 3.41a, b). Next, the partially molten material spreads beneath the crustal base (underplating) for about 20 Myr, and a heated low-viscosity pad, which spreads over about 300 km, is formed (Fig. 3.41c–e). The region of the partial melt (shown in black) represents a feeding channel, extending from the base of the lithosphere, and a flat sill-like body, spreading horizontally along the crust–mantle boundary. Finally, there comes a point of time when the developed convective flow moving in the opposite direction stretches the crust, which thins out (Fig. 3.41f, g). The deformation of the crust (necking) is localized exactly beneath the plume head; the final configuration before rupture of the crust is illustrated in Fig. 3.41g. After this moment, the calculations ceased due to the numerical difficulties; therefore, the stage of local faulting and crustal destruction was modeled within the frame of a different problem (Fig. 3.37).

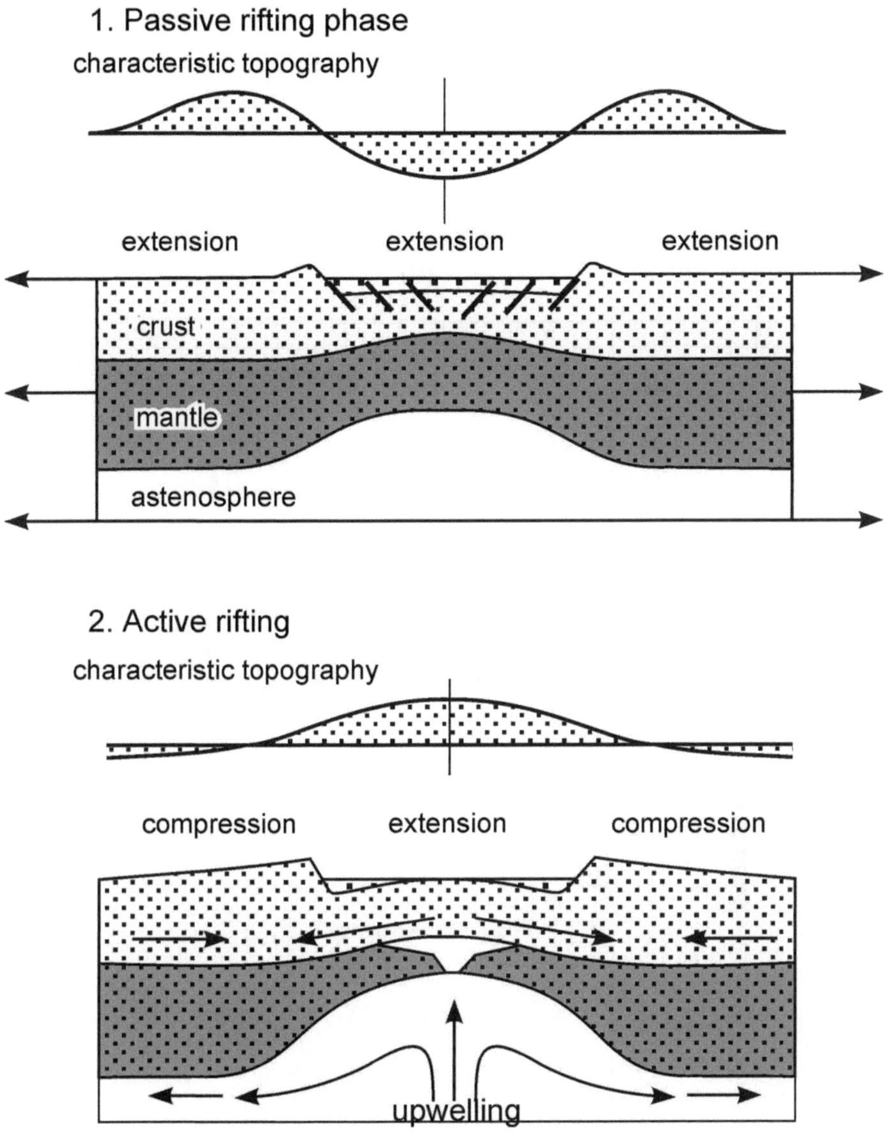

Fig. 3.40 End-member models of rifting after Sengor and Burke (1978). The transition from plate-mode passive extension (1) to diapiric-mode active extension (2) might take place in some basins (Huismans et al. 2001)

As shown by our modeling, the sublithospheric heat source, which is continuously functioning during 20–30 Myr, can heat the mantle, reduce the viscosity and, thus, weaken a segment of the lithosphere. At a high degree of melting [30–40% according to the data presented in (Walter 2003; Herzberg et al. 2010)], magmatic

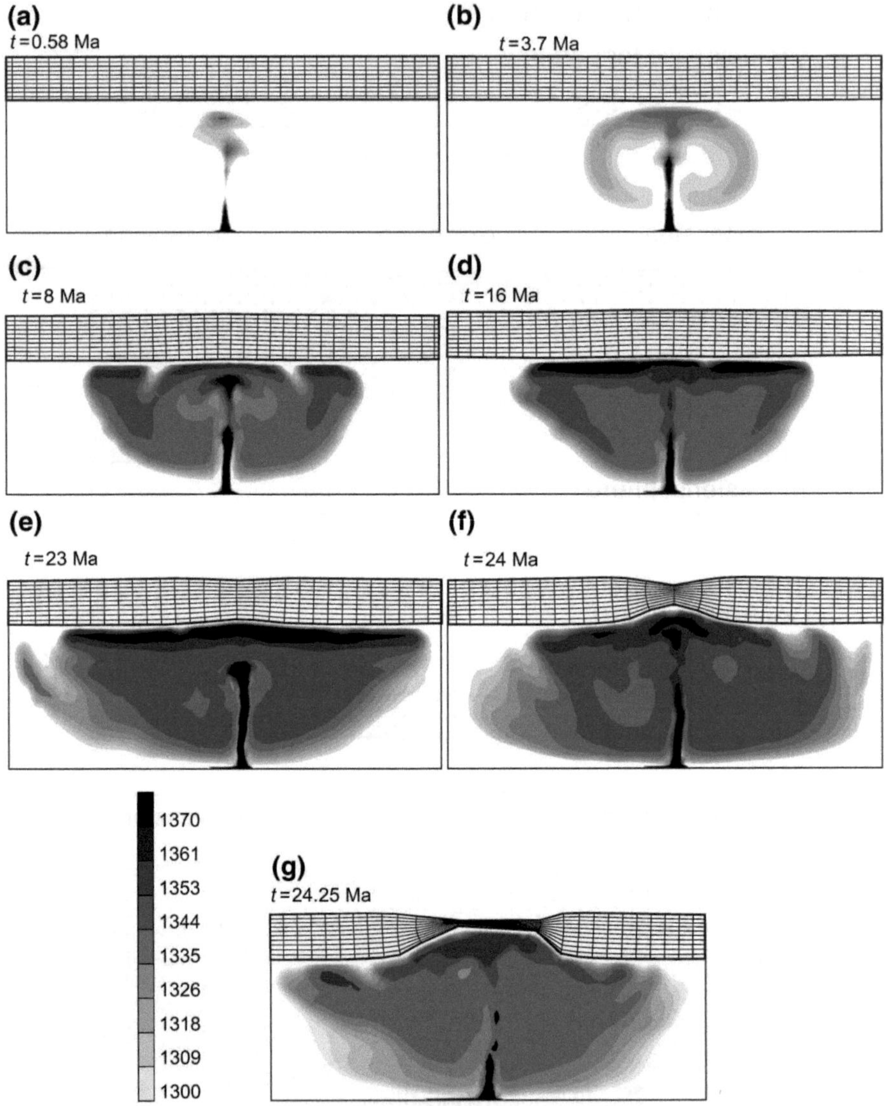

Fig. 3.41 The results for the model of active rifting. The effects of rising of the diapir leading to the thinning and extension of the crust are observed. The mantle rheology corresponds to the properties of the olivine–basalt–water–melt aggregate (Mei et al. 2002). The temperature evolution in the mantle lithosphere around the diapir and the configuration of the crustal layer are shown for different times. The deformed mesh lines are shown in the crust. The temperature scale (g) is in the interval from 1300 to 1370 °C. The width of the area is 450 km

rifting of the thick (up to 200 km) continental lithosphere is possible. According to (Bialas et al. 2010), the characteristic duration of magmatic activity required for oceanic rifting, with a thickness of the suboceanic lithosphere of up to 60–100 km, is 1–10 Myr. The estimates of the duration of thermal action on the cratonic lithosphere obtained in our study (about 25 Myr) are higher than the cited values and correspond to the duration of the magmatic stage of continental rifting. It is quite probable that in the case of intracontinental rifts, the extension does not continue to the total destruction of the crust, as shown in Fig. 3.41g, but, instead, it ceases at an incomplete stage. Our modeling results show that deformation around arising mantle diapir can lead to crustal extension if the density contrast is large enough (at a high degree of melting), as in the case of a mafic composition of diapir. Our results show that the presence of the sublithospheric mantle plume causes the following succession of events: diapirism in the mantle, underplating of the magma, and, with the assumed parameters of melting, extension of the continental crust.

The process of rifting is considered as a combination of stretching and thinning of the lithosphere, upwelling of the hot asthenosphere, isostatic compensation, subsequent cooling and thermal contraction. A generalized geodynamic model of continental rifting was developed by McKenzie (1978), who initially suggested a simple model of "instantaneous" finite extension/stretching of a two-layer lithosphere, which was later modified (Jarvis and McKenzie 1980) by including the duration parameter of finite extension of the lithosphere. Here, it was assumed that lithospheric extension accelerated exponentially and then ceased instantaneously after attainment of finite stretching. Sheplev and Reverdatto (1994) proposed and examined a third version of this model involving finite extension at a constant rate. As it turned out that at the same initial parameters and spreading (extension) duration, the subsidence occurs more rapidly at the end of spreading period, in the case of the accelerated lithospheric extension, and at the beginning, in the case of the constant-rate extension. The difference in subsidence dynamics is diminished with the decreasing duration of spreading, the solutions for the above two versions of calculations approach the solution for a simplified instantaneous case. The same relationships can also be applied to the thermal flux evolution.

The one-dimensional rifting model of Mackenzie can be used to study burial metamorphism during rifting. This mechanism implies that the lithosphere undergoing stretching is deformed as a continuous medium without fractures and rupture. If the initial temperature gradient in the lithosphere is taken to be dT/dz, the instantaneous stretching would lead to an increase in the gradient to $\beta dT/dz$. After rifting, time-dependent conductive cooling of the lithosphere eventually results in reequilibration of the thermal gradient to its initial value.

Below we present a description of the method of mathematical modeling of burial metamorphism based on a combination of lithological and stratigraphic data. We consider a one-dimensional (in depth) sedimentation process leading to the deposition of sedimentary strata with a certain number of layers characterized by the following thermomechanical properties: density of the solid phase, thermal conductivity, compaction coefficient, and initial porosity at the earth's surface (Table 3.6). To obtain quasi-stationary geotherms during subsidence, we calculate the one-dimensional depth-dependent conductive distribution of temperature. The thermal conductivities of porous sediments are calculated through the geometric

Table 3.6 Thermomechanical properties of some volcano-sedimentary rocks (Dortman 1984)

Lithology	Surface porosity, φ_0	Compaction parameter, C (*10^{-5}, cm^{-1})	Density, ρ, (10^{-3} kg/m^3)	Thermal conductivity, λ, W/(m K)
Mud	0.30	0.59	1.85	1.65
Mudstone	0.33	0.44	2.77	2.76
Sandstone	0.19	0.39	2.65	2.80
Siltstone	0.16	0.39	2.68	3.14
Limestone	0.24	0.16	2.71	2.93
Shale	0.13	0.51	2.72	2.30
Quartzite	0.02	0.08	2.50	5.02
Chalk	0.25	0.45	2.45	2.59
Tuff	0.37	0.41	1.66	2.08
Basalt	0.03	0.08	2.90	3.20

mean (Horai 1971) as a function of the thermal conductivity of porous water and rock matrix: $\lambda = \lambda_f^{\phi}\,\lambda_s^{(1-\phi)}$.

Let us consider the extension of the two-layer lithosphere, which is heterogeneous by depth and consists of the crust with a thickness of h and a density of ρ_c and the lithospheric mantle with a thickness of H and a density of ρ_m resting on the asthenosphere with a density of ρ_a. The backstripping method includes successive restoration of the paleotemperature and paleodepth of a multilayer sedimentary sequence with known thermomechanical characteristics of porous rocks. Figure 3.42 shows the Mackenzie's model of lithospheric extension. After thinning, the thickness of the crust and mantle lithosphere decreases to values of h/β and H/δ respectively. In addition, the possible extension of the lithosphere by a relative value of γ ($0 < \gamma < 1$) as a result of basalt dike emplacement is considered. The total amount of lithospheric thinning will be $\varepsilon = \frac{H+h}{(h/\beta+H/\delta)} \cdot \frac{1}{1-\gamma}$, where $\beta > 1$ and $\delta > 1$ are the extension coefficients for the crustal and mantle layers, respectively (Friedinger et al. 1991).

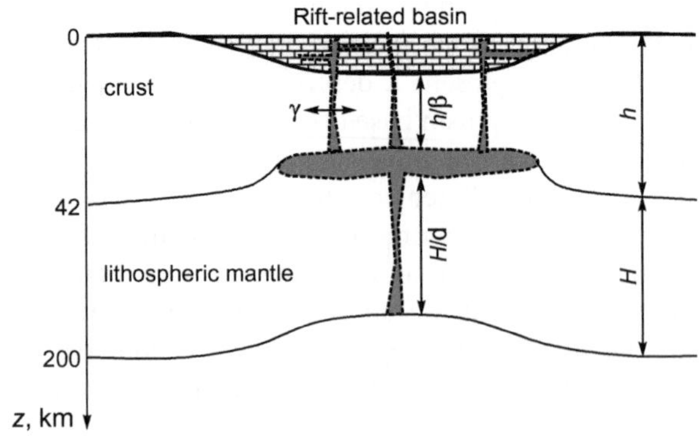

Fig. 3.42 Scheme of the rift mechanism used in backstripping estimations of the extension parameters of the crust and lithospheric mantle. H and h are the pre-rift thicknesses of the lithospheric mantle and crust, respectively; β, δ, γ are the extension coefficients of the crust, lithospheric mantle, and due to dyke intrusion, respectively

The examples shown below illustrate the evolution of burial metamorphism within the Yenisei-Khatanga rift in the north of the Siberian platform, including the Ust-Yenisei, Tsvetkov-Paksa, and Balakhna depressions (Polyanskii et al. 2000). The data on stratigraphy, lithology, and physical properties of rocks were assumed on the basis of stratigraphic sections and well data (Sorokov and Ginsburg 1974). The results of model calculations on burial depths and thermal regimes are presented in Fig. 3.43. The following maximum temperatures were obtained: >400 °C for the Ust-Yenisei depression, >300 °C for the Balakhna depression, and up to 400 °C for the Tsvetkov-Paksa depression. This difference can be explained by varying degrees of crustal extension and thinning, and therefore by upwelling of the asthenospheric mantle. The total amount of extension for these depressions was estimated at 5.4, 2.1, and 1.85, respectively. Crustal rocks beneath the Ust-Yenisei depression are expected to have been subject to higher grades of metamorphism as compared to other rift segments.

The study of rift basins has the following geodynamic implications. The applicability of Mackenzie's model (McKenzie 1978) is confirmed by the close agreement between the model results and the observed records of tectonic subsidence inferred from the structure of sedimentary sequences (with known ages of sedimentary rock layers). The modeling results are consistent with the existence of a positive correlation between depths of rift basins and the amount of spreading in each specific case of rifting (Reverdatto et al. 1992, 1995). There is no need to complicate the subsidence model by the gabbro—eclogite transition at the base of the crust (Artyushkov 1993). The stage of spreading, i.e. the stage of stretching/extension of the lithosphere can be linked to the onset of relatively rapid crustal subsidence, whereas subsequent cooling of the hot asthenosphere was the direct cause of slow subsidence in these depressions over a period of many tens of millions of years. Data from a number of sedimentary basins indicate that the rifting phase lasted up to about 10–20 Myr and was followed by thermal relaxation, which persisted over a period of 50–100 Myr.

In general, crustal extension during rifting amounts to 30–50%, the depth of the basins varies from 6 to 9 km, rocks at the bottom of depressions may reach temperatures characteristic of the prehnite-pumpellyite facies/subfacies and, rarely, greenschists facies. Higher metamorphic temperatures at a depth of about 10–12 km can be attained by additional heat supply from magmatic intrusions emplaced into base of the depressions. However, there are several possible exceptions associated with ancient rifts, where the deepest basins were formed. The reason may be either the more substantial extension of the lithosphere (more than 100%), or eclogitization of the basaltic layer (Artyushkov 1993). However, the eclogitization seems to be a sufficiently exotic reason for a significant crustal subsidence. In some cases, under very obscure conditions, very deep basins may have been formed at different times, such as the Adelaide basin in Australia (Reverdatto et al. 1992; Veevers 1984; Preiss 1987), and the Caspian and Black Sea basins (Artyushkov 1993), where rocks in the lower part of the sedimentary section were metamorphosed under epidote-amphibolite and amphibolite facies conditions. Such basins may have formed as a result of significant, complex and (most

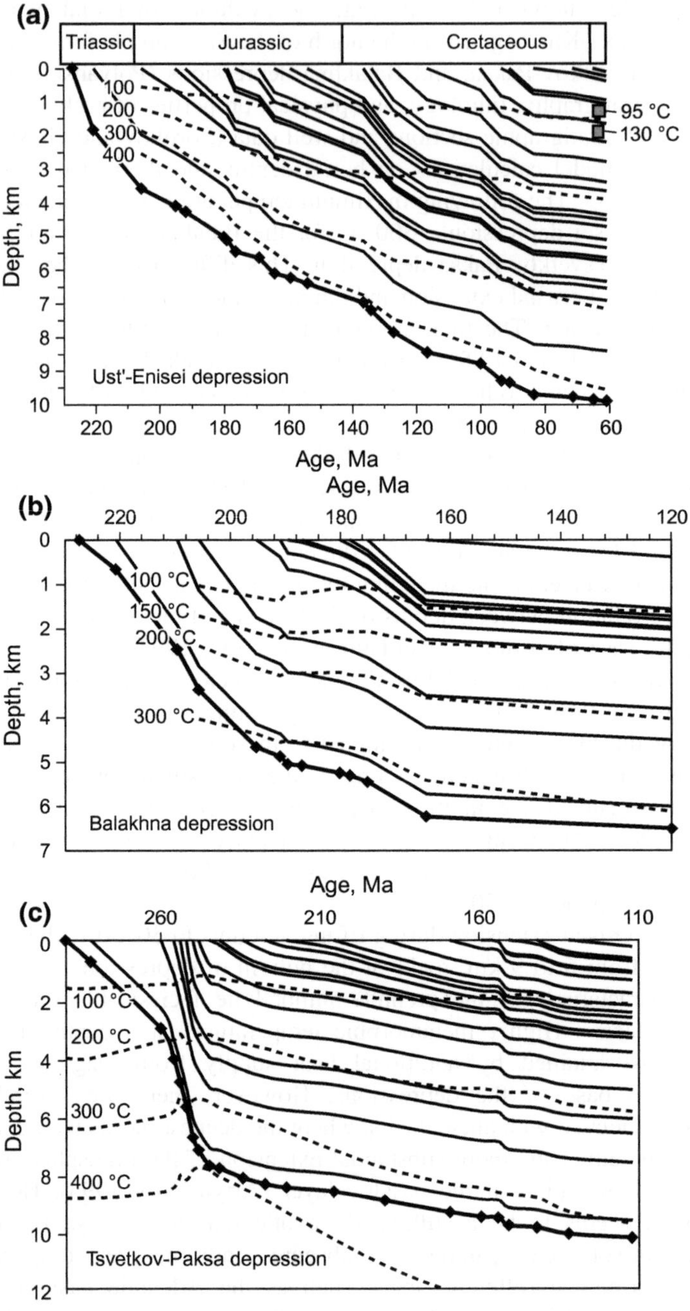

important!) repeated extension of the lithosphere accompanied by mantle upwelling.

◂**Fig. 3.43** Evolution of subsidence of the western (**a**), central (**b**) and eastern (**c**) Yenisei-Khatanga Trough (Ust'-Yenisei, Balakhna and Tsvetkov-Paksa depressions). The squares in (**a**) indicate the determinations of paleotemperature from vitrinite reflectance (Polyanskii et al. 2000). Dashed lines denote a temperature evolution, solid lines denote the sediments subsidence, and line with rhombuses denotes the basin basement subsidence

During the Paleozoic-Mesozoic, metamorphism grade did not exceed greenschists facies in rift basins. Evidently, higher metamorphic temperatures may have been attained only in deeper Precambrian rifts.

3.2.2.2 Metamorphism Caused by Tectonic Stacking in Orogeny

Tectonic stacking is inherently associated with a collision, which results from transpressional stress induced by the colliding plates. Collision of continental slabs and blocks causes compression and crustal thickening. The latter can also lead to horizontal shortening in the rock strata along the strike as a result of folding and thrusting/shearing. Intensive folding and thrust faulting in the collision zones is accompanied by low- to medium-grade metamorphism caused by the rise of isotherms, often cutting across the nappe boundaries. The rise of isotherms is most likely caused either by advective heat transfer induced by the upward movement of hot crustal material along the thrust faults in the lower parts of the continental plates or by deep magmatic intrusions.

The Himalayan belt is the most typical example of metamorphism during tectonic stacking in orogeny. It is characterized by relatively rare manifestations of an inverted metamorphic zonation, generally representing a gradual succession of chlorite to sillimanite isograds, from the base to the top of the unit. The following stages are identified in the Himalayan collisional belt (Fig. 3.44): 1—burial and prograde metamorphism, 2—peak sillimanite-grade metamorphism at 4–5 kbar, 3—partial melting and migmatisation, 4—crystallization and intrusion of leucogranite, 5—extrusion of the Greater Himalayan mid-crustal complexes, 6—recumbent folding of the metamorphic isograds (Searle et al. 2007).

Numerous models have been proposed to explain the inverted metamorphic zoning (Mathew et al. 2013): (a) hot-iron model, (b) isograds deformation by recumbent folding, (c) imbricate thrusting and frictional heating, (d) syn- to post-convergent ductile extrusion (channel flow), (e) critical taper model. Two models of the deformation-metamorphic evolution of the Himalayan crust are most actively discussed: channel flow (Searle et al. 2007; Jamieson and Beaumont 2013) and critical taper models (Kohn 2008). The first model predicts that lateral transport of both heat and material occurs laterally in a mid-crustal channel between two bounding faults, which is extruded onto the surface at the orogenic front (Fig. 3.45a). The extruded channel is bounded by upper normal sense and lower thrust sense shear zones. The critical taper model assumes that the rise of material occurs by means of thrusting at the orogenic front, which leads to the development of a sequence wedges tapered in shape with the thickness decreasing downward

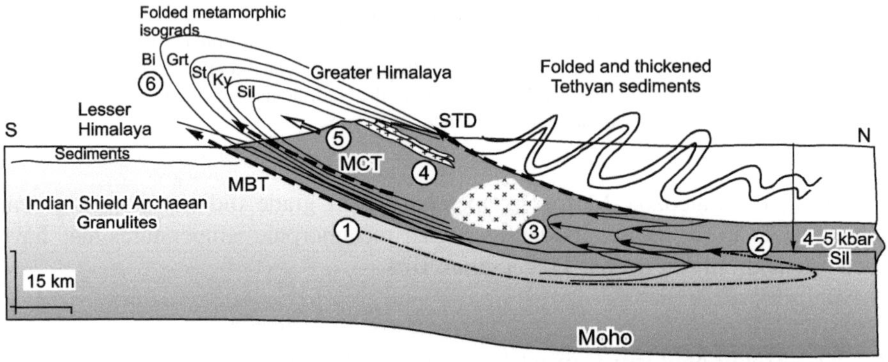

Fig. 3.44 The channel flow model based on regional mapping combined with strain analyses, thermobarometry, and U–Pb geochronology of the Greater Himalayan rocks, after Searle et al. (2007). Stages of metamorphic evolution of the Himalaya (numbers in circles): 1—subsidence and progressive metamorphism, 2—peak sillimanite-grade metamorphism, 3—partial melting to form migmatites and leucogranites, 4—intrusion and crystallization of leucogranites, 5—ductile extrusion of mid-crustal channel. MBT—Main Boundary Thrust, MCT—Main Central Thrust, STD—South Tibetian Detachment

(Fig. 3.45b). In this model, the spatial position of metamorphic zones and *P-T-t* paths suggest that thrusts generally propagate in-sequence and become progressively younger toward the frontal portion of the orogen.

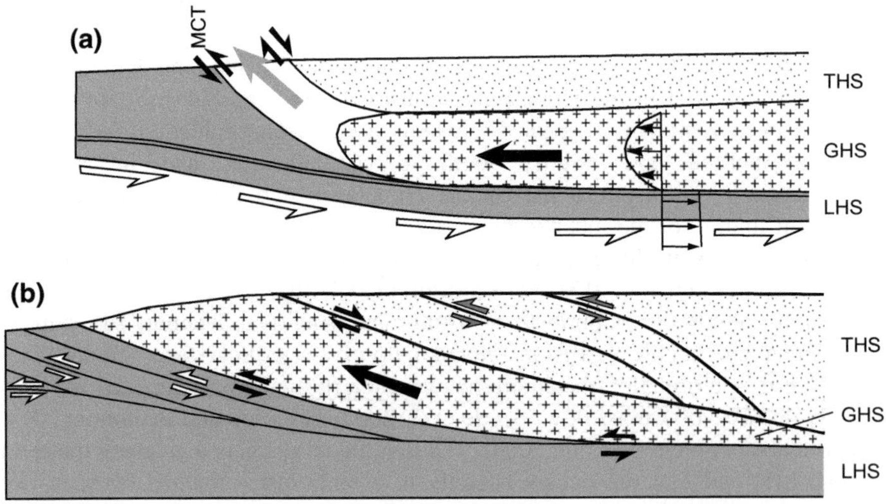

Fig. 3.45 Two end-member models of orogenic front propagation and metamorphic zoning on the example of the Himalayan metamorphic rocks (Jamieson and Beaumont 2013). **a**—channel flow geodynamic model, **b**—critical taper geodynamic model. MCT—Main Central Thrust, THS—Tethyan Himalayan Sequence, GHS—Greater Himalayan Sequence, LHS—Lesser Himalayan Sequence

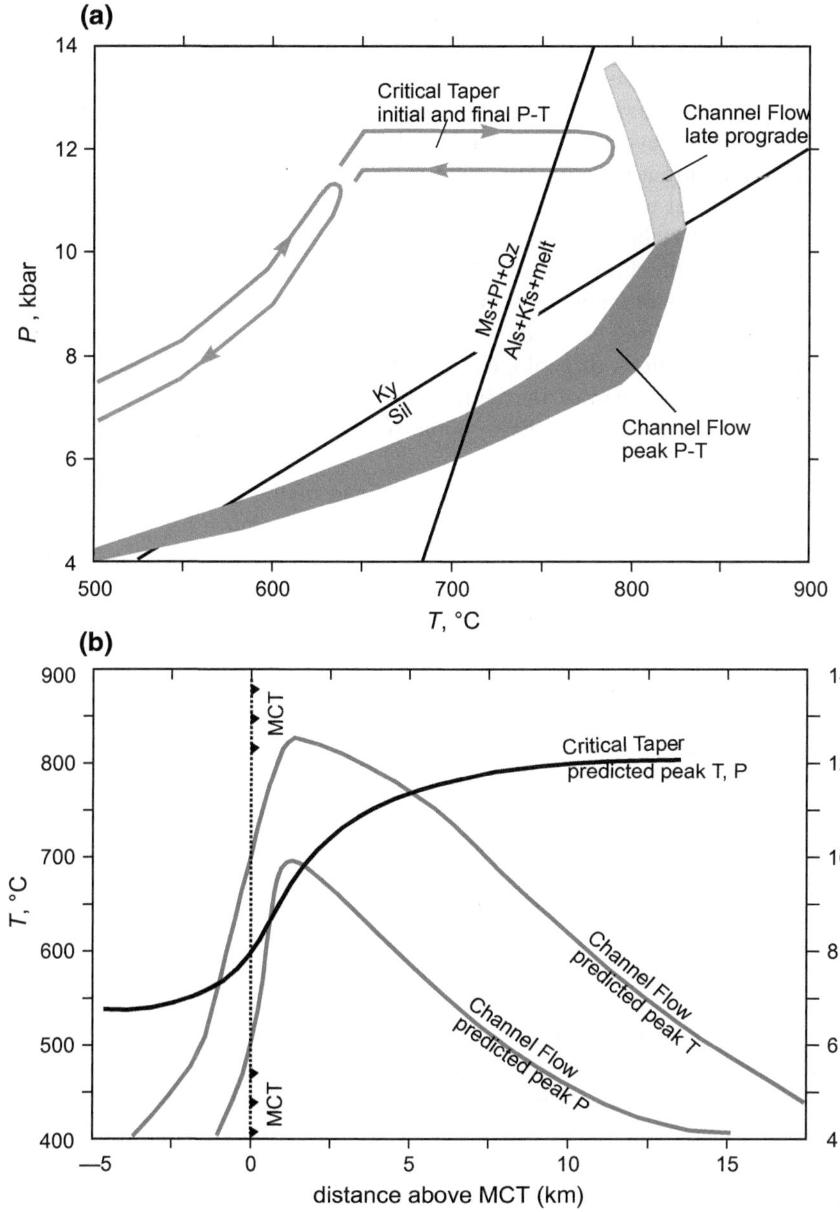

Fig. 3.46 Channel flow and critical taper predictions of *P-T-t* (pressure-temperature-time) conditions and evolution. **a** Initial *P-T* array and peak *P-T* conditions. *T* and *P* versus structural distance from Main Central Thrust according to critical taper (Kohn 2008) and channel flow (Jamieson and Beaumont 2013) models

These two end-members are characterized by different *P-T* parameters, their spatial distribution in the section of the orogen and metamorphic *P-T-t* paths (Kohn 2008). Figure 3.46 shows spatial distributions of *P-T* conditions versus distance from the Main Himalayan Thrust derived from two alternative models. Channel flow *P-T* conditions evolve toward high *T* due to heat advection in the channel, ultimately stabilizing a metamorphic field gradient in the sillimanite stability field (Fig. 3.46a). Critical taper models predict both prograde and retrograde *P-T* paths within the kyanite stability field, generally increasing temperatures and pressures with distance from the thrust toward the back of the orogen. Peak metamorphic *P-T* conditions across the orogenic section are recorded where high-grade metamorphic rocks are brought close to Earth's surface (Fig. 3.46b). Most authors provide conflicting interpretations of geological observations, therefore the mechanism for the inverted metamorphism of the Himalayan sequences still remains a matter of debate (Searle et al. 2007; Kohn 2008; Jamieson and Beaumont 2013; Mathew et al. 2013 and others).

The Central Asian Orogenic Belt (CAOB), one of the largest accretion–collisional belts, is a typical example showing widespread evidence for metamorphism during tectonic stacking: from subduction of the oceanic crust, accretion of island arcs, to collision of continental blocks (Xiao et al. 2004). At the final stage of accretion and collision, the southwestern sector of the Mongolian Altai represented a collage of tectonically juxtaposed and variably metamorphosed terranes (Kozakov et al. 2011). Badarch et al. (2002) suggested that the tectonic framework of this sector of the Mongolian Altai comprises the Erden terrane, Tseel metamorphic block, and Gobi-Altai back-arc terrane, which were amalgamated during the closure of the Mongolian-Okhotsk oceanic basin (Fig. 3.47).

Metamorphism during tectonic stacking in orogeny will be exemplified by the southwestern sector of the Mongolian Altai within the Tseel metamorphic belt (Polyansky et al. 2011; Sukhorukov et al. 2016). This belt extends along the Bulgan fault of the Main Mongolian Lineament, which represents the extension of the of the Irtysh shear zone. The Tseel metamorphic belt comprises the Bodonchin and Tsogt metamorphic blocks, located along the boundary between the Hercynides and Caledonides of the Gobi-Altai zone (Mossakovskii et al. 1993) (Fig. 3.47); structurally, both blocks belong to the Tseel tectonic plate. The Tseel metamorphic belt evolved in several stages (Kozakov et al. 2002). At the early stage, the rocks underwent metamorphism of the andalusite-sillimanite type (384–385 Ma) from amphibolite to granulite facies in places. The second stage (365–371 Ma) was marked by metamorphism of the kyanite-sillimanite type up to amphibolite facies conditions. The final stage was marked by regional metamorphism under greenschist facies conditions.

Sukhorukov et al. (2016) proposed a reconstruction of metamorphic conditions for aluminous schists of the Tsogt metamorphic block of the Tseel metamorphic belt (Fig. 3.47). The block includes a medium-temperature/medium-pressure zonal metamorphic complex, whose metamorphic grade varies from the greenschist to epidote-amphibolite facies. The *P-T* metamorphic parameters estimated by mineralogical geothermometers and geobarometers and by numerical modeling with the

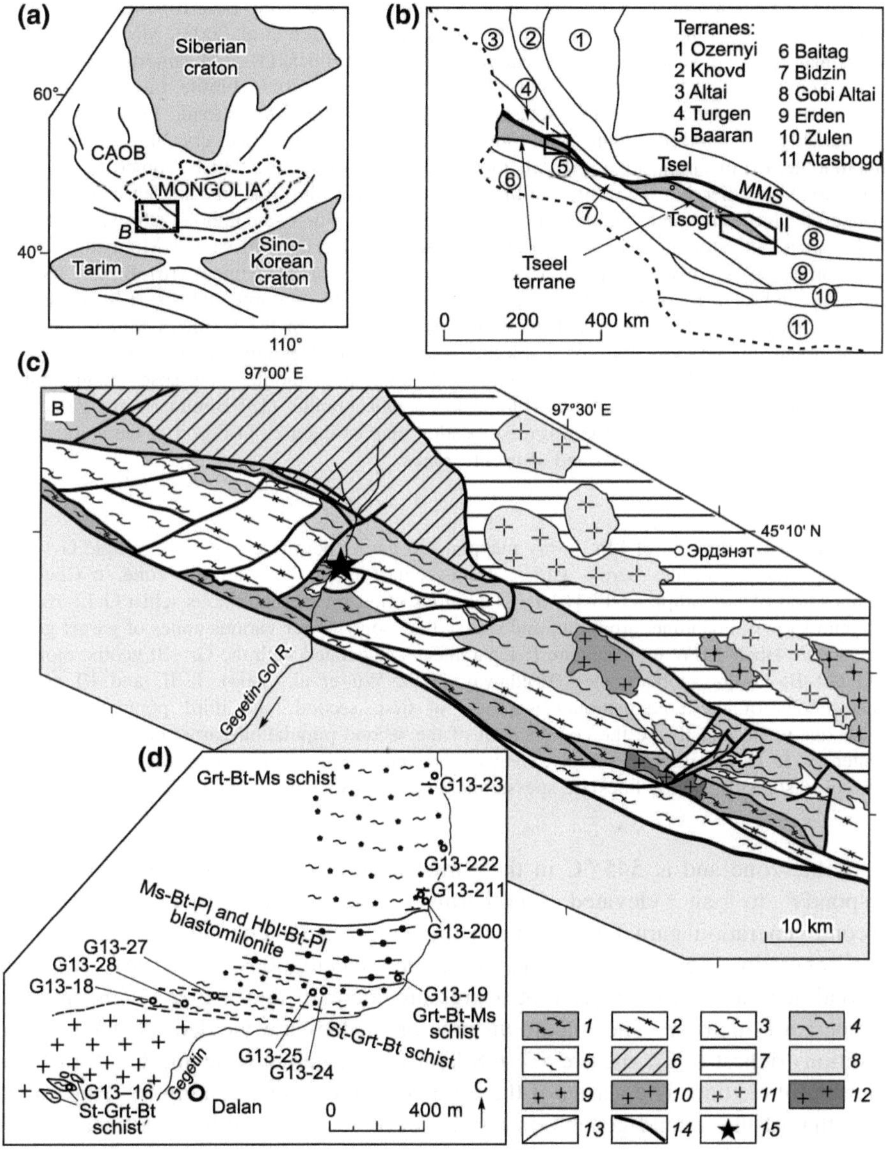

PERPLEX 668 software provide evidence of two successive metamorphic episodes: moderate-gradient (andalusite-sillimanite type, a geothermal gradient of ~40–50 °C/km) and low-gradient (kyanite-sillimanite type, a geothermal gradient of ~27 °C/km).

Three generations of garnet have been recognized in the garnet and garnet–staurolite schists based on differences in their composition and morphology. The data indicate that the first-generation garnet mainly grew at T = 570–575 °C in the

◂**Fig. 3.47** **a** Location of the study area in the Central Asian Orogenic Belt. **b** Schematic tectonic map of the Tsel terrane in southwestern Mongolia (after Badarch et al. 2002). MMS is the Main Mongolian Suture. **c** Schematic geological map of the Tsogt Block. (1–5) Metamorphic complexes of the Southern Altai Metamorphic Belt: (1) amphibolite; amphibole, biotite, biotite–amphibole, and biotite–muscovite crystalline schists and gneisses with garnet and, rarely, pyroxene, (2) variably migmatized biotite and biotite–amphibole gneisses and crystalline schists and amphibolite, (3) metasandstone, metasiltstone, and biotite–muscovite schists, sometimes containing garnet, cordierite, andalusite, sillimanite, and staurolite, (4) metabasalt, metandesite, metatuffite, metasandstone, amphibolite, and biotite and epidote–amphibole schists, sometimes with garnet and staurolite; biotite–sericite and sericite–chlorite schists; (5) Migmatite; (6, 7) Caledonides in the marginal portion of the North Asian paleocontinent: (6) Basalt, basaltic andesite, andesite, rhyolite; chlorite, epidote–, and sericite–chlorite schists; (7) Sandstone, siltstone, clay shale, limestone, and gritstone; (8) Hercynides in the Southern Mongolian Belt; (9) Permian biotite and muscovite–biotite granite; (10) Devonian quartz monzonite and granosyenite; (11) Mid-Devonian biotite and amphibole–biotite gneissose plagiogranite; (12) Devonian biotite and amphibole–biotite leucocratic granite; (13) Boundaries of geological complexes; (14) Tectonic boundaries; (15) location map of the area in Fig. d. **d** Geological map and sampling sites in the garnet and staurolite zones

Fig. 3.48 **a** Distribution of Mn (X-ray mapping) in garnet porphyroblast from sample G-13-16. ▸ **b** Compositional profiles across garnet porphyroblast from the staurolite zone. **c** Chemical composition of the sample G-13-16, **d** *P-T* pseudosection for the aluminous schist G-13-16. The diagram shows almandine, grossular, and spessartine isopleths for various zones of garnet grains. The square labeled HW indicates the *P-T* parameters calculated with the Grt–Bt geothermometer and Grt–Bt–Pl–Qz geobarometer (Holdaway 2000; Wu et al. 2004). I, II, and III are the intersections of the compositional isopleths of first, second, and third population garnet, respectively; IIc and IIi are the composition of the second population garnet closer to the grain center and to its margin, respectively. Heavy solid lines are almandine isopleths, dashed lines are grossular isopleths, dotted lines are spessartine isopleths (Sukhorukov et al. 2016)

staurolite zone and at 545 °C in the garnet zone at $P = 3.1$–3.7 kbar, which corresponds to an elevated geothermal gradient of 40–50 °C/km. The second-generation garnet has a much higher grossular content compared to the first-generation garnet. Thermobarometric calculations indicate that the second-generation garnet grew at a pressure increase and a merely insignificant temperature increase. The maximum pressure values are yielded by garnet compositions richest in Ca and are 5.7–6.8 kbar for the garnet zone and 6.8–8 kbar for the staurolite zone. The garnet of the third generation grew at a higher temperature than that of the second-generation garnet, while the pressure has not considerably changed, which is consistent with the decrease in the Fe#. The maximum temperatures are 560–565 °C for garnet in the garnet zone and 585–615 °C for the staurolite zone. Figure 3.48 shows X-ray compositional map of the third-generation garnet grain with the distribution of Mn (a), concentration profiles across inferred boundaries between zones of the first, second, and third generations of garnet (b), and a *P-T* pseudosection constructed using the PERPLEX 668 software showing the parameters of the second- and third-generation garnet (c).

It should be noted that the composition of the second-generation garnet plots within the stability field of the assemblage Grt + Bt + Ms + Chl + Pl + Qz, and

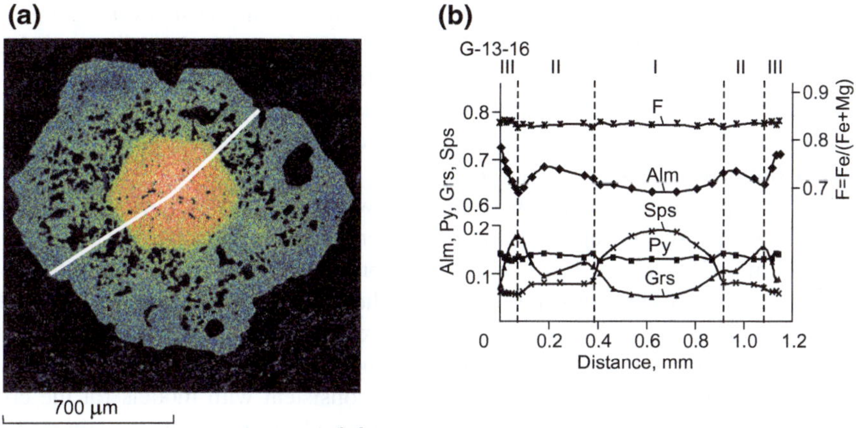

(c)

SiO2	TiO2	Al2O3	FeO	MnO	MgO	CaO	Na2O	K2O
61.52	0.75	17.73	7.19	0.07	3.10	1.66	1.27	3.50

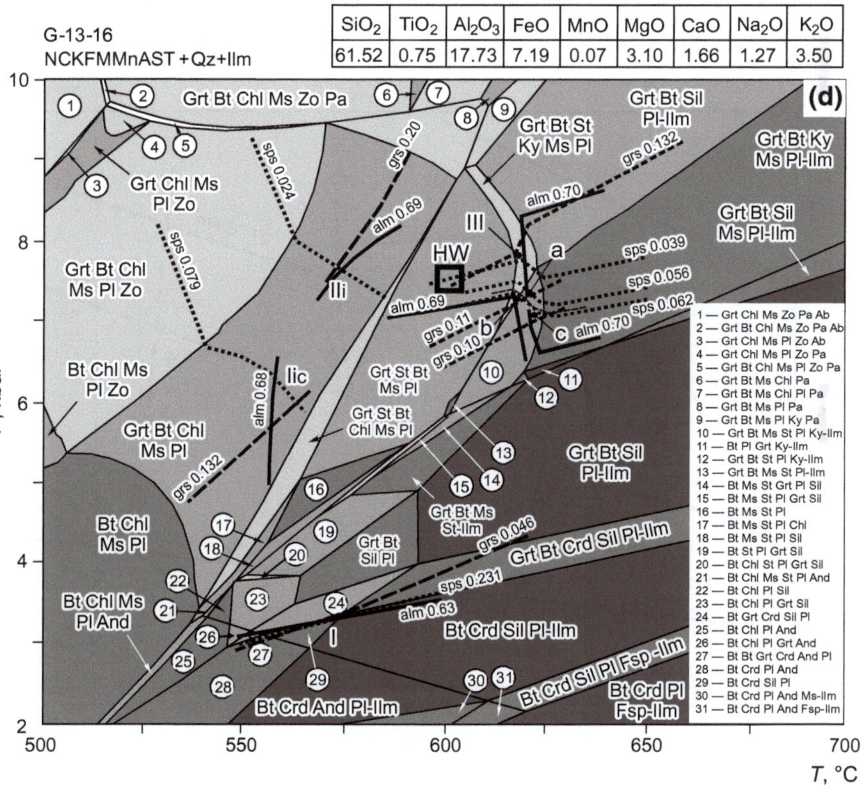

only the third-generation garnet falls into the staurolite stability field. This led us to conclude that staurolite started to grow in the rocks simultaneously with the development of the third-generation garnet rims, during the postkinematic episode. This conclusion is consistent with the observation that mineral inclusions in the garnet grains form S-shaped trails, while inclusion trails in the staurolite grains are linear. The orientation of these trails does not coincide with the foliation, which may suggest that deformation continued when the staurolite grains stopped to grow, but at lower temperatures. This is supported by the presence of muscovite in pressure shadows near staurolite grains. These deformations may have been related to the exhumation of the metamorphic blocks.

The *P-T* parameters of the first episode were 545–575 °C and 3.1–3.7 kbar (Fig. 3.49). The second episode was characterized by zonal metamorphism with peak T = 560–565 °C and P = 6.4–7.2 kbar for the garnet zone and T = 585–615 °C and P = 7.1–7.8 kbar for staurolite zone. The *P-T* metamorphic evolution of the area is described by a clockwise *P-T* path: pressure increase at a nearly constant temperature during the first episode and a temperature increase at a nearly constant pressure during the second one. This trend is consistent with models for the crust thickening during imbrication. The temperature and pressure values estimated by mineral geothermometers and geobarometers are closely similar and often coincide within the errors ($T \pm 50$ °C, $P \pm 1$ kbar) with the estimates derived from

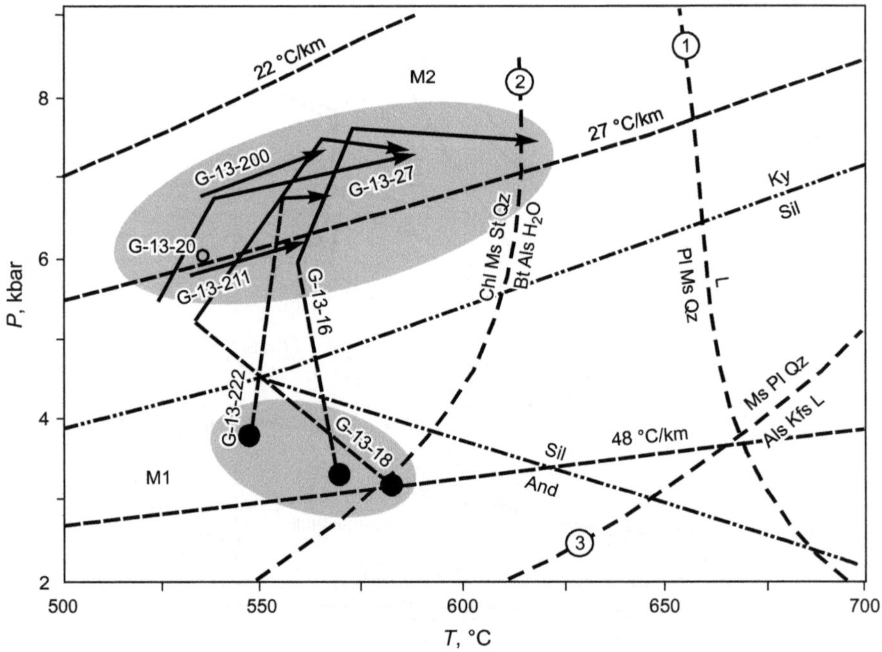

Fig. 3.49 Reconstructed *P-T* metamorphic path of the aluminous schists inferred from geothermobarometric data. M1 is the old high-gradient metamorphism, M2 is the medium-temperature, medium-pressure metamorphism. Numerals indicate mineral equilibria for the metapelite system: (2) Pattison (2001), (1) pelite solidus in a water-saturated system Le Breton and Thompson (1988), (3) Chatterjee and Johannes (1974). The triple point of Al_2SiO_5 polymorphs is according to Pattison (1992). Steady-state geotherms (dashed lines) correspond to the thermal gradient of 22, 27 and 48 °C/km

intersections of compositional isopleths for the rims of garnet grains. The differences may arise due to the use of different thermodynamic data on mineral end members.

The geothermal gradient during the growth of second- and third-generation garnet was about 27 °C/km. The pressure difference between the growth of the first-generation garnet and peak pressure conditions calculated from the second-generation garnet is approximately 4–5 kbar, which corresponds to crustal thickening by 15–18 km at an average density of the continental crust of 2.7 g/cm^3.

The S-shaped inclusion trails in second-generation garnet (with high grossular contents) indicate synchronous deformation. The increase in pressure coeval with deformation may be explained by crustal thickening due to overthrusting of the tectonic slabs. The configuration of the calculated *P-T* path suggests that the rocks were hosted in the footwall of the thrust. The *P-T-t* paths calculated for the compositions of second- and third-generation garnet are in good agreement with the model *P-T* paths for regions of thickened continental crust without additional heat influx (England and Thompson 1984; Spear 1993). In this case, the early evolution is marked by a pressure increase at almost constant temperature due to underplating induced by the overthrusting of a neighboring slab. The direction of the *P-T* path has been eventually changed due to an increase in temperature at nearly constant pressure because of the inertness of heat transfer and re-equilibration of the temperature field.

The exhumation rate of medium- and high-pressure and temperature rocks bears important implications for the preservation of peak metamorphic assemblages. The Bodonchin metamorphic complex of the Tseel belt (Mongolia) was used as an example to illustrate the application of the thermochronological method for the estimation of exhumation rates.

The Bodonchin block extends for 70–80 km from NW to SE and ranges in width from 40 to 10 km, pinching out to the east (Fig. 3.50). The block is truncated by the Paleozoic granitoids massif in the west and is separated from the Lower Devonian mafic and intermediate volcanic rocks by the Bulgan fault in the south. In the north, the Bodonchin block adjoins the Turgen accretionary complex, which is composed of Cambro-Ordvician volcanic and metasedimentary rocks. The Bodonchin metamorphic complex has a zonal structure and exhibits a gradual increase in metamorphic grade from north to south from the greenschists through kyanite-staurolite schists and gneisses with amphibolite interlayers to migmatites. The Bodonchin complex is dominated by mineral assemblages of the second (kyanite-sillimanite) metamorphic stage (Fig. 3.50). Along the northern and southern margins, greenschist facies rocks are separated by tectonic contacts from the main body.

Absolute age determinations and *P-T* estimates using mineral geothermobarometry were used to model metamorphic conditions and tectonic evolution (Polyansky et al. 2011; Sukhorukov et al. 2016).

The estimation of the *P-T* conditions of metamorphism was performed using the THERMOCALC software with the internally consistent thermodynamic dataset (Holland and Powell 1998) and a set of mineral geothermometers and geobarometers: Grt-Bt (Ferry and Spear 1978; Holdaway and Lee 1977; Kleemann and

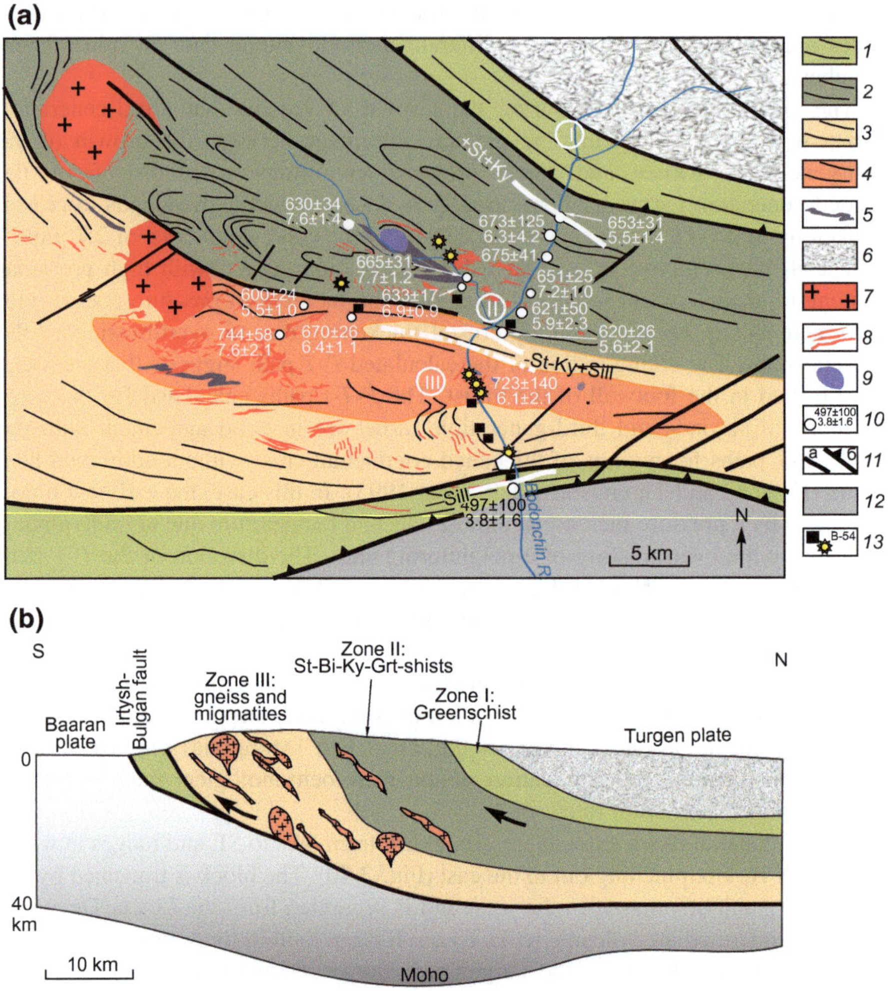

Fig. 3.50 The structure of the Bodonchin metamorphic complex with the sampling points and estimated *P* and *T* parameters of metamorphism (temperature in °C, pressure in kbar) after Polyansky et al. (2011). Zones of metamorphic rocks: I, greenschists, II, staurolite-disthene-garnet-biotite schists, III, gneisses and migmatites. 1—greenschists, 2—staurolite-kyanite schists and amphibolites, 3—garnet-biotite schists and amphibolites, 4—predominantly migmatites, 5—gabbroid rocks, 6—Caledonides of the North-Asian paleocontinent (Turgen plate), 7—late-Devonian granites, 8—syn-deformational granite-pegmatite veins, 9—mafic intrusion, 10—sampling points and T–P conditions (temperature in °C, pressure in kbar), 11—shear faults (**a**) and thrusts (**b**), 12—Hercynides of the South Mongolian Belt (Baaran plate), 13—sample localities of age dating (squares—Ar/Ar, stars—U/Pb). Numbers in circles denote intrusive massfis: 1—Ergiyn-Atgiyn, 2—Alag-Tektin, 3—Ergiyn-Us

Reinhardt 1994), Bt-Ms (Hoisch 1989), Grt-Crd (Holdaway and Lee 1977), Grt-Pl-Ky(Sil)-Qz (Newton and Haselton 1981), Grt-Pl-Ky(Sil)-Qz (Koziol and

Newton 1988), Grt-Ms-Pl-Bt (Ashworth and Evirgen 1985), Grt-Crd (Wells 1979). The temperature varies from 550 to 675 °C at 5.5–7.7 kbar in the kyanite-staurolite zone and from 600 to 720–745 °C at 5.5–7.6 kbar in the migmatite zone. The configuration of the paleogeotherm was reconstructed at the syn-collisional peak of metamorphism (Fig. 3.51). It was shown that the thermal state of the crust was controlled by either a high concentration of radioactive heat sources or high mantle heat flow. Taking the pressure as lithostatic, with a gradient of 1 kbar/3.5 km, we can estimate the paleotemperature gradient in the crust of the Bodonchin complex, using the pressure and temperature conditions determined by mineral geothermometers and geobarometers. In terms of the geothermal gradient, the slopes of the lines correspond to the temperature gradients of $\partial T/\partial z = 25.5$ and 27.2 °C/km (Fig. 3.51). These values are typical of the continental crust in fold belts during tectonic stacking accompanied by burial.

One of the quantitative approaches to reconstruct the evolution of particular crust blocks is the thermochronological method of examination based on the closure temperature concept (Dodson 1973; Hodges 2005). This concept implies the existence of a threshold temperature, below which diffusion processes in isotopic systems virtually cease. The $^{40}Ar/^{39}Ar$ dating reflects a time when a metamorphic mineral cooled through its closure temperature during exhumation to the surface. For dating, we sampled biotite, muscovite, and hornblende monofractions from metamorphic rocks in different parts of the orogenic section. The age estimated from biotite, muscovite, and amphibole varies from 243.9 ± 2.5 to 251.8 ± 2.8 Ma, from 247.8 ± 2.6 to 254.2 ± 2 Ma, and from 261.4 ± 2.9 to 275.1 ± 4.2 Ma, respectively. Using age estimates from isotopic ratios in the

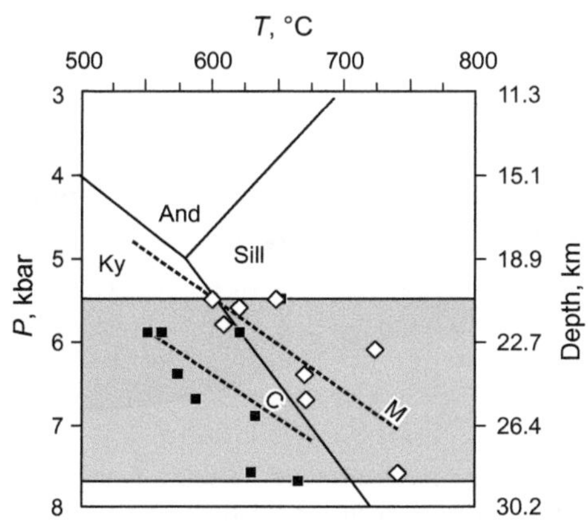

Fig. 3.51 The estimated *P* and *T* parameters of metamorphism in the Bodonchin complex and model geotherms for crust in different geodynamic regimes. Dashed lines are the interpolation geotherms for migmatite (M) and staurolite schists (S) zones. The squares mark the P and T parameters of metamorphism of staurolite schists, and the diamonds mark metamorphism of migmatites and garnet-biotite schists. The coordinates of the triple point and the line of univariant equilibria of aluminosilicates are given after Holland and Powell (1985)

minerals (biotite, muscovite, amphibole), we constructed a temperature—time evolutionary path for a particular point. Using the known temperature for calculating the depth, one can reconstruct the history of burial/uplift of a particular crustal block. To do this, it is necessary to know a temperature change with depth, i.e., the shape of the paleogeotherm for a given location. These data can be obtained by the independent estimation of the temperature and pressure conditions of metamorphism, using geothermobarometers, as described above. The model used in the calculation of the exhumation rate is shown in Fig. 3.52. The exhumation rate can be calculated based on three $^{40}Ar/^{39}Ar$ ages determined for three minerals at the same point, whose uplift trajectory is shown by the curve M (migmatites) or the curve S (schists) in Fig. 3.52. The depth distance between the 350, 380, and 550 °C isotherms corresponding to the closure temperatures of biotite, muscovite, and amphibole (Reiners 2009) can be determined from the geothermal gradient of 25.5–27.2 °C/km estimated using mineral thermometers. Then the vertical rate of exhumation will be calculated as $V = (T^{amf}_{closure} - T^{bi}_{closure})/gradT/\Delta t$, where $T^{amf}_{closure}, T^{bi}_{closure}$ are the closure temperatures of the argon system in the corresponding minerals (amphibole and biotite), *grad T* is the temperature gradient, and Δt is the age difference (Ma) between different minerals. The estimates rates of

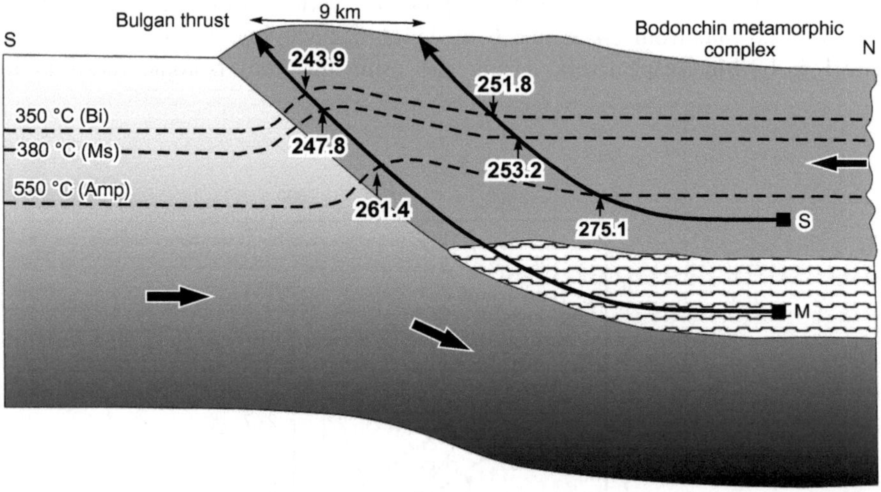

Fig. 3.52 Scheme illustrating the mechanism of cooling of migmatites and staurolite-kyanite schists of the Bodonchin complex used in the interpretation of thermochronological data. M, migmatites, S, staurolite-kyanite schists. The scheme explains the way of estimation of the ascent rate. Dashed lines are isotherms, and solid lines with arrows are the assumed trajectories of the rock ascent at the analyzed points. The intersections of the ascent paths with the isotherms (closure temperatures of the corresponding minerals) correspond to the $^{40}Ar/^{39}Ar$ ages (see text for explanation). Black arrows show the assumed collision and obduction of the Bodonchin complex. Grey and ornament zones mark schists and migmatites, respectively; gradient fill marks Hercynides of the South Mongolian belt

exhumation as a result of thrusting in the Bulgan fault zone vary from 315 to 470 m/Myr (Polyansky et al. 2011).

The mechanisms of formation of accretionary-collisional orogens are of particular interest, because they are regions of growth of new continental crust. Alternative mechanisms of collision include thickening of the crust (lithosphere) or the subduction–obduction model of plate thrusting (Johnson and Harley 2012).

To substantiate the amount of crustal thickening, which was obtained above using geothermobarometry data on metapelitic rocks of the Tseel metamorphic belt of the Mongolian Altai, we suggested a mathematical model of the process on the basis of equations of the deformable solid mechanics (Polyansky et al. 2015). We considered the problem of the crust deformation and distribution of the *P-T* parameters of metamorphism in the Tsogt block during convergence of the Edren and Gobi–Altai terranes.

Model setup, the geometry of the model area, the boundary, and the initial conditions are shown in Fig. 3.53. We set motion of the distant margin of the left plate at various rates (V = 1.0, 3.3, and 6.0 cm/year) for the time period of 33, 10, and 5 Myr, respectively, in which convergence was the same (330 km) for each variant of the model. The rates of plate convergence were selected on the basis of the geological data for the Alpine and Andean orogens (1–2 cm/year) and from 5 to <13 cm/year for the Karakorum–Tibet—Himalayan orogen (Johnson and Harley 2012). We solved equations of the equations of the deformable solid mechanics in the two-dimensional formulation: mechanical equilibrium equations and heat transfer equations were completed by the constitutive relation, which describes the rheological properties of the material. The formulation of equations, description of the mathematical method, and the rheology of the lithospheric material under the typical *P-T* parameters are given in the Appendix. A simplified model of the lithosphere structure composed of the crust of the dioritic and diabase compositions and the peridotitic mantle was considered. The Tseel block has a higher density due to the presence of garnet amphibolite and granulite (Kozakov et al. 2011). Each block of the model crust is characterized by its own rheology. It was suggested that the Edren rigid island-arc terrane could be described by an elastic rheology, whereas other areas can be described by a visco–elastoplastic rheology (Hansen and Carter 1982). It was assumed that the Tseel metamorphic block is characterized by the properties of wet diorite, the Gobi–Altai back-arc terrane is described by the properties of wet diabase, and the mantle is described by the properties of wet dunite.

The numerical experiment included (a) the study of variations in configurations of the deformable plates with different rheological properties upon collision, (b) estimation of the magnitude of thermal anomalies and the degree of crust thickening in the thrust zone as a function of the plate convergence rate. It is evident from the experiments that the viscoplastic block is obducted on the more rigid plate. This results in the formation of a low-angle thrust along the contact of the rigid plate with the viscoplastic Tseel block (Fig. 3.54). The underplating rigid slab was bent under the influence of the additional weight of the thrusted material. The strain was localized in the lower crust along the detachment, from the side of a hanging

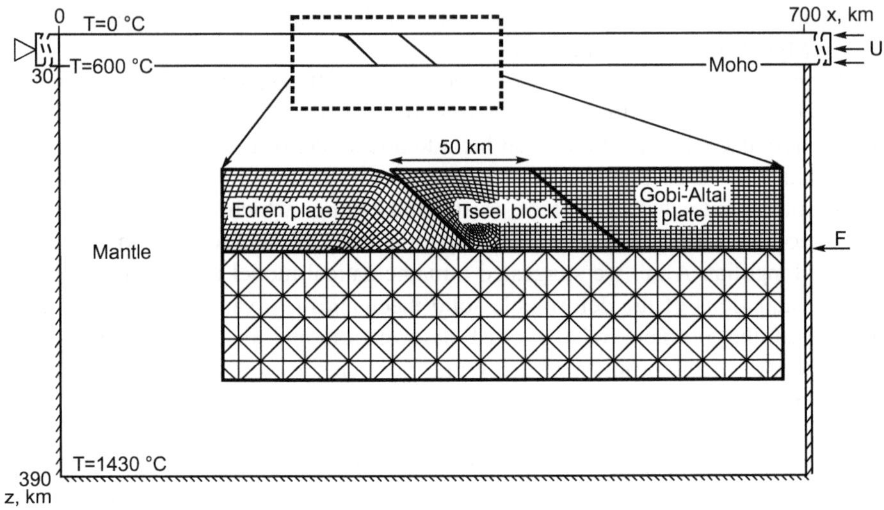

Fig. 3.53 Initial configuration of the numerical model of collision of the island arc and back arc terranes separated by a metamorphic block. The inset shows the detailed structure of the model crust surrounding the Tseel metamorphic block with initial computational mesh. F is the force equivalent to the lithostatic pressure; U is the prescribed plate velocity

block. The low-viscosity Tseel block was subjected to the strong deformation: its compression was accompanied by an upward extrusion of the material with the formation of a low-angle thrust and subduction of the lower crustal material, leading to thickening. Given the amount of plate convergence of 330 km, the central block was compressed by almost two times for all convergence rates. The right plate was deformed to a lesser degree and mainly underwent thickening. Total thrusting of the Tseel block over the rigid plate was 50 km per 10 Myr while the margin of the Gobi–Altai terrane moved a distance of 344 km. The main deformation involved thickening of the Tseel block and crust of the Gobi–Altai terrane. The maximum thickness of the crust in the collision zone during the period of plate convergence (10 Myr) was ~55 km, corresponding to a 1.8-fold thickening. The model predicts heterogeneous deformation of the Tseel block: the upper crustal material was transported almost horizontally; the middle and lower crustal material underwent burial to 15–16 km, corresponding to the lithostatic pressure of 4–4.5 kbar.

Three variants of the model with various plate convergence rates (1.0, 3.3, and 6.0 cm/year) were investigated. The results are shown in Fig. 3.55 at different periods, but for the same amount of convergence. The results of the calculation show that the thrust zones were characterized by changes in rock temperatures. The high rates of convergence (3.3 and 6.0 cm/year) led to the development of the inverted metamorphic zonation, where the hanging wall rocks were hotter (Fig. 3.55b, c). A comparison of these models shows that inverted isograds in the

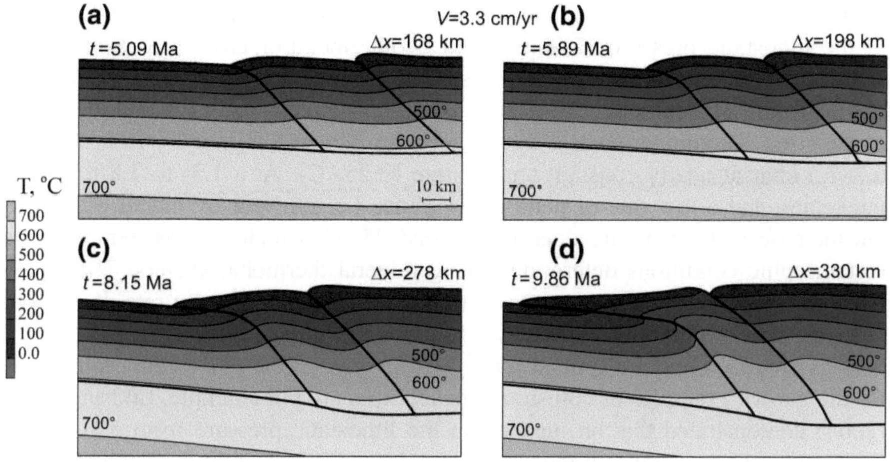

Fig. 3.54 Results of modeling of collision at a right plate convergence rate of 3.3 cm/year. The center of the model area corresponding to the inset in Fig. 3.62 is shown. Thick solid lines indicate the boundaries of the blocks and crust–mantle. **a–d** Temperature evolution (in range of 0–700 °C) and collisional orogen configuration in different stages of the convergence. The right plate displacement is shown in each figure (Δx)

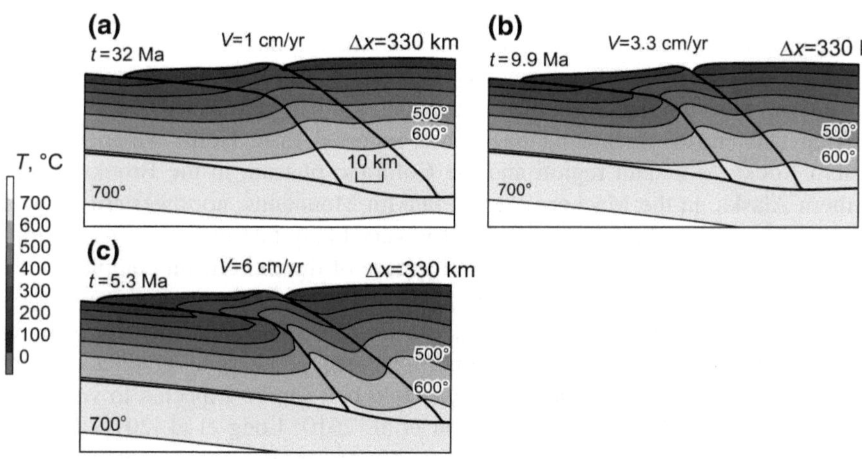

Fig. 3.55 Comparison of temperature fields for the models with relatively slow (1 cm/year), intermediate (3.3 cm/year), and fast (6 cm/year) rates of plate convergence. **a–c** Data are shown for the equal displacement of the right plate margin (330 km) during the different time periods

thrust zone are produced at a convergence rate of 3–6 cm/year and are almost absent at a rate of 1 cm/year.

The model predicts the development of a high-temperature field in the block bounded by thrust planes at rates of plate convergence of not less than 3 cm/year;

the temperature in the hanging block is higher than in the footwall block. Estimates of the lithostatic pressure obtained from geothermobarometric data (Sukhorukov et al. 2016) indicate an increase in pressure up to 4–4.5 kbar, which suggests burial of 15–18 km. The numerical simulation (Fig. 3.55) predicts a ~1.5-fold thickening of the crust. Thermobarometric data suggest an increase in pressure from 3–3.5 to 6.5–7.5 kbar at nearly constant temperature (±25 °C). At a 1.5- to 1.8-fold crust thickening and a low rate of plate convergence (<1 cm/year by model estimates), the increase in temperature does not exceed 25 °C, which is consistent with the metamorphic conditions determined using mineral thermobarometers. The rate of plate convergence controls the rate of thrusting: at a convergence rate of ~1 cm/year, the rate of thrusting will be equal to ~0.15 cm/year.

This approach may be applied for reconstructions of the *P-T* metamorphic paths for the known examples of collisional metamorphism. For example, Likhanov et al. (2004) demonstrated that an increase in the lithostatic pressure from 3.5–4 to 6–7 kbar at an almost constant temperature of 540–600 °C within the Panimba thrust of the Yenisei Ridge can be explained by tectonic thickening of the crust. The thrusting rate of 300 m/Myr (0.03 cm/year) was determined by comparing the geological and mineralogical data on a near-isothermal pressure increase during the evolution of the metamorphic complex near the thrust with the results of numerical modeling (Korobeinikov et al. 2006). The model predicts the thrusting rate of 1.5 mm/year within the Tseel block, i.e., five times higher than that in the zone of the Panimba thrust of the Yenisei Ridge.

The proposed model may serve as a basis for the reconstruction of the formation of zoned metamorphic complexes in orogenic areas, such as Tien Shan, the Himalayas, etc. Similar examples of metamorphism associated with tectonic stacking and crustal thickening have been reported from North America in the Eastern Rocky Mountain region and the Colorado plateau, in the Brooks Range, northern Alaska, in the Mackenzie and Franklin Mountains, northwestern Canada, and possibly also in the Caucasus, etc. It is very likely that metamorphism during tectonic stacking in these orogens, as in the case of the Bodonchin complex of the western Mongolian Altai, occurred along a normal (average continental) geothermal gradient, ranging from ~20 to ~40 °C/km. Despite significant fluctuations in the estimates and highly variable *P-T* conditions, the geothermal gradients during tectonothermal evolution of the Himalayan fold belt can be expected to vary from 15–20 to 40–50 °C/km (Kohn 2008; Yin et al. 2010; Long et al. 2012; Mathew et al. 2013; Mukherjee 2013 and others). Metamorphism was accompanied by intensive deformation and was caused by shearing and thrusting.

3.2.2.3 Metamorphism Associated with Archean Crust Formation

The oldest crustal fragments composed mainly of two-pyroxene gneisses (granulite facies) are reported from all continents. They form the largest shields (Aldan, Anabar, Baltic, Ukrainian, Canadian, Guiana, African, Australian and Antarctic). These gneisses are also found in many median massifs and in some places in the central parts of fold belts. At higher structural level on ancient platforms and in median massifs, the granulites and amphibolite-facies rocks (biotite-sillimanite gneisses) are widespread and usually reflect retrograde alteration and low-temperature metamorphic overprint (Dobretsov et al. 1973). The geodynamic conditions and mechanisms of formation of the Archean-Proterozoic continental crust are still not clearly understood. The geologic history and the causes for the granulite facies metamorphism in ancient shields remain rather controversial. Nevertheless, it can be assumed that Archean and Early Proterozoic geological history, tectonics, and magmatic activity were qualitatively different from Phanerozoic events. Although Earth already had an atmosphere and a hydrosphere in the Archean, it was dominated by widespread volcanism and magmatism while sedimentary processes played a sharply subordinate role. The atmosphere and hydrosphere took part in the destruction of primary volcanic products, as indicated by the composition of the oldest sedimentary rocks. Investigation of detrital zircons from metaconglomerates of Western Australia suggested that the formation of the continental crust began by ~4.4 Ga (Wilde et al. 2001).

The major crystalline rocks of shield areas are the so-called "grey gneisses" of tonalite-*t*rondhjemite-granodiorite affinity (TTG suite) (Martin 1994). The mechanism of their formation is understood completely. One mechanism suggests that the sialic protocrust was produced directly from the primitive mantle during fine-cell convection in a partially molten layer and segregation of the "light material" onto the surface (Garagash and Ermakov 2004). Another, more widely accepted opinion is that grey gneisses are products of repeated re-melting and transformation of the primary komatiite-basalt crust (Rozen et al. 2008). Melting was facilitated by convection of hot subcrustal material and subsidence of fragments of komatiite-basalt rocks into the mantle. At that time, the temperature of the mantle was probably higher than now by about 300 °C (Condie 2001). The radiogenic heat production in the Archean was 2–3 times higher than at present (Patchett 1992; Pollack 1997), but in the Pre-Archean it was probably even much higher (Rozen et al. 2008). It is assumed that Archean magmatic activity was governed predominantly by plume tectonics (Condie 1981; Glukhovskoy and Moralev 1996; Nenakhov 2001; Condie 2001), some argue that plate tectonics already operated in Archean times (Conrad and Hager 1999; Korenaga 2003). Nevertheless, the possible existence of subduction in the Archean is doubtful, because glaucophane schists are not documented in the Archean Eon (Brown 2008). In the Proterozoic, the formation of island-arc systems, arc accretion and subduction processes are assumed to have occurred (Rozen et al. 2008; Mohan et al. 2013; Sajeev et al. 2013).

The grey gneiss basement complex of intermediate composition is overlain by two rock series: (1) granitic gneisses which consist of plagiogneiss, mafic schist, amphibolite, granulite, granulite-migmatite, charnokite, enderbite, mangerite, anorthosite, granite, and granitic gneiss, as well as metavolcanics, metasedimentary rocks, etc., and (2) "greenstone belts" represented by metamorphosed ultramafic, mafic, and intermediate volcanics, chert and sandy-argillaceous deposits. The rocks of these series were formed in several stages over billions of years throughout the

Archean and during the Proterozoic. The rocks of the first series were probably derived from grey gneiss basement, volcanic and magmatic products, and sediments. They experienced tectonic stacking, progressive deformation under plastic flow conditions and several pulses of metamorphism (polymetamorphism). At the prograde metamorphic stage, the rocks reached granulite facies conditions with temperatures probably exceeding 1000 °C and pressures of ~12 kbar (see, for example, Sommer and Kröner 2013); the retrograde stage gave rise to the formation of amphibolite- and low-temperature facies mineral assemblages.

It is believed that granite gneisses may have been formed with the involvement of metasomatic processes, i.e., granitization, and a fluid phase that produced changes in the rock composition was CO_2–H_2O-rich. The greenstone belts were formed on the granite-gneiss basement complexes.

In the Archean-Proterozoic, granite-gneiss and greenstone complexes formed large areas lying at the base of the 30–40 km thick continental crust. The Archean-Proterozoic granite gneiss complex is characterized by a slight variation in *P-T* conditions of metamorphism over large areas. This can be explained, as shown below, by the spread of a thermal source, namely, the convective mantle magma over large areas at the base of the Earth's crust. Another explanation invoking heat transfer from multiple mafic-ultramafic plutons also appears to be plausible, but it would require assumption of the huge size of intrusions that were sequentially emplaced within a relatively small time interval.

Multiple metamorphic stages that took place during a long evolution of the Archaean-Proterozoic continental crust are difficult to distinguish, but it is obvious that Precambrian crust evolved in substantially different thermodynamic regimes (Harley 1989). It is likely that the most widespread types of metamorphism which occured during that period include collision-related metamorphism due to plate collision, burial metamorphism, as well as magmatic diapirism. metamorphism around magmatic intrusions, etc. It appears that during the Late Proterozoic, the metamorphism was characterized by Phanerozoic-style features (Rozen et al. 2008). Clastic material formed by erosion has accumulated in water bodies and may have been subjected to recycling. Intensive multiple transformations of the rocks largely destroyed all evidence for different types of metamorphism. As a result, the general geothermal gradient was close to a normal value at the prograde metamorphic stage (Brown 2008), and may have been deflected at the retrograde stage depending on the tectonic regime.

The geological structure of the shield fragments and the diversity of their constituent rocks can be illustrated using the examples of the Baltic shield, Ladoga region (Fig. 3.27a), Pilbara craton, Western Australia (Fig. 3.28b), and the Adirondack massif as parts of the Grenville Province in North America (Fig. 3.56). The Adirondack massif represents the southern tip of the Canadian shield and is composed predominantly of 0.9–1.3 Ga metamorphic rocks. The massif is dominated by banded metasedimentary and metavolcanic gneisses and schists, metaplutonic granite gneisses, charnokites, anorthosites, olivine gabbros, with subordinate marbles (metamorphosed limestones). The *P-T* parameters of metamorphism are estimated at 8–10 kbar and 700–900 °C. The rocks in the

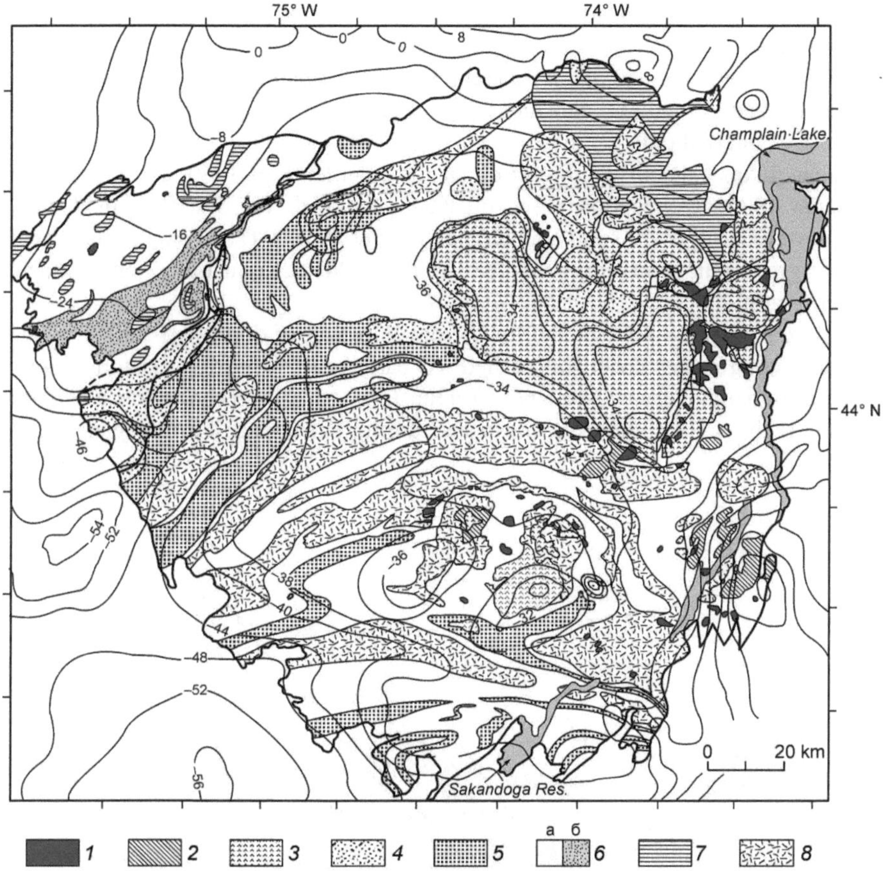

Fig. 3.56 Schematic geological map of Adirondack district, New York, USA (McLelland and Isachsen 1985). 1—metagabbro, 2—charnockite and mangerite gneisses, 3—metaanorthosite, 4—monzonite and syenite gneiss, 5—hornblende granite gneiss, 6—**a** undivided metasedimentary rocks: marbles, quartzites and metapelites, **b** biotite-plagioclase quartz schists and migmatites, 7—leucogranite gneisses, 8—charnockites and granite gneisses. Thin isolines characterize the structure of gravitation field (in milligals) with a strong negative anomaly in the area of anorthosite

Adirondacks were intensely deformed during the Grenville orogeny. Plastic deformation events gave rise to very complex folds of different styles and scales, while subsequent brittle deformation led to the development of fractured zones, numerous faults and mylonites.

A distinctive feature of the Karelian province of the Baltic shield is the widespread development of granite-gneiss domes, hundreds and thousands of kilometers across, which may have been formed by the process of diapirism, with upwelling of relatively less dense granitic material (see Sect. 3.2.1.3). Greenstone belts of the second complex may occur either as extensive linear features or show dome-and-keel patterns, representing the ruptures in the ancient basement and synclinal keels interspersed with domes; they are composed of low-temperature facies rocks.

Two models are invoked to explain the formation of granite-greenstone dome-and-keel structures: subsidence of crustal rocks and thermal effects at the expense of internal radiogenic heat sources (van Kranendonk et al. 2004), as well as the heat fluxes related to convection in the mantle (Fig. 3.57) (Artyushkov and Batsanin 1984). Conception based on the first model was described in detail by Choukroune et al. (1997) and Van Kranendonk et al. (2004) and exemplified by the granite-greenstone complexes of the Archean Pilbara (Australia) and Dharwar (India) cratons.

Figure 3.58 provides a model for the development of dome-and-keel structures. The following stages are distinguished: (I) early crust formation; accumulation of mafic-ultramafic volcanic sequences (greenstone cover) overlying the TTG basement; (II) the TTG complexes are heated during subsidence by radiogenic and/or

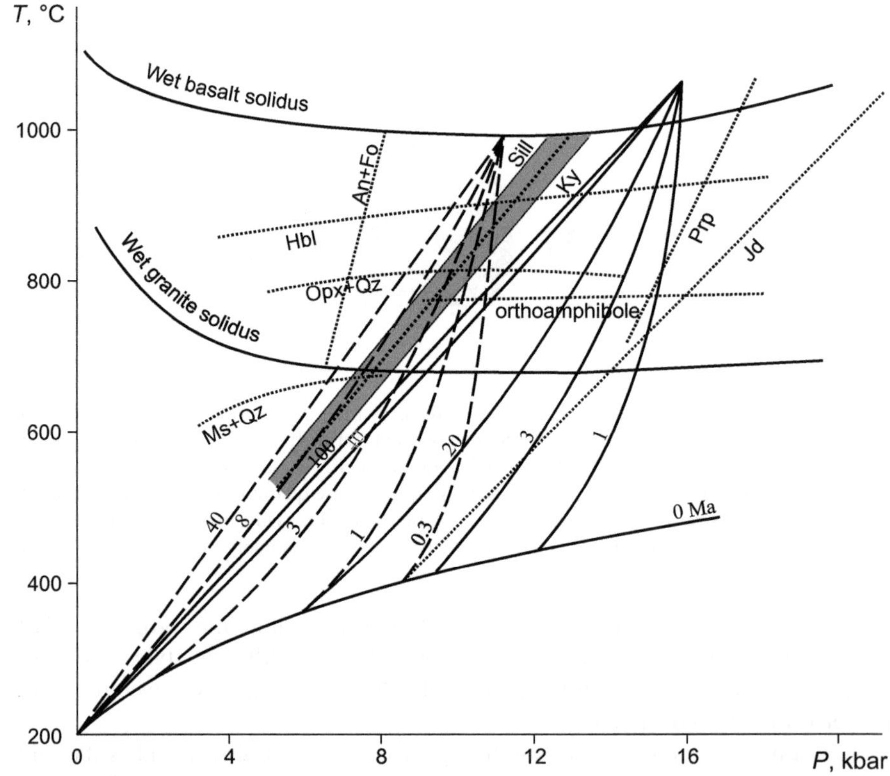

Fig. 3.57 Calculated temperature evolution (two versions) in a two-layer earth's crust when a thermal source with constant temperature is brought to its lower boundary. Numbers on curves (isochrons) indicate the time in million years. The *P-T* diagram displays the stability boundaries of some mineral and mineral assemblasges (Dobretsov et al. 1972; Spear 1993). Shaded field is the zone of uncertainty (from experimental data) between the stability fields of sillimanite and kyanite

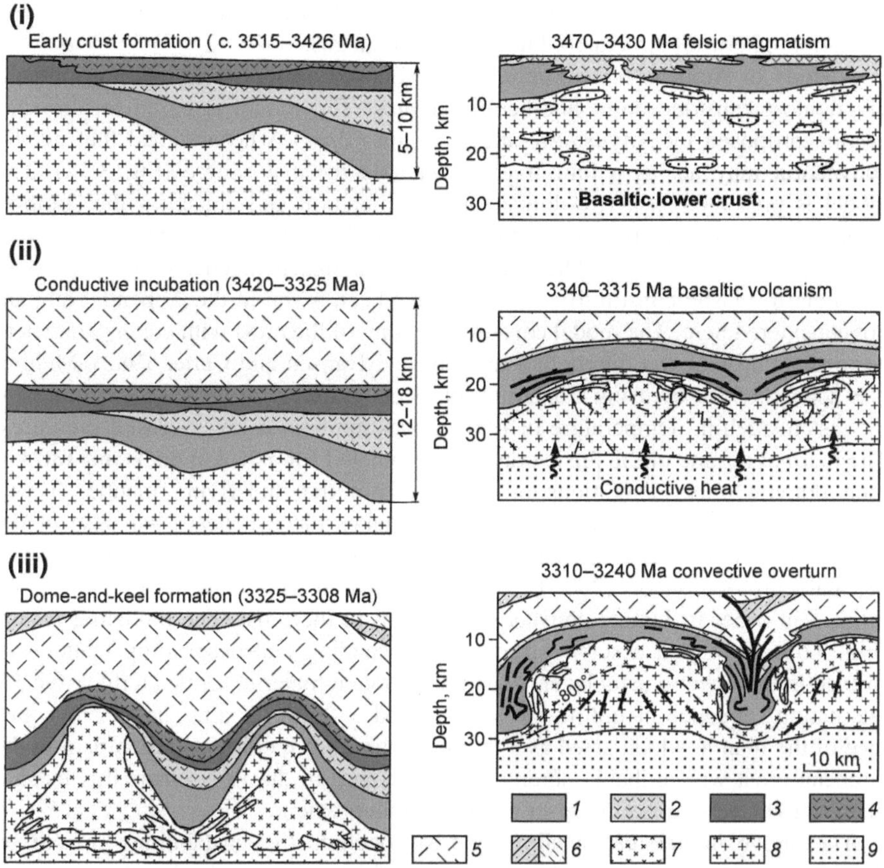

Fig. 3.58 Schematic model of partial convective overturn for the Archean East Pilbara Granite-Greenstone Terrane showing the development of dome-and-keel geometry through time after (Sandiford et al. 2004) (left) and Van Kranendonk et al. (2004) (right). 1–4—greenstones of the Pilbara Supergroup; 5—Euro Basalt; 6—Wyman Formation and Budjan Creek Formation; 7—voluminous granitoids of the Mount Edgar and Corunna Downs granitoid complexes; 8—thermally weakened middle crust; 9—underplated mafic lower crust

mantle heat sources (a conductive incubation period when the viscosity of the TTG decreases); (III) density inversion results in the partial overturn of denser greenstones and a less dense sequence and formation of TTG domes and synclines filled with greenstone rocks. Such a density inversion is explained by two processes. The first one is partial melting caused either by radiogenic heat sources (Sandiford et al. 2004) or a heat flux related to convection in the mantle (Artyushkov and Batsanin 1984). The second process is the inverted density stratification of the ancient crust where less dense TTG complexes become overlain by a denser greenstone succession. This process resulting from Rayleigh-Taylor gravitational instabilities was described by Polyansky and Volkov (1990) and Robin and Bailey (2009) using numerical modeling. It is therefore likely that both factors may contribute to the formation of crustal structures, but at different stages.

A study of Archean metamorphism was focused on modeling the development of granite-greenstone dome-and-keel structures. The model is based on the suggestion proposed by Artyushkov and Batsanin (1984), who assumed that mantle convection may act as a long-lived and stable thermal source (Fig. 3.57). The model is presented in a two-dimensional formulation and considers the comprehensive system of equations of motion, heat transfer and mass conservation, as was described in Sect. 3.2.1.3. The equations of the thermomechanics coupled with visco-elasto-plastic constitutive equation have been given in the Appendix and are not considered here. The model considers heating of the crust with creep parameters corresponding to a wet quartz diorite rheology (Hansen and Carter 1982). Melting was accounted for by using a diagram for "wet" granite. For the sake of simplicity, a constant melting temperature of 650 °C was applied at lower-crustal pressures. The thermal regime of the felsic crust was simulated for a 30 km thick and 180 km long domain under conditions of heating from the underlying mantle. It was assumed that continuous mantle convection takes place beneath the base of the crust, maintaining a constant temperature of 1200 °C. In a 20 km wide region, a thermal perturbation of 1250 °C was prescribed in the center of the subcrustal domain in order to initiate convective overturn in the crust. The modeling results (Fig. 3.59) show that a stable structure of ascending and descending matter was formed in the lower crustal layer. A partially molten material formed a rhythmically alternating sequence of uplifts (domes) located $\sim$25 km apart. The spacious surface of the mantle which acted as a thermal source for melting and diapirism in the Archean crust was probably shrank with time, and in the Proterozoic this mechanism must have operated only in areas of limited extent.

The above model results allow one to conclude that instabilities created by density inversion during partial melting of the felsic crust can lead to the formation of a system of uplifts and troughs, and this result is applicable to the formation of dome-shaped structures during the Precambrian. It can be assumed that the presence of basement TTG complex and overlying greenstones in the upper crust created an additional driving force for granite-gneiss dome formation.

3.2.3 Collisional Metamorphism

Collisional metamorphism accompanying large-scale deformation of the crust has a typical range of temperature and pressure conditions from 250 to 750 °C and from 4 to >15 kbar (Korikovsky 1995). These processes produce a combination of low-, medium-, and high-pressure facies series. Some deeply buried (in subduction zones or at convergent plate margins) rock complexes may reach diamond-bearing peridotite and eclogites facies conditions at temperatures up to $\sim$1000 °C and pressures >40 kbar. The formation of high/ultrahigh-pressure (HP/UHP) rocks (coesite- and diamond-bearing) in different parts of the world can be unequivocally

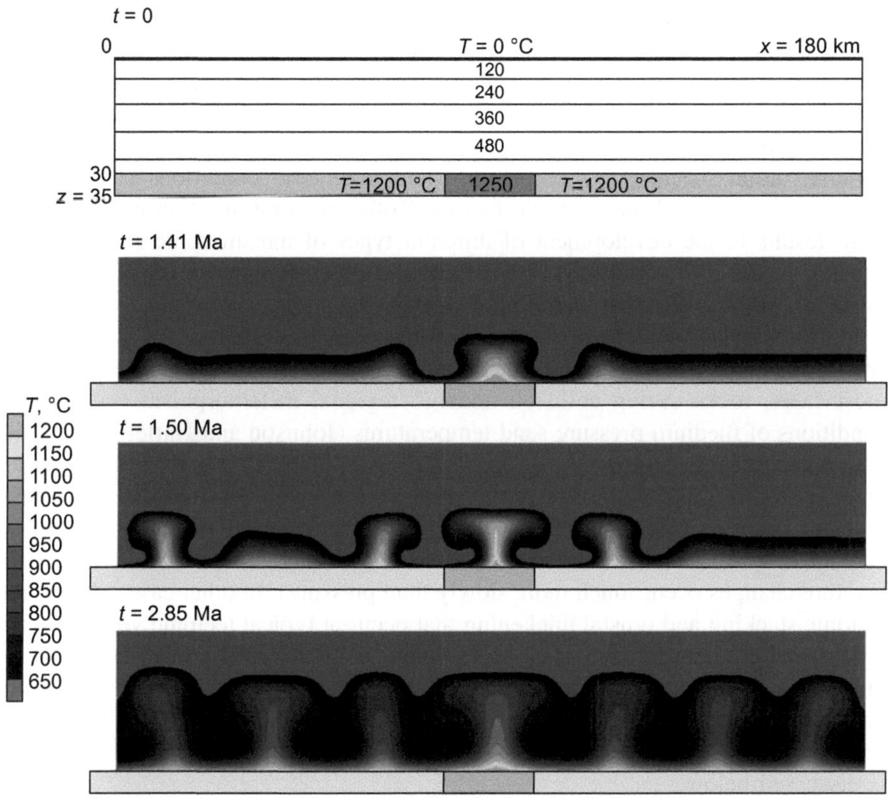

Fig. 3.59 Upper panel: setting up the problem of the Archean granite-greenstone "dome-and-keel" structures formation in the lower crust. Shown are the modeling areas, parameters, boundary and initial temperature. Mafic magma (light and dark grey) with initial temperature of 1200/1250 °C intruded undeformable lower crust. Central and lower panels: Temperature fields in the range of 650–1200 °C during the upwelling of the material, described by creep rheology. Three sequential stages of dome-and-keel structures formation are shown. Scale is shown on the left. Grey field depict the area with the temperature below wet granite solidus (650 ° C)

explained by deep burial of the Earth's crust during intracontinental collision. Coesite-bearing eclogites have been found, in particular, in the Himalayas. These finds provide a close link between the syn-collisional metamorphism associated with thrusting and nappe-forming in the continental crust and subduction of plates at convergent margins. Thrusting of crustal blocks at relatively shallow depths during collision can also lead to metamorphism, but at lower *P-T* conditions than in the Himalayas (Korobeinikov et al. 2006; Likhanov et al. 2004). This can result in the development of a specific metamorphic zoning characterized by either an almost isothermal increase in pressure (Spear et al. 1990; Likhanov et al. 2004), or by the inverted *P-T* gradient (Pecher 1989).

From a standpoint of mechanics, both subduction/convergence of plates and overthrusting/underthrusting of smaller crustal blocks can be considered as the deformation processes in solids, which differ mainly in the scales of interaction, duration, thermomechanical properties of materials and *P-T* parameters. Collision metamorphism during overthrusting/underthrusting of blocks takes place under variable initial mechanical, temperature and geometric conditions caused by a number of geological and tectonic factors. Both structural and thermal rearrangement results in the development of different types of metamorphic zoning which depend on the *P-T* conditions, thermophysical properties and rates of motion of interacting blocks (Ruppel and Hodges 1994).

A characteristic feature of most collisional orogens is heating of rocks during crustal thickening, which causes metamorphism and, in some cases, anatexis. In most cases, rocks during collision undergo regional metamorphism occurring in conditions of medium pressure sand temperatures (Johnson and Harley 2012). The magnitude of the geothermal gradient during collisional metamorphism varies greatly. During rapid underthrusting of one plate beneath another, the geothermal gradient becomes lower than the average continental crustal value, which can be explained by inaccessible thermal equilibrium of interacting blocks (since temperature changes occur much more slowly than pressure). In other cases, during the tectonic stacking and crustal thickening that occur at typical tectonic velocities, the geothermal gradient remains close to the average value, while elevated contents of radiogenic heat sources in the thickened granite crust may cause a significant increase in the geothermal gradient at the postcollisional stage. The relationship between the rates of overthrusting and collision and the thermal regime of the crust is discussed below.

The pioneering works of Miyashiro (1961, 1973), Zwart (1969), and Ernst (1973, 1975) in the 1960s and 1970s established the relationship between metamorphism in collisional orogens and tectonic processes. In these works, the orogens were typified by the character of metamorphism based on metamorphic facies series. These ideas were amplified by England and Richardson (1977) and fully developed by England and Thompson (1984), who by applying thermal modeling techniques used the concept of *P-T-t* path (described either as "clockwise" or "counterclockwise") and distinguished the stages of isobaric heating/cooling and isothermal (de-)compression. The next step in the development of collisional zone metamorphic and tectonic models involved two-dimensional numerical thermomechanical simulations based on the experimentally derived rheological behavior of materials over a wide range of temperatures and pressures (Gerya 2010; Burov et al. 2014 and others). In most recent models, density changes are accounted for by recalculating the equilibrium mineral assemblages for specific *P-T* parameters at every time step. For this purpose, specific thermodynamic algorithms for calculating equilibrium mineral assemblage based on Gibbs' free energy minimization e.g., THERIAK (de Capitani and Petrakakis 2010) or PERPLEX (Connolly 2005) software, are built into the thermomechanical block of the model. Unfortunately, changes in the rheological propoerties of metamorphic rocks cannot be implemented in the same way as density changes, due to the absence of relevant experimental data (Burov et al. 2014).

3.2.3.1 Metamorphism Associated with Overthrusting

Collision between continents leads to significant crustal thickening and produces manifestations of collisional metamorphism. This problem was analyzed using examples of metamorphism on the Yenisei Ridge in Siberia. In this region, collisional metamorphism is typically localized in the vicinity of thrusts in zones of deep faults and results in the development of new mineral assemblages and deformation structures (Likhanov et al. 2006, 2014). We recognized several narrow zones of medium-pressure kyanite-sillimanite metamorphism (2.5–7 km in width), which exhibit a gradual increase in pressure (from 4 to 6.5–8 kbar) in the vicinity of thrusts. A small variation in temperatures suggests a very low metamorphic gradient of 7–14 °C/km (Likhanov et al. 2008a, b). This type of metamorphism is typical of the Teya Complex of the Transangarian Yenisei Ridge.

The study area is located between the Yeruda and Tchirimba Rivers at the eastern margin of the middle part of the Transangarian Yenisei Ridge (Fig. 3.60). A geological sketch map of this area shows two different tectonic units divided by a thrust zone. The Korda lower plate (KLP) is located SW of the Panimba thrust fault, tectonically underlying the Early Proterozoic Penchenga upper plate (PUP) to the NE. The KLP represents regionally metamorphosed low-pressure, andalusite-bearing metapelites of Middle Proterozoic age that were overprinted by Late Proterozoic medium-pressure kyanite-bearing assemblages and regional metamorphic textures (Likhanov et al. 2012). A sheet of medium-pressure rocks consists of an up to 7 km wide zone restricted in the NE by the Panimba thrust fault (Fig. 3.60). From approximately 7–8 km to the southwest of the Panimba thrust fault (Fig. 3.60), the degree of medium-pressure metamorphic overprint on the andalusite-bearing rocks and the intensity of deformation in the metapelites increase toward the fault (Likhanov et al. 2000). Four distinct metamorphic zones in the metapelites of the Korda Formation can be recognized. Rocks that are unaffected by the overprint comprise Zone I. The boundary between Zone I and II coincides with the first appearance of kyanite toward the Panimba thrust fault. All rocks of Zones III and IV are characterised by similar mineral assemblages and differ only in the degree and style of deformation.

The results of geothermobarometry for rocks suggest a progressive pressure increase toward the overthrust from 3.5 to 4.0 kbar in Zone I (metapelites of the andalusite-sillimanite type) through 4.5–5.0 kbar in the outer Zone II (metapelites of the kyanite-sillimanite type) to 5.5–6.0 kbar in the Zone III and 6.2–6.7 kbar in the inner Zone IV with a small increase in temperature (from 540 to 600 °C). *P-T* path calculations for all metapelites from the Korda Formation document an increase in pressure up to 2.2 kbar from west to east associated with only minor heating (20 ± 15 °C) (Likhanov et al. 2001).

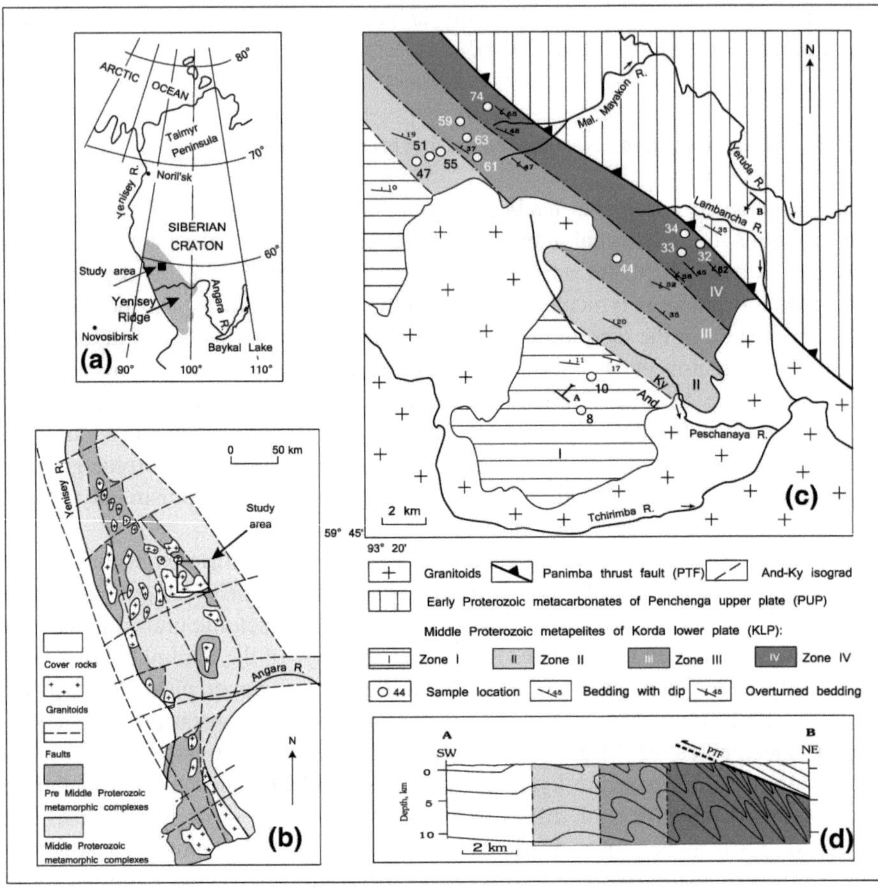

Fig. 3.60 **a** Location of the study area (the shaded area is the Yenisei Ridge); **b** general geological situation of the Yenisei Ridge; **c** geological sketch map of metamorphism at the eastern margin of the middle part of the Transangarian region in the vicinity of Panimba thrust (interfluve between Eruda and Chirimba Rivers, Teya metamorphic complex) and **d** schematic cross section across the Panimba thrust showing relative placement of metamorphic zones. **a–b** in Fig. 3.69c shows the location of cross section

Geochronological data coupled with *P-T* path calculations derived from garnet zoning patterns records two superimposed metamorphic events. The first stage occurred as a result of the Grenville-age orogeny during late Meso-early Neoproterozoic (~970 Ma) and was marked by low-pressure zoned metamorphism at *c.* 3.3–4.0 kbar and 540–560 °C with a metamorphic field gradient of dT/dH = 25–35 °C/km. At the second stage, the rocks experienced middle Neoproterozoic (~850 Ma) collision-related medium-pressure metamorphism at *c.* 4.5–6.7 kbar and 570–600 °C with dT/dH < 10 °C/km as a result of which low-pressure metamorphic effects were overprinted by medium-pressure regional metamorphic mineral assemblages and textures. In all metapelites from the Teya Complex, the available data document an increase in pressure up to 2.2 kbar from west to east associated with only minor heating (Likhanov et al. 2001), which is indicative of near-isothermal subsidence of rocks along a low metamorphic field gradient.

A number of specific features (an abrupt increase in Grs and a concomitant decrease in Sps, and minor variations in X_{Fe}, Prp, and Alm in garnet close to the boundary with the middle zone, the large volume and small entropy effects of the calculated reactions) and extremely low metamorphic gradient are typical of collision-related metamorphism during overthrusting of continental blocks and suggest a near-isothermal loading in accordance with the transient emplacement of thrust sheets. Similar garnet zoning patterns and *P-T* paths were derived from rocks in overthrust zones from mobile belts of the Scandinavian Caledonides, the Western Alps, the Appalachians, the Cordilleras and the Himalayas. These results are also consistent with metamorphic recrystallization that occurred during thrust emplacement of the tectonically overlying plate. Detailed mass-transfer analysis showed that the large volume and small entropy effects characterize the calculated reactions. The detailed mass-transfer analysis revealed that the prograde evolution of chemical and modal compositions of minerals during collisional metamorphism was controlled by the gradual increase of pressure (at nearly constant temperature) accompanied by deformation within the bulk composition of the protolith (Likhanov and Reverdatto 2002).

In this processes, the low geothermal gradients were associated with relatively short-lived events and the lack of thermal equilibrium between the blocks of rock at the respective depths because of a strong thermal inertia relative to pressure. Based on the kinetics for the reversible reactions, this would require enhanced exhumation rates under fluid-absent conditions during the early stages of rapid exhumation (Sklyarov 2006), which can explain the preservation of medium-pressure prograde assemblages in regions with thrusting tectonics. *P-T* paths calculated from metamorphic rocks of the Teya Complex agree well with the *P-T* conditions of the metamorphic rocks from other collisional orogens of the world where the prograde And → Ky transition was explained by rapid crustal thickening with subsequent rapid exhumation and erosion (Fig. 3.61).

In general, most rocks that witnessed a significant pressure increase with only minor heating during the later tectonometamorphic stage are characterized by a low *P-T* gradient of <12 °C/km. Although diverse tectonic mechanisms have been proposed to explain near–isothermal pressure increase for crustal thickening and extension, the major tectonic stress is correlated to isostatically compensated loads and from plate boundary forces (Likhanov et al. 2018 and references therein). We briefly address below some of the salient models that explain the pressure increase in the study area.

(1) The initially horizontal or sub-horizontal low-pressure metamorphic rock sequences were folded and preserved either as a fold limb or as a monocline. So, in the eastern part of the structure, these rocks would represent either the core of a fold or the most deep-seated part of a monocline. However, geothermobarometry

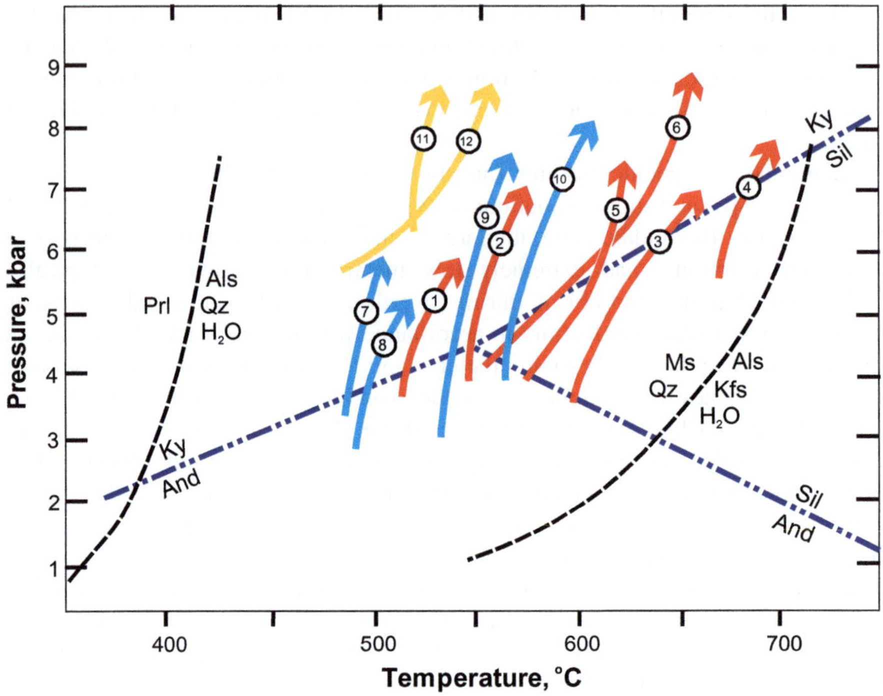

Fig. 3.61 *P-T* diagram summarizing the results of *P-T* path calculations for evolution of collision metamorphism in overthrust terranes. Arabic numerals in arrow body of *P-T* paths correspond to study areas: (red color) 1—Angara area; 2—Mayakon area; 3—Chapa area; 4—Teya area; 5—Polkan area, and 6—Garevka area, Yenisei Ridge (Likhanov and Reverdatto 2011); (blue color) 7—Bellows Falls area (Spear et al. 2002); 8—Mascoma-Orfordville area (Kohn et al. 1992); 9—Nason terrain and Southern British Columbia area (Whitney et al. 1999); 10—Piedmont Plateau area (Crawford and Mark 1982). The calculated prograde segment of modeling *P-T* trajectory (yellow arrows) using two-dimensional thermal-mechanical model during continental collision consistently show increases in pressure with little or no heating [11—(Jamieson et al. 2002]; 12—(Huerta et al. 1999). The coordinates of aluminum silicate triple point and univariant equilibrium curves correspond to those in Fig. 2.3

indicates a pressure difference of 1.5–2.5 kbar over a horizontal distance of 7 km at nearly constant temperature. In this model, the temperature would also have to increase toward the thrust zone (i.e. from west to east) by at least 70–140 °C, if a thermal gradient of 10–20 °C/km is assumed (e.g. Ruppel and Hodges 1994).

(2) The revealed increase in pressure in the vicinity of the thrust could have been a result of separate shifting of crustal slices which were sequentially pushed up from different depth levels during thrusting process (e.g. Beaumont et al. 2001). However, this model of "strike-slip structural pattern" lacks the support of geological evidence for the presence of tectonic contacts and high lateral temperature gradients between adjacent metamorphic zones.

(3) The pressure increase could be related to plate boundary forces such as ridge push (e.g. Kusznir and Park 1984). Petrini and Podladchikov (2000) investigated the pressure distribution with depth in regions undergoing horizontal shortening and experiencing crustal thickening. Their two-dimensional modeling results show that, in compressive tectonic settings, the pressure values range from one to two times the lithostatic values. However, in most cases, the magnitude of non-hydrostatic corrections to pressure estimates is smaller than the uncertainty of geobarometers. Laboratory experiments on rheology of the rocks show that the average stress level at a depth of 15–20 km within a continental plate arising from such sources is likely to lie within the range 0.2–0.3 kbar, if distributed over the whole lithospheric plate at a geologically realistic strain rate of 10^{-17}–10^{-14} s^{-1} (Strehlau and Meissner 1987). Such stress is not sufficient to achieve overpressures of as much as 1.5–2 kbar. Despite this experimental constrains, recent evaluation of pressure deviation from lithostatic, i.e. tectonic overpressure due to deviatoric stress and deformation, in lithospheric tectonics and metamorphism (Gerya 2015 and references therein) suggests that non-lithostatic phenomena are common on different time and space scales. These are considered to be responsible for some of the processes during plate convergence. Magnitudes of overpressure are strongly variable and may potentially reach up to 100% of the lithostatic pressure (Pleuger and Podladchikov 2014), depending mainly on the rheology of deforming rocks and on the nature of related tectonic processes. Rheological heterogeneity of deforming rock units has a tendency to enhance overpressure (e.g., Mancktelow 2008; Schmalholz et al. 2014). Large overpressure can typically be expected in rheologically strong (dry) bending rock units, in particular in the mantle lithosphere (e.g. Burov and Yamato 2008; Faccenda et al. 2009). However, rheological weakness of rocks and small local deviatoric stresses do not necessary endorse the absence of large overpressures in these rocks (Vrijmoed et al. 2009; Schmalholz and Podladchikov 2013; Schmalholz et al. 2014; Li et al. 2010). Therefore, the influence of significant tectonic overpressure cannot be excluded for any metamorphic complex a priori but should be instead tested by exploring realistic thermomechanical models for tectono-metamorphic scenarios. We therefore assume that such models, which involve dynamic stresses deviating from lithostatic pressure, could be a possible alternative to account for the pressure variations in rocks within the shear zones (Likhanov et al. 2018).

(4) The increase in lithostatic pressure could be due to loading by overlying magma (e.g. Brown and Walker 1993). The magma loading model is based in part on the observation that pressures in metapelitic rocks decrease away from granitoid plutons. As noted above, however, comparable intermediate pressures are recorded in rocks irrespective of their proximity to plutons. Furthermore, magmatism plays a particularly sensitive role in models of crustal thickening, because of the higher heat content of magma in comparison with the base of a thrust sheet, and the significant thermal contribution of the heat of crystallization of magma (typical values for latent heat are orders of magnitude larger than heat production due to radioactive decay). In addition, depending on intrusion rates and volume, advective heating by magmatism is one to several orders of magnitude faster than advective heating by

thrusting (e.g., compare rates in Ruppel and Hodges (1994) to those in Paterson and Tobisch (1992).

Since these models are inconsistent with petrological evidence, it remains to be discussed whether the observed *P-T* paths are consistent with a model of thrust emplacement (England and Thompson 1984), and, if so, what they reveal about the tectonic evolution of the study region. All *P-T* paths observed suggest nearly isothermal loading, in accordance with the emplacement of a thrust sheet, such that the change in pressure corresponds to a change in depth of a given sample. Our preferred model for the tectonic evolution of the rocks is presented in Fig. 3.62, in an attempt to integrate the results of this study into a coherent model. Prior to thrusting, the pressures recorded in the low-pressure, andalusite-bearing rocks suggest a depth of approximately 15–17 km. During thrusting, these rocks experienced a pressure increase of 1.5–2 kbar, equivalent to an increase in burial depths of 5–7 km. The post-thrusting pressures (4.5–6.7 kbar) imply the depth of rocks was on the order of approximately 20–24 km, which is a measure of the total tectonic overburden. This could be explained by tectonic crustal thickening due to the southwestward thrusting of the Penchenga upper plate (PUP) onto the Korda lower plate (KLP). In addition, it is clear that about 15 km of the upper part of the KLP directly beneath the overthrust must have been tectonically removed in order to juxtapose at the present erosion surface, as presently observed, Zone IV rocks with the Panimba thrust (Fig. 3.60).

The major difficulty with such models is the near-isothermal nature of the loading process. One of the possible mechanisms is the rapid thrusting with subsequent fast uplift and erosion in collisional orogens.

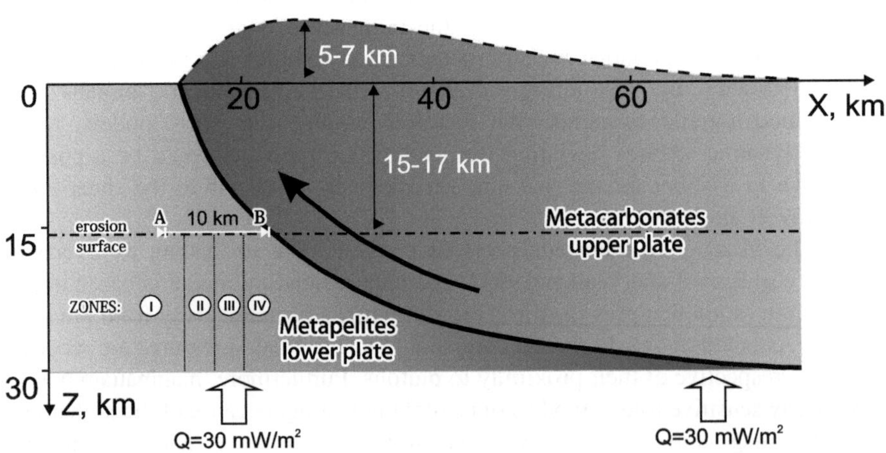

Fig. 3.62 Model of crustal thickening in the vicinity of the Panimba overthrust in the Teya complex. Thrust fault plane is shown by heavy black line; direction of motion is shown by arrow. Mantle heat flow (*Q*) is assumed to remain constant. White line A-B corresponds to the position of the cross section through A-B line, zones corresponds to the metamorphic zones in Fig. 3.69

According to this idea, the subducted/underthrust slab fails to equilibrate thermally before the rocks are brought back near the surface. Thermal-mechanical modelling provide constraints on likely mechanism of loading because *P-T* paths inferred from thermobarometric results can be compared with paths predicted by models calculated from heat transfer equations for crustal thickening. Early models for the thermal effects of thrust emplacement in the vicinity of major thrust zones (e.g. England and Thompson 1984) relied on assumption of instantaneous thrust emplacement and one-dimensional heat conduction. Later two-dimensional modeling of Ruppel and Hodges (1994) incorporated the effects of radiogenic heat and isostasy, and these authors examined the effects of transient thrusting with variable thrust rates, thrust dips, and exhumation rates on *P-T* paths. Jamieson et al. (1998) used thermal models to examine the effects of tectonic redistribution of crust rich in heat-producing elements on the thermal and deformational history of small convergent orogens. Beaumont et al. (2001) and Jamieson et al. (2002) investigated the thermal and mechanical evolution, and resulting metamorphic response of large convergent orogens that are driven by subduction of lithosphere. Most of the burial *P-T* paths produced by these models are very steep (Fig. 1.19), a style observed in many overthrust terranes. Shi and Wang (1987), Karabinos and Ketchman (1988), and Spear et al. (1989) have presented two-dimensional thermal models of thrusting that demonstrate that rocks of the lower plate of a thrust system cannot undergo isothermal loading during thrusting unless the rate of thrusting is unreasonably fast (for example, on the order of meters per year). Spear et al. (1989) did demonstrate, however, that isothermal loading could occur in the middle plate of a multi-plate system if both bounding thrusts move simultaneously. In such a middle plate, heating from above is balanced by cooling from below, so that the temperature remains relatively constant. For this model to apply to the Transangarian region, however, would require that the rocks beneath the KLP be allochthonous and underlain by another thrust, in contradiction to the geological situation in the study area.

The two-dimensional nature of above mentioned models requires that not only the vertical extent, but also the horizontal tectonic motion, of thrust faulting lie within physically realistic limits (Ruppel and Hodges 1994). The use of such models for our region is limited by the lack of the necessary input parameters, such as thrusting and erosion rates, thrust geometry, to calculate model. Therefore, we propose an alternative interpretation of the observed tectono-metamorphic evolution of the study area. The simplified model used here makes no attempt at reproducing these effects of horizontal fault geometry, as in real settings thrust faults do not have constant dip and often propagate stepwise. Unlike in previous models of continental crustal thickened by thrusting (Peacock 1989; Huerta et al. 1999), we suggest that the two plates (KLP and PUP) on both sides of the thrust have different thermal conductivities and heat generation properties, in order to account for the lack of significant heating of the underthrust rocks. Previous models imply that, after fast underthrusting of slabs with similar steady-state geotherms, the lower-plate rocks should be heated by 150 °C (Ruppel and Hodges 1994). However, the significant difference in the type of rock on both sides of the Panimba thrust suggests that differences in radioactive heat sources, thermal conductivity coefficients and, therefore, steady-state geotherms must be accounted for.

In order to demonstrate the existence of an overthrusting mechanism, the problem of conductive heating when moving the hanging wall (PUP) relative to the footwall (KLP) along the steeply dipping thrust fault needs to be examined (Fig. 3.62). The one dimensional lateral heating model is used to evaluate the temperature distribution along the horizontal section A-B. Holding the overthrusting velocity constant along the fault leads to temperature increasing at point B, which depends on temperature gradient and exhumation rate in the upper plate. Prediction of the uplift velocity under which nearly isothermal loading condition is obeyed would be useful.

The temperature evolution is described by the one-dimensional heat transfer equation in the horizontal direction:

$$\frac{dT}{dt} = \kappa \frac{\partial^2 T}{\partial x^2}, \tag{3.14}$$

where t is the time, T is the temperature, and κ is the thermal diffusivity.

Boundary conditions of zero horizontal heat flow at the left boundary (point x = A) and linearly increasing temperature in time at the right boundary (point x = B) were imposed:

$$\frac{\partial T(\mathrm{A}, t)}{\partial x} = 0; \ T(\mathrm{B}, t) = T_0 + V\frac{\partial T}{\partial z}t, \tag{3.15}$$

where z is the depth and V is the uplift velocity.

Initial condition is uniform temperature T_0 = 500 °C at the depth of 15 km (the present erosion surface):

$$T(x, 0) = T_0 \tag{3.16}$$

In dimensionless form, the solution of Eq. (3.14) with boundary (3.15) and initial condition (3.16) may be expressed as (Carslaw and Jaeger 1959):

$$\theta = \left\{1 + erf\left(\frac{x'}{2\sqrt{\tau}}\right)\right\} + Pe \cdot a \cdot \tau \left\{\left(1 + \frac{x'^2}{2\tau}\right) erf^*\left(\frac{x'}{2\sqrt{\tau}}\right) - \frac{x'}{\sqrt{\pi\tau}} \exp\left(-\frac{x'^2}{4\tau}\right)\right\}, \tag{3.17}$$

where $Pe = VL/\kappa$ is the Peclet number, L is the characteristic length scale, $a = \partial T/\partial z \sin\alpha$ is the reduced dimensionless geothermal gradient, α is the dip angle of fault at point B, τ is the dimensionless time, $x' = x/L$, $\theta = T/T_0$, $\tau = L^2/\kappa t$, $erf^*(x) = 1 - erf(x)$—complementary error function.

Since maximum temperature changes during metamorphism were small, and increased toward the thrust on average by 30 °C, dimensionless thrust velocity corresponds to Pe = 0.09. Knowing the Peclet number means it is possible to estimate the overthrusting velocity by using the total width of metamorphic zones of 10 km for the characteristic length scale L, 500 °C for temperature scale, reduced dimensionless geothermal gradient of 0.7 $\approx$ sin45° for a, and 10^{-6} m^2/s for thermal diffusivity κ. Under these conditions, the thrust velocity must be less then $V \leq Pe\kappa/L = 0.09 \times 10^{-10}$ (m/s) $\approx$ 300 m/Myr. This value is less than independently constrained estimates of uplifting and accretion rates in the orogenic collision events. For example, within the Himalayan thrust belt the rates of erosion and accretion derived from the evolution of metamorphic parameters, apatite fission-track data and $^{40}Ar/^{39}Ar$ analysis of detrital K-feldspar and muscovite, range from 0.5 to 1.8 ± 0.7 km/Myr (Huerta et al. 1999; Copeland and Harrison 1990). The exhumation of ultrahigh pressure metamorphic rocks related to continent-continent collision occurred at the rate of over 13 km/Myr (Coleman and Wang 1995). The thrusting velocity determined here of ~300 m/Myr compares well with that obtained using *P-T-t* path reconstruction and numerical models in the overthrust terranes (England and Thompson 1984; Peacock 1989; Spear et al. 1991).

In order to estimate the possible temperature increase in the lower plate, the thermal evolution of two juxtaposed plates with different radioactive heat sources and thermal conductivities was considered. It is assumed that the evolution of the geotherm is controlled by the initial distribution of radioactive heat-generating elements in the overthrust and underthrust plates and by the heat conduction through the crust from the mantle. The instantaneous juxtaposition of a hot thrust nappe on a cold footwall leads to initiation of the "sawtooth" geotherm and progressive heating of lower plate (Spear 1993).

Analytical and numerical solutions to the differential equations concerning thermal conduction with radiogenic heat production have been used to substantiate the thrust model. One-dimensional, conductive heat transfer equation in a horizontal infinite plate H km thick can be represented by:

$$\frac{dT}{dt} = \frac{\lambda}{\rho C_p}\frac{\partial^2 T}{\partial z^2} + \frac{A}{\rho C_p}, \tag{3.18}$$

where t (s) is the time, T (°C) is the temperature, λ (W/m K) is the thermal conductivity, z (m) is the depth, ρ (kg/m^3) is the density, Cp (kJ/kg) is the isobaric specific heat, and A (W/m^3) represents radiogenic heat sources.

It is appropriate to assume that the heat production due to radioactive elements decreases exponentially with depth (Turcotte and Schubert 1982):

$$A = A_0 \exp(-z/D_r). \tag{3.19}$$

The upper boundary is maintained at T_{surf} = 0 °C, and the lower boundary condition is a specified basal heat flux, Q = 30 mW/m^2. Integration of Eq. (3.18) with appropriate boundary conditions gives the steady-state geotherm (Spear 1993):

$$T = T_{surf} + \frac{Q\,z}{\lambda} + \frac{A_0 D_r^2}{\lambda}\left(1 - e^{-z/D_r}\right), \tag{3.20}$$

where A_0 is the surface radiogenic heat production and $Dr = 10^4$ m is the depth constant.

As indicated above, the average thermal gradient in the rocks during peak metamorphism is lower than the gradients in many Precambrian terranes. However, there is an increasing awareness that the distribution of radiogenic heat production can influence the thermal structure of the crust (e.g. Huerta et al. 1999). Using the concentrations of heat-producing elements in the KLP and PUP rocks (Table 3.7), heat generation in these rocks can be calculated on the basis of the Ljubimova's model (1968). In the case of five KLP metapelites, the value averages 2.04 μW/m^3 and for the PUP metacarbonates it is an order of magnitude lower, with A_0 = 0.375 μW/m^3 (Table 3.7). The calculated geotherms for the PUP and KLP before and after thrusting are shown in Fig. 3.63. The curves in Fig. 3.63a are steady-state geotherms for a model crust with maximum values for the observed range of radioactive heat sources in the KLP and PUP. Thermal conductivities of metapelites and metacarbonates are assumed to be 1.5 and 2.5 W/m K, respectively, based on literature data for thermal conductivities of limestone and shale (Clark 1966). The thrusting resulted in the KLP rocks being buried in depths of 20–24 km, as shown in Fig. 3.63b by the relative positions of the grey rectangles. The re-equilibrated steady-state geotherms after instantaneous thrusting of the 5–7 km-thick PUP are shown in Fig. 3.63b. These curves show a range of temperatures from a minimum range (curve 4) to maximum heat production (curve 6) for KLP and PUP rocks. The modelled results show that heat production in the KLP is the most important parameter. The steady-state geotherm (curve 5) constructed using the mean values of heat production for PUP and KLP rocks also passes through the *P-T* estimates of the medium-pressure rocks of zones II-IV in the vicinity of the thrust. It is evident from Fig. 3.63 that there is only a small temperature increase if when comparing the position of the maximum steady-state geotherms (curves 2 and 6) relative to the grey rectangle. This small increase is in good agreement with the *P-T* paths shown in Fig. 3.61, and lends support to the validity of the proposed model.

The transient one-dimensional heat conduction in two-layered crust can be simulated using the finite difference method. Time integration of the heat flow equation was carried out using a second order accurate implicit Crank-Nicolson scheme. Figure 3.64 shows the calculated geotherms at four different stages of evolution and the equilibrium steady-state geotherm in the case of no erosion. It is assumed that the proportions and distribution of radioactive heat sources remain unchanged during thrusting in both of the plates and are described by Eq. (3.19), and the boundary conditions are the same. The maximum temperature increase in the KLP metapelites is only observed near the interface boundary between plates (Fig. 3.64). After thrusting, the lower plate is significantly heated up to 90 °C in the depth range from 7 to 20 km. Below this depth, the difference between initial and

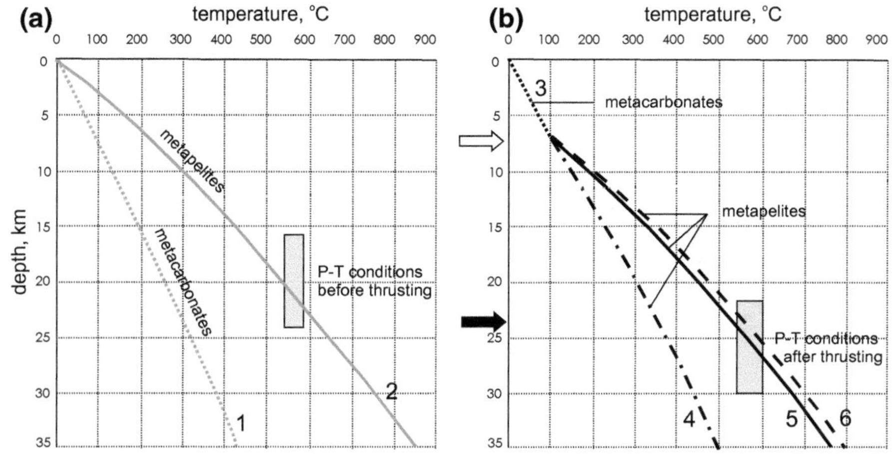

Fig. 3.63 Calculated steady-state geotherms for the upper plate metacarbonates (1, 3) and lower plate metapelites (2, 4, 5, 6) before (**a**) and after (**b**) thrusting. The grey rectangle indicates the depth-temperature coordinates determined for the medium-pressure rocks of the II–IV zones. White arrow shows upper plate/lower plate boundary, black arrow indicates assumed present erosion surface. Thermal conductivities 1.5 W/m K in metapelites and 2.5 W/m K in metacarbonates were used to calculate steady-state geotherms (curves 1–6). The following heat production parameters were used for metapelites: 2.4 μW/m^3 (curves 1–3), 1.86 μW/m^3 (4), 2.03 μW/m^3 (5) and 2.38 μW/m^3 (6); for metacarbonates: 0.439 μW/m^3 (1–3, 6), 0.311 μW/m^3 (4), and 0.375 μW/m^3 for curve 5

Table 3.7 Concentrations of heat-producing elements and heat production (*H*) in metapelites of the lower plate and metacarbonates of the upper plate

	U, ppm	Th, ppm	K, %	*H*, W/kg	*H*, W/m^3
Metapelites					
51	2.0	16.1	2.70	7.2×10^{-10}	1.93×10^{-6}
59	2.1	14.7	2.42	6.8×10^{-10}	1.83×10^{-6}
63	2.4	14.7	2.00	6.9×10^{-10}	1.86×10^{-6}
34	3.1	17.0	1.65	8.1×10^{-10}	2.17×10^{-6}
74	3.3	18.6	2.08	8.8×10^{-10}	2.38×10^{-6}
Mean H					2.03×10^{-6}
Metacarbonates					
1	1.4	0.8	0.3	1.6×10^{-10}	4.39×10^{-7}
2	0.4	2.1	0.6	1.2×10^{-10}	3.11×10^{-7}
Mean H					3.75×10^{-7}

final, steady-state temperature at a depth of over 20 km is less than 10 °C. Such a small increase is due to low radioactive heat generation in the PUP as compared with the KLP. In the case of erosion of the upper plate, an increase in temperature will be smaller because of heat removal due to decreasing plate thickness.

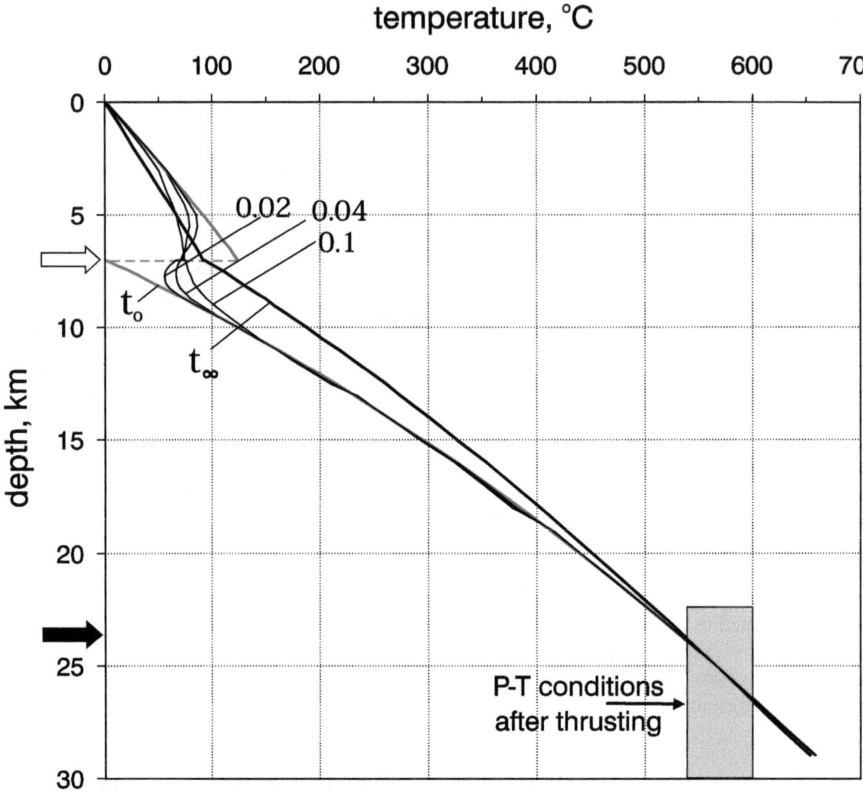

Fig. 3.64 Temperature-depth plot showing thermal evolution of thickened crust in the overthrust model. Curve labeled t_o represents initial (saw-tooth) geotherm characteristic of instantaneous overthrusting. Curves labeled with time in millions of years after thrusting. Curve labeled t_∞ represents steady-state geotherm approached after tens of millions of years. Other symbols are the same as in Fig. 3.63

The proposed model for tectono-metamorphic evolution of the study area, calculated within a framework of crustal thickening by southwestward thrusting, explains a number of features (e.g. the origin of kyanite after andalusite, common increase in garnet grossular content from core to rim, the gradual change in recorded pressure between low- and medium-pressure rocks, small temperature increase, etc.) associated with this tectonic phenomenon and confirms the possibility of nearly isothermal loading during overthrusting. A minor temperature increase (up to 20 ± 15 °C) of the upper part of the overthrust plate is explained by specific behaviour of steady-state geotherms computed using the low radioactive heat production of metacarbonates as compared with the metapelites. The suggested thermal-mechanical model corresponds well with *P-T* paths inferred from obtained thermobarometric data and correlates satisfactorily with *P-T* trajectories predicted by two-dimensional thermal models calculated for different crustal thickening and exhumation histories (e.g. Jamieson et al. 2002; Beaumont et al. 2001). This model can operate if the rock blocks on each side of the fault differ significantly in their thermo-mechanical and heat-producing properties, which is also at variance with the geological history in the present case. This model can be used for interpretation of metamorphic evolution in overthrust terranes with collisional tectonics.

3.2.3.2 Metamorphism Associated with Underthrusting and Subduction

The variety of types of metamorphism in convergent settings is dependent on the different nature of the tectonic interaction processes (Sklyarov et al. 2001). Different types of interactions in a convergent setting may arise between an island arc and a Japanese subduction zone, an Andean-type continental margin and a Chilean-type subduction zone of shallow dip, Himalayan collision zone, etc.

Several major factors that control the processes in subduction zones are as follows (Dobretsov 2000): (1) the thermal structure of the subducted oceanic plate, which depends on the age of the rocks, (2) subduction rate, (3) plate dip and the curvature of the subduction zone, (4) viscous flow in an accretionary wedge and at the interface between the oceanic and continental plates; (5) heat generated in the friction zone, (6) induced convection in the mantle wedge above the subduction zone; and (7) distribution of radioactive elements in the subducting slab and in the mantle wedge. In accordance with this, Khain and Lomize (1995) distinguish several types of subduction zones, each of which is characterized by the inherent distribution of earthquake foci along the seismic focal zone, the specific endogenous processes and geodynamics.

The Mariana (West-Pacific)-type subduction zone is characterized by a relatively low rate of convergence, an older, cold (and, correspondingly, heavier) subducting lithosphere, the motion of both slabs in one or more directions (but at different velocities), the steep dip of the Benioff zones, the predominance of extensional conditions at the margins of both interacting slabs, the active development of back-arc basins with high heat flow and, sometimes, basalt-rhyolite bimodal volcanism. Under extensional conditions, the subducting oceanic slab often bends to form grabens, acting as traps for oceanic sediments erosion products from continental slopes. This is one of the possible explanations for the absence of lateral accretion features at the front of active margins in the western Pacific region, where accretionary prisms are absent or poorly developed. Such grabens can be traced on seismic profiles across the Japan arc.

The Sunda- and Japanese-type subduction can be accounted for by identical mechanisms of formation and is identified at the periphery of the Pacific Ocean in an ocean-continent subduction setting. The Sunda type is characterized by the thinning of the continental crust due to the absence of significant compressive stresses; the Japanese type is characterized by the opening of marginal seas with a newly formed oceanic crust.

The Chilean (Andean)-type subduction is marked by high convergence rates, a younger and warmer subducting slab, the opposite direction of slab motion, and the rapid overthrusting of a gently-dipping plunging slab by an overriding continental slab. The dominance of compressive regimes results in the absence of active back-arc basins. At a rapid convergence rate, the magma beneath the arc has no enough time to be extensively contaminated by crustal material, so that granite-granodiorite intrusions and latite-andesitic volcanic suites predominate. This type of subduction is not unique to the Andes, it can also be observed in the west of Myanmar and in the western half of the Sunda arc (Sumatra, Java), in the Alpine zones of the Middle East and Europe.

The Cordilleran-type subduction is characterized by a relatively moderate rate of convergence, a younger and warmer subducting plate, an oblique convergence angle for both slabs. Unlike the other types, this type is characterized by a relatively immature overriding continental slab, which was dominated, before the onset of subduction, either by a passive margin or arc–continent collision setting. IN a compressive regime, partial melting in the mantle wedge occurs above the Benioff zone, which does not lead to the significant crustal enrichment of lithophile elements because of a small thickness and immaturity of the crust. Voluminous granite-granodiorite batholiths predominate in this zone. In the later stages, the continental slab usually overrides the subducting oceanic slab along its spreading axis, which is followed by a specific magmatism and extension along continental margins. The examples are the settings dominated in the west of North America during the Late Mesozoic-Cenozoic and in the Indochina block at the late Permian-Early Triassic.

The Malayan type subduction is typified by the relatively slow subduction, a fixed position of the continental margin, the moderate dip of the Benioff zone (no more than 45°), no significant difference in slab age, repeated recycling of crustal material and the formation of palingenic granites. The examples include the Mesozoic Burma-Malayan, Cathaysian, Mongol-Okhotsk and Okhotsk-Chukotka belts, and some segments in the back-arc parts of the modern Andean and Cordilleran belts of America.

Depending on the contribution of the above factors controlling the processes in subduction zones, Cloos (1993) proposed four main scenarios describing temperature evolution in the metamorphic processes, which can be reflected in:

(1) a geotherm corresponding to ultrahigh temperatures at the front of the upwelling magma directly beneath the island-arc;
(2) an average continental geotherm corresponding to a "hot" subduction setting beneath Andean-type continental margins;
(3) a warm subduction geotherm corresponding to a slow subduction at an average velocity of 3–5 cm/year and/or subduction of younger and warmer plates in incipient subduction zones;
(4) a cold subduction geotherm, characteristic of long-tern rapid subduction (>10 cm/year) of older and cooler oceanic lithosphere.

The realistic metamorphic *P-T* estimates obtained for the eclogite-glaucophane belts, which were formed as a result of metamorphic transformation of the subducted plates, correspond to scenarios (3) and (4). In this case, the curve of the "cold" subduction and (4) does not fall below the thermal gradient of 5 °C/km. This suggests that even in the case of subduction of the older and colder oceanic slab it retains much of its original heat to form eclogite and glaucophane schists belonging to a high-pressure type of metamorphism.

High-pressure glaucophane schist (T = 250–500 °C and P = 6–20 kbar) and eclogite-glaucophane schist (T $\geq$ 500 °C and P $\geq$ 12 kbar) complexes are among the most important indicators of the evolution of subduction-accretionary settings at convergent plate boundaries (Dobretsov 1974, 2000; Ernst 1988; Peacock 1993). As the subduction zone evolves, these complexes can be replaced by collision- subduction complexes (Platt 1986). Subsequent accretion and collision result in the formation of an orogenic "collage", containing fragments of island arcs, collisional complexes, and oceanic plateaus that were successively accreted to continental margins. The main elements of accretionary-collision complexes are ophiolitic nappes and lenses, associated eclogites and glaucophane schists, olistostrome and mélange complexes. They typically form spatially associated rock complexes: eclogitess and glaucophane schists occur at the base of ophiolitic nappes as isolated sheets or blocks in the mélange zone, while rocks in the olistostrome unit can be metamorphosed under glaucophane schist facies conditions (Dobretsov 1974, 1981).

Therefore, contrasting types of metamorphism (Sklyarov et al. 2001) are manifested in the system subduction-suprasubduction zone. High-pressure metamorphism (eclogite-glaucophane schist complexes) recorded in the subducting oceanic plate is due to rapid subduction of a relatively cold plate to great depths. The inverted metamorphic zoned complexes may develop at the contact of the subducting plate and directly at the hanging wall contact. This unusual inverted zoning pattern is formed during cooling of a hot hanging wall by a cold oceanic plate. And, finally, a distinctive feature of the suprasubduction zone is heating by injected magmas during melting in the subducting plate. The upward displacement of isotherms results in the formation of low-pressure zonal metamorphic complexes (of andalusite-sillimanite type). Within the continental arcs (Andean type active margins), the emplacement of large volumes of magmas of different composition at depths of 10–15 km results in the formation of specific granulites at shallow levels (P = 4–6 kbar at $\sim$800 °C).

High-pressure metamorphic rocks as an indicator of subduction processes are characterized by a wide range of *P-T*-conditions, a specific evolution of *P-T* parameters and composition (Zhang et al. 2015). Knowledge of the composition of the protolith of high-pressure metamorphic complexes is crucial for understanding tectonic processes. For example, the protoliths for glaucophane schists from continental collisional orogens (type A according to Maruyama et al. (1996)) are rocks of passive margins, e.g., platform-type sedimentary carbonates, bimodal volcanics, high-aluminous sediments. These metamorphic complexes were formed in a collisional setting from continental crustal rocks and contain high-pressure and ultrahigh-pressure assemblages. On the other hand, glaucophane schists from active continental margins were formed by subduction of paleoceanic plates and have an oceanic protolith (type B according to Maruyama et al. (1996)) represented by metamorphosed siliceous sediments, mid-ocean ridge and ocean island basalts, reef limestones, and greywackes.

Metamorphic *P-T-t* paths of rocks play a key role in elucidating the tectonic regime, responsible for the formation and exhumation of subduction-related high-pressure metamorphic complexes (Ernst 1988). In this case, the resultant *P-T-t* paths show two different scenarios for the metamorphic history, depending on different convergent tectonic settings. These are an Alpine-type subduction/collision setting and an East Pacific-type intra-oceanic subduction setting (Ibid). The most important differences in the inferred metamorphic history are associated with retrograde *P-T-t* trajectories: the Alpine-type subduction is characterized by near-isothermal decompression, whereas the East Pacific-type shows a simultaneous decrease in temperature and pressure (Ibid).

Another typical example would be different metamorphic paths reconstructed for three blocks of high-pressure eclogite and glaucophane schist facies rocks in a serpentine mélange of the Rio San Juan Complex, Dominican Republic (Krebs et al. 2008). These blocks record different stages of the subduction zone evolution: a high-temperature counterclockwise (CCW) *P-T-t* path is interpreted to be related to incipient subduction, while a low-temperature clockwise (CW) *P-T-t* path responds to a more mature stage (Ibid). The rocks that have been buried and exhumed within a younger and warmer subduction zone generally follow counterclockwise *P-T-t* paths (Lazaro et al. 2009).

Eclogites that formed in subduction zones differ considerably from those formed in a collisional setting. Subduction-related eclogites are found in close spatial association with glaucophane schists and formed at relatively low temperatures of 450–650 °C. Unlike collision-related eclogites, they may contain lawsonite and glaucophane, more Fe-rich garnet, amphibole represented mainly by glaucophane, omphacite with lower jadeite component, etc. Typical examples include the Franciscan complex of California, the Sambagawa belt of Japan, and the Maksyutov Complex of the Southern Urals.

Eclogites formed at collision zones are widely found in gneiss complexes of mountain belts in many parts of the world (western Norway, Front Range of the Greater Caucasus, the Polar Urals, and western White Sea). These eclogites are crustal metamorphic rocks which were formed at moderate temperatures and over a wide range of pressures. Garnets are typically compositionally zoned and contain mineral inclusions inherited from the earlier paragenesis.

The formation of high- and ultrahigh-pressure (HP/UHP) metamorphic rocks can be explained only by deep subduction of oceanic and continental crust. This problem of primary geodynamic and petrological interest has been extensively studied in a number of monographs and review papers (Dobretsov et al. 1989; Agard et al. 2009; Guillot et al. 2009 and others). Good examples of such metamorphism have been reported from convergent regions: the UHP Western Gneiss Region of Norway, Kokchetav Massif of Kazakhstan, Dabie Shan in China, Dora

Maira Massif of the Western Alps, Bohemian Massif of Central Europe, etc. More than 60 worldwide occurrences of HP and UHP massifs have been described in Guillot et al. (2009) and their number tends to increase. HP to UHP metamorphic rocks of continental or oceanic origin occur in convergent zones, while the reconstructed *P-T-t* paths suggest their subduction and subsequent return to the earth's surface (exhumation). Proposed mechanisms for exhumation include: (1) subduction channel flow (Cloos 1993), (2) return flow in an accretionary wedge (Dobretsov 2000), (3) compression and extrusion of metamorphic rocks from a soft zone between two rigid blocks (Thompson et al. 1997), and other numerous modifications thereof.

Based on the thermal gradient typical for subduction settings, the rocks in subduction zones fall into the lowest gradient (<10 °C/km) region IV (Fig. 3.1). Geodynamically, the continental subduction was generally preceded by the oceanic subduction, which has its own characteristics such as: (1) relatively fast convergence rates (5–15 cm/year), (2) negative buoyancy, (3) high strength of the plate, (4) plastic flow in the plate bending zone, (5) hydration of the mantle wedge, (6) serpentinization at the crust-mantle boundary, (7) shear heating (Burov et al. 2014). Many of these factors are absent in the case of continental subduction, which is retarded by the positive buoyancy of continental crust, slow convergence rates (up to a few millimeters per year), and the absence of a serpentinite lubrication zone at the subduction interface. Continental subduction is favored by several factors such as metamorphic reactions leading to mechanical softening of the subduction interface, the UHP eclogitization of the basalt part of the crust, leading to negative buoyancy of the plate with a bulk effect of the reaction of up to 15–17%, and a low yield stress of the middle and lower crust.

The isochemical phase diagrams (pseudosections) for rocks formed in collisional and subduction zones can be calculated from the corresponding mineral equilibria (Connolly 1990). A version of the generalized phase diagram for metabasites with MORB compositions calculated with THERMOCALC (Powell and Holland 1994) using results of Hacker et al. (2003) is shown in Fig. 3.65. The calculated stability fields of mineral assemblages and reaction lines are consistent with actual bulk compositions of HP and UHP rocks and minerals reported from various geological objects. This phase diagram also shows changes in the *P-T* parameters on the surface of the subducting plate, which were calculated for different natural examples of subduction zones. The age of the plate and subduction rate in these examples are as follows: Cascadia subduction zone (Juan de Fuca plate)—40 mm/year and 5–9 Ma; Nankai subduction zone (Philippine plate, southwest Japan)—45 mm/year and 15 Ma; Costa-Rica subduction zone (Cocos plate)—87 mm/year and 18 Ma; Tohoku subduction zone (Pacific plate, northeast Japan)—91 mm/year and 130 Ma. The calculated *P-T-t* paths for the above examples show that the geothermal gradient was 4–7 °C/km, while melting in the subducting slab would occur only in the case of the slowest subduction (curve 4).

Geophysical observations in modern subduction zones reveal the difference in the geometry of the subducting plates at the 100–150 km depth in the upper mantle (Jarrard 1986). There is a general increase in dip with depth and the descent angle

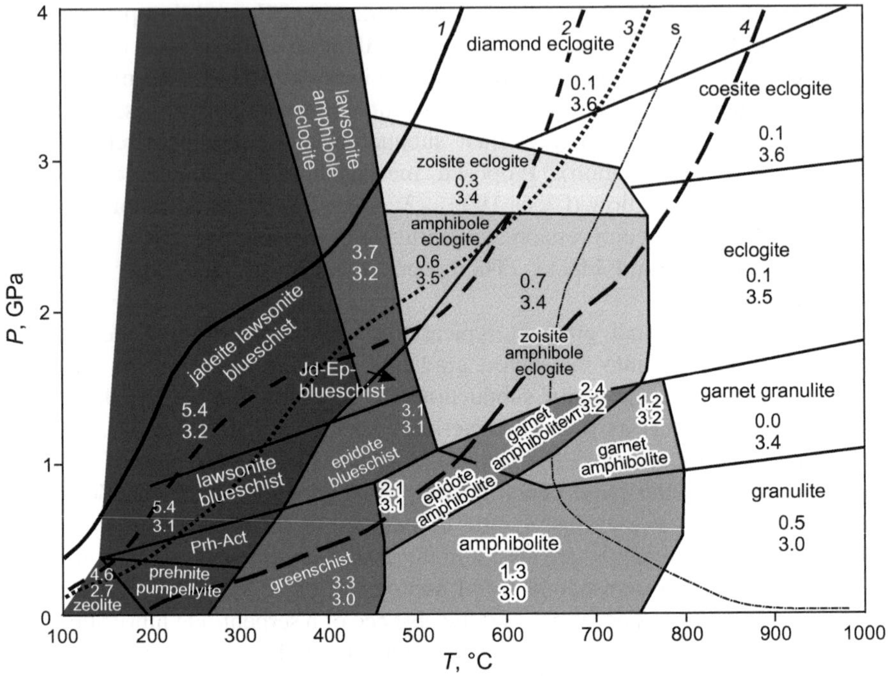

Fig. 3.65 Phase diagram for metabasites constructed using THERMOCALC software (Powell and Holland 1994) agreed with data on natural mineral parageneses in HP/UHP rocks (Hacker et al. 2003). Dark to light grey shades indicate the increase in water content. The dash-dotted line (s) indicates the wet basaltic solidus. Lines 1–4 show the *P-T* paths along the subducting slab-mantle wedge interface for Tohoku (1), Costa Rica (2), Nankai (3), and Cascadia (4) subduction zones. The numbers in the stability fields indicate the water content (wt%)/density (g/cm^3)

becomes closer to vertical (e.g., Mariana subduction zone). However, the plate may be flattened beneath the continent (e.g., Andean subduction zone) to cause flat subduction or underthrusting.

Models for the subduction wedge and exhumation of subduction complexes using analytical expressions and analog experiments were proposed by Peacock (1992), Dobretsov (1991), Dobretsov and Kirdyashkin (1991), and Chemenda et al. (1995). Further evolution of models describing subduction and exhumation of high and ultrahigh-pressure metamorphic rocks is associated with the development of high-resolution numerical simulations to explore the mechanisms of recycling material in a wedge-shaped subduction channel (Gerya 2002; Burov et al. 2014). Gerya (2002) show that the burial and exhumation of high-pressure metamorphic rocks depend on the progress of hydration of the mantle wedge, which controls the shape and circulation pattern of a subduction channel with a forced return flow of material and a progressive upward widening of the mantle wedge. Both clockwise and counterclockwise *P-T-t* trajectories are possible depending on specific conditions.

To understand the causes of the onset of subduction, we have produced models (Korobeinikov et al. 2008; Korobeynikov et al. 2009; Polyansky et al. 2010b) controlled by two factors: eclogitization (or densification associated with growth of high-density phases) in a subducting slab and (2) the presence of a keel at the slab edge. The model rheology is an elastoplastic crustal material, as opposed to a viscoplastic mantle rheology, which is assumed for incompressible material. The model takes into account an increase in the weight of the slab due to gabbro–eclogite transition within the depth interval of 35–40 km with a density increment of 500 kg/m^3 (from 3000 to 3500 kg/m^3). We consider the motion of two converging slabs at a rate of 1.4 mm/year with the same thickness of 30 km and the length of over 300 km, one of which was set immobile and rigid. The slabs interact along the contact inclined at 45°, reproducing an accretionary wedge and defining a subduction zone. We used two configurations of the subducting slab, one with a smooth or flat bottom and the other having an initial bulde at its edge at the beginning of the collision.

The results of numerical modeling are presented as configurations of the deformed subducted slab at some time after the onset of subduction. The images show enlarged fragments of the central model area of the slab–slab interface. The first series of numerical experiments was performed for a model of a deformable slab with a flat lower surface. Figure 3.66a shows the slab geometry 8.5 Myr after the beginning of collision. The deformable slab underthrusts the immobile one along the contact and does not tend to simply sink into the mantle. Figure 3.66b shows the results of numerical simulation for the case of the initial heterogeneity developing as a 15 km thick bulge (plate thickening of up to 1.5 times) at the slab edge. The results depict a downward bending of the slab with an observable increase in the dip angle, from 45° at the contact to the near vertical dip. The model predicts the formation of the highest plastic strain domain on the upper bent portion of the slab, which is in contact with the mantle wedge. Figure 3.66c shows configuration of the subducted slab 6.5 Myr after the onset of subduction. It can be seen that the slab thins out below the gabbro–eclogite phase transition zone. The thickness of the slab, in the narrowest portion, is 18 km at the initial thickness of 30 km. This portion also concentrates the equivalent plastic strains with a maximum in the vicinity of the upper slab–mantle interface.

The proposed model geometries for the subducted slab and the distribution of deformations suggest two critical factors that control the character of subduction: (1) slab rheology, which depends on the composition and basalt-eclogite transition transition, and (2) initial slab geometry at the convergence zone. The computations reproduced deep sinking only for the case of initial heterogeneity in a deformable slab (keel formation or thickening) and considering eclogitization. Otherwise, the slabs may underthrust and imbricate as in the case of the Andean-type subduction zone (Jarrard 1986). Using an elastoplastic rheology enables us to properly explain the thinning of the plate as the initial phase of slab detachment. Another outcome of numerical simulations with different models was the recognition of a high plastic strain near the upper slab surface at the contact with the mantle wedge.

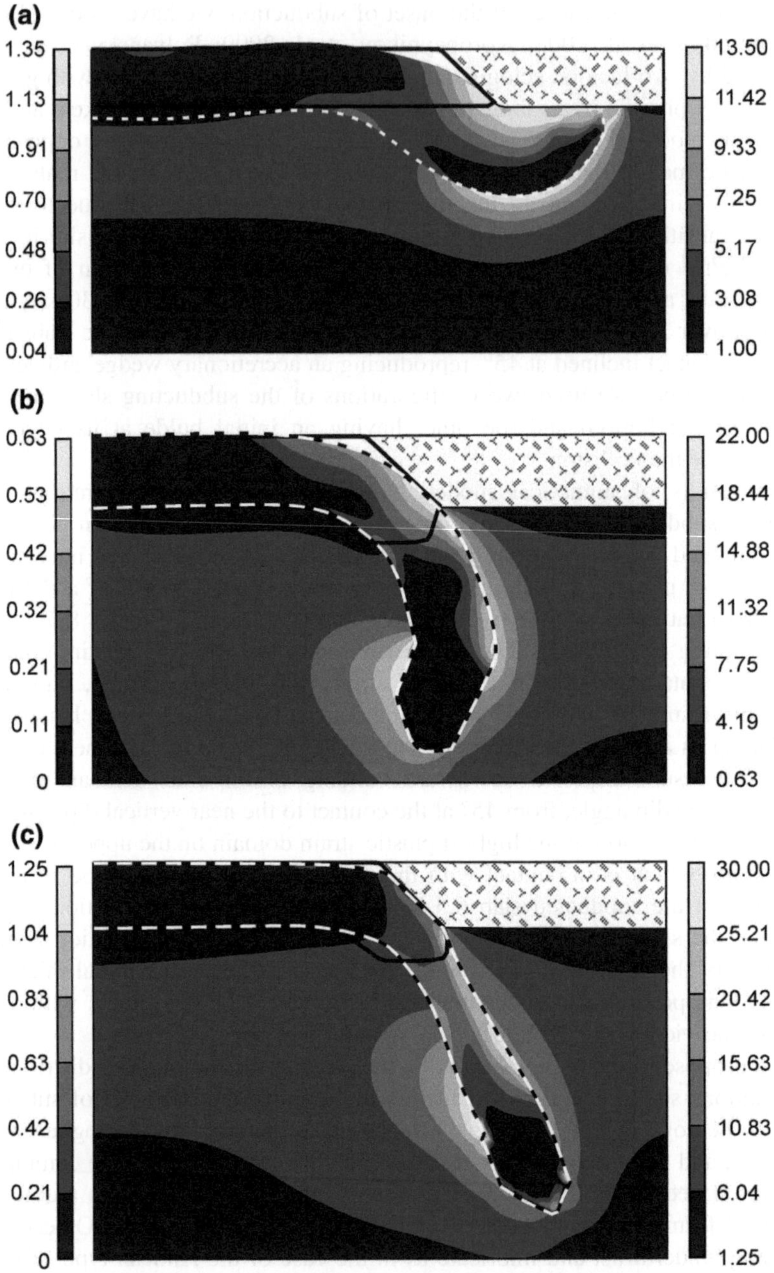
(a)
1.35
1.13
0.91
0.70
0.48
0.26
0.04
13.50
11.42
9.33
7.25
5.17
3.08
1.00
(b)
0.63
0.53
0.42
0.32
0.21
0.11
0
22.00
18.44
14.88
11.32
7.75
4.19
0.63
(c)
1.25
1.04
0.83
0.63
0.42
0.21
0
30.00
25.21
20.42
15.63
10.83
6.04
1.25

◂**Fig. 3.66** **a** Underthrusting of a slab with a smooth lower surface using the linear Mohr–Coulomb model. Slab position and distribution of the plastic strain field at 8.5 Myr after the beginning of subduction. Here and in **b** and **c** light color depicts maximum deformations; deformation scales for slab (left) and mantle (right); slab boundary is marked with a white dash. **b** Subduction of a slab with initial heterogeneity using the linear Mohr–Coulomb model; **c** Subduction of a slab with initial heterogeneity using the parabolic Mohr–Coulomb model

The structure and deformation at the plate contact area in subduction zones control the mechanisms of exhumation of HP and UHP rocks. Based on the rock mineralogy, peak *P-T* parameters and the exhumation patterns of metamorphic rocks, Guillot et al. (2009) distinguished three types of subduction: accretionary, serpentinite-subduction channel and continental collision type (Fig. 3.67a).

During accretionary-type subduction, the exhumation of metasediments to the surface is driven by a return flow in the accretionary prism. The maximum pressures recorded in the exhumed rocks of this type vary from 0.7 to 2 kbar, the geothermal gradients are 5–14 °C/km, which corresponds to modern subduction zones. In some cases, eclogites were formed at a depth of at least 75 km, which is much greater than the expected depth of present-day accretionary wedges (20–40 km). Another common feature of this type of subduction is low exhumation velocities of 1–5 mm/year (Agard et al. 2009).

The serpentinite-subduction channel exhumes high and ultrahigh-pressure rocks. A 1–10 km thick serpentinite channel originates from both subducted abyssal peridotites and hydrated mantle wedge. The exhumation velocity ranges from 3 to

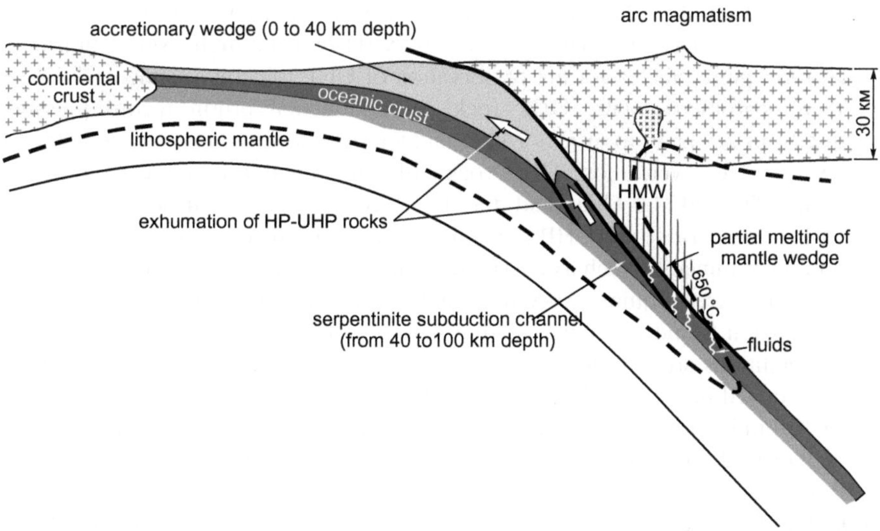

Fig. 3.67 Schematic model showing the relationship between different conditions of subduction and exhumation of HP-UHP/LT rocks, modified after Chemenda et al. (1995), Gerya (2002), Guillot et al. (2009), and Burov et al. (2014). HMW, hydrated mantle wedge

10 mm/year, which is somewhat higher than that for accretionary-type subduction. The exhumation is controlled by the positive buoyancy (a low density of 2600 kg/ m^3), a high Poisson ratio (0.29), low shear modulus (36–54 GPa) and a low viscosity of serpentinites (4×10^{19} Pa s). These physical properties allow serpentinites to be highly ductile to lubricate subduction planes.

In paleosubduction zones, serpentinites are commonly considered as fragments of the oceanic crust. The boundary of the serpentinite subduction channel is defined by the 650 °C isotherm which corresponds to the stability field of serpentinite in *P*-*T* space of the phase diagram (Hacker et al. 2003). It forms a ~60 km long (from 40 to 100 km depth) channel between the dry (rigid) subducted oceanic lithosphere and the dry (rigid) mantle wedge. It contains exotic blocks of metabasalys, metasediments, and metagabbros, exhumed within subduction zones of this type and incorporated in a soft matrix of deformed serpentinites. Due to the low viscosity and low density of serpentinite mineral and the triangle shape of the serpentinite channel, the dowgoing material is progressively entrained upward.

Regarding the metamorphic conditions, most eclogitic blocks reached HP between 1.8 and 2.5 GPa and relatively low temperatures [500–700 °C, as is the case of the Maksyutov eclogite—blueschist complex (Dobretsov 2000)], which defines paleogeothermal gradients lower than 10 °C/km. Two localities provide evidence for deeper *P-T* conditions, at 3.2 and 4 GPa, and temperatures at 800 and 1550 °C, respectively, in the Rio San Juan complex in Dominican Republic and Sanbagawa belt of Japan, as deduced from garnet peridotite blocks embedded in the serpentinite mélange.

The continental-type subduction exhumes rocks of continental origin that were buried down to a depth between 100 and 200 km along cold geotherms. UHP rocks are exhumed rapidly ($20 < V < 60$ mm/year) over a period <10 Myr. Structural, mineralogical and isotope data (Dobretsov 1991) indicate high exhumation rates, which form prerequisites for the preservation of UHP mineral assemblages. Based on thermobarometric data, UHP rocks of continental origin record pressures ranging from 2.5 kbar up to $P > 7$ kbar and temperatures varying between 500 and ~1335 °C. However, most UHP rocks of continental origin record pressures between 2.50 and 4.0 GPa, and the UHP conditions are mostly recorded in garnet-bearing peridotites. UHP minerals are rarely preserved and mostly occur as relict in other minerals, such as coesite in garnet or omphacite or diamond in zircon. Their occurrence requires specific conditions, such as rapid cooling during decompression, rapid exhumation, fluid-absent condition during the exhumation. These conditions are only locally attained so that the evidence for UHP conditions is only retained in lenses. UHP metamorphism records low geotherms ranging from 5 to 6 °C/km in in the Kokchetav massif (Sobolev and Shatsky 1987), and down to 3.5 °C/km in the Forbidden Zone in China (Liou et al. 2000). The *P-T-t* paths of UHP rocks are characterized by isothermal decompression until crustal depths (1.0 to 0.5 GPa). The absence of significant heat loss during the exhumation indicates their rapid exhumation, up to 2.5–7 cm/year at the initial stages of exhumation in the Kokchetav massif (Dobretsov 2000) and up to 8 cm/year in the Alps and the Himalaya (Guillot et al. 2009).

It is noteworthy that an accretionary wedge and a serpentinite subduction channel may coexist in a single subduction zone but differ in their burial and exhumation depths, as observed in the Franciscan complex of California, the northern subduction complex in Dominican Republic, and the Western Alps. The Pennine Alps appears to be a good example of coexistence of all three types of subduction: an accretionary wedge (the Schistes Lustés unit), a serpentinite channel (the Monviso unit) and a continental subduction unit (the Dora Maira unit) (Agard et al. 2009).

Glaucophane schists and eclogites mark the position of ancient subduction zones and form part of accretionary-subduction complexes where they occur as tectonic plates, nappes, lenses or exotic blocks in a mélange of serpentinite. This serpentinite mélange with inclusions of high-pressure rocks is documented in many accretionary-subduction complexes and is interpreted to be associated with the recycling of the subducted oceanic plate into the accretionary prism. The serpentinite mélange is formed by disintegration of ophiolites, and also due to tectonic juxtaposition of various oceanic complexes (Volkova et al. 2008).

Most glaucophane schist complexes at the southern margin of the Siberian Craton were formed in an intra-oceanic forearc environment, characterized by tectonic accretion, subduction, underplating and obduction of mafic terranes (Volkova and Sklyarov 2007). The next stage in the formation of glaucophane schist complexes in the Central Asian fold belt involved island arc accretion and the formation of rock associations which included fragments of the deformed oceanic crust, glaucophane schists, metagreywackes, and ophiolitic sheets. Subsequent accretionary collision events within this belt led to the formation of an orogenic "collage" consisting of fragments of island arcs, collisional complexes and oceanic plateaus that were successively accreted to the southern edge of the Siberian craton.

Another example is HP rocks in a serpentinite mélange of the Chara zone, northeastern Kazakhstan (Volkova et al. 2011). The *P-T* conditions of metamorphism (Volkova et al. 2008) of the glaucophane and garnet-barroisite schists estimated with THERMOCALC (Holland and Powell 1998) indicate a relatively narrow temperature range (500–570 °C) and a wide range of pressures (5–13 kbar) (Fig. 3.68). The metamorphic conditions of eclogites in the Chara zone were estimated at 650–700 °C and 15–19 kbar. A large difference in the metamorphic pressures may reflect the different depth of subduction of the protoliths (oceanic basalts and siliceous sediments). The line in Fig. 3.68 approximating the *P-T* parameters is close to the geotherm corresponding to warm subduction, which indicates subduction of a relatively young and cool oceanic crust at the final stage of closure of a paleooceanic basin. On the other hand, this line can be considered as a metamorphic path of exhumed HP rocks. The high-pressure segment of this line connecting the *P-T* estimates for eclogite and garnet-barroisite schists with well-preserved eclogite paragenesis represents a retrograde trajectory of high-pressure metamorphism.

In discussing the mechanisms of formation of HP metamorphic complexes, we need to know whether these complexes are fragments of oceanic or continental crust. Data on trace element and REE contents provide invaluable information on their protolith. The study of element mobility in oceanic basalts and subducted

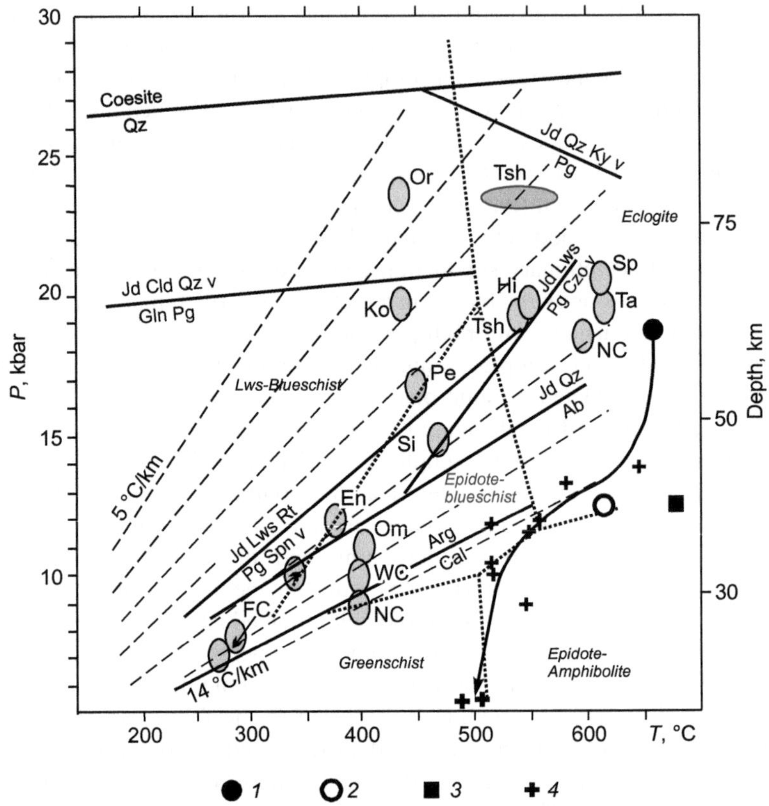

Fig. 3.68 Peak *P-T* conditions of blueschist and eclogite facies metamorphism for selected areas. The dotted lines are the facies boundaries after Evans (1990). The data sources are: Franciscan Complex, California (FC); western Crete (WC) and Peloponnese (Pe), Greece; Oman; Sifnos, Greece (Si); New Caledonia (NC); Tauern Window, Alps (Ta); Engadine Window (En); Himalaya (Hi); Tianshan (Tsh); Spitsbergen (Sp); Kocasu (Ko), Orhaneli (Or), Turkey (compiled by Okay (2002)). The *P-T* uncertainties are usually larger than that indicated by the ellipses. Solid line shows the *P-T* metamorphic path calculated for high-pressure rocks of the Chara zone (Kazakhstan). (1)—eclogite; (2)—eclogite assemblage "preserved" in garnet–barroisite schist; (3)—garnet clinopyroxenite; (4)—garnet–barroisite schists, after Volkova et al. (2008) and Simonov et al. (2008)

sediments (Bebout 2007) reveals obvious enrichments or depletions of only the most fluid-mobile elements in HP metamorphic rocks as compared to their protolith, whereas concentrations of some high-field-strength elements (HFSE) and rare-earth elements (REE) remain close to those of the protoliths. This may provide the basis for determining the geochemical nature of the protoliths for HP rocks and geodynamic conditions for their formation.

The geochemistry of the Dzhebash Group glaucophane schists of the Kurtushibinsky range of the Western Sayan showed that their protoliths were oceanic basalts comparable to some enriched MORB varieties (Volkova et al.

2009). The REE distribution patterns in glaucophane shales show a slight enrichment of LREE over HREE: $Ce_n = 34.3–73.2$, $(Ce/Yb)_n = 2.5–6.5$, $(La/Sm)_n > 1$. The multi-element diagrams for glaucophane schists have a slight negative slope, due to elevated concentrations of U, Th, Nb, Ta, and LREE in some samples, depletion in Rb and Ba, and pronounced negative K and Sr anomalies (Fig. 3.69). Such patterns are typical of some oceanic plateau basalts of the E-MORB and P-MORB types (Klein 2003). A comparison of the chemical and trace element compositions of the glaucophane schists with weakly altered basalts from the Kurtushibinsky Formation showed that they are the almost indistinguishable with respect to many fluid-immobile elements, such as Ti, P, Zr, Hf, Y, MREE, and HREE. The spatial association and identical trace element patterns in these rocks allow us to suppose that the basalts of the Kurtushibinsky Formation and the protoliths of glaucophane schists were derived from an enriched mantle source in an oceanic plateau setting. These data also suggest that the glaucophane schists facies metamorphism was essentially isochemical and resulted only in minor changes in the chemical composition of the subducted oceanic basalts, namely, significant loss of K (up to 90–95%).

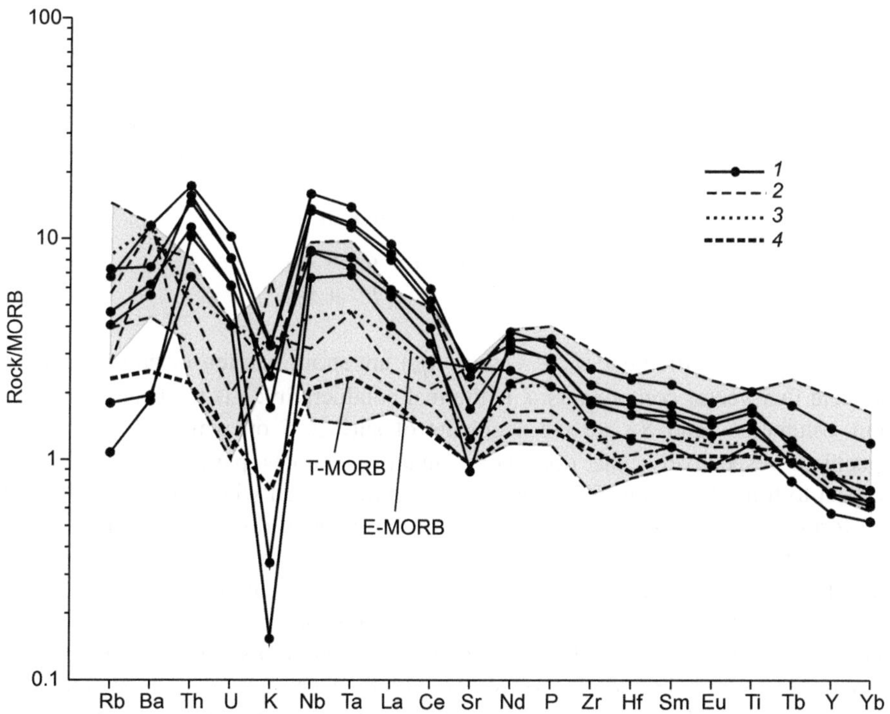

Fig. 3.69 Comparison of N-MORB-normalized contents of trace elements in the (1) blueschists of the Dzhebash Group and (2) basalts of the Kurtushibinsky Formation. The shaded area shows the range of compositional variations for the basalts of the Kurtushibinsky Formation. Also shown are the compositions of (3) E-MORB and (4) T-MORB after Klein (2003)

Appendix

The formulation is based on a coupled thermomechanical model of planar deformation. For twice continuously differentiable fields of the velocity vector *ui*, the equilibrium equations in differential form, together with the boundary conditions, have the form (here and below, the indices *i* and *j* run through 1, 2) (Korobeynikov 2000):

$$\begin{aligned} \sigma_{ij,j} + \rho g_i &= 0 \text{ in } V, \\ n_i \sigma_{ij} &= f_j^* \text{ at } S_f, \\ u_i &= u_i^* \text{ at } S_u. \end{aligned} \tag{3.21}$$

Here, σ_{ij} are components of the symmetric Cauchy stress tensor ($i, j = 1, 2$); g_i the components of the gravity vector; ρ is the current mass density of the material; V is the domain occupied by the body in its current configuration; S is the closed boundary of domain V; S_u and S_f are the segments of surface S on which the components of velocity vector u_i and Cauchy stress vector $n_i\sigma_{ij}$ are specified; n_i are the vector components of the external unit normal vector to surface $S_{f;}$ and summation goes over repeated indices from 1 to 2; the comma denotes partial derivative with respect to the corresponding coordinate, and the asterisk denotes a prescribed variable value.

For twice continuously differentiable temperature fields T, the heat balance equation may be expressed in the form (Landau and Lifshitz 1986):

$$\begin{aligned} \rho c_p \dot{T} &= \sigma_{ij} d_{ij} + kT_{,ii} + r \text{ in } V, \\ kT_{,i} n_i &= q_n^* \text{ at } S_q, \\ T &= T^* \text{ at } S_T. \end{aligned} \tag{3.22}$$

Here, c_p is specific heat; $d_{ij} \equiv \frac{1}{2}\left(u_{i,j} + u_{j,i}\right)$ are strain rate tensor components; q_i are heat flux vector components; k is thermal conductivity; r is the heat source in a unit volume; S_q and S_T are the segments of surface S on which q_n^* and T^* are specified, respectively; the dot above quantity denotes partial derivative with respect to time. Equation (3.22) should be complemented with initial conditions in the form of temperature fields specified at each point of domain V at initial time.

Note that the differential Eqs. (3.21) and (3.22) require that the functions contained in them be highly smooth. If we ignore this requirement (for instance, in the case of discontinuous fields of stress tensor components), we should introduce the internal interfaces and specify the discontinuity conditions for functions or their derivatives on these boundaries. However, with the weak formulations of equations, the internal boundaries are not required because the equations will be satisfied irrespective of these discontinuities. The finite element method, which is applied in this study for numerical modeling of geodynamical processes, is based on the weak forms of thermomechanical equations. The equilibrium equations in weak form correspond to the principle of balance of virtual works of the internal and external forces:

$$\int_V \sigma_{ij}\delta d_{ij}dV = \int_V \rho g_i \delta u_i dV + \int_{S_f} f_i^* \delta u_i dS \quad \forall \delta u_i (\delta u_i = 0 \text{ at } S_u), \tag{3.23}$$

while the weak form of heat transfer equation is

$$\int_V (\rho c \dot{T} \delta T + k T_{,i} \delta T_{,i}) dV = \int_V (\sigma_{ij} d_{ij} + r) \delta T dV + \int_{S_q} q_n^* \delta T dS \quad \forall \delta T \; (\delta T = 0 \text{ at } S_T), \tag{3.24}$$

where $\delta(\,), \forall(\,)$ mean variation and universal quantifier in corresponding variable.

The computational domain, including the sections of the earth's crust and mantle lithosphere, is presented in Fig. 3.30. In the same figure, boundary conditions are given both for solving the mechanical problem (Fig. 3.30a) and for the thermal problem (Fig. 3.30b). Figure 3.30b shows the initial temperature distribution that corresponds to the stationary geothermal lithosphere of the craton with a mantle heat flux Q = 17 mW m^{-2} and the thermophysical parameters of the crust and mantle specified in Table 3.8. The following boundary conditions for the mechanical problem are employed: the top boundary is a free surface, the bottom boundary and lateral boundaries of the lithospheric mantle are free slip, lithostatic pressure is assumed on the lateral boundaries of the crust. The latter condition means that all lateral boundaries of the crust move horizontally. The crust and the mantle represent two bodies in contact, and Coulomb friction with the coefficient of friction $\mu = 0.9$ is assumed for the surface of the deformable contact. For the thermal problem, we set up the boundary conditions of the isothermal upper surface (T = 0 °C), heat-insulated lateral boundaries, the constant thermal flux at the base of the lithosphere outside the plume region and the constant temperature (1450, 1550 and 1650 °C) in the 100-km zone of the sublithospheric upper mantle plume (Fig. 3.30).

The model accounts for interaction of a partially molten material with the mantle at subsolidus temperature, which requires consideration of contrasting rheological properties of the rocks. The main strain mechanism in mantle under high temperatures and pressures is described by dislocation creep (Karato and Wu 1993). In the subsolidus and partially molten state, the medium is described by non-Newtonian nonlinear viscosity. A common approach to modeling the mantle flows consists in the use of constitutive relations for the non-Newtonian viscous incompressible fluid:

$$\sigma_{ij} = -p\delta_{ij} + s_{ij}; \quad s_{ij} = 2\eta d_{ij}, \tag{3.25}$$

Table 3.8 Thermomechanical model parameters of the crust and mantle

Parameter, symbol (units)	Crust	Mantle
Rheological parameters		
Yield strength, σ_y (MPa)	50	1
Density, ρ_0 (kg/m^3)	2820	3250
Young's modulus, *E* (GPa)	50	50
Poisson ratio, **v**	0.25	0.25
Friction angle, ϕ	30°	
Quartzite–H_2O (Kronenberg and Tullis 1984)		
Pre-exponential constant, **A** (Pa^{-n} s^{-1})	4.0×10^{-21}	
Activation energy, **H** (KJ/mol)	134	
Power-law exponent, **n**	2.6	
Olivine–basalt–H_2O–melt (Mei et al. 2002)		
Pre-exponential constant, **A** (Pa^{-n} s^{-1})		3.9×10^{-21}
Activation energy, **H** (KJ/mol)		470
Power-law exponent, **n**		3.5
Thermal parameters		
Thermal conductivity, **k** (J/(m s K))	2.5	3.5
Heat capacity, $\mathbf{C_P}$ (J/(kg K))	1250	1250
Thermal diffusivity, **K** (10^{-6} m^2/s)	0.99	0.86
Radioactive heat, **r** (J/(m^3 s))	4.5×10^{-7}	0
Mantle heat flux, **Q** (mW/m^2)		17
Thermal expansion, $\boldsymbol{\alpha}$ (K^{-1})	1.0×10^{-5}	3.1×10^{-5}
Melt fraction, ϕ		0.1–0.5
Density contrast, $\boldsymbol{\rho_s}$–$\boldsymbol{\rho_m}$ (kg/m^3)		50–250
Solidus temperature, $\mathbf{T_s}$ = T(P), (°C)		1360–920
Temperature at the base of the lithosphere (°C)		1350
Temperature at Moho (initial) (°C)	625	

where p is the lithostatic pressure; s_{ij} are the components of the Cauchy deviator stress tensor; η is the nonlinear *P-T*-dependent viscosity expressed as (Ranalli 1995):

$$\eta = \exp\left(\frac{-c\phi}{n}\right) A^{-\frac{1}{n}} [\dot{\varepsilon}_{II}]^{\frac{(1-n)}{n}} \exp\left(\frac{H + pV_a}{nRT}\right), \; \dot{\varepsilon}_{II} \equiv \frac{1}{2} d_{ij} d_{ij}, \qquad (3.26)$$

where $\dot{\varepsilon}_{II}$ is the second invariant of the strain rate tensor; T is temperature; p is pressure; A is the coefficient before the exponent; n is the power-law exponent (n = 1 for the Newtonian liquid); ϕ is the melt fraction; H and Va are activation energy and activation volume, respectively; c is a given parameter in the interval from 30 to 45; and R is the universal gas constant.

In this model we used complete constitutive equations of thermal elastoplasticity with allowance for creep deformations. With the combined rheological model, the components of strain rate tensor are cast as the sum of rates of elastic d^e_{ij}, thermal d^T_{ij}, plastic d^p_{ij} and creep strains d^c_{ij} (Korobeynikov 2000):

$$d_{ij} = d^e_{ij} + d^T_{ij} + d^p_{ij} + d^c_{ij}. \tag{3.27}$$

It is suggested that non-elastic components of the strain rate tensor correspond to the conditions of incompressibility, i.e., for the plain strain conditions, they can be written as $d^p_{11} + d^p_{22} = 0$ and $d^c_{11} + d^c_{22} = 0$.

In order to describe the rates of plastic and thermal strains, we used standard approach described in the works (Korobeynikov 2000; Polyansky et al. 2010a). As was shown by experiments on the rock strength, the tangential stresses never exceed some value corresponding to the yield stress, σ_Y. In this relation, our modeling was based on the rheological model of medium involving both plastic and viscous strains. The plastic strains in mantle (in nearly liquid state) can be described by the model of material with the Huber—Mises yield surface, with sufficiently low value of yield stress (Gerya and Burg 2007; Polyansky et al. 2010b). The computations show that the yield stress varied from 100 to 400 MPa for crustal material and of 0.1 to 1 MPa for mantle plume.

For the creep components of the strain rate tensor, we use the Norton steady-state creep law (Korobeynikov 2000):

$$d^c_{ij} = \frac{3}{2}\frac{\dot{\bar{\varepsilon}}^c}{\bar{\sigma}} s_{ij}, \tag{3.28}$$

where $\bar{\sigma} = \sqrt{3/2 s_{ij} s_{ij}}$ is the effective stress and $\dot{\bar{\varepsilon}}^c = \sqrt{2/3 d^c_{ij} d^c_{ij}}$ is the effective rate of creep strain. According to (Ranalli 1995), the equation for effective rate of creep deformations has the form

$$\dot{\bar{\varepsilon}}^c = 3^{-\frac{n+1}{2}} 2^{1-n} A \exp(c\phi)\bar{\sigma}^n \exp\frac{-(H + pV_0)}{RT} \tag{3.29}$$

The experimental data on the strains in partially molten and aqueous olivine–basalt aggregates suggest the exponential decay of viscosity with increase of melt fraction: $\eta/\eta_0 = \exp\left(\frac{-c\phi}{n}\right)$ where $c \approx 45$ (Mei et al. 2002).

At such parameters, the ratio of mixture viscosity to that in the absence of melt was taken within the range from 0.2 to 10^{-3}. Geochemical data on mantle peridotites from xenoliths within the ancient cratons such as Siberian, Slave and Kaapval (Walter 2003), indicate a degree of melting and extraction of the melt in the range of 30–50% with an average value of 45%. Figure 3.70 shows the temperature dependence of viscosity of wet and dry mantle rocks according to (3.26) at the strain rate of 10^{-13} s^{-1} based on experimental data of Chopra and Patterson (1984), Karato and Wu (1993).

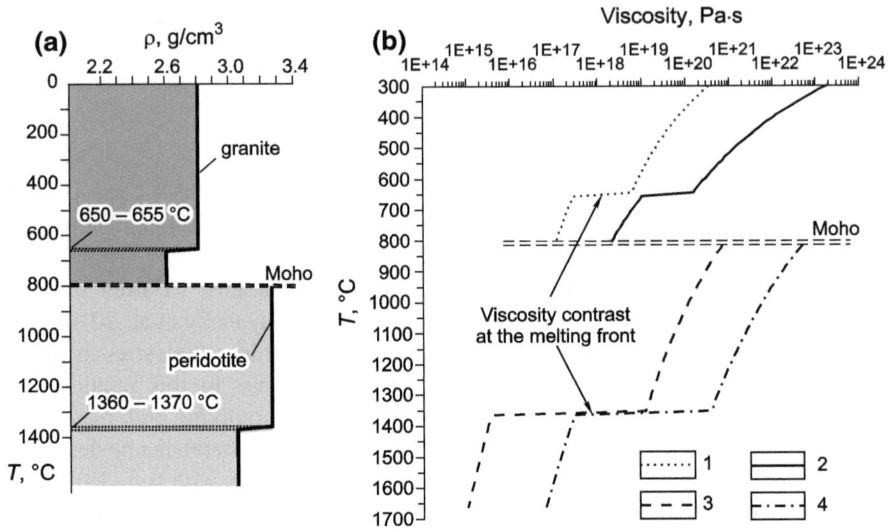

Fig. 3.70 Density and viscosity distribution as a function of temperature and phase transition during melting, accepted in models. Moho depth is provisional. 1, wet quartzite (Kronenberg and Tullis 1984); 2, diabase (Carter and Tsenn 1987); 3, wet dunite (Chopra and Patterson 1984); 4, molten wet olivine-basalt (Mei et al. 2002)

The model was underlain by the reference values of parameters, which corresponded to the continental felsic crust (Ranalli 1995; Gerya and Burg 2007) and mantle with rheology of wet dunite (Chopra and Patterson 1984), dry olivine (Karato and Wu 1993) or olivine–basalt–water–melt aggregate (Mei et al. 2002). Rheological parameters of crustal and mantle rocks are shown in Table 3.8.

The equation of state is assumed in the form of the dependence of density ρ on thermal expansion α and melt fraction ϕ:

$$\rho = \rho_s\left(1-\alpha T - \frac{\rho_s - \rho_m}{\rho_s}\phi\right), \tag{3.30}$$

where ρ_m and ρ_s are the densities of the melt and the solid matrix, respectively. Beyond the melt region, $\phi = 0$ and density variations are only due to thermal expansion; in the melt region, the melt fraction is assumed to be constant. Figure 3.70 shows the temperature dependence of density accounting for phase transition during melting of the granite crust and peridotite mantle.

References

Agard P, Yamato P, Jolivet L et al (2009) Exhumation of oceanic blueschists and eclogites in subduction zones: timing and mechanisms. Earth Sci Rev 92:53–79

Anan'ev VA (1999) Kontaktoviy metamorfizm Ayu-Daga (Contact metamorphism on the Ayu-Dagh mountain). Candidate of Science Dissertation, Novosibirsk

Anan'ev VA, Polyansky OP, Lepezin GG et al (2003) Metamorphic zoning of the Tongulak mountain range, Altai: mathematical modeling. Russ Geol Geophys 44(4):297–304

Aranovich LY, Podlesskii KK (1983) Phase conformity in the system: cordierite–garnet–sillimanite– quartz In: Marakushev AA (ed) The biotite – garnet – cordierite equilibria and the evolution of metamorphism Nauka, Moscow, pp 89–123

Arndt NT (2013) Formation and evolution of the continental crust. Geochem Perspectiv 2(3):405–533

Artemieva IM (2003) Lithospheric structure, composition and thermal regime of the East European Craton: implications for the subsidence of the Russian platform. Earth Planet Sci Lett 213:431–446

Artyushkov EV (1993) Fizicheskaya tektonika (Physical tectonics) Nauka, Moscow

Artyushkov EV, Batsanin SF (1984) On the thermal regime variations in the Earth crust connected with the approach to its lower boundary of anomalous mantle. Izvest Acad Sci USSR. Phys Solid Earth 12:3–9

Ashworth JR, Evirgen MM (1985) Plagioclase relations in pelites, Central Meuderes Massif, Turkey. II. Perturbation of garnet-plagioclase geobarometers. J Metamorph Geol 3:219–229

Babichev AV, Polyansky OP, Korobeynikov SN et al (2014) Mathematical modeling of magma fracturing and dike formation. Dokl Earth Sci 458(2):1298–1301

Badarch G, Cunningham WD, Windley BW (2002) A new terrane subdivision for Mongolia: implications for the Phanerozoic crustal growth of central Asia. J Asian Earth Sci 21:87–110

Baltybaev ShK, Glebovitskii VA, Shul'diner VI et al (1996) The Meyeri thrust: the main element of the suture at the boundary between the Karelian Craton and the Svecofennian belt in the Ladoga region of the Baltic Shield. Dokl Acad Nauk 348(4):581–584

Baltybaev ShK, Levchenkov OA, Levskii LK (2009) Svekofennskiy poyas Fenno-skandii: prostranstvenno-vremennaya korrelyatsiya ranneproterozoyskikh endogennykhprotsessov (The Svecofennian belt of the Fennoscandia: spatio-temporal correlation of the early Proterozoic endogenous processes), Nauka, St. Petersburg

Barber JP, Yardley BWD (1985) Conditions of high grade metamorphism in the Dalradian of Connemara, Ireland. J Geol Soc London 142:87–96

Barton MD, Ilchik RP, Marikos MA (1991a) Metasomatism. In: Kerrick DM (ed) Contact metamorphism. Reviews in mineralogy, vol 26, Book Crafters Inc, Chelsea, Michigan, pp 321–350

Barton MD, Staude J-M, Snow EA et al (1991b) Aureole systematics In: Kerrick DM (ed) Contact metamorphism. Reviews in mineralogy, vol 26, Book Crafters Inc, Chelsea, Michigan, pp 723–847

Baumgartner L, Ferry JM (1991) A model for coupled fluid-flow and mixed-volatile mineral reactions with applications to regional metamorphism. Contr Mineral Petrol 106:273–285

Beaumont C, Jamieson RA, Nguyen MH et al (2001) Hymalayan tectonics explained by extrusion of a low-viscosity crustal channel coupled to focused surface denudation. Nature 414:738–742

Bebout GE (2007) Metamorphic chemical geodynamics of subduction zones. Earth Planet Sci Lett 260:373–393

Berg JH (1977a) Dry granulite mineral assemblages in the contact aureoles of the Nain complex, Labrador. Contr Mineral Petrol 64:33–52

Berg JH (1977b) Regional geobarometry in the contact aureoles of the anorthositic Nain complex, Labrador. J Petrol 18:399–430

Bialas RW, Buck WR, Qin R (2010) How much magma is required to rift a continent? Earth Planet Sci Lett 292:68–78

Birth F, Schairer JF, Spicer HC (1942) Handbook of physical constants. Geol Soc Amer, Spec Paper 36:325

Bittner D, Schmeling H (1995) Numerical modelling of melting processes and induced diapirism in the lower crust. Geophys J Int 123:59–70

Bohlen SR, Montana A, Kerrick DM (1991) Precise determinations of the equilibria kyanite – sillimanite and kyanite – andalusite and a revised triple point for Al_2SiO_5 polymorphs. Am Mineral 76:677–680
Bowers JR, Kerrick DM, Furlong KP (1990) Conduction model for the thermal evolution of the Cupsuptic aureole, Maine. Am J Sci 290:644–665
Brown M (2008) Characteristic thermal regimes of plate tectonics and their metamorphic imprint throughout earth history: when did earth first adopt a plate tectonics mode of behavior? In: Condie KC, Pease V (eds) When did plate tectonics begin on planet earth? Geol Soc Am, Spec Paper, vol 440, p 97–128
Brown M (2013) Granite: from genesis to emplacement. Geol Soc Am Bull 125:1079–1113
Brown EH, Walker NW (1993) A magma-loading model for Barrovian metamorphism in the Southeast Coast Plutonic Complex, British Columbia and Washington. Geol Soc Am Bull 105:479–500
Buck WR (1986) Small-scale convection induced by passive rifting: the cause of uplift of rift shoulders. Earth Planet Sci Lett 77:362–372
Buck WR (1991) Modes of continental lithospheric extension. J Geophys Res 96(B12):20161–20178
Buntebarth G (1984) Geothermics—an introduction. Springer, Berlin, Heidelberg, New York
Buntebarth G (1991) Thermal model of cooling. In: Voll G, Töpel J, Pattison DRM, Seifert F (eds) Equilibrium and kinetics in contact metamorphism. The Ballachulish igneous complex and its aureole. Springer, Berlin, Heidelberg, pp 379–402
Burke K, Dewey JF (1973) Plume-generated triple junctions. Key indicators in applying plate tectonics to old rocks. J Geol 81:406–433
Burov E, Cloetingh S (2009) Controls of mantle plumes and lithospheric folding on modes of intraplate continental tectonics: differences and similarities. Geophys J Int 178:1691–1722
Burov E, Yamato P (2008) Continental plate collision, P-T–t–z conditions and unstable vs. stable plate dynamics: insights from thermo-mechanical modeling. Lithos 103:178–204
Burov E, Jaupart C, Guillou-Frottier L (2003) Ascent and emplacement of buoyant magma bodies in brittle-ductile upper crust. J Geophys Res 108:2177
Burov E, Francois Th, Yamato Ph et al (2014) Mechanism of continental subduction and exhumation of HP and UHP rock. Gondwana Res 25:464–493
Carslaw HS, Jaeger JC (1959) Conduction of heat in solids, 2nd edn. Clarendon Press, Oxford
Carter NL, Tsenn MC (1987) Flow properties of continental lithosphere. Tectonophysics 136:27–63
Chatterjee ND, Johannes WS (1974) Thermal stability and standard thermodynamic properties of synthetic 2M1- muscovite, KAl2Al3Si3O10(OH)2. Contrib Mineral Petrol 48:89–114
Chemenda AL, Mattauer M, Malavieille J et al (1995) A mechanism for syn-collisional deep rock exhumation and associated normal faulting: results from a physical modelling. Earth Planet Sci Lett 132:225–232
Chopra PN, Patterson MS (1984) The role of water in the deformation of dunite. J Geophys Res 89:7861–7876
Choukroune P, Ludden JN, Chardon D et al (1997) Archaean crustal growth and tectonic processes: a comparison of the Superior Province, Canada and the Dharwar Craton, India. Geol Soc Spec Publ 121:63–98. Blackwell, London
Christensen UR (1992) An Eulerian technique for thermomechanical modeling of lithospheric extention. J Geophys Res 97:2015–2036
Clark SP (1966) Handbook of physical constants. Revised edition. Geol Soc Amer Memoir 97
Cloos M (1993) Lithospheric buoyancy and collisiona l orogenesis: sulbduction of oceanic p lateaus, continenta l margins, is land arcs, spreading ridges, and seamounts. Geol Soc Am Bull 105:715–737
Coleman RG, Wang X (eds) (1995) Ultrahigh pressure metamorphism. Cambridge University Press, Cambridge, New York, Port Chester, Melbourne, Sydney
Compton RR (1960) Contact metamorphism in the Santa Rosa Range, Nevada. Geol Soc Amer Bull 71:1383–1416

Condie KC (1981) Archean greenstone belts. Elsevier, Amsterdam
Condie KC (2001) Mantle plumes and their record in earth history. Cambridge University Press, Cambridge
Connolly JAD (1990) Multivariable phase-diagrams—an algorithm based on generalized thermodynamics. Am J Sci 290:666–718
Connolly JAD (2005) Computation of phase equilibria by linear programming: a tool for geodynamic modeling and its application to subduction zone decarbonation. Earth Planet Sci Lett 236:524–541
Conrad CP, Hager BH (1999) The thermal evolution of an earth with strong subduction zones. Geophys Res Lett 26:3041–3044
Copeland P, Harrison TM (1990) Episodic rapid uplift in the Himalaya revealed by 40Ar/39Ar analysis of detrital K-feldspar and muscovite, Bengal Fan. Geology 18:354–357
Crawford ML, Mark LE (1982) Evidence from metamorphic rocks for overthrusting. Pennsylvania Piedmont, USA. Can Mineral 20:333–347
De Capitani C, Petrakakis K (2010) The computation of equilibrium assemblage diagrams with Theriak/Domino software. Am Mineral 95:1006–1016
De Yoreo JJ, Lux DR, Guidotti CV (1989a) A thermal model for Carboniferous metamorphism near the Sebago batholoth in western Maine. In: Tucker RD, Marvinney RG (eds) Igneous and metamorphic geology. Maine Geol Sur V, Augusta, ME, vol 3, pp 19–34
De Yoreo JJ, Lux DR, Guidotti CV et al (1989b) The Acadian thermal history of western Maine. J Metamorph Geol 7:169–190
De Yoreo JJ, Lux DR, Guidotti CV (1991) Thermal modelling in low-pressure/high-temperature metamorphic belts. Tectonophysics 188:209–238
Delaney PT (1988) FORTRAN 77 programs for conductive cooling of dikes with temperature-dependent thermal properties and heat of crystallization. Comput Geosci 14:181–212
Dobretsov NL (1974) Glaukofanslantsevyye i eklogit-glaukofanslantsevyye kompleksy SSSR (Glaucophane schists and eclogite-glaucophane schists complexes of the USSR). Nauka, Novosibirsk
Dobretsov NL (1981) Global'nyye petrologicheskiye protsessy (Global petrological processes). Nedra, Moscow
Dobretsov NL (1991) Blueschist and eclogites: a possible plate tectonic mechanism for the emplacement from the upper mantle. Tectonophysics 186:253–268
Dobretsov NL (2000) Collision processes in Paleozoic foldbelts of Asia and exhumation mechanisms. Petrology 8(5):403–427
Dobretsov NL (2010) Global geodynamic evolution of the Earth and global geodynamic models. Russ Geol Geophys 51(6):592–610
Dobretsov NL, Kirdyashkin AG (1991) Dynamics of subduction zones: models of accretionary wedge formation and exhumation of glaucophane schist and eclogites. Soviet Geol Geophys 32 (3):9–20
Dobretsov NL, Kirdyashkin AG (1994) Glubinnaya geodinamika (Deep-seated geodynamics). GEO, Novosibirsk
Dobretsov NL, Kirdyashkin AG (1994) Blueescltist belts of North Asia and models of subduction-accretion wedge. In: Coleman RG (ed) Reconstuction of Paleo-Asian ocean. VSP International Science Publication, Netherland, pp 91–106
Dobretsov NL, Khlestov VV, Reverdatto VV et al (1972) The facies of metamorphism. Australian National University, Canberra, Publ, p 214
Dobretsov NL, Khlestov VV, Sobolev VS (1973) The facies of regional metamorphism at moderate pressure. Australian National University, Canberra, Publ, p 236
Dobretsov NL, Sobolev VS, Sobolev NV et al (1975) The facies of regional metamorphism at high pressure. Australian National University, Canberra, Publ, p 266
Dobretsov NL, Sobolev NV, Shatsky VS (1989) Eklogity i glaukofanovyye slantsy v skladchatykh oblastyakh (Eclogites and glaucophane schists in folded areas). Nauka, Siberian Branch, Novosibirsk

Dodson MH (1973) Closure temperature in cooling geochronological and petrological systems. Contrib Mineral Petrol 40:259–274
Dortman MB (ed) (1984) Phisicheskie svoistva gornyh porod i poleznyh iskopaemyh (petrophisica) (Physical properties of rocks and minerals (petrophysics). Nedra, Moscow
Droop GTR, Brodie KH (2012) Anatectic melt volumes in the thermal aureole of the Etive Complex, Scotland: the roles of fluid-present and fluid-absent melting. J Metamorph Geol 30. https://doi.org/10.1111/j.1525-1314.01001.x.2012
Duchkov AD, Lysak SV, Balobaev VT et al (eds) (1987) Teplovoe pole nedr Sibiri (Thermal field of Siberia interiors). Nauka, Novosibirsk
Dudarev AN, Kudryavtsev VA, Melamed VG et al (1972) Teploobmen v magmatogennykh protsessakh (Heat transfer in magmatic processes). Nauka, Siberian Branch, Novosibirsk
Elan R (1985) High grade contact metamorphism at the Lake Isabella north shore roof pendant, southern Sierra Nevada, California. Dissertation, University of Southern California
England PC, Richardson SW (1977) The influence of erosion upon the mineral facies of rocks from different metamorphic environments. J Geol Soc London 134:201–213
England PC, Thompson AB (1984) Pressure-temperature-time paths of regional metamorphism: heat transfer during the evolution of regions of thickened continental crust. J Petrol 25:894–928
Englebrecht JP (1990) Contact metamorphism processes related to the aureole of the Bushveld complex in Marico district, western Transvaal, South Africa. Geology 93:339–349
Ernst WG (1973) Blueschist metamorphism and P-T regimes in active subduction zones. Tectonophysics 17:255–272
Ernst WG (1975) Systematics of large-scale tectonics and age progressions in Alpine and circum-Pasific blueschist belts. Tectonophysics 26:229–246
Ernst WG (1988) Tectonic history of subduction zones inferred from retrograde blueschist P-T paths. Geology 16:1081–1084
Eskola P (1948) The problem of mantled gneiss domes. Quart J Geol Soc London 104:461–476
Evans BW (1990) Phase relations of epidote-blueschists. Lithos 25:3–23
Faccenda M, Gerya TV, Burlini L (2009) Deep slab hydration induced by bending related variations in tectonic pressure. Nature Geosci 2:790–793
Ferry JM, Spear FS (1978) Experimental calibration of the partitioning of Fe and Mg between biotite and garnet. Contrib Mineral Petrol 66:113–117
Friedinger PJ, Reverdatto VV, Polyansky OP (1991) Metamophism of subsiding sediments in rift structures of the Earth's crust (approach on the basis of model). Sov Geol Geophys 32(9):71–79
Friedrich AM, Hodges KV, Bowring SA et al (1999) Geochronological constraints on the magmatic, metamorphic and thermal evolution of the Connemara Caledonides, western Ireland. J Geol Soc London 156:1217–1230
Fullagar D, Butler JR (1979) 325 to 265 Ma-old granitic plutons in the Piedmont of the south-eastern Appalachians. Am J Sci 279:161–185
Furlong KP, Hanson RB, Bower JR (1991) Modeling thermal regimes. In: Kerrick DM (ed) Contact metamorphism. Reviews in mineralogy, vol 26, Mineral Soc Am, Book Crafters Inc., Chelsea, Michigan, pp 437–505
Gaiduk VV (1988) Vilyuyskaya srednepaleozoyskaya riftovaya sistema (The Middle Paleozoic Vilyui Rift System). YaF SO AN SSSR, Yakutsk
Garagash IA, Ermakov VA (2004) A probable geodynamic model of the early earth. Dokl Earth Sci 394(1):73–77
Gerya TV (2002) P-T-trendy i model' formirovaniya granulitovykh kompleksov dokembriya (P-T trends and model of formation of Precambrian granulite complexes). Doctor of Science Dissertation. Moscow State University, Moscow
Gerya TV (2010) Introduction to numerical geodynamic modelling. Cambridge University Press, Cambridge
Gerya TV (2015) Tectonic overpressure and underpressure in lithospheric tectonics and metamorphism. J Metamorph Geol 33:785–800

Gerya TV, Burg J-P (2007) Intrusion of ultramafic magmatic bodies into the continental crust: numerical simulation. Phys Earth Planet Inter 160:124–142

Gibson RG, Speer JA (1986) Contact aureoles as constraints on regional P-T trajectories, an example from the northern Alabama Piedmont, USA. J Metamorph Geol 4:285–308

Glukhovskoy MZ, Moralev VM (1996) Geotektonika plyumov rannego dokembriya na primere evolyutsii Sunnaginskogo enderbitovogo kupola (Aldanskiy shchit) (Geotectonics of plumes of the Early Precambrian on the example of the evolution of the Sunnagin enderbite dome (Aldan Shield)). Geotectonika 6:81–93

Golubev VS, Sharapov VN (1974) Dinamika endogennogo rudoobrazovaniya (Dynamics of endogenous ore formation). Nedra, Moscow

Green DH (1967) High-temperature peridotite intrusions. In: Wyllie PJ (ed) Ultramafic and related rocks. Wiley, New York, pp 212–222

Griffin WL, Sturt BA, O'Neill CJ et al (2013) Intrusion and contamination of high-temperature dunitic magma: the Nordre Bumandsfjord pluton, Seiland, Arctic Norway. Contrib Mineral Petrol 165:903–930

Grigor'eva LV, Shinkarev NF (1981) Usloviya obrazovaniya kupol'nykh struktur v Priladozh'ye (Emplacement Conditions of Dome-Shaped Structures of the Ladoga region). Izvestiya Akad Nauk SSSR, Ser Geol 3:41–50

Guillot S, Hattori K, Agard Ph et al (2009) Exhumation processes in oceanic and continental subduction contexts: a review. In: Lallemand S, Funiciello F (eds) Subduction zone geodynamics. Springer, Berlin, pp 175–205

Hacker BR, Abers GA, Peacock SM (2003) Subduction factory 1. Theoretical mineralogy, densities, seismic wave speeds, and H_2O contents. J Geophys Res 108(B1):2029. https://doi.org/10.1029/2001jb001127

Hansen FD, Carter NL (1982) Creep of selected crustal rocks at 1000 MPa. Trans Amer Geophys Union 63:437

Harley SL (1989) The origin of granulites: metamorphic perspective. Geol Mag 126:215–247

Harwood DS (1973) Bedrock geology of the Capsaptic and Arnold Pond quadrangles, west central Maine. US Geol Surv Bull 1346:90

Hasenclever J, Morgan JP, Hort M et al (2011) 2D and 3D numerical models on compositionally buoyant diapirs in the mantle wedge. Earth Planet Sci Lett 311:53–68

Hayba DO, Ingebritsen SE (1997) Multiphase groundwater flow near cooling plutons. J Geophys Res 102:12235–12252

He B, Xu Y-G, Paterson S (2009) Magmatic diapirism of the Fangshan pluton, southwest of Beijing, China. J Struct Geol 31:615–626

Hermes OD, Murray DP (1988) Middle Devonian to Permian plutonism and volcanism in the N. American Appalachians. In: Harris AL, Fettes DJ (eds) The Caledonian Appalachian Orogeny, vol 38. Geological Society Special Publication, Blackwell, London, pp 559–571

Herzberg C, Condie K, Korenaga J (2010) Thermal history of the earth and its petrological expression. Earth Planet Sci Lett 292:79–88

Hodge DS (1974) Thermal model for origin of granitic batholiths. Nature 251:297–299

Hodges KV (2005) Geochronology and thermochronology in orogenic system. In: Rudnick RL (ed) The crust. Holland HD, Turekian KK (eds) Treatise on geochemistry, vol 3. Elsevier, Oxford, pp 263–292

Hoisch TD (1989) A muscovite-biotite geothermometer. Am Mineral 74:565–572

Holdaway MJ (2000) Application of new experimental and garnet Margules data to the garnet-biotite geothermometer. Am Mineral 85:881–892

Holdaway MJ, Lee SM (1977) Fe-Mg cordierite stability in high-grade pelitic rocks based on experimental, theoretical and natural observations. Contrib Mineral Petrol 63:175–198

Holland TJB, Powell R (1985) An internally consistent thermodynamic dataset with uncertainties and correlations: 2. Data and results. J Metamorph Geol 3:343–370

Holland TJB, Powell R (1990) An enlarged and updated internally consistent thermodyna- mic dataset with uncertainties and correlations: the system K2O–Na2O–CaO–MgO–MnO–FeO–Fe2O_3–Al_2O_3–TiO_2–SiO_2–C–H_2–O_2. J Metamorph Geol 8:89–124

Holland TJB, Powell R (1998) An internally consistent thermodynamic data set for phases of petrological interest. J Metamorph Geol 16:309–343
Holness MB (1999) Contact metamorphism and anatexis of Torridonian arkose by minor intrusions of the Rhum Igneous Complex, Inner Hebrides, Scotland. Geol Mag 136:527–542
Horai K (1971) Thermal conductivity of rock-forming minerals. J Geophys Res 76:1278–1308
Huerta AD, Royden LH, Hodges KV (1999) The effects of accretion, erosion and radiogenic heat on the metamorphic evolution of collisional orogens. J Metamorph Geol 17:349–366
Huismans RS, Podladchikov YY, Cloetingh SAPL (2001) Transition from passive to active rifting: relative importance of asthenospheric doming and passive extension of the lithosphere. J Geophys Res 106:11271–11292
Ingersoll LR, Zobel OJ (1913) An introduction to the mathematical theory of heat conduction with engineering and geological applications. Ginn and Co, Boston
Jaeger JC (1957) The temperature in the neighbourhood of a cooling intrusive sheet. Am J Sci 255:305–318
Jaeger JC (1959) Temperatures outside a cooling intrusive sheet. Am J Sci 257:44–54
Jaeger JC (1964) Thermal effect of intrusions. Rev Geophys 2:443–466
Jaeger JC (1968) Cooling and solidification of igneous rocks. In: Hess HH and the late Poldervaart A (eds) Basalts. The Poldervaart treatise on rocks of basaltic composition. Interscience Publisher, New York-London-Sidney, pp 503–536
Jamieson RA, Beaumont C (2013) On the origin of orogens. Geol Soc Am Bull 125:1671–1702. https://doi.org/10.1130/B30855.1
Jamieson RA, Beaumont C, Fullsack P et al (1998) Barrovian regional metamorphism: where's the heat? In: Treloar PJ, O'Brien PJ (eds) What drives metamorphism and metamorphic reactions? special publication, vol 138. Geological Society, London, pp 23–45
Jamieson RA, Beaumont C, Nguyen MH et al (2002) Interaction of metamorphism, deformation and exhumation in large convergent orogens. J Metamorph Geol 20:9–24
Jansen JBH, Schuiling RD (1976) Metamorphism on Naxos: petrology and geothermal gradients. Am J Sci 276:1225–1253
Jarrard RD (1986) Relations among subduction parameters. Rev Geophys 24:217–284
Jarvis GT, McKenzie DP (1980) Sedimentary basin formation with finite extension rates. Earth Planet Sci Lett 48:42–52
Jellinek AM, Kerr RC (1999) Mixing and compositional stratification produced by natural convection. 2. Applications to the differentiation of basaltic and silicic magma chambers and komatiite lava flows. J Geophys Res 104:7203–7218
Johannes W (1984) Beginning of melting in the granite system Qz–Or–Ab–An–H_2O. Contrib Mineral Petrol 86:264–273
Johnson MRW, Harley SL (2012) Orogenesis: the making of mountains. Cambridge University Press, Cambridge
Karabinos P, Ketchman R (1988) Thermal structure of active thrust belts. J Metamorph Geol 6:559–570
Karato S, Wu P (1993) Rheology of the upper mantle: a synthesis. Science 260:771–778
Katz RF, Weatherley SM (2012) Consequences of mantle heterogeneity for malt extraction. Earth Planet Sci Lett 335–336:226–237
Khain VE, Lomize MG (1995) Geotektonika s osnovami geodinamiki (Geotectonics with the basics of geodynamics). Moscow State University, Moscow
Kirdyashkin AA, Kirdyashkin AG (2013) Interaction of a thermochemical plume with free convection mantle flows and its influence on mantle melting and recrystallization. Russ Geol Geophys 54(5):544–554
Kiselev AI, Yarmolyuk VV, Ivanov AV et al (2014) Middle Paleozoic basaltic and kimberlitic magmatism in the northwestern shoulder of the Vilyui Rift, Siberia: relations in space and time. Russ Geol Geophys 55(2):144–152
Kleemann U, Reinhardt J (1994) Garnet-biotite thermometry revisited: The effect of AlVI and Ti in biotite. Eur J Mineral 6:925–941

Klein EM (2003) Geochemistry of the igneous oceanic crust. In: Rudnick RL (ed) Treatise on geochemistry. Crust, vol 3, pp 433–463
Kohn MJ (2008) *P-T-t* data from central Nepal support critical taper and repudiate large-scale channel flow of the Greater Himalayan Sequence. Geol Soc Amer Bull 120:259–273
Kohn MJ, Orange DL, Spear FS et al (1992) Pressure, temperature, and structural evolution of west-central New Hampshire: hot thrusts over cold basement. J Petrol 33:521–556
Kolobov VY, Likhanov II, Reverdatto VV (1992) Kontaktovo-metamorficheskiye porody (Contact-metamorphic rocks). In: Dobretsov NL, Bogatikov OA, Rosen OM (eds) Classification and nomenclature of metamorphic Rocks Nauka. Siberian Branch, Novosibirsk, pp 77–97
Korenaga J (2003) Energetics of mantle convection and the fate of fossil heat. Geophys Res Lett 30:1437–1440
Korikovsky SP (1995) Kontrastnyye modeli progradno-retrogradnoy evolyutsii meta-morfizma fanerozoyskikh skladchatykh poyasov v zonakh kollizii i subduktsii (Contrast models of the prograde-retrograde evolution of metamorphism of the Phanerozoic folded belts in zones of collision and subduction). Petrologiya 3(1):45–63
Koritnig S (1955) Die Blaue Kuppe bei Eschwege mit ihren Kontacterscheinungen. Heidelberg Beitr Mineral Petrogr 4:504–521
Korobeinikov SN, Polyansky OP, Likhanov II et al (2006) Mathematical modeling of overthrusting fault as a cause of andalusite–kyanite metamorphic zoning in the Yenisei Ridge. Dokl Earth Sci 408(4):652–656
Korobeinikov SN, Polyansky OP, Sverdlova VG et al (2008) Computer modeling of underthrusting and subduction under conditions of gabbro–eclogite transition in the mantle. Dokl Earth Sci 421(5):724–728
Korobeynikov SN (2000) Nelineynoye deformirovaniye tverdykh tel (Nonlinear Deformation of Solids). SO RAN Press, Novosibirsk
Korobeynikov SN, Reverdatto VV, Polyanskii OP et al (2009) Computer simulation of undrthrusting and subduction due to collision of slabs. Num Anal Applic 2(1):58–73
Korsman K, Korja T, Pajunen M et al (1999) The GGT/SVEKA transect—structure and evolution of the continental crust in the Paleoproterozoic Svecofennian orogen in Finland. Int Geol Rev 41:287–333
Kozakov IK, Glebovitsky VA, Bibikova EV et al (2002) Hercynian granulites of Mongolian and Gobian Altai: geodynamic setting and formation conditions. Dokl Earth Sci 386(7):781–785
Kozakov IK, Didienko AN, Azimov PY et al (2011) Geodynamic settings and formation conditions of crystalline complexes in the South Altai and South Gobi metamorphic belts. Geotectonics 45(3):174–194
Koziol AM, Newton RC (1988) Redetermination of the garnet breakdown reaction and improvement of the plagiclase-garnet-Al_2SiO_5-quartz geobarometer. Am Mineral 73:216–223
Krebs M, Maresch WV, Schertl H-P et al (2008) The dynamics of intra-oceanic subduction zones: A direct comparison between fossil petrological evidence (Rio San Juan Complex, Dominican Republic) and numerical simulation. Lithos 103:106–137
Kremenetsky AA, Milanovsky SY, Ovchinnikov LN (1989) A heat generation model for the continental crust based on deep drilling in the Baltic Shied. Tectonophysics 159:231–246
Kronenberg AK, Tullis J (1984) Flow strength of quartz aggregates: grain size and pressure effects due to hydrolytic weakening. J Geophys Res 89:4281–4297
Kruckenberg SC, Vangervaeghe O, Ferre EC et al (2011) Flow of partially molten crust and the internal dynamics of a migmatite dome, Naxos, Greece. Tectonics 30. https://doi.org/10.1029/2010tc002751
Kusznir NJ, Park RG (1984) The strength of intraplate lithosphere. J Phys Earth Planet Int 36:224–235
Kuzmin MI, Yarmolyuk VV, Kravchinsky VA (2010) Phanerozoic hot spot traces and paleogeographic reconstructions of the Siberian continent based on interaction with the African large low shear velocity province. Earth-Sci Rev 102:29–59

Lachenbruch AH, Sass JH (1978) Models of an extending lithosphere and heat flow in the Basin and Range Province. Cenozoic Tectonics und Regional Geophysics of the Western Cordillera. Geol Soc Am Mem 152:209–250

Landau LD, Lifshitz EM (1986) Gidrodinamika (Hydrodynamics), 3rd edn. Nauka, Moscow

Lazaro C, Garcia-Casco A, Rojas Agramonte Y et al (2009) Fifty-five-million-year history of oceanic subduction and exhumation at the northern edge of the Carribean plate (Sierra del Convento mélange, Cuba). J Metamorph Geol 27:19–40

Le Breton N, Thompson AB (1988) Fluid-absent (dehydration) melting of biotite in metapelites in the early stages of crustal anataxis. Contrib Mineral Petrol 99:226–237

Lee MK (1986) A new gravity survey of the Lake district and three-dimensional model of the granite batholith. J Geol Soc London 143:425–435

Leonov YuG (2004) Kharakternyye osobennosti stroyeniya i razvitiya nekotorykh tipov osadochnykh basseynov (Characteristic features of the structure and development of some types of sedimentary basins). In: Leonov YuG, Volozh YuA (eds) Sedimentary basins: methods of study, structure and evolution. Nauch Mir, Moscow, pp 38–60

Lepezin GG (1972) Metamorfizm fatsii epidotovıh amfibolitov (Metamorphism of the facies of epidote amphibolites). Nauka, Moscow

Li ZH, Gerya TV, Burg P (2010) Influence of tectonic overpressure on P-T paths of HP-UHP rocks in continental collision zones: Thermomechanical modelling. J Metamorph Geol 28:227–247

Likhanov II, Reverdatto VV (2002) Mass transfer during andalusite replacement by kyanite in Al- and Fe-rich metapelites in the Yenisei Range. Petrology 10(5):479–494

Likhanov II, Reverdatto VV (2011) Neoproterozoic collisional metamorphism in overthrust terranes of the Transangarian Yenisei Ridge, Siberia. Int Geol Rev 53:802–845

Likhanov II, Reverdatto VV, Memmi I (1994) Short-range mobilization of elements in the biotite zone of contact aureole of the Kharlovo gabbro intrusion (Russia). Eur J Mineral 6:133–144

Likhanov II, Ten AA, Reverdatto VV et al (1996) Low-grade metamorphism of clays at the contacts of andesite plugs in Turkmenistan (western Badkhys). Transactions (Doklady) Earth Sci 346(1):78–81

Likhanov II, Polyanskii OP, Kozlov PS et al (2000) Replacement of andalusite by kyanite with increasing pressure at a low geothermal gradient in metapelites of the Enisei Ridge. Dokl Earth Sci 375(9):1411–1416

Likhanov II, Polyanskii OP, Reverdatto VV et al (2001) Metamorphic evolution of high-alumina metapelites near the Panimba overthrust (Yenisei Range): mineral associations, P-T conditions, and tectonic model. Geol Geofiz 42(8):1205–1220

Likhanov II, Polyansky OP, Reverdatto VV et al (2004) Evidence from Fe- and Al- rich metapelites for thrust loading in the Transangarian Region of the Yenisei Ridge, eastern Siberia. J Metamorph Geol 22:743–762

Likhanov II, Kozlov PS, Popov NV et al (2006) Collision metamorphism as a result of thrusting in the Transangara region of the Yenisei Ridge. Dokl Earth Sci 411(1):1313–1317

Likhanov II, Reverdatto VV, Kozlov PS et al (2008a) Collision metamorphism of Precambrian complexes in the Transangarian Yenisei Range. Petrology 16(2):136–160

Likhanov II, Reverdatto VV, Verschinin AE (2008b) Fe- and Al-rich metapelites of the Teya sequence, Yenisei Range: geochemistry, protoliths and the behavior of their matter during metamorphism. Geochem Int 46(1):17–36

Likhanov II, Reverdatto VV, Kozlov PS (2011) The Teya polymetamorphic complex in the Transangarian Yenisei Ridge: an example of metamorphic superimposed zoning of low- and medium-pressure facies series. Dokl Earth Sci 436(2):213–218

Likhanov II, Reverdatto VV, Kozlov PS (2012) U-Pb and ^{40}Ar-^{39}Ar evidence for Grenvillian activity in the Yenisei ridge during formation of the Teya metamorphic complex. Geochem Int 50(6):551–557

Likhanov II, Nozhkin AD, Reverdatto VV et al (2014) Grenville tectonic events and evolution of the Yenisei Ridge at the western margin of the Siberian craton. Geotectonics 48(5):371–389

Likhanov II, Régnier J-L, Santosh M (2018) Blueschist facies fault tectonites from the western margin of the Siberian Craton: implications for subduction and exhumation associated with early stages of the Paleo-Asian Ocean. Lithos 304–307:468–488

Liou JG, Hacker BR, Zhang RY (2000) Into the forbidden zone. Science 287:1215–1216

Lister GS, Etheridge MA, Symonds PA (1991) Detachment models for the formation of passive continental margins. Tectonics 10:1038–1064

Little TA, Hacker BR, Gordon SM et al (2011) Diapiric exhumation of Earth's youngest (UHP) eclogites in the gneiss domes of the D'Entrecasteaux Islands, Papua New Guinea. Tectonophysics 510:39–68

Ljubimova EA (1968) Termika Zemli i Luny (Thermics of Earth and Moon). Nauka, Moscow

Logatchev NA, Rogozhina VA, Solonenko VP et al (1978) Deep structure and evolution of the Baikal rift zone. In: Ramberg IB, Neumann ER (eds) Tectonisc and geophysics of continental rifts. Norwell MaS Reidel, pp 49–62

Long SP, McQuarrie N, Tobgay T et al (2012) Variable shortening rates in the eastern Himalayan thrust belt, Bhutan: insights from multiple thermochronologic and geochronologic data sets tied to kinematic reconstructions. Tectonics 31(5). https://doi.org/10.1029/2012tc0 03155

Loosveld RJH, Etheridge MA (1990) A model for low-pressure facies metamorphism during crustal thickening. J Metamorph Geol 8:257–267

Lovering TS (1935) Theory of heat conduction applied to geological problems. Geol Soc Amer Bull 46:69–94

Lovering TS (1936) Heat conduction in dissimilar rocks and the use of thermal models. Geol Soc Amer Bull 47:87–100

Lovering TS (1955) Temperatures in and near intrusions. Econ Geol 50:249–281

MacKenzie DB (1960) High-temperature alpine-type peridotite from Venezuela. Geol Soc Amer Bull 71(3):303–318

Mancktelow NS (2008) Tectonic pressure: theoretical concepts and models. Lithos 103:149–177. In: Abstracts of the international symposium "Large Igneous Provinces of Asia, Mantle Plumes and Metallogeny", Novosibirsk, 13–16 Aug 2007. Publ. House of SB RAS, Novosibirsk, pp 39–42

Martin H (1994) The Archean grey gneisses and the genesis of continental crust. In: Condie KC (ed) Archean crustal evolution. Elsevier, Amsterdam, pp 205–259

Maruyama S, Liou JG, Terabayashi M (1996) Blueschists and eclogites of the world and their exhumation. Int Geol Rev 38:485–594

Masaitis VL (2007) Devonian basalts of Siberian platform, and their heterogeneous mantle sources. In: Abstracts of the international symposium "Large Igneous Provinces of Asia, Mantle Plumes and Metallogeny", Novosibirsk, 13–16 August 2007. Publishing house of SB RAS, Novosibirsk, p 39–42

Masaitis VL, Mikhailov MV, Selivanovskaya TV (1975) Vulkanism i tektonika Patomsko-Viluyiskogo srednepaleozoiskogo avlakogena (Volcanism and tectonics of the middle Paleozoic Patom-Vilyui aulacogen). Nedra, Moscow

Mason R (1978) Petrology of the metamorphic rocks. George Allen and Unwin, London

Mathew G, De Sarkar S, Pande K et al (2013) Thermal metamorphism of the Arunachal Himalaya, India: Raman thermometry and thermochronological constraints on the tectono-thermal evolution. Int J Earth Sci 102:1911–1936

McCarthy TS, Fripp REP (1980) The crystallization history of a granitic magma, as revealed by trace element abundances. J Geol 88:211–224

McLelland JM, Isachsen YW (1985) Geological evolution of the Adirondack mountains: a review. In: Tobi AC, Touret JLR (eds) The deep Proterozoic crust in the North Atlantic provinces. NATO ASI series (Series C. Mathematical and physical sciences). vol 158. Springer, Dordrecht, pp 175–215

McKenzie D (1978) Some remarks on the development of sedimentary basins. Earth Planet Sci Lett 40:25–32

McKenzie D, Bickle MJ (1988) The volume and composition of melt generated by extension of lithosphere. J Petrol 29:625–679

Mei S, Bai W, Hiraga T et al (2002) Influence of melt on the creep behavior of olivine-basalt aggregates under hydrous conditions. Earth Planet Sci Lett 201:491–507

Milliard Y (1959) Les massifs metamorphiques et ultrabasiques de la zone paleozoique interne du Rif. Notes du service Geol du Maroc 18(147):125–160

Mints MV, Glaznev VN, Konilov AN et al (1996) Ranniy dokembriy severo-vostoka Baltiyskogo shita (Early Precambrian of the northeast Baltic shield: paleogeodynamics, structure and evolution of the continental crust). Nauchnyi Mir, Moscow

Miyashiro A (1961) Evolution of metamorphic belts. J Petrol 2:277–311

Miyashiro A (1973) Metamorphism and metamorphic belts. Allen and Unwin, London

Mohan MR, Satyanarayanan M, Santosh M, Sylvester PJ, Tubrett M, Lam R (2013) Neoarchean suprasubduction zone arc magmatism in southern India: geochemistry, zircon U-Pb geochronology and Hf isotopes of the Sittampundi anotthosite complex. Gondwana Res 23:539–557

Montel J-M, Kornprobst J, Vielzeuf D (2000) Preservation of old U-Th-Pb ages in shielded monazite: example from the Beni Bousera Hercynian kinzigites (Marocco). J Metamorphic Geol 18:335–342

Montelli R, Nolet G, Dahlen FA et al (2004) Finite- frequency tomography reveals a variety of plums in the mantle. Science 303:338–343

Moretti I, Froidevaux C (1986) Thermomechanical models of active rifting. Tectonics 5:501–511

Morozov YuA, Gaft DE (1985) O prirode granitogneisovyh kupolov Severnogo Priladozya (On the nature of granite-gneiss domes of the Northern Ladoga region). In: Ez VV (ed) Structure and petrology of the precambrian complexes. IFZ AN SSSR, Moscow

Mossakovskii AA, Ruzhentsev SV, Samygin SG et al (1993) Centralnio-asiatskiy skladchtyi poyas: geodinamicheskaya evoluciya i istoriya formirovaniya (Central Asian fold belt: geodynamic evolution and the history of formation). Geotectonika no 6:3–33

Mukherjee S (2013) Channel flow extrusion model to constrain dynamic viscosity and Prandtl number of the Higher Himalayan Shear Zone. Int J Earth Sci 102:1811–1835

Nabelek PI, Hofmeister AM, Whittington AG (2012) The influence of temperature-dependent thermal diffusivity on the conductive cooling rates of plutons and temperature–time paths in contact aureoles. Earth Planet Sci Lett 317(318):157–164

Nakada M (1994) Convective coupling between ductile lower crust and upper mantle and its tectonic implications. Geophys J Int 118:579–603

Nell J (1985) The Bushveld metamorphic aureole in the Potgietersrus area, evidence for a two-stage metamorphic event. Econ Geol 80(4):1129–1152

Nenakhov VM (2001) Geodinamicheskie osobennosti rannego arheya (Geodynamic features of the early Archean). Geotectonika no 1:3–15

Newton RC, Haselton HT (1981) Thermodynamics of the garnet-plagioclase-Al_2SiO_5- quartz geobarometer. In: Newton RC, Navrotsky A, Wood BJ (eds) Advances in physical geochemistry, vol 1. Springer, New-York, pp 131–147

Norlander BH, Whitney DL, Teyssier Ch et al (2002) Partial melting and decompression of the Thor Odin dome, Shuswap metamorphic core complex. Lithos 61:103–125

Norton D, Knight J (1977) Transport phenomena in hydrothermal systems: cooling plutons. Amer J Sci 277:937–981

Nozhkin AD, Krendelev FP, Myronov AG (1975) Radioaktivniye elementy v dokembrii Eniseyskogo kriazha (Radioactive elements in Precambrian of Yenisei Range). In: Kuznetsov VA (ed) Radioactive elements in rocks. Nauka, Novosibirsk, pp 183–189

Nozhkin AD, Turkina OM, Bibikova EV et al (1999) Riphean granite-gneiss domes of the Yenisei Ridge: geological structure and U-Pb isotope age. Geol Geofiz 40(9):1284–1292

Obata M (1980) The Ronda peridotite: garnet-, spinel-, and plagioclase – lherzolite facies and the *P-T* trajectories of a high-temperature mantle intrusion. J Petrol 21:533–572

Okay AI (2002) Jadeite–chloritoid–glaucophane–lawsonite blueschists in north-west Turkey: unusually high *P*/*T* ratios in continental crust. J Metamorphic Geol 20:757–768

Oxburgh ER, Turcotte DL (1974) Membrane tectonics and the East African Rift. Earth Planet Sci Lett 22:l33–140

Parker EC, Davis PM, Evans JR et al (1984) Upwarp of anomalous asthenosphere beneath the Rio Grande Rift. Nature 312:354–356

Patchett PJ (1992) Isotopic studies of Proterozoic crustal growth and evolution. In: Condie KC (ed) Proterozoic crustal evolution. Developments in Precambrian geology, vol 10. Elsevier, Amsterdam, p 481–508

Paterson SR, Tobisch OT (1992) Rates and progress in magmatic arcs: implications for the timing and nature of pluton emplacement and wall rock deformation. J Struct Geol 14:291–300

Paterson SR, Vernon RH, Fowler TK (1991) Aureole tectonics. In: Kerrick DM (ed) Contact metamorphism. Reviews in mineralogy, vol 26. Mineral Soc Amer, Washington, pp 673–722

Pattison DRM (1991) P-T-a(H_2O) conditions in the thermal aureole. Equilibrium and kinetics in contact metamorphism. In: Voll G, Töpel J, Pattison DRM et al (eds) The Ballachulish igneous complex and its aureole. Springer, Heidelberg, pp 327–350

Pattison DRM (1992) Stability of andalusite and sillimanite and the Al_2SiO_5 triple point: constraints from the Ballachulish aureole, Scotland. J Geol 100:423–446

Pattison DRM (2001) Instability of Al_2SiO_5 « triple point » assemblages in muscovite + biotite + quartz – bearing metapelites, with implications. Am Mineral 86:1414–1422

Pattison DRM, Tracy RJ (1991) Phase equilibria and thermobarometry of metapelites. In: Kerrick DM (ed) Contact metamorphism. Reviews in mineralogy, vol 26. Mineral Soc Amer, Washington, pp 105–206

Peacock SM (1989) Numerical constraints on rates of metamorphism, fluid production, and fluid flux during regional metamorphism. Geol Soc Amer Bull 101:476–485

Peacock SM (1992) Blueschist facies metamorphism, shear heating and P-T-t paths in subduction shear zones. J Geophys Res 97:17693–17707

Peacock SM (1993) The importance of blueschist-eclogite dehydration reactions in subducting oceanic crust. Geol Soc Amer Bull 105:684–694

Pearson DG, Davies GR, Nixon PH (1995) Orogenic ultramafic rocks of UHP (diamond facies) origin. In: Coleman RG, Wang X (eds) Ultrahigh pressure metamorphism. Cambridge University Press, Cambridge, pp 456–510

Pecher A (1989) The metamorphism in the central Himalaya. J Metamorphic Geol 7:31–41

Pekhovich AI, Zhidkikh VM (1976) Raschety teplovogo regima tverdykh tel (Calculations of the thermal regime of solids). Energiya, Leningrad

Persikov ES (1984) Vyazkost magmaticheskih rasplavov (Viscosity of igneous melts). Nauka, Moscow

Pertsev NN (1977) Vysokotemperaturnyi metamorfizm i metasomatoz karbonatnykh porod (High-temperature metamorphism and metasomatism of carbonate rocks). Nauka, Moscow

Petford N (1996) Dykes or diapirs? Trans Royal Soc Edinburgh Earth Sci 87:105–114. https://doi.org/10.1017/S0263593300006520

Petrini K, Podladchikov Yu (2000) Lithospheric pressure-depth relationship in compressive regions of thickened crust. J Metamorphic Geol 18:67–77

Pitcher WS, Berger AR (1972) The geology of Donegal. A study of granite emplacement and unroofing. Wiley, New York

Pitcher WS, Sinha RC (1958) The petrochemistry of the Ardara aureole. Quart J Geol Soc London 113:393–408

Platt JP (1986) Dynamics of orogenic wedges and the uplift of high-pressure metamorphic rocks. Bull Geol Soc Amer 97:1037–1053

Pleuger J, Podladchikov YY (2014) A purely structural restoration of the NFP20-East cross section and potential tectonic overpressure in the Adula nappe (central Alps). Tectonics 33:656–685

Pollack HN (1997) Thermal characteristics of the Archaean. In: de Wit MJ, Ashwal LD (eds) Greenstone belts. Oxford monograph on geology and geophysics, vol 35. Oxford University Press, New York, pp 223–232

Polyansky OP, Efremov VN (1989) Diagnostics of dome-like structures of the Northern Ladoga region on the basis of thermodynamic data and tectonophysical analysis. Sov Geol Geophys 30 (4):36–39

Polyansky OP, Reverdatto VV (2002) Fluid convection in sediment-hosted reservoirs due to thermal action of dikes and sills. Russ Geol Geophys 43(1):25–39

Polyansky OP, Volkov PK (1990) Model of metamorphism at advection processes. Geologiya i Geofizika no 2:29–36

Polyanskii OP, Reverdatto VV, Anan'ev VA (2000) Evolution of the rift sedimentary basin as an indicator of geodynamic setting (on example of the Enisei-Khatanga depression). Dokl Akad Nauk SSSR 370(1):71–75

Polyanskii OP, Reverdatto VV, Sverdlova VG (2002) Convection of two-phase fluid in a layered porous medium driven by the heat of magmatic dikes and sills. Geochem Int 40(S1):S69–S81

Polyansky OP, Reverdatto VV, Khomenko AV et al (2003) Modeling of fluid flow and heat transfer induced by basaltic near-surface magmatism in the Lena-Tunguska petroleum basin (eastern Siberia, Russia). J Geochem Explor 78–79:687–692

Polyansky OP, Babichev AV, Reverdatto VV et al (2009) Computer modeling of granite magma diapirism in the Earth's crust. Doklady Earth Sci 429(8):1380–1384

Polyansky OP, Babichev AV, Korobeynikov SN et al (2010a) Computer modeling of granite gneiss diapirism in the Earth's crust: Controlling factors, duration, and temperature regime. Petrology 18(4):432–446

Polyansky OP, Korobeynikov SN, Sverdlova VG et al (2010b) The influence of crustal rheology on plate subduction based on numerical modeling results. Doklady Earth Sci 430(2):158–162

Polyansky OP, Sukhorukov VP, Travin AV et al (2011) Tectonic interpretation of the thermochronological data and P-T conditions of rock metamorphism in the Bodonchin zone complex (Mongolian Altai). Russ Geol Geophys 52(9):991–1006

Polyansky OP, Korobeynikov SN, Babichev AV et al (2012) Formation and upwelling of mantle diapirs through the cratonic lithosphere: numerical thermomechanical modeling. Petrology 20 (2):120–137

Polyansky OP, Prokop'ev AV, Babichev AV et al (2013) The rift origin of the Vilyui basin (East Siberia), from reconstructions of sedimentation and mechanical mathematical modeling. Russ Geol Geophys 54(2):121–137

Polyansky OP, Korobeinikov SN, Babichev AV et al (2014) Numerical modeling of mantle diapirism as a cause of intracontinental rifting. Izvestiya, Physics Solid Earth 50(6):839–852

Polyansky OP, Babichev AV, Sukhorukov VP et al (2015) A thermotectonic numerical model of collisional metamorphism in the Mongolian Altai. Doklady Earth Sci 465(1):1164–1167

Polyansky OP, Reverdatto VV, Babichev AV et al (2016) The mechanism of magma ascent through the solid lithosphere and relation between mantle and crustal diapirism: numerical modeling and natural examples. Russ Geol Geophys 57(6):843–857

Polyansky OP, Prokopiev AV, Koroleva OV et al (2017) Temporal correlation between dyke swarms and crustal extension in the middle Palaeozoic Vilyui rift basin, Siberian platform. Lithos 282–283:45–64

Powell R, Holland TJB (1994) Optimal geothermometry and geobarometry. Am Mineral 79:120–133

Preiss WV (1987) The Adelaide geosyncline: late Proterozoic stratigraphy, sedimentation, palaeontology and tectonics. Geol Surv South Australia Bull no 53:411–426

Ranalli G (1995) Rheology of the Earth. Chapman & Hall, London

Reiners PW (2009) Nonmonotonic thermal histories and contrasting kinetics of multiple thermochronometers. Geochim Cosmochim Acta 73(12):3612–3629

Reverdatto VV (1970) Ob izochemicheckoy prirode kontaktovogo metamorfisma (On the isochemical nature of contact metamorphism). Geologiya i Geofisika no 5:53–63

Reverdatto VV (1973) The facies of contact metamorphism. Australian National University Publ no 233, Canberra

Reverdatto VV (1973b) Velichiny geotermicheskih gradientov pri regionalnom metamorfizme (The values of geothermal gradients in regional metamorphism). Geologiya i Geofisika no 8:36–43

Reverdatto VV Kalinin AS (1989a) Two-dimensional models of metamorphism and anatexis in folded regions of the crust. 1. Model of magmatic intrusive. Geologiya i Geofizika 30(6):54–58

Reverdatto VV, Kalinin AS (1989b) Two-dimensional models of metamorphism and anatexis in folded regions of the Earth's crust. 2. Model of fluid flow. Geologiya i Geofizika 30(8):37–42

Reverdatto VV, Kalinin AS (1990) Two-dimensional models of metamorphism and anatexis in folded belts of Earth's crust. 3. Combined fluid-magmatic model, comparison with other models, and analysis of the problem. Geologiya i Geofizika 31(6):1–8

Reverdatto VV, Melenevskij VN (1983) Magmatic heat as a factor of hydrocarbon generation; the case of the basaltic sills. Geol Geofiz 24(6):13–21

Reverdatto VV, Polyansky OP (1992) Evolution of PT-parameters in the alternative models of metamorphism. Dokl Akad Nauk SSSR 325(5):1017–1020

Reverdatto VV, Polyansky OP (2004) Modelling of the thermal history of metamorphic zoning in the Connemara region (western Ireland). Tectonophysics 379:77–91

Reverdatto VV, Sheplev VS (1998) Geodynamic factors of metamorphism and their modeling: review and analysis of the problem. Russ Geol Geophys 39(12):1664–1677

Reverdatto VV, Sharapov VN, Melamed VG (1970) The controls and selected peculiari- ties of the origin of contact metamorphic zonation. Contrib Mineral Petrol 29:310–337

Reverdatto VV, Sharapov VN, Slobodskoy RM (1972) Some questions of analytical simulation of contact metamorphism. Contrib Mineral Petrol 36:195–206

Reverdatto VV, Sharapov VN, Lavrent'ev Y et al (1974) Investigations in isochemical contact metamorphism. Contrib Mineral Petrol 48:287–299

Reverdatto VV, Melenevskii VN, Melamed VG (1982) Contact-metamorphism of the rocks containing dispersed organic-matter - time and temperature as factors of hydrocarbon generation in the case of parallel basaltic sills. Dokl Akad Nauk SSSR 266(4):952–955

Reverdatto VV, Polyansky OP, Ananyev VA (1992) Model estimates of paleotemperatures and burial metamorphism during rifting processes. Dokl Akad Nauk SSSR 323(5):921–924

Reverdatto VV, Sheplev VS, Polyansky OP (1995) Burial metamorphism and evolution of rift troughs—a model approach. Petrology 3(1):31–37

Rey PF, Teyssieur C, Whitney DL (2009) Extension rates, crustal melting, and core complex dynamics. Geology 37:391–394

Robin CMI, Bailey RC (2009) Simultaneous generation of Archean crust and subcratonic roots by vertical tectonics. Geology 37(6):523–526

Rosenberg CL, Handy MR (2005) Experimental deformation of partially melted granite revisited: implications for the continental crust. J Metamorph Geol 23:19–28

Ross DC (1985) Mafic gneissic complex (batholithic root?) in the southernmost Sierra Nevada, California. Geology 13:288–291

Rozen OM, Shchipansky AA, Turkina OM (2008) Geodinamika ranney Zemli: evoluciya i ustoychivost' geologicheskih processov (ofiolity, ostrovnye dugi, cratony, osadochnye basseiny) (Geodynamics of the Early Earth: the evolution and stability of geological processes (ophiolites, island arcs, cratons, sedimentary basins)). Nauchnyi Mir, Moscow

Ruppel C (1995) Extensional processes in continental lithosphere. J Geophys Res 100 (B12):24187–24215

Ruppel C, Hodges KV (1994) Pressure-temperature-time paths from two-dimensional thermal models: prograde, retrograde and inverted metamorphism. Tectonics 13:17–44

Sajeev K, Windley BF, Hegner E et al (2013) High-temperature, high-pressure granulites (retrogressed eclogites) in the central region of the Lewissian, NW Scotland: crustal-scale subduction in the Neoarchaean. Gondwana Res 23:526–538

Sams DB, Saleeby JB (1988) Geology and petrotectonic significance of crystalline rocks of the southernmost Sierra Nevada, California. In: Ernst WG (ed) Metamorphism and crustal evolution, Western United States. Rubey vol 7. Prentice-Hall, Englewood Cliffs NJ, pp 865–893

Sandiford M, van Kranendonk MJ, Bodorkos S (2004) Conductive incubation and the origin of dome-and-keel structure in Archean granite-greenstone terrains: a model based on the eastern Pilbara Craton, Western Australia. Tectonics 23. https://doi.org/10.1029/2002tc001452

Schumacher JC, Schumacher R, Robinson P (1989) Acadian metamorphism in central Massachusetts and southwestern New Hampshire: evidence for contrasting *P-T* trajectories

In: Daly JS, Cliff RA, Yardley BWD (eds) Evolution of metamorphic belts, vol 43. Geol Soc Spec Publ, pp 453–460

Schmalholz SV, Podladchikov YY (2013) Tectonic overpressure in weak crustal-scale shear zones and implications for the exhumation of high pressure rocks. Geophys Res Lett 40:1984–1988

Schmalholz SM, Duretz T, Schenker FL et al (2014) Kinematics and dynamics of tectonic nappes: 2-D numerical modelling and implications for high and ultra-high pressure tectonism in the Western Alps. Tectonophysics 631:160–175

Searle PM, Stephenson B, Walker J et al (2007) Restoration of the Western Himalaya: implications for metamorphic protoliths, thrust and normal faulting, and channel flow models. Episodes 30 (4):242–257

Sengor AMC, Burke K (1978) Relative timing of rifting and volcanism on the earth and its tectonic implications. Geophys Res Lett 5:419–421

Sharapov VN, Averkin YuA (1990) Dinamika teplo- i massoperenosa v ortomagmaticheskih fluidnyh sistemah (Dynamics of heat and mass transfer in orthomagmatic fluid systems). Nauka, Novosibirsk

Sharapov VN, Akimtsev VA, Dorovsky VN et al (2000) Dinamika razvitiya rudno-magmatitcheskih sistem zon spredinga (Dynamics of development of ore-magmatic systems of spreading zones). Publishing house of SB RAS, Novosibirsk

Sheplev VS, Reverdatto VV (1994) The investigation of a model of rifting process. Dokl Akad Nauk SSSR 334(1):103–105

Shi Y, Wang C (1987) Two-dimensional modeling of the P-T-paths of regional metamorphism in simple overthrust terraines. Geology 15:1048–1051

Shimizu M (1986) The Tokuwa batholith, central Japan, – an example of occurrence of ilmenite—series and magnetite—series granitoids in a batholith. The University Museum Bull no 28, The University of Tokyo, Tokyo

Shmonov VM, Vitovtova VM, Zharikov AV (2002) Fluidnaya pronitsaemost' porod zemnoi kory (Fluid permeability of rocks of the Earth's crust). Nauchniy Mir, Moscow

Siegesmund S, Becker JK (2000) Emplacement of the Ardara pluton (Ireland): new constraints from magnetic fabrics, rock fabrics and age dating. Int J Earth Sci 89(2):307–327

Simonov VA, Sakiev KS, Volkova NI et al (2008) Conditions of formation of the Atbashi Ridge eclogites (South Tien Shan). Russ Geol Geophys 49(11):803–815

Sklyarov EV (2006) Exhumation of metamorphic complexes: basic mechanisms. Russ Geol Geophys 47(1):68–72

Sklyarov EV, Gladkochub DP, Donskaya TV et al (2001) Metamorfizm i tektonika (Metamorphism and Tectonics). Internet Engineering, Moscow

Sobolev NV, Shatsky VS (1987) Inclusions of carbon minerals in garnets from metamorphic rocks. Geologiya i Geofisika 28(8):1–18

Sobolev AV, Sobolev SV, Kuzmin DV et al (2009) Siberian meimechites: origin and relation to flood basalts and kimberlites. Russ Geol Geophys 50(12):999–1033

Sommer H, Kröner A (2013) Ultra-high temperature granulite-facies metamorphic rocks from the Mozambique belt of SW Tanzania. Lithos 170–171:117–143

Sorokov DS, Ginsburg GD (eds) (1974) Yenisei-Khatangskaya neftegazonosnaya oblast' (Yenisei-Khatanga oil and gas province). Sci Res Inst Geol Arctic, Leningrad

Spear FS (1993) Metamorphic phase equilibria and pressure-temperature-time paths. Mineralogical Society of America Monograph, Washington

Spear FS, Kohn MJ, Harrison TM (1989) A thermal model for west-central New Hampshire. Geol Soc Am, 1989 Annual Meeting 21:67–68

Spear FS, Hickmott DD, Selverstone J (1990) Metamorphic consequences of thrust emplacement, Fall Mountain, New Hampshire. Geol Soc Am Bull 102:1344–1360

Spear FS, Kohn MJ, Cheney JT et al (2002) Metamorphic, thermal and tectonic evolution of Central New England. J Petrol 43:2097–2120

Spear FS, Peacock SM, Kohn MJ et al (1991) Computer programs for petrologic P-T-t path calculations. Am Mineral 76:2009–2012

Spry AH, Solomon M (1964) Columnar buchites at Apsley, Tasmania. Quart J Geol Soc London 120:519–545

Strehlau J, Meissner R (1987) Estimation of crustal viscosities and shear stresses from an extrapolation of experimental steady state flow data. In: Fuchs K, Froidevaux C (eds) Compositions, structure and dynamics of the lithosphere-astenosphere system, Geodyn Ser, vol 16. AGU, Washington, pp 69–87

Stüwe K (2002) Geodynamics of the lithosphere. Springer, Heidelberg

Sukhorukov VP, Polyansky OP, Krylov AA et al (2016) Reconstruction of the metamorphic P-T path from the garnet zoning in aluminous schists from the Tsogt Block, Mongolian Altai. Petrology 24(4):409–432

Tapponnier P, Molnar P (1979) Active faulting and Cenozoic tectonics of the Tian Shan, Mongolia and Baikal regions. J Geophys Res 84(B7):3425–3459

Toé W, Vanderhaeghe O, André-Mayer A-S et al (2013) From migmatites to granites in the Pan-African Damara orogenic belt, Namibia. J African Earth Sci 85:62–74

Thompson A, Schulmann K, Jezek J (1997) Extrusion tectonics and elevation of lower crustal metamorphic rocks on convergent orogens. Geology 25:491–494

Thybo H, Artemieva IM (2013) Moho and magmatic underplating in continental lithosphere. Tectonophysics 609:605–619

Turcotte DL, Emerman SH (1983) Mechanism of active and passive rifting. Tectonophysics 94:39–50

Turcotte DL, Schubert G (1982) Geodynamics. Applications of continuum physics to geological problems. Wiley, New York

Turner FJ (1968) Metamorphic petrology. Mineralogical and field aspects. McGraw-Hill, New York

Turner FJ (1973) Buoyancy effects in fluids. (Cambridge monographs on mechanics and applied mathematics). Cambridge University Press, Cambridge. https://doi.org/10.1017/cbo9780511608827

Turner FJ, Verhoogen J (1951) Igneous and metamorphic petrology, 2nd edn. McGraw-Hill, New York

Van der Wal D, Vissers RLM (1993) Uplift and emplacement of upper mantle rocks in the western Mediterranean. Geology 21:1119–1122

Van Kranendonk MJ, Collins WJ, Hickman A et al (2004) Critical tests of vertical vs. horizontal tectonic models for the Archaean East Pilbara Granite-Greenstone Terrane, Pilbara Craton, Western Australia. Precambrian Res 131:173–211

Vanderhaeghe O (2004) Structural development of the Naxos migmatite dome. In: Whitney DL, Teyssier C, Siddoway CS (eds) Gneiss domes in orogeny. Geol Soc Am Special Paper, vol 380. pp 211–228

Veevers JJ (1984) Phanerozoic earth history of Australia. Oxford geol sci ser no 2. Clarendon Press, Oxford

Volkova NI, Sklyarov EV (2007) High-pressure complexes of Central Asian Fold Belt: geologic setting, geochemistry, and geodynamic implications. Russ Geol Geophys 48(1):83–90

Volkova NI, Tarasova EN, Polyanskii NV et al (2008) High-pressure rocks in the serpentinite melange of the Chara zone, Eastern Kazakhstan: Geochemistry, petrology, and age. Geochem Intern 46(4):386–401

Volkova NI, Stupakov SI, Babin GA et al (2009) Mobility of trace elements during subduction metamorphism as exemplified by the blueschists of the Kurtushibinsky Range, Western Sayan. Geochem Intern 47(4):380–392

Volkova NI, Travin AV, Yudin DS (2011) Ordovician blueschist metamorphism as a reflection of accretion-collision events in the Central Asian orogenic belt. Russ Geol Geophys 52(1):72–84

Voll G, Töpel J, Pattison DRM et al (1991) Equilibrium and kinetics in contact metamorphism. The Ballachulish igneous complex and its aureole. Springer, Heidelberg

Vrijmoed JC, Podladchikov YY, Andersen TB et al (2009) An alternative model for ultra-high pressure in the Svartberget Fe-Ti garnet-peridotite, Western Gneiss Region, Norway. Eur J Mineral 21:1119–1133

Walter MJ (2003) Melt extraction and compositional variability in mantle lithosphere. In: RW Carlson (ed) Treatise in geochemistry. The mantle and core, vol 2. Holland HD, Turekian KK (eds) Elsevier-Pergamon, Oxford, pp 363–394

Watson JV (1978) Precambrian thermal regimes. Philosoph transact Royal Soc London 288:431–440

Weinberg RF, Podladchikov Y (1994) Diapiric ascent of magmas through power crust and mantle. J Geophys Res 99(B5):9543–9559

Weissel JK, Karner GD (1989) Flexural uplift of rift flanks due to mechanical unloading of the lithosphere during extension. J Geophys Res 94(B10):13919–13950

Wells PRA (1979) Chemical and thermal evolution of Archean sialic crust, southern West Greenland. J Petrol 20:187–226

Wernicke B (1985) Uniform-sense normal simple shear of the continental lithosphere. Can J Earth Sci 22:108–125

Wernicke B, Burchfiel BC (1982) Modes of extensional tectonics. J Struct Geol 4:105–115

White R, McKenzie D (1989) Magmatism at rift zones: the generation of volcanic continental margins and flood basalts. J Geophys Res 94(B10):7685–7729

Whitney DL, Miller RB, Paterson SR (1999) *P-T-t* evidence for mechanisms of vertical tectonic motion in a contractional orogen: north-western US and Canadian Cordillera. J Metamorphic Geol 17:75–90

Wilde SA, Valley JW, Peck WH et al (2001) Evidence from detrital zircons for the existence of continental crust and oceans on the Earth 4.4 Gyr ago. Nature 409:175–178

Wu CM, Zhang J, Ren LD (2004) Empirical garnet-biotite-plagioclase-quartz (GBPQ) geobarometry in medium- to high-grade metapelites. J Petrol 45:1907–1921

Xiao W, Windley BF, Badarch G et al (2004) Paleozoic accretionary and convergent tectonics of the southern Altaids: implications for the growth of Central Asia. J Geol Soc London 161:339–342

Yardley BWD, Barber JP, Gray JR (1987) The metamorphism of the Dalradian rocks of western Ireland and its relation to tectonic setting. Philosoph Transact R Soc London 321:243–270

Yarmolyuk VV, Kozlovsky AM, Kuzmin MI (2016) Zoned magmatic areas and anorogenic batholith formation in the Central Asian Orogenic Belt (by the example of the Late Paleozoic Khangai magmatic area). Russ Geol Geophys 57(3):357–370

Yin A, Dubey CS, Kelty TK et al (2010) Geologic correlation of the Himalayan orogen and Indian craton: part 2. Structural geology, geochronology and tectonic evolution of the Eastern Himalaya. Geol Soc Amer Bull 122:360–395

Zen E, White WS, Hadley JB et al (1968) Studies of Appalachian geology, northern and marine. Wiley, New York

Zhang J, Wei C, Chu H (2015) Blueschists metamorphism and its tectonic implication of Late Paleozoic-Early Mesozoic metabasites in the mélange zones, central Inner Mongolia, China. J Asian Earth Sci 97:352–364

Zlobin VA, Kulikov AA, Bobrov VA (1975) Zakonomernosti raspredeleniya radioaktivnykh elementov v dokembriyskih otlozheniyah Eniseyskogo kryazha (Objective laws of distribution of radioactive elements in Precambrian suits of Yenisei Range). In: Kuznetsov VA (ed) Radioactive elements in rocks. Nauka, Novosibirsk, pp 198–203

Zorin YuA (1981) The Baikal rift: an example of the intrusion of asthenospheric material into lithosphere as the cause of disruption of lithospheric plates. Tectonophysics 73:91–104

Zwart HJ (1969) Metamorphic facies series in the European orogenic belts and their bearing on the causes of orogeny. Geol Assoc Canada Special Paper 5:7–16

Chapter 4
Metamorphic Processes in Rocks

4.1 Pressure-Temperature-Time (*P-T-t*) Paths as a Result of Metamorphic Evolution

There are two basic approaches to quantitative geothermobarometry of metamorphic rocks. The first one is the conventional "absolute" thermobarometry, which focuses on the reconstruction of the pressure and temperature conditions at which a rock equilibrated at some point in its history. A second approach, the "relative" thermobarometry is exemplified by the determination of changes in physical conditions experienced by a rock through analysis of mineral zoning and reaction microtextures (Spear 1989).

Information obtained with these two approaches has different applications. The evaluations of the thermodynamic conditions by means of "absolute" thermobarometry may be applied to quantify the depths and thermal structures at which the metamorphic complexes were produced. The output of the "relative" thermobarometry is the *P-T-t* paths, which can be used to develop a model of rock evolution with time (Spear and Peacock 1989). Although the *P-T-t* paths provide more information than the peak *P-T* parameters of metamorphism for estimating the mechanisms of geodynamic processes, a complete characteristic of a metamorphic rocks should involve the application of both approaches.

There are several stages of changing thermodynamic conditions that most metamorphic rocks passed through during their evolution. These transformations are preserved as mineral relics, microtextures and chemical zoning within individual grains, indicative of previous events. On *P-T* diagrams, these transformations are reflected in *P-T-t* paths (directed lines or loops), which represent a record of coherent changes in temperature and pressure during different stages of rock evolution. The differences in *P-T* paths usually suggest distinct tectonic processes. Each *P-T-t* path or its section is given specific reference to a time variable, and the overall trajectory represents a different geotectonic setting. Different regional metamorphic conditions and/or events can be recognized within single or multiple stages of

V. V. Reverdatto et al., *The Nature and Models of Metamorphism*,
Springer Geology, https://doi.org/10.1007/978-3-030-03029-2_4

tectogenesis, and the time lag between different metamorphic events may be significant. Therefore, the reconstruction of *P-T-t* paths requires the application of a set of petrological methods, including geothermobarometry and distinctive mineral assemblages, coupled with high-precision ages of metamorphic stages. Despite a number of difficulties that complicate the analysis of *P-T-t* paths, they provide an effective tool for understanding the tectonothermal processes taking place during regional metamorphism and are crucial for solving geodynamic problems.

P-T-t paths describe possible mechanisms leading to burial and subsequent exhumation under different geological conditions. During such tectonic transport, any rock follows its individual and unique path in space and time. Each rock unit may experience loss or gain of heat, and changes in its position in different crustal levels result in variations in the lithostatic pressure load on the rock. Figure 4.1 shows a model of destructive plate margin with position of a rock unit at the different depth level h in the crust as a function of time during a continent-continent collision (a) and corresponding paths followed by the rock unit in *P-T* space (b–e). The situation here depicts a continent-continent collision with the formation of continental crust twice its normal thickness (Bucher and Grapes 2011). This model is used to explain the relationship between tectonics and metamorphism during collisional orogeny, which was first recognized by England and Thompson (1984). In Fig. 4.1a at t_1 (0 Ma) there is a rock unit (indicated by an open square) at depth h, the position of which changes depth with time (t_1 through to t_6 at 30 Ma). In Fig. 4.1b–e, the position of the rock unit (filled squares) during tectonic transport is shown in terms of *P-T* space. The time slices are arbitrary and have been chosen in accordance with time scales of the formation of Alpine-type orogenic belts. At t_1 the rock unit lies on a stable steady-state geotherm. At t_2 (10 Ma) tectonic transport moves the crust together with the rock unit beneath another continental crust of normal 35 km thickness. Increasing depth of the rock unit is accompanied by increasing pressure. At the same time, the rock unit begins to receive more heat than at its former position at t_1. However, because heat transport is a slow process compared with tectonic transport, dP/dT tends to be much steeper (Fig. 4.1c) than the corresponding dP/dT slope of the initial steady-state geotherm (Fig. 4.1b). Between t_1 and t_2, the rock unit has traveled in *P-T* space along a path that is on the high-pressure side of the initial steady-state geotherm. The rock unit is now on a transient geotherm that changes its shape as time progresses. At t_3 the crust is twice its normal thickness (about 70 km), which is about the maximum thickness in continent–continent collision zones. The rocks have reached their maximum depth and consequently their maximum pressure of about 2 GPa (Fig. 4.1c). Continued plate motion does not increase the thickness of the crust and pressure remains constant as long as underthrusting is going on. On the other hand, heat transfer to the rock unit in question increases the temperature as shown in Fig. 4.1d (t_4). From this time, a number of feasible mechanisms may control the path of the rock unit. Continued tectonic transport may return slices and fragments of rock to shallower levels in a material counter current (dashed arrow in Fig. 4.1a), or simply, after some period of time, plate convergence stops, e.g. because frictional forces balance the force moving the plate. The thickened crust starts to uplift, and erosion restores the crust to its original thickness. By this mechanism the rock unit may return to its original depth position and, given enough time, the stable steady-state geotherm will be re-established. The path between t_4 and t_5 is characterized by decompression (transport along the h-axis). If initial uplift rates are slow compared with heat transport rates, the rocks will experience a continued temperature increase during uplift as shown in Fig. 4.1e. However, at some stage along the path the rocks must start to lose more heat to the surface than they receive from below and consequently cooling begins. The point t_5 in Fig. 4.1e represents the maximum temperature position of the path traveled by the rock unit. At t_6 the rock has returned to its former position on the steady state geotherm (Fig. 4.1a).

The consequences for pressure and temperature of the geologic process illustrated in Fig. 4.1a–e are summarized in Fig. 4.1f. A rock at depth *h* follows a clockwise pressure–temperature loop (CW) (P_{max} before T_{max}). Such clockwise *P-T-t* paths are a characteristic feature of orogenic metamorphism and have been documented from such diverse mountain belts as the Scandinavian Caledonides, Western Alps, Appalachians, and Himalayas. Very often "normal" orogenesis is characterized by the following sequence of *P-T-t* path sections: isothermal thickening, isobaric heating, isothermal decompression and isobaric cooling. In detail, clockwise *P-T-t* loops may show a number of additional complications and local

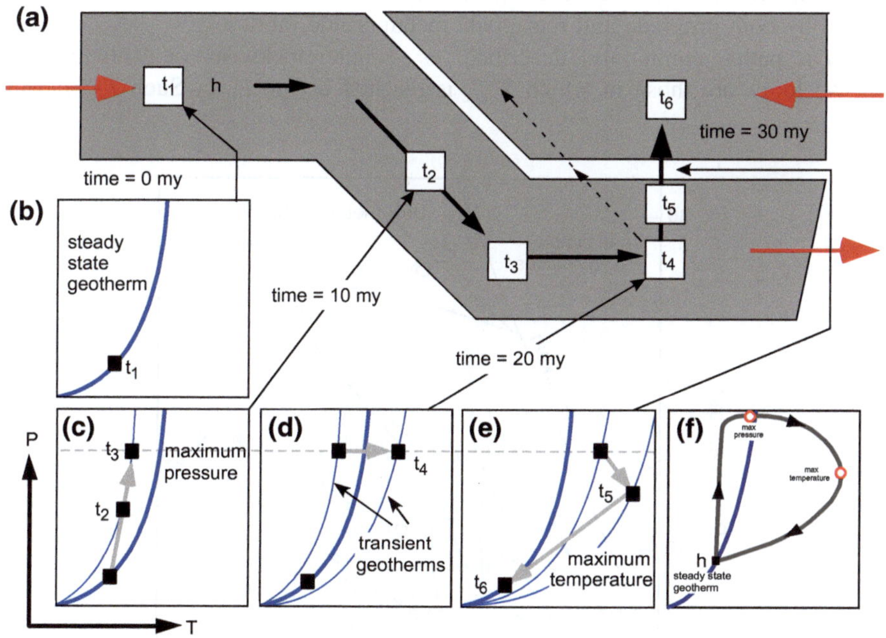

Fig. 4.1 Schematic diagram showing the position of a rock unit at the different depth level *h* in the crust as a function of time during a continent-continent collision (**a**) and corresponding paths followed by the rock in *P-T* space (**b–e**) after Bucher and Grapes (2011). The rock depicts a clockwise pressure-temperature loop (**f**). Direction of plate motion is shown by heavy red arrows

features. A similar evolution of *P-T* metamorphic parameters can be inferred for cool subduction zones. *P-T-t* paths for different types of metamorphism differ mainly in *dT/dP* slopes for the prograde and retrograde segments of the metamorphic evolution and in the length of recorded *P-T* history, which correspond to subduction (low *dT/dP*) and collisional (moderate *dT/dP*) conditions. For example, some rocks in convergent settings witness a significant pressure increase with only minor heating during the later tectonometamorphic stage, suggesting a low *dT/dP* gradient. One of the possible mechanisms of such process is the rapid subduction/thrusting with subsequent fast uplift and erosion (Huerta et al. 1999; Jamieson et al. 2002; Korobeinikov et al. 2006). According to this model, the subducted/underthrust slab fails to equilibrate thermally before the rocks are brought back near the surface. In other words, the low *dT/dP* gradients are associated with relatively short-lived events and the lack of thermal equilibrium between the blocks of rock at the respective depths because of a strong thermal inertia relative to pressure. As seen in Fig. 4.1f, the maximum temperature point along the *P-T-t* path followed by a metamorphic rock does not necessarily coincide with the maximum pressure point of the path. This means that maximum pressure and maximum temperature will be generally diachronous.

In Fig. 4.2 shows a generalized *P-T* diagram of the evolution of metamorphic complexes in various geodynamic settings representing different types of metamorphism. It was constructed using data on well-studied natural objects characterized by both prograde and retrograde metamorphic fabrics.

P-T-t paths commonly described as counterclockwise or anticlockwise (CCW) loops are those in which T_{max} is reached before P_{max}. Such *P-T-t* paths

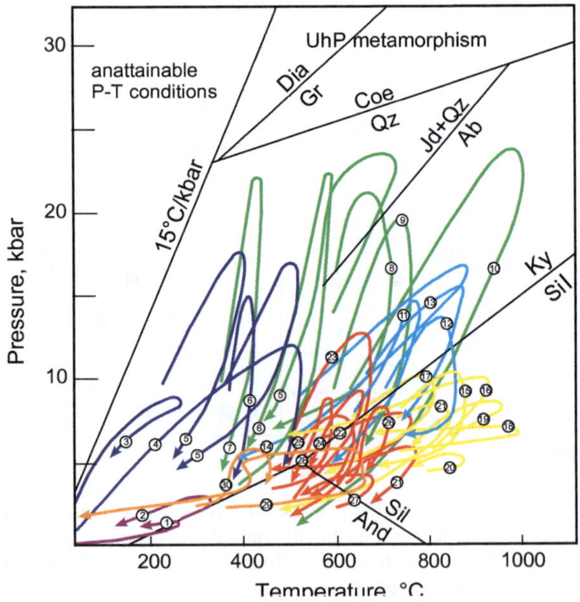

◀**Fig. 4.2** *P-T* diagram showing the diversity of reconstructed *P-T-t* paths for geological complexes of the different types of metamorphism. Low-temperature metamorphic rocks (purple color of loop): 1—Dniepr-Donetsk basin (Reverdatto and Polyansky 1992) CCW; 2—Welsh basin (Robinson and Beavins 1989) CCW. Blueshists (dark blue color of loop): 3—mélange of Franciscan type of subduction, western California (Ernst 1988) CW; 4—mélange of Alpine type of subduction, western Alps (Ernst 1988) CW; 5—serpentinite mélange in Rio San Juan Complex, Dominican Republic (Krebs et al. 2008) CW; 6—northern Caribbean subduction–accretionary complex (Escuder-Viruete and Pérez-Estaún 2013) CW. Eclogites (green color of loop): 5—serpentinite mélange in Rio San Juan Complex, Dominican Republic (Krebs et al. 2008) CCW; 6—northern Caribbean subduction–accretionary complex (Escuder-Viruete and Pérez-Estaún 2013) CCW/CW; 7—central Qiangtang, northern Tibet, China (Zhai et al. 2011) CW; 8—metabasites from northeastern Sardinia, Italy (Cruciani et al. 2011) CW; 9—western Odenwald Crystalline Complex, Variscan Mid-German Crystalline Rise, southern Germany (Will and Schmadicke 2003) CW; 10—northern North China Craton (Wan et al. 2015) CW. High-pressure granulites of eclogite-granulite type (light blue color of loop): 11—northeastern segment of the Tarim Craton (Dunhuang block) (He et al. 2014) CW; 12 and 13—mafic and pelitic granulites, respectively, the Jiaobei massif in the Jiao-Liao-Ji Belt, North China Craton (Tam et al. 2012a, b) CW; 14—Dulan area, North Qaidam Mountains, northwestern China (Yu et al. 2011) CW. Granulites, including ultrahigh-temperature granulites (yellow color of loop): 15—Daqingshan Complex of the Khondalite Belt, North China Craton (Cai et al. 2014) CW; 16—north Qinling-Tongbai orogen, Central China (Xiang et al. 2012) CCW; 17—Limpopo Belt, southern Africa (Tsunogae and van Reenen 2006) CW; 18—Ongole domain of the Eastern Ghats Belt, India (Sarkar and Schenk 2014) CCW; 19—Angara-Kan block, Yenisei Ridge, western margin of the Siberian Craton, Russia (Likhanov et al. 2015a, 2016) CCW; 20—Musgrave Block, central Australia (White et al. 2002) CCW; 21—Lapland Granulite Belt, northern Finland (Cagnard et al. 2011) CW. Amphibolites of overthrust terranes, including migmatites, gneisses and schists (red color of loop): 21—Lapland Granulite Belt, northern Finland (Cagnard et al. 2011) CW; 22—Feiran–Solaf region, Egypt (Abu-Alam and Stuwe 2009) CW; 23—Slavonian Mountains, Tisia Mega-Unit, Croatia (Balen et al. 2015) CW; 24—Garevka Complex, Transangarian Yenisei Ridge, Russia (Likhanov et al. 2015b) CCW; 25—generalized evolution of the several Al_2SiO_5 tripple point localities: Boehls Butte area, Idaho, Mt. Moosilauke area, New Hampshire, and areas in north-central New Mexico (Pattison 2001) CW; 26—Teya Complex, Transangarian Yenisei Ridge, Russia (Likhanov and Reverdatto 2011b) CW; 27—western Odenwald Crystalline Complex, southern Germany (Will and Schmadicke 2003) CCW; 28—accretionary mélange between Kunlun and Karakorum, Aghil Range, southwestern Sinkiang, China (Groppo and Rolfo 2008) CCW; 29 and 30 (orange color of loop): different thermal history in the two structural levels —upper-plate (CCW) and lower-plate (CW) of the Fall Mountain nappe, southwest New Hampshire, USA (Spear et al. 1990)

occur when rocks undergo burial and exhumation within a young and hot subduction zone (Lazaro et al. 2009). They may also occur in terrains that have experienced an initial phase of crustal extension. Counter-clockwise *P-T-t* paths have been reported from granulite facies terrains where an event of heating from igneous intrusions precedes crustal thickening. The same path may be followed by rocks undergoing low-temperature burial metamorphism or it may be recorded by high-temperature granulites associated with crustal extension.

The largest and most complex changes in the thermal structure of the crust occur at convergent plate margins. Collision can generate clockwise or counterclockwise metamorphic *P-T-t* paths, according to the relative rates of thickening and heat transfer (Brown 2007). Collision-related *P-T-t* paths are commonly described as clockwise. Several examples of such collision complexes with different *P-T-t* paths, but resulted from the same geodynamic event—crustal thickening—have been described by Likhanov et al. (2010a, b, 2011a, b, c). Most rocks that have experienced collisional metamorphism under amphibolite facies conditions (7–9 kbar/550–700 °C) with subsequent retrograde decompression (4–5 kbar/450–500 °C) would follow clockwise *P-T-t* paths (Fig. 4.2). However, there are several well-documented examples of metamorphic belts, which have counterclockwise *P-T-t* paths (Likhanov et al. 2013a, b; Reinhardt and Rubenach 1989; Bohlen 1991; Collins and Vernon 1991; Hand et al. 1992; Rubenach 1992; Johnson and Vernon 1995; Brown 2001; Johnson et al. 2003; Perchuk et al. 2006). The *P-T-t* paths inferred in these studies correlate with *P-T* trajectories predicted by thermomechanical models calculated for different crustal thickening and exhumation histories, accounting for different subsidence mechanisms and exhumation rates (Beaumont et al. 2001; Jamieson et al. 2002; Gerya 2014). A good example of such counterclockwise *P-T* trajectory is metapelites of the Garevka complex (Yenisei Ridge), which documents initial prograde low pressure heating followed by near isothermal medium-pressure compression and post-peak retrograde syn-exhumation decompression and cooling. During the late metamorphic stage the mid-crustal amphibolite-facies rocks were exhumed to upper-crustal levels (accompanied by a total pressure and temperature decrease by 3 kbar and 120 °C, respectively). However, such counterclockwise near isothermal decompression and cooling path also suggests that retrograde metamorphism may have occurred within an overall extensional tectonic setting (Johnson and Harley 2012; Tong et al. 2014). In the sequence of tectonic events associated with the Precambrian evolution of the Yenisei Ridge, the late stages of orogenesis were marked by collision-related medium-pressure metamorphism as recorded in the pelitic gneisses (Likhanov et al. 2009), followed by the development of rift-related bimodal dike swarms, as a result of Neoproterozoic extensional processes along the western margin of the Siberian craton and the onset of Rodinia breakup. This is supported by the tectonic and age relationships between the gneisses and schists of the Garevka complex ($\sim$800 Ma) and the cross-cutting felsic and mafic dikes with ages of 797–792 Ma, which are interpreted to postdate the peak of collisional metamorphism in the region (Likhanov and Santosh 2017).

Within deep thrust or subduction zones, where metamorphic rocks show an inverted metamorphic zonation, the hanging wall or footwall near the shear plane can also be characterized by an opposite *P-T-t* path with different metamorphic field gradients (England and Molnar 1993). Contrasting metamorphic *P-T-t* paths from the upper (CCW) and lower (CW) plates of the nappe, New England, USA were considered in detail by Spear et al. (1990) (Fig. 4.2; curves 29 and 30). Different metamorphic paths reconstructed from eclogite and glaucophane schist block in the serpentine mélange of the Rio San Juan complex, Dominican Republic are characteristic of subduction zones (Fig. 4.2; curve 5). These blocks represent different stages of subduction: the highest temperature CCW path corresponds to the incipient subduction, while the low-temperature CW path corresponds to a more mature stage (Krebs et al. 2008).

The interpretation of *P-T-t* paths for regions experienced different metamorphic events can be controversial (Sklyarov et al. 2001; Perchuk et al. 2015; Likhanov et al. 2018). Contrasting types of metamorphism in the system subduction-suprasubduction zone can be represented by low- to medium- to high-pressure eclogite-glaucophane facies rocks to medium-pressure zonal andalusite-sillimanite and granulite facies rocks. Under these conditions, the construction of *P-T-t* paths reflecting different scenarios of metamorphic rock formation in different tectonic settings is significantly hampered (Brown 2007), and differences can be recognized only from the retrograde segments of *P-T-t* paths. For example, the Alpine-type subduction is characterized by a decrease in pressure at an almost constant temperature (Fig. 4.2, curve 4), whereas the rocks in the East Pacific (Cordilleran) type subduction zones usually record evidence for synchronous cooling and decreasing pressure (Fig. 4.2, curve 3) (Ernst 1988; Gao and Klemdt 2003). More complex *P-T-t* paths can also be observed. They can differ even within the same type of metamorphism and record several cycles of burial and exhumation. For example, the precise *P-T-t* path suggests that the eclogites of the Akeyazi terrane in Chinese Tianshan underwent a polycyclic evolution including two burial–exhumation cycles during convective flow in a subduction channel with two *P-T* loops indicated by the occurrence of polyphase garnet and mutually replaced and regrown amphibole (Li et al. 2016).

In general, cases of fast exhumation are generally related to the collision of thick and/or buoyant oceanic or continental lithosphere with the trench-fore arc–arc system, which is normally associated with the end of subduction. Cases of slow exhumation, on the other hand, appear to be typical of near-steady-state subduction. In this scenario, fragments of the subducted slab are incorporated into the overlying subduction channel, where the blocks are generally slowly exhumed (Gerya 2002 and references therein). This is in contrast to fast exhumation where discrete tectonic events of collision and subsequent extension/collapse cause deep-seated rocks to rapidly ascend to the surface. Observations on natural rocks have also established a variety of *P-T-t* paths for subducted material (Ernst 1988), in agreement with geophysical calculations and modelling (Gerya 2002). Thus oceanic rocks exhumed in syn-subduction scenarios normally show cold exhumation paths (clockwise cooling during decompression or counterclockwise cooling at depth). Whereas the corner flow model provides an explanation for the existence of counterclockwise *P-T-t* paths (Cloos 1982), it appears that this type of path is not as general as the clockwise type. Counterclockwise *P-T-t* paths are considered to be characteristic of rocks subducted in juvenile subduction scenarios and, consequently, to document a transient thermal state during onset of subduction (e.g., Gerya 2002).

An increasing number of very high temperature terranes or occurrences are being recognized, both on the basis of distinctive mineral assemblages and geothermobarometry. Granulite facies terranes provide excellent opportunities for the accurate definition of *P-T-t* paths because of the availability of geothermobarometers and experimental data through which mineral zoning may be interpreted. Despite these advantages, the interpretation of granulite *P-T-t* paths is not simple because high cation diffusion rates generally preclude the preservation of early, prograde or maximum temperature segments of the paths. Usually, we are forced to infer tectonic processes based only on the later, post-peak or retrograde segments of the *P-T-t* paths, coupled with definitive structural and timing relationships and pertinent geochemical data on the nature and sources of the preserved rock types. The only way to restore a prograde segment of the *P-T-t* paths appears when porphyroblastic garnet or pyroxene contain mineral inclusions which can be used to estimate *P-T* conditions of pre-peak metamorphic stage. In this case mineral inclusions are attributed to the prograde evolution, whereas abundant and well-preserved reaction textures yield evidence for a multistage retrograde evolution (e.g., Tong et al. 2014; Likhanov et al. 2016; Nozhkin et al. 2018).

It is significant that a diversity of *P-T-t* paths is preserved in granulite terranes. Through the constraints of geothermobarometry, mineral assemblage evolution and related observation of textural relationships, it is apparent that many granulites have experienced near-isothermal decompression subsequent to their apparent thermal maximum (ITD paths, Fig. 4.2) whilst other terranes have undergone near-isobaric cooling following their thermal maxima (IBC paths, Fig. 4.2). Some granulite terranes exist where both types of paths are preserved in sequence and either related to one event (e.g. decompression followed by cooling) (e.g., Scourie Complex, Scotland) or resulting in overprinting of earlier basement gneisses by subsequent unrelated metamorphic event (e.g. shear zones in Napier Complex, Antarctica).

Certain types of mineral reaction textures, usually developed on the retrograde or post-peak segments of a *P-T* evolution, and mineral assemblage relationships within well-defined compositions are diagnostic of either near-isothermal decompression (ITD) or near isobaric cooling (IBC) in granulite facies terranes. For a detail review we refer our readers to an excellent work "The origin of granulites: a metamorphic perspective" by Harley (1989).

Under normal conditions, the exhumation of rock complexes from the metamorphic zone to the surface is characterized by a simultaneous decrease in pressure and temperature. Recently obtained evidences indicate that the exposure of metamorphic complexes whose genesis was related to collision process was usually an integrated effect of several tectonic mechanisms: tectonic denudation, i.e. the tectonic removal of overlying complexes in the process of large-amplitude extension, at a significant role of erosion denudation (Sklyarov 2006). However, the concept of the predominantly erosion-controlled exhumation of metamorphic complexes from depths of $\sim$25 km poses the question of the location of large volumes of sedimentary rocks of corresponding age, which are absent from nearby territories. The problem of these sediments was most convincingly solved in application of the exposure of some granulite and eclogite complexes. It was assumed that the uplift of granulite complexes and the simultaneous rock subsidence of the greenstone complexes were controlled by the gravity-driven redistribution of rocks in the Precambrian crust via crustal diapirism (Gerya and Maresch 2004; Perchuk et al. 2001). Exhumation of high-pressure rocks to the surface during subduction is possible with a viscous return flow over the upper contact of the subsucted plate, where the highly plastic hydrated rocks—serpentinites—occurred (Gerya 2010, 2014). Thus, the effect of individual exhumation mechanisms may vary at different stages of rock evolution, and metamorphic complexes may present examples of various combinations of these mechanisms, which are considered most exhaustively by Teyssier and Whitney (2002).

ITD *P-T-t* paths determined from granulite terranes are characterized by some cooling, generally through 50–110 °C concominant with decompression (Fig. 4.2). Most examples record decompression intervals of between 2 and 4 kbar, although the granulites of the Gruf Complex (Droop and Bucher-Nurminen 1984) may record decompression through some 10 kbar. Whilst most of the ITD paths are found in granulites metamorphosed in the range 6–9 kbar and 700–850 °C, most of their ITD paths define a common array with typical *dP*/*dT* gradient of 2–3 kbar/100 °C.

ITD granulites are interpreted to have formed in crust thickened by collision, with magmatic additions being an important extra heat source. Decompressional paths may, however, also be generated in extensional settings within the footwall to a low-angle extensional datachment zone (Wernicke 1985), provided that magmas accrete simultaneously onto the base of extending crust. This type of setting would result in quite hot but shallow level (2–5 kbar) ITD path such as those found in some Cordilleran core complexes. Erosion alone is not, however, considered to be the dominant post-collision thinning process. Instead, the ITD paths are generated during more rapid thinning (1–2 mm/year exposure) related to tectonic exhumation during moderate-rate or waning extension.

ITD paths in mafic granulites result in the removal of garnet from higher-pressure assemblages as decompression proceeds. Similar reaction textures involving the formation of orthopyroxene-plagioclase symplectites or fringes at the expense of garnet are produced in both quartz-bearing and quartz-free rock types. In quartz-bearing mafic granulites the reactions may also lead to the mantling of clynopyroxene by orthopyroxene. In quartz-absent mafic granulites initially containing garnet, spinel may be an additional phase in the orthopyroxene-plagioclase symplectites. Additional equilibria involving garnet and amphibole may also lead to orthopyroxene-plagioclase symplectites in mafic granulites following ITD paths. Pelitic and felsic granulites may also preserve good textural evidence for ITD. In felsic granulites reaction involving the Grt+Opx+Pl+Qz will produce plagioclase as moats and rims on garnet and between garnet-orthopyroxene on decompression. In very high-temperature granulites, textures indicative of decompression typically involve the overprinting of garnet- and sapphirine-bearing assemblages by cordierite-spinel-orthopyroxene-bearing one. Garnet breakdown textures are often responsible for spectacular orthopyroxene-cordierite symplectites and pseudomorphs replacing resorbed garnet. Other textures indicative of ITD in pelitic granulites include sillimanite rimming or enclosing garnets which have only sparse sillimanite needles, rutile grains in the cores, coupled with matrix ilmenite replacing rutile external to garnet, and plagioclase coronas, moats and rinds between garnet and nearby sillimanite. An important feature consistent with ITD paths is the appearance of late syn-metamorphic melts in pelitic and mafic granulites.

IBC paths usually occur in granulites at deep-crustal levels of 7–10 kbar, and at mid-crustal level of 4–7 kbar. Isobaric cooling usually does not involve some drop in pressure, through 0.5–2 kbar, but typical *dP*/*dT* gradients are only 0.3–0.5 kbar/100 °C.

IBC granulites may have formed in a variety of settings. Those which show anticlockwise *P-T-t* histories are interpreted to have formed in and beneath areas of voluminous magmatic accretion, with or without additional crustal extension. IBC granulites at shallow levels (<5 kbar) may also be formed during extension of normal thickness crust, but deeper-level IBC requires more complex models. Many granulites exhibiting IBC at deep crustal levels (>7.5 kbar) may have formed in thickened crust which underwent very rapid (5 mm/year) extensional thinning subsequent to collision. It is suggested that the preservation of IBC paths rather than ITD paths in many granulites is primarily related to the rate and timescale of extensional thinning of thickened crust, and that hybrid ITD to IBC paths should also be observed.

Mafic granulites on IBC paths usually develop coronitic textures with the production of secondary garnet as rims, lamellae or granules on orthopyroxene-plagioclase contacts or as overgrowths on earlier garnet; clinopyroxene overgrowths on orthopyroxene in such coronites. In essentially pyroxenitic rocks types the reaction high-Al pyroxene = low-Al pyroxene + garnet leads to textures such as garnet lamellae or grains within pyroxene; garnet nucleated at kink-bands or on exsolution lamellae in pyroxene; or garnet as grains or rims on pyroxene grain boundaries, often with additional recrystallized pyroxene neoblasts. Textures indicative of IBC in pelitic to felsic (e.g. charnochitic) granulites include those where secondary garnet is produced as overgrowth on earlier garnet, or as grains and rim lamellae on early orthopyroxene, through continuous equilibria involving a decrease in the Al-content of pyroxene or reaction with other aluminous phases such as plagioclase. In addition, there are numerous corona and intergrowth textures involving sapphirine, spinel, cordierite, sillimanite, garnet and orthopyroxene, which have been used to discriminate IBC paths. Granulite terranes with IBC paths should also be characterized by the lack of kyanite and general prelevance of sillimanite throughout metamorphic histories.

As noted previously, some examples exist (e.g., Scourie Complex, Scotland—Sills and Rollinson 1987; Serre, Italy—Schenk 1984; Angara-Kan, Siberia—Likhanov et al. 2016) where isobaric cooling is preceded by some decompression at temperatures in the range 800–850 °C. These terranes may represent cases where post-thickening extension was not rapid enough to sustain high temperatures through the whole decompressional stage.

Another possible variant of the general IBC path is one which involves some (1–2 kbar) increase in pressure with cooling (e.g. Namaqualand, S. Africa—Waters 1986) a feature which might be consistent with thermal subsidence and sedimentation onto a thinned continental crust (e.g., Sandiford and Powell 1991).

In ultrahigh-temperature granulites (UHT) of the Angara-Kan block (South Yenisei Ridge), the early retrograde segment of the *P-T-t* path also can be characterized by limited extent of near-isothermal decompression preceding isobaric cooling. This is documented by the specific reaction textures, including the cordierite + orthopyroxene symplectites, which are commonly developed during a rapid decompression event (Harley 1989). Similar successive retrograde *P-T* paths with initial decompression followed near isobaric cooling were reported for metapelitic granulites from the Kan complex (Perchuk 1989). These paragneisses show rare reaction textures such as replacement of cordierite by garnet, sillimanite and quartz, sillimanite coronas around spinel and corundum, and plagioclase rims around garnet. These data are consistent with a counterclockwise *P-T-t* path characterized by a high metamorphic field gradient of dT/dP = 100–200 °C/kbar involving initial prograde heating and post-peak retrograde decompression followed by near isobaric cooling. The near isobaric cooling and counterclockwise P-T path suggest that UHT metamorphism likely occurred in an overall extensional tectonic setting with associated underplating of mantle-derived mafic magma. This is supported by the close spatial and temporal association between ultrahigh-temperature metamorphic rocks and products of anorogenic magmatism and bimodal volcanism in the region, whereas a large gravity high above the South Yenisei Ridge indicates the presence of large volumes of dense mafic masses at lower-crustal depths, which may be sourced from synchronous giant radiating dike swarms linked to the Vilyuy mantle plume as part of the Trans-Siberian LIP (Ernst et al. 2008; Gladkochub et al. 2010).

The application of *P-T-t* paths for distinguishing tectonic settings and establishing the types of metamorphism is significantly hampered without additional geological information. This is illustrated in Fig. 4.2, which shows that the evolution of *P-T* parameters of the same type of metamorphism can be characterized by both clockwise (CW) and counterclockwise (CCW) trajectories. The principal differences in the direction of the retrograde segments defining the final trajectory of *P-T-t* paths are mainly controlled by the mechanisms of exhumation in various geodynamic settings of crustal compression or extension. The tectonic settings for the formation of granulites are best explored. The other types of rocks, especially those formed in tectonically active areas at convergent plate margins the situation, requires further investigation. Nevertheless, the *P-T-t* paths followed by rocks during their metamorphic history provide one of the most effective tools for studying tectonothermal processes during metamorphism. The interpretation of the results of the *P-T-t* evolution of rocks coupled with thermomechanical numerical simulation of lithospheric plate interaction accounting for varying velocities and mechanisms of the burial and exhumation of rocks (Gerya 2010, 2014) has potential for providing correct solution for a broad spectrum of geodynamic problems.

4.2 Mass Transfer During Metamorphism

4.2.1 Coronites and Models for Zoning Growth

4.2.1.1 Korzhinskii–Fisher–Joesten Model

The transport of matter during metamorphism is undoubtedly limited by the low rock permeability, which is usually less than 10^{-18} m^2 and may be reduced to 10^{-20}–10^{-22} m^2 with increasing temperature and lithostatic pressure; the minimum value of permeability is of the order of 10^{-23} m^2. The permeability of metamorphic rocks is generally comparable to the permeability of unaltered and undeformed igneous rocks, such as granites, diorites, basalts, etc. (Oelkers 1996; Shmonov et al. 2002).

Low permeability values of rocks suggest that the transfer of matter and the change in chemical composition during metamorphic reactions are controlled by diffusion. The diffusion of substance occurs down its concentration gradient and is efficient only for short distances; in the presence of a fluid (in the form of intergranular fluid films) the diffusion proceeds faster and over a longer distance than in the absence of a fluid. Given this limited mass transfer with increasing P and T, metamorphism can be treated as isochemical process (within a small volume), which involves virtually no change in the bulk chemistry of the rock almost impermeable to infiltrating fluids. Only in the case of high permeability, when it is possible for the substance to be introduced in the intergranular space by infiltrating fluids (as noted in Sect. 1.2), there are prerequisites for significant chemical change in the bulk composition of the original rock (in a large volume), which leads to metasomatism. The mass transfer by fluid infiltration is controlled by a hydrodynamic gradient, which appears to be a more effective mode of transport of substance and metasomatism than diffusion.

The interaction in the originally heterogeneous rock units undergoing metamorphism under changing P-T conditions takes place at the contact between the chemically contrasting (incompatible) rock units, which are not in equilibrium. The interaction occurs either between adjacent layers of the rocks or mineral inclusions (grains, concretions, segregations, xenoliths, etc.) with the composition differing from that of the surrounding rocks. Exchange and diffusion of chemical components result in the development of a reaction zoning (corona structure). Following Korzhinskii (1955), it can be referred to as "bimetasomatic". The diffusion zone formed during metamorphism over a geological timescale, i.e. 10^3–10^7 years, may range in width from a few tenths of millimeter, a few millimeters or centimeters to >10 cm. Zoning typically shows well-defined transverse banding with sharp boundaries. Such structure represents a frozen incomplete reaction, involving both the reactant and product phases. In other words, zoning is an intermediate product of the subsolidus transformation approaching overall equilibrium conditions. Reaction zoning in metamorphic rocks has been widely studies as it provides important information about the transport of matter and formation of metamorphic minerals.

The mechanism of formation of corona textures is still obscure. The use of mineral equilibria with participation of a fluid phase for solving this problem is not a simple task because of the lack of thermodynamic data and information on the amount and composition of the fluid phase. In a simplified form, the problem of the reaction rim formation must include information on the dissolution of initial mineral grains, the transfer of dissolved substances in intergranular fluid, and the growth of new solid phases. To describe the dynamics of this process, it is necessary to know the composition of the fluid phase, the solubility of minerals, the mobility (migration rate) of dissolved substances in intergranular spaces, and the growth kinetics of new minerals. Because of the unavailability of these data, the problem of dynamics is very complex and its solution requires a number of simplifications and assumptions.

Most recent models for corona growth are based on the ideas of Korzhinskii (1955, 1957, 1962, 1973, 1982), Fisher (1973, 1977, 1978; Fisher and Elliott 1974) and Joesten (1977, 1991; Joesten and Fisher 1988), who introduced the concept of local equilibrium between minerals or mineral assemblages in metamorphic rocks. As noted above, the latter is caused by low permeability of rocks, a large surface of mineral grains, differential mobility of chemical components. It is assumed that the concentration of chemical components in the intergranular fluid during metamorphism remains approximately constant within a certain local equilibrium volume, and the diffusion transport of matter occurs in a stationary (quasi-stationary) regime (Korzhinskii 1962; Fisher and Elliott 1974; Weare et al. 1976). Since the mass of solid phases is much greater than the mass of the solution (fluid) and the mineral assemblage in the local volume buffers the composition of intergranular fluid, the variance of the metamorphic system (rocks) will be low.

The Korzhinskii–Fisher–Joesten model is based on a few fundamental ideas about statistically interpreted transport of particles of substance treated as a continuum. The formation of corona textures can be described within the framework of thermodynamics of irreversible processes (Prigogine 1961; De Groot and Mazur 1962; Prigogine and Defay 1962). The assumption of local equilibrium makes it possible to use the Gibbs-Duhem equation in combination with mass balance equation for a chemical reaction and linear equations relating the fluxes and chemical potential gradients. Assuming the phenomenological coefficients in the equations relating the fluxes and forces, i.e. chemical potential gradients, are constant, the problem of determining the fluxes of substances diffusing in the intergranular fluid and the reaction rates at the interface within a bimetasomatic zone reduces to solving a closed system of linear equations.

Suppose that two rocks initially not in equilibrium and composed of different minerals are brought into contact. During metamorphism at constant pressure and temperature at the contact between these two rocks, the interaction and diffusion of chemical components through an intergranular fluid result in the formation of zoning consisting of layers of newly formed minerals. It is assumed that all diffusing components are linearly independent, i.e. the chemical composition of any of them cannot be obtained by a linear combination of other substances.

Let us consider a one-dimensional model, where mass transfer occurs perpendicularly to the growing bimetasomatic zoning consisting of layers. The leading role in this model, according to Joesten (1977) and Ashworth and Sheplev (1997), is assigned to three basic Eqs. (4.1, 4.2 and 4.4).

A linear relation exists between the fluxes of components J_i and chemical potential gradients $d\mu_i/dx$:

$$J_i = -\sum_{j=1}^{S} L_{ij} \frac{d\mu_j}{dx}, \tag{4.1}$$

where L_{ij} is the phenomenological diffusion coefficient (Onsager coefficient) in a multicomponent system, x is the coordinate along the direction of transport of matter, μ_j is the chemical potential of the j-th component in fluid, S is the number of components, $i = 1 \ldots S$, $j = 1 \ldots S$. The Onsager coefficient directly relates the diffusion of component i to the chemical potential gradient i at $i = j$; in the case of $i \neq j$, the "off-diagonal" components L_{ij} connect the diffusion of component i with the gradient of potential of component j, i.e. allowing for the interaction between different diffusion fluxes. Neglecting the "off-diagonal" coefficients simplifies the equation.

In a layer where the mineral m is present, the gradients are related to each other in the form

$$\sum_{i=1}^{S} n_{im} \frac{d\mu_i}{dx} = 0, \tag{4.2}$$

where n_{im} is the number of moles of the component i in 1 mol phase m. This follows from the constancy (for constant T and P) of the molar free Gibbs energy G_m for the mineral m:

$$G_m = \sum_{i=1}^{S} n_{im}\mu_i \tag{4.3}$$

For small deviations from equilibrium, variations in the Onsager coefficient can be neglected.

It is assumed that the mineral reactions are confined only to the boundaries of the layers, the mineral within them are not formed, while the fluxes of diffusing components and the potential gradients within the layers are assumed to be constant (Ashworth and Sheplev 1997; Sheplev et al. 1998). The reaction at the first boundary between the layers is the summation over the mass for each component i over all minerals k:

$$\sum_{k=1}^{\Phi} v_k n_{ik} = 0, \tag{4.4}$$

where Φ is the number of phases, v_k is the stoichiometric coefficient of the mineral k in the overall reaction.

At the boundary r, the material balance condition is fulfilled:

$$J_i^r = J_i^{r-1} + \frac{d\xi'}{dt}\sum_{k=1}^{\Phi} n_{ik}v_k^r, \tag{4.5}$$

where v_k^r is the stoichiometric coefficient of the phase k in the reaction at the boundary r, the sign of v_k^r assuming to be positive if the phase disappears, t is the time, ξ is the development of the complete reaction, $\xi' = \xi/a$ is the total reaction on the unit area of zoning structure, a is the area of the layer in the plane perpendicular to the flux of substance.

The system is closed if

$$J^0 = J^{K+1} = 0, \tag{4.6}$$

where K is the number of boundaries between layers.

From the above equations it follows that the growth of zoning obeys a simple relationship: ξ' proportional to $\sqrt{t}$. A dependence of this kind is typical for metamorphic mineral formation by diffusion-controlled mass transfer (Fisher 1978; Korzhinskii 1982).

The model under consideration assumes that the concentrations of components at the layer boundaries remain almost constant with time, i.e. a stable or quasi-stationary state is achieved during growth of zoning (Fisher 1973, 1978; Fisher and Elliott 1974; Lichtner 1988; Ashworth and Sheplev 1997; Sheplev et al. 1998). With respect to the intergranular fluid occupying a very small part of the metamorphosed rock volume, this means that the concentrations of dissolved components undergo only minor and relatively short-term changes.

In a closed system, it is assumed that the number of minerals (Φ) in zoning is one more than the number of components (S): $\Phi = S + 1$. This restriction looks somewhat artificial, but the simplifications that follow from it are important and convenient. The simplification of combining (uniting) or eliminating certain components leads to the fact that during the formation of zoning a single general reaction can be written for all interacting phases. By solving Eqs. (4.1), (4.2) and (4.5) for this reaction, we can calculate different variants of the distributions of minerals and mineral assemblages along the layers in the composition of zoning. The calculation results are used to obtain stable sequences of layers of different mineral composition, while the ratios of Onsager coefficients L are used as variables.

For a specific zoning, the coefficients v_k^q on the boundary q are calculated using Eqs. (4.1), (4.2) and (4.5). The summation over Eq. (4.5) from boundary 1 to boundary q allows us to calculate the flux of component i in the layer q:

$$J_i^q = \frac{d\xi'}{dt}\sum_{k=1}^{\Phi} n_{ik}\sum_{r=1}^{q} v_k^r \tag{4.7}$$

Taking into account $L_{ij} = 0$ for $i \neq j$, Eq. (4.1) is transformed into:

$$\left(\frac{d\mu_i}{dx}\right)^q = \frac{-J_i^q}{L_{ii}}. \tag{4.8}$$

The unknown values of fluxes J_i^q and chemical potential gradients $(d\mu_i/dx)^q$ are eliminated in Eqs. (4.2), (4.7) and (4.8). The substitution $(d\mu_i/dx)^q$ in Eq. (4.2) gives the expression:

$$\sum_{i=1}^{S} n_{im}\frac{J_i^q}{L_{ii}} = 0 \tag{4.9}$$

for $m = 1, \ldots, \Phi^q$, where Φ^q is the number of minerals in the layer q. By replacing J_i^q in Eq. (4.9) by the expression from (4.7), we obtain:

$$\sum_{i=1}^{S}\frac{n_{im}}{L_{ii}}\sum_{k=1}^{\Phi} n_{ik}\sum_{r=1}^{q} v_k^r = 0, \tag{4.10}$$

where $m = 1, \ldots, \Phi^q$. By solving these equations awe obtain the cumulative reaction coefficients in zoning based on the known phase compositions n_{ik} for all i, k and the selected values of coefficients L_{ii}.

For practical purposes, instead of the absolute values of L_{ii}, it is advisable to use relations of the type L_{ii}/L_{SiSi} (Joesten 1977) or the like. When calculating the sequences of layers (mineral assemblages) in zoning, it is convenient to derive the value of A_{kl} (Ashworth and Birdi 1990) for each pair of minerals k and l:

$$A_{kl} = \sum_{i=1}^{S}\frac{L_{SiSi}}{L_{ii}} n_{ik} n_{il}. \tag{4.11}$$

Then the key Eq. (4.10) is transformed into:

$$\sum_{k=1}^{\Phi} A_{mk}\sum_{r=1}^{q} v_k^r = 0 \tag{4.12}$$

for each mineral m in the layer q.

Summation over the phases from $k = 1$ to Φ can be divided into two parts. One part for Φ^q minerals is represented in the layer q (whose coefficients are unknown); the other is the remaining phases (some of them may be completely absent between boundaries 1 and q and not participate in the summation). The phases to the left of the boundary q, but not in the layer q (they are denoted as Φ^{left}), are represented on the right side of the equation:

$$\sum_{k=1}^{\Phi^q} A_{mk} \sum_{r=1}^{q} v_k^r = - \sum_{k=\Phi^q+1}^{\Phi^q+\Phi^{left}} A_{mk}(v_k')^q, \tag{4.13}$$

where $(v_k')^q$ is simple v_k, if the minerals Φ^{left} are only to the left of the layer q. However, Sheplev et al. (1991) showed that in the calculation of zoning some kind of mineral can reappear in the layers to the right. The sequence of layers of the type … ‖ A + B ‖ B ‖ C ‖ A + C ‖ …, in which A occurs repeatedly, illustrates exactly such a case: to check how realistic it is we shall use the criterion of thermodynamic stability—the criterion for choosing the sequence of layers (see below, and also Sheplev et al. 1991; Ashworth and Sheplev 1997).

An example of calculating the zoning is the sequence of layers developed at the boundary between plagioclase and olivine in the metamorphosed troctolite (Fig. 4.3) (Ashworth and Birdi 1990; Ashworth and Sheplev 1997). Equation (4.13) is solved for a certain sequence of layers with given ratios L_{SiSi}/L_{ii} included in A_{mk}. However, the computed set v_k^q for some sequence of layers may not indicate that this sequence is formed for given L-ratios, therefore another sequence of layers must be calculated. In some cases, the calculated quantities of minerals in the layers can satisfy the "transfer" (of matter) condition across the boundaries. The transfer occurs when the layer contains more than one "consumable" mineral (Joesten 1977). If after such transfer a sequence of layers is theoretically possible, then it is possible to calculate both the molar amount of each mineral in each layer and the model layer thickness h^{r^*}, using molar volumes of minerals. Assuming that h^r is the absolute thickness of the layer r, then $h^{r^*} = h^r/\xi'$. It follows from Eq. (4.5) that the model flux $J_i^{r^*}$ can be expressed by:

$$J_i^{r^*} = -\frac{L_{SiSi}}{d\xi'/dt} \tag{4.14}$$

Hence, using Eq. (4.8), it can be shown (Ashworth and Sheplev 1997) that:

$$\left(\frac{d\mu_i}{dx}\right)^{r^*} = \frac{L_{SiSi}}{d\xi'/dt}\left(\frac{d\mu_i}{dx}\right)^{r} = -\frac{L_{SiSi}}{L_{ii}} J_i^{r^*}. \tag{4.15}$$

Possible sequences of layers in zoning can be either stable or unstable. If the mineral k is in local equilibrium with fluid, the chemical potential of each component in the mineral is equal to the chemical potential of corresponding component in the fluid. The free energy of this mineral G_k can be calculated from Eq. (4.3). A similar value defined as

$$G_k^f = \sum_{i=1}^{S} n_{ik}\mu_i, \tag{4.16}$$

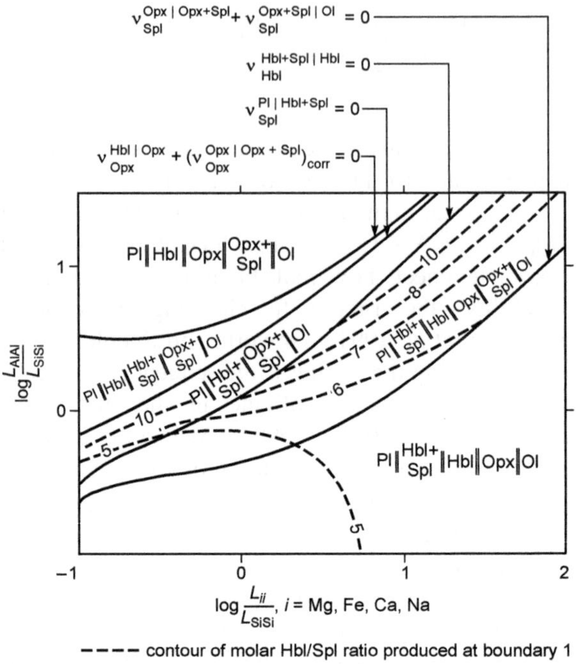

Fig. 4.3 A model example illustrating the variants of zonal sequence of mineral layers in the corona structure (Ashworth and Birdi 1990). Stability fields of different layers for the overall reaction at the contact between plagioclase and olivine are calculated. In computations carried out with *L*-ratios in the range shown here, five different stability fields are found. Each field is labelled with its distinctive layer sequence containing different minerals and mineral assemblages. Each curve separating fields is labelled to show how it is recognized by a quantity going to zero and thus requiring a change in layer sequence, on approaching the curve from one side (the side where stable layer sequence contains the layer boundary or boundaries named in the curve's label). In two cases a single coefficient of reaction for one phase at one boundary v_k^r goes to zero. Other cases are less simple in that combination of coefficients must be examined to find the quantity going to zero (subscript "corr" indicates a coefficient corrected for transfer of a mineral across a boundary, in the case Opx from Opx + Spl to the Opx layer).The sequences of layers satisfy the condition of thermodynamic stability criterion (Ashworth and Sheplev 1997). The configuration of boundaries and fields shown in the figure is determined by the range of *L*-ratios which seem to correspond to what is observed in nature. The dashed lines refer to the molar ratio Hbl/Spl in the course of reaction with the participation of Pl of variable composition (the molar ratios stated are based on 24-oxygen formula units). One of the sequences of layers in the corona structure is Pl $\|$ Hbl + Spl $\|$ Hbl $\|$ Opx $\|$ Ol with Hbl/Spl ratio ≈ 6.0 (Ashworth and Birdi 1990): the ruled ornament indicates the range *L*-ratios within which the modeling produces. The compositions of minerals are: Opx (Si = 7.900, Al = 0.200, Fe = 1.340, Mg = 6.540, Ca = 0.020), Spl (Al = 12.000, Fe = 2.400, Mg = 3.600), Hbl (Si = 6.200, Al = 2.780, Fe = 0.720, Mg = 3.530, Ca = 1.850, Na = 0.660), Pl (Si = 6.680, Al = 5.320, Ca = 2.320, Na = 0.680), Ol (Si = 6.000, Fe = 2.230, Mg = 9.770)

is used to estimate the fluid saturation with respect to the mineral k, where G_k^f is the molar free energy of the phase k dissolved in fluid. Assuming local equilibrium between the mineral and the fluid, $G_k^f = G_k$, *if* $G_k^f > G_k$, the fluid must be supersaturated with respect to k, but if k is absent, then $G_k^f < G_k$. This relationship is a criterion of stability.

Variations of G_k^f across the layered (zoning) structure are related to its thickness and chemical potential gradients. The chemical potential μ_i^q at the boundary q is expressed as:

$$\mu_i^q = \mu_i^1 + \sum_{r=1}^{q-1} h^r \left(\frac{d\mu_i}{dx}\right)^r, \tag{4.17}$$

so that $(G_k^f)^q$ on this boundary is:

$$\left(G_k^f\right)^q = \sum_{i=1}^{S} n_{ik} \left[\mu_i^1 + \sum_{r=1}^{q-1} h^r \left(\frac{d\mu_i}{dx}\right)^r\right]. \tag{4.18}$$

Model calculations allow us to determine the quantities h^{r*} and $(d\mu_i/dx)^{r^*}$, which, respectively, are proportional to h^r and $(d\mu_i/dx)^r$. From here it is possible to calculate the difference in the amount of free energy $\left(G_k^f\right)^*$ across zoning:

$$\left[\left(G_k^f\right)^q - \left(G_k^f\right)^1\right]^* = \sum_{i=1}^{S} n_{ik} \sum_{r=1}^{q-1} h^{r^*} \left(\frac{d\mu_i}{dx}\right)^{r^*}. \tag{4.19}$$

The relationship between this model value and the corresponding estimate in real zoning can be written as:

$$\left[\left(G_k^f\right)^q - \left(G_k^f\right)^1\right]^* = \frac{L_{SiSi}}{C}\left[\left(G_k^f\right)^q - \left(G_k^f\right)^1\right], \tag{4.20}$$

where the constant $C = \xi' \frac{d\xi'}{dt}$. Equation (4.2) defines a constant value $\left[\left(G_k^f\right)^q - \left(G_k^f\right)^1\right]^*$ in all layers of zoning where the mineral k is present. The stability criterion is satisfied if the solution of the Eq. (4.19) gives a lower value $\left[\left(G_k^f\right)^q - \left(G_k^f\right)^1\right]^*$, than a constant value $\left[G_k - \left(G_k^f\right)^1\right]^*$, at all boundaries where the mineral k is absent. Subtraction of a constant value makes it possible to obtain $\left[\left(G_k^f\right)^q - G_k\right]^*$. If the value obtained is less than or equal to zero, then the sequence of layers in zoning is stable under the given conditions, and the model result is considered as real.

In particular, the sequence of layers Pl ‖ Hbl + Spl ‖ Hbl ‖ Opx ‖ Ol in Fig. 4.3 was investigated in detail by Ashworth and Sheplev (1997); along with other options, the difference in free energy values was calculated for use of the stability criterion. In the case of the sequence Pl ‖ Hbl + Spl ‖ Hbl ‖ Opx ‖ Ol, the value $\left[\left(G_k^f\right)^q - G_k\right]^* \leq 0$, and this indicates its stability for certain ratios L_{AlAl}/L_{SiSi} and L_{ii}/L_{SiSi}, where i = Mg, Fe, Ca, and Na.

The affinity of a chemical reaction (Prigogine and Defay 1962) in a closed system is expressed as:

$$(-\Delta G) = \sum_{k=1}^{\Phi} \nu_k G_k. \tag{4.21}$$

With respect to zoning comprising a sequence of layers, the affinity of the reaction is determined by the chemical potential gradients that control the diffusion of matter through the layers, while the energy spent in the formation of crystal grain boundaries, on deformation during crystal growth and on the kinetics of formation or dissolution of phases is ignored. The local equilibrium between coexisting phases and components at any point means that the affinity is zero. But affinity is not equated to zero for zoning as a whole, since not all the phases in it are in contact.

Using Eqs. (4.3), (4.18) and (4.21), the affinity of the reaction can be written as:

$$(-\Delta G) = \sum_{k=1}^{\Phi} \nu_k \sum_{i=1}^{S} n_{ik} \left[\mu_i^1 + \sum_{r=1}^{q^k - 1} h^r \left(\frac{d\mu_i}{dx}\right)^r\right], \tag{4.22}$$

where q^k is the boundary of layer on which the phase k appears for the first time. In this equation, the sums inside the square brackets are different for different minerals k, but the chemical potential is constant and eliminated [according to the stoichiometry of the reaction, see Eq. (4.4)], since:

$$\mu_i^1 \sum_{k=1}^{\Phi} \nu_k n_{ik} = \mu_i^1 \cdot 0 = 0 \tag{4.23}$$

for all i. Thus, it turns out that

$$(-\Delta G) = \sum_{k=1}^{\Phi} \nu_k \sum_{i=1}^{S} n_{ik} \sum_{r=1}^{q^k - 1} h^r \left(\frac{d\mu_i}{dx}\right)^r. \tag{4.24}$$

The affinity of the reaction is a function of the difference in chemical potentials across the layers of zoning, rather than chemical potentials at the last boundary.

The model affinity $(-\Delta G)^*$ is calculated as:

$$(-\Delta G)^{*} = \sum_{k=1}^{\Phi} v_k \sum_{i=1}^{S} n_{ik} \sum_{r=1}^{q^k-1} h^{r^*} \left(\frac{d\mu_i}{dx}\right)^{r^*}. \tag{4.25}$$

The relationship between (–ΔG), (–ΔG)* and the constant rate $C = \xi' \frac{d\xi'}{dt}$ (see above) is expressed as:

$$C = \xi' \frac{d\xi'}{dt} = \frac{\xi\prime^2}{2t} = L_{SiSi} \frac{(-\Delta G)}{(-\Delta G)^{*}}. \tag{4.26}$$

In reality, ξ' is a ratio of measured thickness to the model layer thickness in zoning, i.e. h^r/h^{r^*}. If $L_{SiSi} \cdot t$ is known, where t is the reaction duration, then the affinity of the reaction (–ΔG) can be found from (–ΔG)*. Evaluations were performed by Ashworth and Sheplev (1997).

The concentration of Al at two outer boundaries of hornblende layer in the retrograde zoning structure was quantified by Ashworth (1993) using microprobe analysis. The results show that the concentration limits range from $\sim 1.3 \times 10^3$ mol m^{-3} to $\sim 1.3 \times 10^4$ mol m^{-3}. This allowed to determine the value of $D_{Al} \cdot t$ in the interval from $\sim 2.0 \times 10^{-11}$ m^2 to $\sim 8.9 \times 10^{-10}$ m^2, where $D_{Al} \cdot t$ can be correlated with $L_{AlAl} \cdot t$ using the equation (Katchalsky and Curran 1965):

$$L_{ii} = D_i \cdot c_i / RT, \tag{4.27}$$

where D_i is the Fick's diffusion coefficient for component i, c_i is the concentration of component i at the grain boundaries; c_{Al} in the layer of hornblende is estimated in the range of (0.56–1.89) × 10^4 mol m^{-3} according to Ashworth (1993). Combining the expression $h^{r^*} = h^r/\xi'$ (see above) with Eqs. (4.26) and (4.27), we obtain:

$$\frac{-\Delta G}{RT} = \left(\frac{h^r}{h^{r^*}}\right)^2 \frac{(-\Delta G)^{*}}{2(L_{SiSi}/L_{AlAl}) \cdot c_{Al} \cdot D_{Al} \cdot t} \tag{4.28}$$

Using a specific example and the proposed model the observed relative thicknesses of layers in the zoning can be estimated from Fig. 4.3 in the range of ratios L_{SiSi}/L_{AlAl}. Using Eq. (4.28) it is possible to calculate the affinity (–ΔG) for the sequence of layers Pl ‖ Hbl + Spl ‖ Hbl ‖ Opx ‖ Ol. The model thickness of hornblende layer h^{r^*} (r = 2) is approximately equal to 3.8 × 10^{-5} m^3 mol^{-1}, the observed thickness is h^r = (10 ± 2) × 10^{-6} m, (– ΔG)* = $\sim 1.32 \times 10^{-4}$ m^3 mol^{-1} at $L_{SiSi}/L_{AlAl} = 0.1$ and so forth (Ashworth and Sheplev 1997). Then, taking into account Eq. (4.28), the required value (–ΔG) is in the interval between 1.7 RT (associated with a minimum estimate for Al concentration gradient) to 580 RT.

For the reaction Pl + Ol using the thermodynamic data of Holland and Powell (1990) at a temperature of 600–700 °C and P ≈ 5 kbar it is possible to approximately determine that (–ΔG) > 10 kJ mol^{-1} (at mole of the consumable Pl, the composition of which is presented in formula units with 24 oxygen atoms); the deviation from the equilibrium temperature is at least ~100 °C (Ashworth and Birdi 1990; Ashworth and Sheplev 1997). Using the information obtained, we can estimate the diffusion coefficient for Al: $D_{Al} > 10^{-25}$ m^2 s^{-1}, if $t = 100$ million years, and $D_{Al} > 10^{-23}$ m^2 s^{-1}, if $t = 1$ million years (Ashworth 1993; Ashworth and Sheplev 1997). It is appropriate to note here that in studies on the experimental determination of diffusion coefficients of silicon along the grain boundaries, D_{Si} is found to be in the range between 10^{-19} and 10^{-17} m^2 s^{-1} (Liu et al. 1997; Yund 1997), however these results are difficult to apply to the conditions of coronite formation in nature (Ashworth and Sheplev 1997). For the case of garnet-pyroxene-quartz coronites formed in the Precambrian metabasite granulites of the Yenisei ridge in Siberia, the diffusion coefficients for Fe, Mg, and Ca are estimated in the range between 9×10^{-23} and 5×10^{-20} m^2 s^{-1} (Ashworth et al. 1998).

The local production rate of entropy σ in a small volume is a function of fluxes and their controlling gradients (Fisher and Lasaga 1981):

$$T\sigma = -\sum_{i=1}^{S} J_i \frac{d\mu_i}{dx}. \tag{4.29}$$

Integrating this equation over the volume of the system, we obtain the total production rate of the internal entropy, d_iS/dt for the entire sequence of layers in zoning:

$$T\frac{d_iS}{dt} = -a\sum_{r=1}^{K-1} h^r \sum_{i=1}^{S} J_i^r \left(\frac{d\mu_i}{dx}\right)^r. \tag{4.30}$$

The affinity (–G) can be identified with $T(d_iS/d\xi)$ (Prigogine and Defay 1962):

$$(-G) = T\frac{(d_iS/dt)}{d\xi/dt} = -\frac{1}{d\xi'/dt}\sum_{r=1}^{K-1} h^r \sum_{i=1}^{S} J_i^r \left(\frac{d\mu_i}{dx}\right)^r. \tag{4.31}$$

An analogous relation for the model quantities J_i^{r*}, $(d\mu_i/dx)^{r^*}$, $(-G)^*$ and h^{r^*} is as follows:

$$(-G)^* = -\sum_{r=1}^{K-1} h^{r^*} \sum_{i=1}^{S} J_i^{r^*} \left(\frac{d\mu_i}{dx}\right)^{r^*}. \tag{4.32}$$

There is a relation between model affinity, entropy production and criterion of stability. Assuming that one out of the two computed sequences of layers in zoning for the same overall reaction and at the same set of L_{ii}/L_{SiSi} satisfies the stability criterion, and the other does not. Empirically, it turns out that the stable sequence has a smaller value $(-G)^*$. This is interpreted as follows. The actual affinity (–G) for given P and T is fixed by the free energies of the phases and by the stoichiometry of overall reaction (see Eq. 4.21). A smaller $(-G)^*$ means a faster reaction rate $(d\xi'/dt)$ at given ξ' and L_{SiSi} in Eq. (4.26). The stable reaction is the one that proceed faster than non-stable alternatives under the same conditions. Larger $d\xi/dt$ also implies a greater rate of entropy production (d_iS/dt) for given affinity ($-\Delta G$) (see Eq. 4.31). Combining Eqs. (4.26) and (4.31) gives:

$$T\frac{d_iS}{dt} = a(-\Delta G)^{3/2}\left[\frac{L_{SiSi}}{2t(-\Delta G)^*}\right]^{1/2}, \tag{4.33}$$

i.e., for fixed values ($-\Delta G$), L_{SiSi} and t the rate of entropy production is inversely proportional to the square root of $(-\Delta G)^*$.

The principle of maximum entropy production is generally valid for near-equilibrium irreversible reaction: a thermodynamically stable reaction is one that produces entropy faster (Sheplev et al. 1991). This principle should not be confused with, and does not conflict with, Prigogine's theorem of minimum rate of entropy production (Prigogine and Defay 1962), which indicates that any quasi-stationary state is stable against small perturbations. Thus, the objects of application of these two principles are different. The new principle (Sheplev et al. 1991) of maximum entropy production is used to distinguish a stable quasi-stationary state from a metastable one.

The application of principle of maximum entropy production rate can be shown in the following simple example. In experiments (duration of 124 days using natural rock samples) on investigation of the formation of bimetasomatic zoning in the $CaO–MgO–SiO_2–H_2O–CO_2$ system at the contact between calcite and serpentine at $T = 500$ °C, $P_{H_2O+CO_2} = 200$ MPa and $X_{CO_2} = 0.2–0.4$, the occurrence of two layers (in the direction from calcite to serpentine) was established: diopside and diopside-forsterite (Shvedenkov et al. 2006). In the three-component system $CaO–MgO–SiO_2$, the reaction is: 1Cal + 3Srp = 1Di + 4Fo (without water). In zoning containing two newly formed phases, the following combinations are possible (Sheplev et al. 1991; Ashworth and Sheplev 1997; Sheplev et al. 1998): Cal ‖ Di ‖ Fo ‖ Srp; Cal ‖ Fo ‖ Di + Fo ‖ Srp; Cal ‖ Di + Fo ‖ Srp; Cal ‖ Di + Fo ‖ Fo ‖ Srp; Cal ‖ Fo ‖ Di ‖ Srp; Cal ‖ Di ‖ Di + Fo ‖ Srp; Cal ‖ Di + Fo ‖ Di ‖ Srp. The free energy of the mineral i (G_i) in the layer must be less than or equal to the free energy of this mineral in the equilibrium fluid G_{if}, i.e. $\frac{dG_{if}}{dx} \geq 0$, where x is the spatial coordinate across zoning, i is Cal, Di, Fo, and Srp. This function should not decrease in the direction from the layer without i-mineral, to the layer containing i-mineral. The calculations of this function for all minerals show that the sequence Cal ‖ Di ‖ Di + Fo ‖ Srp is thermodynamically stable over the entire range of Onsager coefficients, and it satisfies the stability criterion (Sheplev et al. 1991; Ashworth and Sheplev 1997). All other sequences among the above-mentioned do not meet the principle of maximum entropy production rate.

In the case of several independent reactions, when $\Phi > S + 1$, each of them has its own model affinity; by analogy with Eq. (4.25):

$$(-\Delta G)_s^* = \frac{L_{SiSi}}{C}(-\Delta G)_s = \sum_{k=1}^{\Phi} R_{ks} \sum_{i=1}^{S} n_{ik} \sum_{r=1}^{q^k-1} h^{r^*} \left(\frac{d\mu_i}{dx}\right)^{r^*}, \tag{4.34}$$

where R_{ks} is the stoichiometric coefficient of phase k in the reaction s. For two reactions, the ratio of model affinity is the same as the ratio of actual affinity:

$$\frac{(-\Delta G)_1^*}{(-\Delta G)_2^*} = \frac{(-\Delta G)_1}{(-\Delta G)_2}. \tag{4.35}$$

If a diagram similar to that shown in Fig. 4.3 is adopted as isobarothermal, then the affinity values and their ratios will be constant. But in variants of calculations, a varying affinity relationship may arise if the reaction remains the same, while the L-ratios change. Constant affinity relations are possible only if the relative reaction rates presented in the form ξ_1'/ξ_2', vary in the same way as the L-ratios. Calculations of this kind, with fixed affinity ratio and variable reaction coefficients v_k, were performed by Sheplev et al. (1992a).

In addition to the above-described case of formation of corona metamorphic structures at the contact between plagioclase and olivine (Fig. 4.3), it is also interesting to consider an example of zoning in the ternary system of SiO_2–MgO–CaO, occurring between quartz and dolomite involving pore fluid (Kuznetsova et al. 1992; Sheplev 1998; Sheplev et al. 1998). Newly-formed minerals include diopside, calcite, tremolite, forsterite, and talc. Condition of zoning formation is as follows: temperature is 600 °C, pressure is 2 kbar, mole fraction of CO_2 in water—carbon dioxide fluid is 0.5. The values of free energy of minerals, their change at the dissolution in fluid and the affinity of independent reactions are calculated on the basis of information provided by Frantz and Mao (1979), Holland and Powell (1990), and Skippen (1974). The method for determining the stability field of mineral sequences was discussed above. With the participation of listed minerals in the formation of zoning, a large number of reactions can be written, of which only four are linearly independent: 2Qz + Dol = Di, 8Qz + 5Dol = 3Cal + Tr, Qz + 2Dol = 2Cal + Fo, 4Qz + 3Dol = Tlc + 3Cal (H_2O and CO_2 are omitted). Any other reaction is their linear combination. In the system, for the given P-T parameters, the reactions of formation of tremolite and diopside are also realized with the participation of quartz, calcite, and talc: 3Tr = 5Tlc + 4Qz + 6Cal и 5Di = Tr + 2Qz + 3Cal (H_2O and CO_2 are omitted).

In Fig. 4.4, in the coordinates of chemical potentials of components, the saturation surfaces of minerals in pore fluid are constructed using Eqs. (4.3) and (4.16). The profiles of change in the potentials of solutes across the zoning are shown in Fig. 4.5 by thick broken lines ("reaction paths"), lying on the surface-planes of saturation of minerals or at the boundaries of their intersection; the line begins at the boundary with quartz and ends at the boundary with dolomite. The potential of SiO_2 along the profile decreases (does not increase), and the potentials of MgO and CaO increase (do not decrease). Figure 4.6 shows the variants of stable zoning sequences

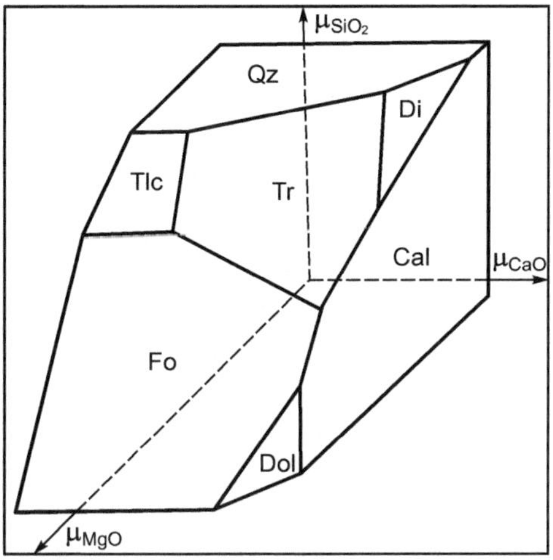

Fig. 4.4 The saturation surfaces of minerals in the space of chemical potentials: μ_{SiO_2}, μ_{CaO} и μ_{MgO}

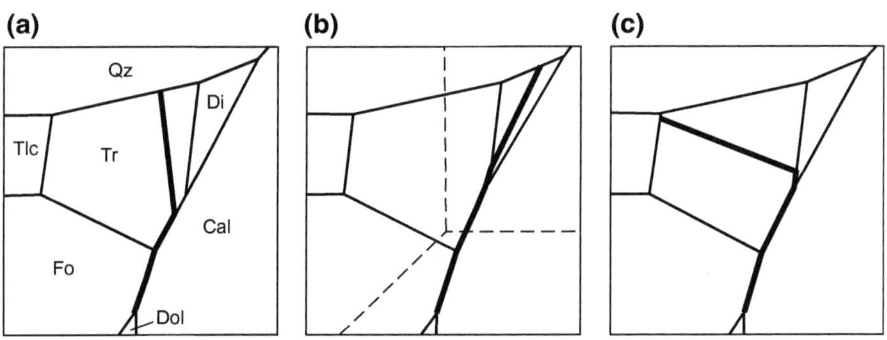

Fig. 4.5 The reaction paths (thick lines) in the space of chemical potentials illustrating different mineral sequences at the contact between dolomite and quartz: **a** Qz ‖ Tr ‖ Tr + Cal ‖ Fo + Cal ‖ Dol, **b** Qz ‖ Di ‖ Tr ‖ Tr + Cal ‖ Fo + Cal ‖ Dol, **c** Qz ‖ Tr + Tlc ‖ Tr ‖ Di ‖ Di + Cal ‖ Tr + Cal ‖ Fo + Cal ‖ Dol

of mineral layers, each of which has its own field in the coordinates $L_c = L_{CaO-CaO}/L_{SiO_2-SiO_2}$ and $L_m = L_{MgO-MgO}/L_{SiO_2-SiO_2}$. An increase in one of the L-ratios, for example, L_m, means alignment along the sequence of potentials of corresponding component μ_m; the decrease in L_c acts in the same direction as the increase in L_m, and vice versa. It follows from this that, in particular, in the coordinates L_c–L_m sequences containing talc and not containing diopside should be located in the upper left corner and the sequence containing diopside and free of talc should be located in the lower right corner in Fig. 4.6. Forsterite is present in all sequences, but only in association with calcite on the border with dolomite

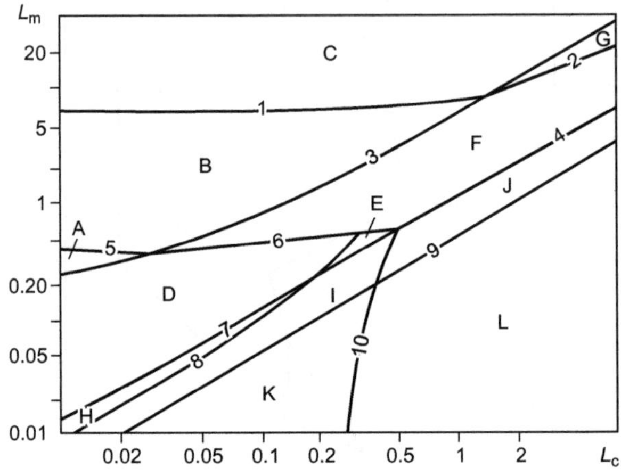

Fig. 4.6 The variants of zoning sequence of mineral layers. Stability fields are shown for the layers in corona structure that appeared at contact between quartz and dolomite. Calculations performed with *L*-ratios in the range of characteristic values refer to 12 fields in the coordinates: $L_c = L_{CaO-CaO}/L_{SiO_2-SiO_2}$ and $L_m = L_{MgO-MgO}/L_{SiO_2-SiO_2}$. In each field, there is a stable sequence of layer differing in their minerals and mineral assemblages: A: Qz ‖ Tlc + Tr ‖ Tr ‖ Di ‖ Di + Cal ‖ Tr + Cal ‖ Fo + Cal ‖ Dol. B: Qz ‖ Tlc + Tr ‖ Tr ‖ Tr + Cal ‖ Fo + Cal ‖ Dol. C: Qz ‖ Tlc + Tr ‖ Tr ‖ Cal ‖ Fo + Cal ‖ Dol. D: Qz ‖ Tr ‖ Di ‖ Di + Cal ‖ Tr + Cal ‖ Fo + Cal ‖ Dol. E: Qz ‖ Tr ‖ Tr + Di ‖ Tr + Cal ‖ Fo + Cal ‖ Dol. F: Qz ‖ Tr ‖ Tr + Cal ‖ Fo + Cal ‖ Dol. G: Qz ‖ Tr ‖ Cal ‖ Fo + Cal ‖ Dol. H: Qz ‖ Tr + Di ‖ Di ‖ Di + Cal ‖ Tr + Cal ‖ Fo + Cal ‖ Dol. I: Qz ‖ Tr + Di ‖ Tr + Cal ‖ Fo + Cal ‖ Dol. J: Qz ‖ Tr + Di ‖ Tr ‖ Tr + Cal ‖ Fo + Cal ‖ Dol. K: Qz ‖ Di ‖ Tr + Di ‖ Tr + Cal ‖ Fo + Cal ‖ Dol. L: Qz ‖ Di ‖ Tr ‖ Tr + Cal ‖ Fo + Cal ‖ Dol

(Fo + Cal layer); layers made up of pure forsterite or pure talc (monolayers) are absent. The greatest variety of types of zoning occurs in the central part of Fig. 4.6; just at that spot the field F, where the characteristic sequence Qz ‖ Tr ‖ Tr + Cal ‖ Fo + Cal ‖ Dol is disposed, in which there is neither talc nor diopside. Its appearance is controlled only by two independent reactions: 8Qz + 5Dol = 3Cal + Tr and Qz + 2Dol = 2Cal + Fo.

Sequences of the type Qz ‖ Tr + Tlc ‖ … are located above the field F, and the sequences containing diopside are to the left or below the field F. The truncated sequence Qz ‖ Tr ‖ Tr + Cal ‖ Dol is stable throughout the range of parameters L_c–L_m, if the supersaturation of pore fluid with respect to talc, diopside and forsterite is not taken into account. However, forsterite can be excluded from consideration, since it enters all types of sequences in the same way, i.e. only as part of the last layer … ‖ Fo + Cal ‖ … The truncated sequence is decisive for the reproduction of the main border lines and their possible types in Fig. 4.6. The saturation of fluid with respect to talc may occur initially at the boundary of Qz ‖ Tr, and with diopside—at the boundary of Qz ‖ Tr and Tr ‖ Tr + Cal.

The formation of zoning with non-constant Onsager coefficients *L*, under conditions of deviation from equilibrium, was considered by Kuznetsova et al. (1992).

The procedure for calculating the stability fields of mineral sequences for non-constant values of L was described by Sheplev et al. (1992b). In the case of significant deviation from equilibrium in the system SiO_2–MgO–CaO, new mineral sequences may appear and be different from those shown in Fig. 4.6. In particular, the sequences where forsterite occurs in the layers preceding the last layer … ‖ Fo + Cal ‖ … may be formed and the appearance of calcite monolayer on the Tr + Cal ‖ Fo + Cal boundary etc. becomes possible. If variations in L during zoning are accompanied by change in the relative amounts of phases A and B inside a layer of the … ‖ A + B ‖ …type, then the disappearance of phase B in this layer can lead to the appearance of a sequence … ‖ A ‖ B ‖ …

Finally, it is of interest to consider concentric zoning structures, i.e. segregation, in heterogeneous metapelites. They occur in gneisses and crystalline schists and consist of high-alumina central nuclei composed of sillimanite or kyanite and surrounded by spinel-cordierite, corundum-cordierite, spinel-cordierite-magnetite, etc. symplectittic rims, which are separated from the rock matrix by biotite—K-Na-feldspar borders. The segregations are similar to those described by Carmichael (1969) and were mentioned earlier in Sect. 1.2. As an example, we use the sillimanite-plagioclase-cordierite biotite-quartz gneisses containing zoning segregations, which are developed in the Kuray Range, Gorny Altai (Kuznetsova et al. 1994). In these gneisses, in the contact aureole near the Terandzhik gabbro-monzonite massif (Reverdatto et al. 1974), K-Na-feldspar and spinel appear around sillimanite segregations, which are the product of a reaction involving biotite. The spinel-cordierite zone at the boundary with sillimanite is represented by a symplectite. For a given segregation within the five-component system SiO_2–Al_2O_3–FeO–MgO–K_2O, the compositions of minerals participating in the zoning are determined: Sil—Al_2SiO_5, Crd—$Mg_{1.46}Fe_{0.48}Al_{4.0}Si_{5.0}O_{18}$, Spl—$Mg_{0.22}Fe_{0.89}Al_{1.92}O_4$, Kfs—$K_{0.83}Al_{1.02}Si_{2.98}O_8$, Bt—$K_{0.93}Mg_{1.21}Fe_{1.06}Al_{1.63}Si_{2.71}O_{10}$, Qz—$SiO_2$. Calcium, sodium and water were not taken into account, therefore for these minerals the only reaction is written in the following form: 0.879Sil + 0.451Bt + 1.0Qz = 0.319Crd + 0.506Kfs + 0.366Spl. The volume ratios (in %) of newly formed minerals calculated from this net reaction are as follows: Crd—52, Spl—10, Kfs—38, which roughly corresponds to proportions observed in segregation. Plagioclase and cordierite in the gneiss matrix were not taken into account in the calculations. Consequently, the balance of chemical components is preserved in the system SiO_2–Al_2O_3–FeO–MgO–K_2O, and their redistribution occurs only within segregation. The study of layer sequences is performed using the following phenomenological coefficients: L_a—$L_{Al_2O_3Al_2O_3}/L_{SiO_2SiO_2}$; L_f—$L_{FeOFeO}/L_{SiO_2SiO_2}$; L_m—$L_{MgOMgO}/L_{SiO_2SiO_2}$; L_k—$L_{K_2OK_2O}/L_{SiO_2SiO_2}$. Analysis of the zoning of Sil ‖ Crd + Spl ‖ Crd ‖ Bt ‖ Kfs ‖ Qz shows that this zoning is unstable at $L_f = L_m \leq 1.2$ over the entire range of parameters L_a and L_k. Its stability field is shown in Fig. 4.7 at $L_f = L_m = 2$. With predetermined L_f and L_m the sequence Sil ‖ Crd + Spl ‖ Crd ‖ Bt ‖ Kfs ‖ Qz is stable for $L_a < 0.25$ and $L_k > 0.25$. The lines limiting its stability field correspond to the supersaturation of pore fluid: (1) relative to K-feldspar at the boundary Crd ‖ Bt and (2) relative to cordierite at the boundary Bt ‖ Kfs (see curves in Fig. 4.7). For $L_k = 10$ and $L_a = 0.1$, the volume fraction of spinel in the layer ‖ Crd + Spl ‖ is 33.3%. The ratio of thicknesses of newly formed layers in the zoning is as follows: ‖ Crd + Spl ‖ – 30.9%, ‖ Crd ‖ – 30.9% and ‖ Kfs ‖ – 38.2%. If L_k and L_a are varied by several orders of magnitude, no significant changes occurs with respect to thickness of layers or the volume fraction of spinel in the symplectite ‖ Crd + Spl ‖ within the stability field of this sequence. If L_f and $L_m > 5$, the calculated fraction of spinel in the symplectite considerably exceeds the ratios observed in the segregation.

When the outer layer in the zoning is represented not only by quartz, but also by a biotite-quartz aggregate, the sequence Sil ‖ Crd + Spl ‖ Crd ‖ Kfs ‖ Bt + Qz appears. It is stable within the system SiO_2–Al_2O_3–FeO–MgO–K_2O with the following parameters: L_f and L_m values should be within the range of 2–5, $L_a < 0.2$, and $L_k > 2.5$. The line bounding its stability field corresponds to a supersaturation of pore fluid relative to cordierite at the boundary Kfs ‖ Bt + Qz. The ratio of thicknesses of the newly formed mineral layers and the volume fraction of spinel in symplectite ‖ Crd + Spl ‖ within the stable growth range of this sequence for given L_f and L_m vary insignificantly and correspond to the proportions observed in actual zoning.

It should be noted in conclusion that, despite some assumptions, the simple theory formulated here based on the principle of local equilibrium is very useful in interpreting coronitic structures in minerals. Using this theory can help us to obtain some expressions for calculating the growth rate of mineral sequence layers, the thicknesses of layers in the mineral sequence, the rate of entropy production and stability of layer boundaries. The solution has the form of observed reaction kinetics between spatially separated minerals, as if they were in direct contact. The role of reaction constants is played by complexes including the thicknesses of layers and the values of the phenomenological coefficients. In the study of metamorphic zoning, the relative mobility of chemical components becomes apparent, with aluminum, titanium and, to some extent, silicon among the slow-moving ones, and iron, magnesium and calcium are among the more mobile, and alkalis are the most mobile. This issue has been discussed above in Sect. 1.2. Using a special criterion

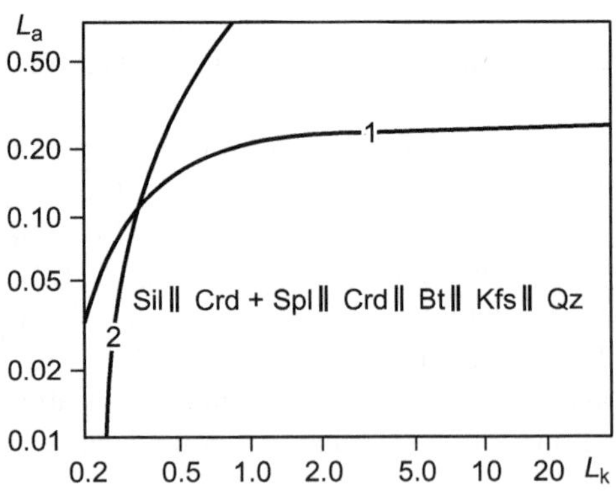

Fig. 4.7 The field of stability of zoning sequence Sil ‖ Crd + Spl ‖ Crd ‖ Bt ‖ Kfs ‖ Qz in metapelites at $L_f = L_m = 2$

of stability makes it possible to ascertain which of the model sequences of mineral layers can actually exist. A metastable equilibrium exists in the rejected sequences. The diversity and peculiarities of zoning sequences permit us using model quantities for estimating the actual differences in chemical potentials and, consequently, the affinity of the reactions. The diffusion coefficients can be determined with great approximation and only tentatively, because of the uncertainty in the duration of reactions.

4.2.1.2 Mineral Thermometry with the Use of Zoning Structures

In the previous Sect. 4.2.1.1, we described a model of bimetasomatism according to Korzhinskii (1955, 1957, 1962, 1973, 1982), Fisher (1973, 1977, 1978; Fisher and Elliott 1974) and Joesten (1977, 1991; Joesten and Fisher 1988). It is based on the assumption about local equilibrium and the quasi-stationarity of processes in the bimetasomatic zoning growing at constant P and T between two initially nonequilibrium minerals or mineral assemblages. The model is still an accurate representation of reality and, at the same time, enables a more thorough analysis. The scientific significance of coronitic zoning structures is driven by the fact that the reactants and reaction products, as well as the quantitative relationships between them are retained in spatially separated layers with distinct boundaries.

The system of linear equations which were used to describe the Korzhinskii–Fisher–Joesten model in Sect. 4.2.1.1 can be represented in a more compact form by using matrix algebra.

Let K be the number of boundaries ($K - 1$ is the number of layers formed within the zoning structure), ξ is the coordinate along the direction of substance transport, N^k is the matrix of the component composition of phases of k-th layer, N is the same for all phases within zoning, μ^k is the vector of chemical potentials of pore fluid components, J^k is the vector of flux of components, G^k is the vector of Gibbs energies of participating phases, h^k is the thickness of k-layer, w^k is the phase formation rates in a unit volume of k-th layer, w_k is the phase formation rates at the k-th boundary, and L is the matrix of the Onsager phenomenological coefficient; superscript T is the transposition symbol. The dynamics of quasi-stationary growth of zoning is defined by the set of equations:

$$\frac{dJ^k}{d\xi} - N^k w^k = 0 \tag{4.36}$$

$$J^k = -L\frac{d\mu^k}{d\xi} \tag{4.37}$$

$$(N^k)^T \mu^k = G^k, \tag{4.38}$$

$$J^k - J^{k-1} = -N_k w_k, \tag{4.39}$$

$$J^0 = J^{k+1} = 0, \tag{4.40}$$

which reflect the balance of components in the k-th layer (4.36), the linear relationship between the vector-flux and vector-gradient of potentials (4.37), the equilibrium between all layer phases and pore fluid of specific composition (4.38), component balances at zone boundaries (4.39), and the closeness of the system to all components (4.40).

Equation (4.38) is satisfied for all phases of the kth layer including boundaries. Since it is used below in a differential form (Gibbs–Duhem), therefore, it is sufficient that finite relationships would be satisfied only in one (for definiteness, the extreme right) point of this layer. The number of layers, their phase compositions, phase components, the order of layer alternation, and component potentials and their flows are unknown. The boundary conditions are laid not only at the outer but also at the inner boundaries of each layer, whose number and arrangement are unknown beforehand.

The reactions taking place in the growing zones are defined by matrix R, whose rows are composed of stoichiometric coefficients of phases in a corresponding reaction. The way of matrix construction is described in a number of manuals (for example, Denbig 1966), and, for a close system, is governed by the condition of conservation $NR^T = 0$.

It was shown in previous Sect. 4.2.1.1 (see also, Ashworth and Sheplev 1997; Sheplev et al. 1998) that for the above assumptions, minerals are formed or dissolved only at layer boundaries; moreover, the J^k flow in the kth layer is constant and can be expressed in terms of the total rate of formation J_{-k} at all previous boundaries of minerals that are absent in the kth and all next layers:

$$J^k = P_k J_{-k}, \quad P_k = E - N^k \left(N^{kT} L^{-1} N^k\right)^{-1} N^{kT} L^{-1}.$$

The matrix P_k shows the properties of a projection operator:

$$P_k^2 = P_k, \quad P_k N^k = 0.$$

Flow constancy in layers makes it possible to calculate component potentials at the kth boundary by potentials at the right-hand boundary

$$\mu_{K+1} = \mu_k + \sum_{r=k}^{K} h^r (\mu^r)' \tag{4.41}$$

If we multiply this equation by the vector of composition of any phase of k-th layer, then according to (4.38) we shall obtain the Gibbs energy of this phase in place of the first term on the right-hand side of (4.41). On performing this operation for all phases present in the zones (each phase has its own index k), we obtain the set of equations

$$N^T \mu_{K+1} = G + \Delta = G^* \tag{4.42}$$

Here, we omit the explicit form of Δ due to its awkwardness. Below in (4.44), it will be given in a compact form. From (4.42) we find the rates of all independent reactions between phases (see above, and Sheplev et al. 1991).

$$w = \left(RzR^T\right)^{-1} A, z = \sum_{r=1}^{K} h^r D^{rT} L^{-1} D^r,\ D^k = P_k J_{-k} \tag{4.43}$$

The remaining independent equations from (4.42) make it possible to find the potentials of pore fluid components at the right-hand boundary of zoning and then determine the potentials at all other boundaries from (4.41). With the use of (4.43), the right side of (4.42) can be represented as

$$G^* = G - zR^T \left(RzR^T\right)^{-1} A$$

or

$$G^* = P_G G,\ \ P_G = E - \mathbf{z}R^T \left(RzR^{\mathbf{T}}\right)^{-\mathbf{1}} R. \tag{4.44}$$

P_G, as well as P_k, is the projection operator (it projects the free energy vector on the subspace orthogonal to a reaction space)

$$P_G^2 = P_G, \quad RP_G = 0. \tag{4.45}$$

The explicit expression for the reaction rates (4.43) makes it possible to find the rate of entropy production in the zoning

$$T\sigma = -wA.$$

Among all types of zoning admitted by the set of Eqs. (4.36)–(4.40), the zoning with the maximum rate of entropy production (see above, and Sheplev et al. 1991; Ashworth and Sheplev 1997) is realized (4.45). The basic equation follows from (4.44) to (4.45):

$$A^* = -RG^* = 0. \tag{4.46}$$

Comparing this equation with the equation for the affinities of reactions (see above, and Ashworth and Sheplev 1997) gives the following inequality:

$$A = -RG \neq 0.$$

This inequality, which is a necessary condition for the growth of zoning, does not allow us to use equilibrium thermodynamics for temperature and pressure determination. At the same time, it follows from (4.46) that such a possibility nevertheless exists, if we make certain changes in the free energy vector. Then we can use the reactions with any combinations of nonequilibrium components as if they would be at mutual equilibrium.

The physical sense of the introduced free energy transformation (4.44) is as follows: according to the assumption of local equilibrium, all phase components are in equilibrium with a pore fluid at any point of a growing zoning. The fluid composition continuously changes across the zoning, and inevitably there comes a point of time when Eq. (4.38) cannot be satisfied; the fluid becomes undersaturated or supersaturated with respect to a certain phase, and a new layer appears. The introduced corrections (4.42) and (4.44) have the form of summand corresponding to the number of layers separating the k-th layer from the extreme right boundary of zoning. Each summand brings all components of the k-th layer into equilibrium with the potential of fluid components in the next layer, and the transformation (4.44) will bring finally all phase components to equilibrium with the pore fluid of fixed composition at the zoning right boundary. As a result, all components prove to be driven to equilibrium with one another.

The introduced transformation extends Korzhinskii's concept of perfectly mobile components to all components with reduced mobility.

The mathematical sense of the Gibbs energy correction is that all boundary conditions are transferred to one point (we remind that all these conditions correspond to very different zone boundaries). The number of phases driven to equilibrium by means of this transformation may exceed the number of phases governed by the Gibbs' phase rule.

The introduced transformation makes it possible to eliminate the determined parameter error introduced by the state of phase nonequilibrium. It is sound practice to illustrate transformation effectiveness in a model not affected by other factors. For this purpose, we utilize a thermodynamic dataset and activity models (Sobolev and Babeyko 1994). The zoning was calculated for the initial compositions of olivine and plagioclase at fixed temperature and pressure. Then, according to Berman (1991), we calculated all possible states of equilibrium generated by a given mineral assemblage with the use of the correlated thermodynamic data and rules of mixing without or with the correction for a possible state of nonequilibrium. By doing this, we estimate the influence of the state of nonequilibrium on the accuracy of restoration of conditions for zoning growth and the effectiveness of the introduced transformation.

In the calculations we assumed the following parameters: the compositions of plagioclase and plagioclase are $Ab_{44}An_{56}$ and $Fo_{51.5}Fa_{48.5}$, respectively; temperature is 720 °C, pressure is 8 kbar; SiO_2, Al_2O_3, FeO, MgO, CaO and Na_2O are taken as pore fluid components; their mobilities measured as Onsager coefficients relative to SiO_2 (Ashworth and Sheplev 1997) are equal to 1, 0.5, 2, 2, 2, and 5, respectively. All Fe is bivalent. The calculations show that garnet, clinopyroxene and orthopyroxene develop between plagioclase and olivine having the following compositions: Grt = $Prp_{30.3}Alm_{50.1}Grs_{18.6}$, Cpx = $cEn_{2.6}Di_{69.9}cFs_{8.5}Hd_{9.4}Ca–Ts_{2.7}Jd_{6.9}$, Opx = $En_{58.8}oDi_1Fs_{38}oHd_{1.1}Crn_{1.2}$. Six oxides ($SiO_2$, Al_2O_3, FeO, MgO, CaO, Na_2O) and 18 components (Ab, An, Fo, Fa, Prp, Alm, Grs, cEn, Di, cFs, Hd, Ca–Ts, Jd, En, oDi, Fs, oHd, Crn, where c—clino- and o—ortho-polymorphs, Ca–Ts—calcium tschermakite) participate in 780 reactions of which 12 are independent. For any reaction, Eq. (4.46) with allowance for the calculated composition and correction of Gibbs free energies has the form

$$a + bP + cP^2 + dT + ePT + fTlnT = 0 \tag{4.47}$$

with the corresponding numerical values of the coefficients *a*, *b*, *c*, *d*, *e*, *f*.

Figure 4.8a shows the curves (4.47) for the first 30 out of 780 reactions. Figure 4.8b shows the same curves with uncorrected Gibbs free energies. All 780 reactions are displayed in Fig. 4.8c for the sake of completeness. It can be seen that the spread in the temperature and pressure estimates, corresponding to intersections of uncorrected curves, is very wide. After correction, all curves pass through one and the same point (8 kbar, 720 °C), i.e. all 3×10^5 curve intersections coincide and give the only estimate of growth conditions. This is an important internal verification of the introduced transformation validity.

The proposed transformation of Gibbs free energies permit the use of all available geothermobarometers for estimating temperature and pressure from nonequilibrium spatially separated mineral zones in the coronitic structure, as if they would be at equilibrium (Sheplev and Reverdatto 1998).

4.2.2 *Metamorphic Reactions in the Matrix*

4.2.2.1 The Model of Dissolution and Growth of Scattered Mineral Grains

Among a wide variety of reactions in the matrix of metamorphic rocks there are cases where interaction occurs between isolated grains dispersed in the inert medium. These cases satisfy the conditions of the present model. A good example is the regional low-temperature metamorphism in siliceous dolomitic limestones of the Alps (Trommsdorff 1966, 1972). The separate dolomite and quartz grains scattered in the inert calcite mass interact through the fluid yielding new calcite together with talc or tremolite grains depending on the proportions of H_2O and CO_2 (Fig. 4.9). Another example (not quite adequate) is retrograde metamorphic reactions in spurrite-merwinite-melilite marble at the contact zone near the trap intrusion (Reverdatto et al. 1979). The grains of essentially gehlenitic melilite and merwinite scattered in the spurrite-calcite aggregate interacted with each other

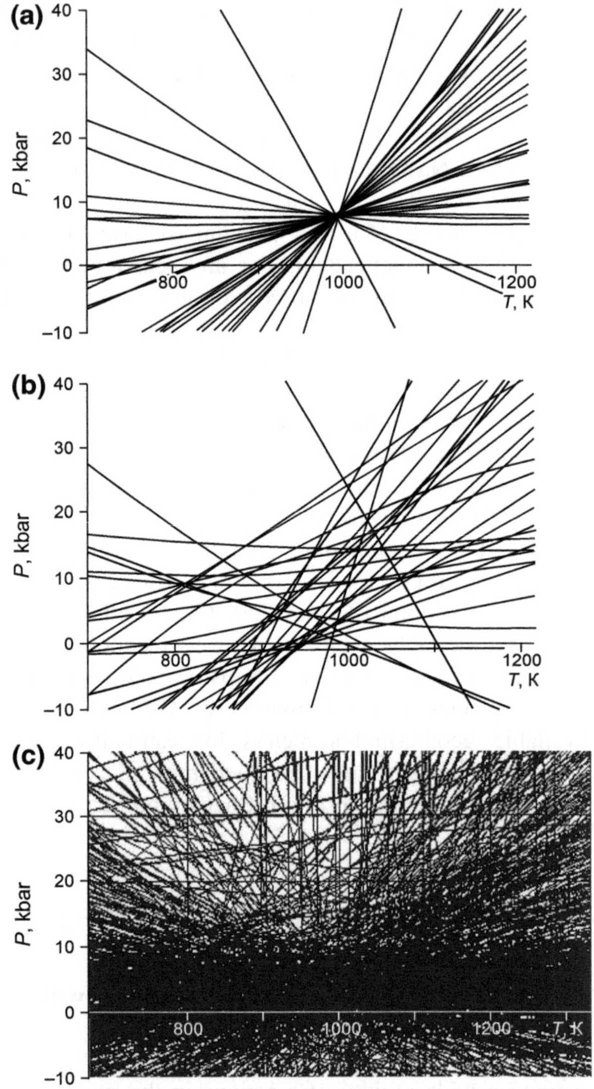

Fig. 4.8 The curves of 30 out of 780 reactions involving Ol, Pl, Grt, Cpx, and Opx, with Gibbs free energy correction (**a**); the same curves without Gibbs free energy correction (**b**); all 780 reactions (**c**)

through the fluid in the presence of CO_2. These reactions produced calcite and new melilite with a high åkermanite content.

It is assumed that a set of intergranular channels uniformly distributed in the rock acting as an effective means of transport and interaction of solutes during metamorphism can be simulated as a continuous unbounded aqueous medium, given that the diffusion-controlled mass transfer and reactions of dissolution and growth of mineral grains proceed at very low rates. Assume that the medium contains a certain number of spherical grains of three types: A, B and C which are randomly distributed. At t (time)= 0, grains A and B begin to dissolve, and the dissolved

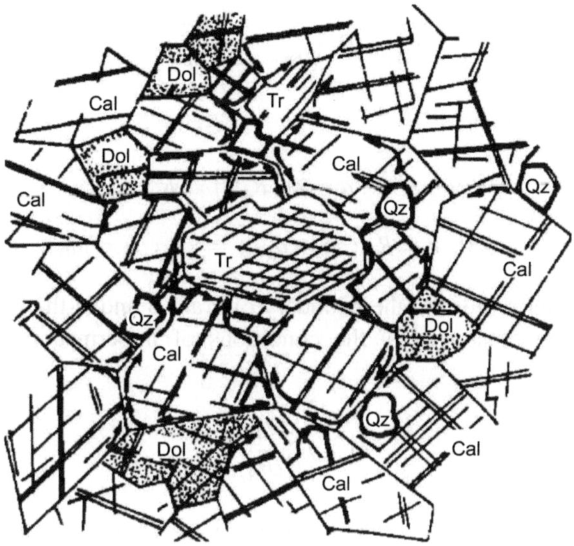

Fig. 4.9 The formation of scattered grains of tremolite and calcite under metamorphism of quartz-bearing dolomitic limestone in accordance with the reaction Dol + Qz + H_2O → Tr + Cal + CO_2. Schematic drawing on thin section. The arrows indicate the directions of mass transfer through fluid films along the grain boundaries

components diffuse through the solution. The irreversible reaction $a + b \rightarrow c$ occurs on the surface of grain C at a rate $\omega(a, b)$. The dynamics of growth of grain C needs to be investigated.

Let us first consider the case with two dissolving grains A and B and one growing grain C (Fig. 4.10). Let us denote the initial and saturation concentrations of grain A by a_o and a_s, respectively, the diffusion coefficient by D_a, the specific volume by V_a and the grain radius by R_a. A similar procedure can be made for grains B and C.

The problem of the dissolution of grain A and the distribution of concentration a (t) reduces to solving the equation:

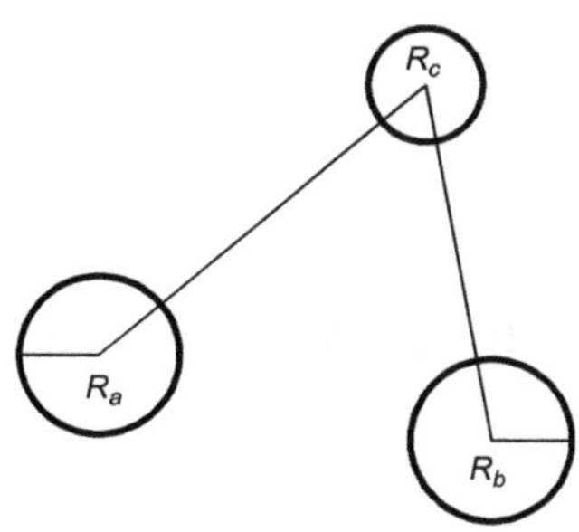

Fig. 4.10 A schematic representation of the model of dissolution and growth of scattered mineral grains during metamorphism. Two grains of large radius R_a and R_b are dissolving, and a grain of smaller radius R_c is growing

$$\frac{\partial a}{\partial t} = D_a\left(\frac{\partial^2 a}{\partial r^2} + \frac{2}{r}\frac{\partial a}{\partial r}\right), \tag{4.48}$$

subject to the conditions

$$t = 0 : a = a_0, R_a(t) = R_a^0; \tag{4.49}$$

$$t > 0;\ r = R_a :\ a = a_s,\ r = \infty :\ a = a_0, \tag{4.50}$$

where r is the outer radius. At the boundary of grain B and at the boundaries of the model region, we must specify the conditions of impermeability, while at the boundary of grain C:

$$\omega = D_a \frac{\partial a}{\partial r}\bigg|_{r=R_c} \tag{4.51}$$

The equation for grain B, is then written in the same way, and with obvious changes for grain C.

The solution of this problems is a challenging task: the spherical symmetry of matter diffusion from one grain is broken by the other grains and by the boundaries of the simulation region. Additional complexity is associated with the moving boundaries of the growing and dissolving grains and with nonlinearity of the reaction at grain C. It is necessary to make substantial simplifying assumptions: (1) the distances between grains are much larger than the grain sizes; by assuming this we can then consider the dissolving grains independently, i.e., assuming spherical symmetry; (2) the rates of dissolution of A and B are considerably greater than the characteristic diffusion rate D/L, in which L is the characteristic dimension; (3) the quasi-stationary approximation is applied for growth of grain C; and (4) the volume of the aqueous solution is unbounded.

Second assumption enables us to solve Eq. (4.48) with boundary (4.49) and initial (4.50) conditions, while neglecting the change in R_a. The solution takes the form (Boltaks 1961; Raychenko 1981):

$$a(r,t) = a_0 + \frac{R_a(a_s - a_0)}{r} erfc\frac{r - R_0}{2\sqrt{D_a t}}. \tag{4.52}$$

At the boundary of grain A the equation is satisfied:

$$\frac{dR_a}{dt} = D_a V_a \frac{\partial a}{\partial r}\bigg|_{r=R_a}. \tag{4.53}$$

By substituting $a(r, t)$ from (4.52) into (4.53), we obtain

$$\frac{dR_a}{dt} = -V_a D_a (a_s - a_0)\left(\frac{1}{R_a} + \frac{1}{\sqrt{\pi D_a t}}\right),\ t = 0:\ R_a = R_a^0 \tag{4.54}$$

Then introduce

$$\rho = \frac{R_a}{R_a^0},\ \tau = \frac{V_a D_a (a_s - a_0) t}{R_a^2},\ \beta = \sqrt{\frac{V_a (a_s - a_0)}{\pi}} \tag{4.55}$$

The function $R_a(t)$ contains five parameters. The conversion to dimensionless variables enables us to reduce, without loss of generality, the amount of parameters to a single dimensionless complex β, which reflects the relation between the capacity of a unit volume of solution and of the grain. Figure 4.11 shows a one-parameter family of curves that contains all the information on the grain dissolution (using these assumptions). The $R(t)$ curves have $dR/dt = \infty$ at the initial time, which can be explained by a concentration jump at the boundary, and at the time of complete dissolution, which corresponds to the unbounded increase in the ratio of grain surface to volume; the curves tend to collapse.

Equation (4.54) can be rewritten as:

$$\frac{d\rho}{d\tau} = -\left(\frac{1}{\rho} + \frac{\beta}{\sqrt{\tau}}\right), \tag{4.56}$$

$$\tau = 0:\ \rho = 1. \tag{4.57}$$

The solution of Eq. (4.56) is obtained by the substitution:

$$\rho = z\sqrt{\tau}, \tag{4.58}$$

$$\tau = \frac{\exp\left[\frac{2\beta}{\sqrt{2-\beta^2}} arctg\left(\frac{z+\beta}{\sqrt{2-\beta^2}} - \frac{\pi}{2}\right)\right]}{z^2 + 2\beta z + 2}. \tag{4.59}$$

Equation (4.59) inexplicitly defines the dependence $\rho(\tau)$. The complete dissolution time τ_c for a grain is derived from (4.59) with the substitution $\rho = z = 0$:

Fig. 4.11 Dissolution of a grain *A* (variations in the ratio R/R_0) as a function of parameter β. The curves correspond to the following values of parameter β: 1—0.2; 2—0.6; 3—1.0; 4—1.3

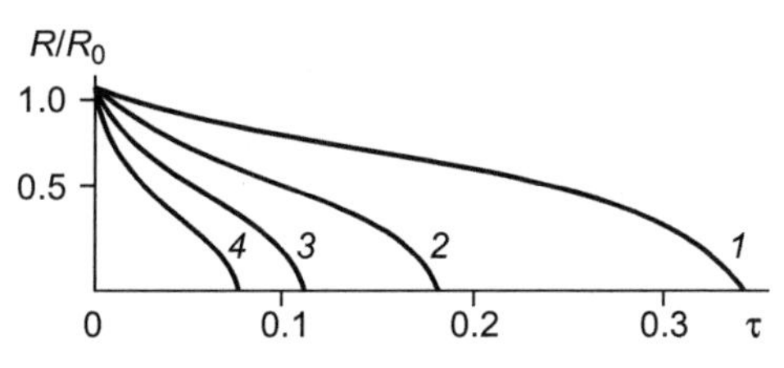

$$\tau_c = \frac{1}{2}\exp\left[\frac{2\beta}{\sqrt{2-\beta^2}} arctg\left(\frac{\beta}{\sqrt{2-\beta^2}} - \frac{\pi}{2}\right)\right]. \tag{4.60}$$

To examine the dissolution of grain A, the following steps need to be done: (1) to obtain dimensionless values of β and τ from (4.55); (2) to calculate τ_c from (4.60); (3) given that $\tau < \tau_c$ to solve Eq. (4.59) for z and derive $\rho = z\sqrt{\tau}$ and $R_a = \rho R_a^0$; and (4) to obtain the concentration a at the desired point from (4.52). The dissolution of grain B is investigated in a similar way. Constructing the curve $\rho(\tau)$ does not require solving the nonlinear Eq. (4.59): using the parametric expressions (4.58) and (4.59) we must specify the value for z and use (4.59) to calculate τ and then find ρ using Eq. (4.58). After changing the value of z, we will derive a new pair of τ and ρ, and so on.

Let us consider the growth of grain C. Assuming that the distance of grain C from the dissolving grains is sufficiently large, a and b near grain C can be derived from (4.59), and the unknown concentrations will be denoted as $\bar{a}$ and $\bar{b}$, respectively. The rate of the reaction on the surface of grain C

$$\omega = k(a \cdot b)\big|_{r=R_c}$$

should be equal to the flux of a and b (4.51):

$$\omega = D_a \frac{\partial a}{\partial r}\bigg|_{r=R_c} \quad D_b \frac{\partial b}{\partial r}\bigg|_{r=R_c}.$$

With the quasi-stationary approximation, the concentration a near grain C follows equation:

$$D_a \frac{\partial}{\partial r}\left(r^2 \frac{\partial a}{\partial r}\right) = 0, \tag{4.61}$$

which is solved as

$$a = \bar{a} - \frac{R_c G}{r}. \tag{4.62}$$

A similar equation is written for the substance b:

$$b = \bar{b} - \frac{R_c H}{r}. \tag{4.63}$$

The coefficients G and H are determined from boundary conditions (4.51). Substitution gives:

$$H = \frac{D_a}{D_b} G; \; D_a \frac{G}{R_c} = k(\bar{a} - G)\left(\bar{b} - \frac{D_a}{D_b} G\right) \tag{4.64}$$

Then

$$G = D_b \left(\rho - \sqrt{\rho^2 - \frac{\bar{a}\,\bar{b}}{D_a\,D_b}} \right); \; \rho = 2\left(\frac{\bar{a}}{D_b} + \frac{\bar{b}}{D_a} + \frac{1}{k\,R_c} \right). \tag{4.65}$$

The smallest root is taken to solve the quadratic equation for G.

Given the above assumption, the change in the radius of grain C can be given in the form of a first-order differential equation whose right-hand side is analytically expressed

$$\frac{dR_c}{dt} = V_c D_a \frac{G}{R_c} = \frac{V_c D_a D_b \left(\rho - \sqrt{\rho^2 - \frac{\bar{a}\,\bar{b}}{D_a\,D_b}} \right)}{R_c} \tag{4.66}$$

and the initial condition is $R_c = R_c^0$. This equation is numerically solved using the Runge-Kutta method. In the case of dissolution of multiple grains, the growing grain C receives substances a and b from many sources. These concentrations of substances a and b are obtained by summing the expressions like (4.52):

$$a(t) = a_0 + (a_s - a_0) \sum_i \frac{R_{a_i}}{r_i} erfc \frac{r_i - R_{a_i}}{2\sqrt{D_a t}} \tag{4.67}$$

A similar formula is applied to substance b.

In the case of complete dissolution of grain A ($t = t_{cd}$) the substance a is redistributed by diffusion (Boltaks 1961; Carslaw and Jager 1964):

$$a(r,t) = \frac{1}{2\sqrt{\pi D_a t}} \int_0^r a(x, t_{cd}) \left\{ \exp\left[-\frac{(x-r)^2}{4 D_a t} \right] + \exp\left[-\frac{(x+r)^2}{4 D_a t} \right] \right\} dx, \tag{4.68}$$

in which t_{cd} is the time of complete dissolution. A similar equation is applied to the concentration of substance b.

The solutions obtained can be applied to the case of grain dissolution in a limited volume. Let the solution volume be bounded by plane faces of the cube or parallelepiped, within which there may be certain numbers of grains A and B at any points. Let us first consider the transport of the concentrations a and b through the faces 11 – 11 and 12 – 12, perpendicular to the x-axis. If another cube (or parallelepiped) is added next to the face 11 – 11 where grains A and B are aligned with the mirror symmetry about the initial cube (or parallelepiped), the transport through this face will be equal to zero because of the symmetry, as if this face is impermeable. To prevent diffusion through 12 – 12 in the initial cube (or parallelepiped) and the respective face in the second cube, we need to include mirror reflections about either side indefinitely. As a result, the faces 11 – 11 and 12 – 12 will appear to be impermeable. This one-dimensional sequence formed by an infinite number of mirror reflections about the faces perpendicular to the y-axis can thus be expanded to a two-dimensional array. In this case, the faces of the initial cube (or parallelepiped) perpendicular to the y-axis will also appear to be impermeable. This is also true for the z-axis. Then solute diffusion is restricted mainly to region inside the initial cube (or parallelepiped), with arbitrary number and orientation of grains *A* and *B*.

The concentration a at a given point is found by the formula obtained by solving (4.52):

$$a(t) = a_0 + (a_s - a_0)\sum_{i=0}^{N_a} R_{ai} \sum_{j=-\infty}^{\infty} \sum_{k=-\infty}^{\infty} \sum_{l=-\infty}^{\infty} erfc\left(\frac{r_{ijkl} - R_i}{2\sqrt{D_a t}}\right)\frac{1}{r_{ijkl}}, \tag{4.69}$$

where R_i is the current radius of the *i*-th grain *A*, N_a is the number of grains, r_{ijkl} is the distance from the *i*-th grain *A* inside the cube (or parallelepiped) whose *j* number can be counted along the x-axis, *k* number along the y-axis, and *l* number along the z-axis.

Let the dimensions of the cube (or parallelepiped) along the *X*, *Y*, *Z* axes be h_x, h_y, h_z, the coordinates of grain *A* in the initial cube (parallelepiped) be a_{xi}, a_{yi}, a_{zi}, and the coordinates of grain *C* be $c_{x_i}, c_{y_i}, c_{z_i}$.

Then

$$r_{ijkl} = \sqrt{(r_x - c_x)^2 + (r_y - c_y)^2 + (r_z - c_z)^2} \tag{4.70}$$

where, for example:

$$r_x = (j - 1/2)h_x + q \cdot dx; \; dx = (a_{x_i} - h_x/2), \tag{4.71}$$

and $q = 1$, if *j* is even (mirror reflection); $q = -1$, if *j* is odd.

Therefore, the limitations imposed by the assumption on an unbounded volume can be relaxed. Equations (4.66)–(4.68) can be used to simulate the dynamics of grain growth.

Assume that grains *A* and *B* whose number may be up to 40 are randomly distributed in a cube with side L_1, while grain *C* is in a cube with side L_2. One cube is placed in another. The calculations of growth dynamics of grain *C* were repeated many times with other random positions of grains of all types, and the results were averaged. The average of six realizations (a large number did not influence the results) was calculated. The calculation results for many cases are shown in Figs. 4.12 and 4.13. Figure 4.12 illustrates the growth of tremolite during metamorphism in siliceous dolomitic limestones in the Alps (Trommsdorff 1966, 1972). Figure 4.13 shows the case of retrograde metamorphism in a marble near the trap

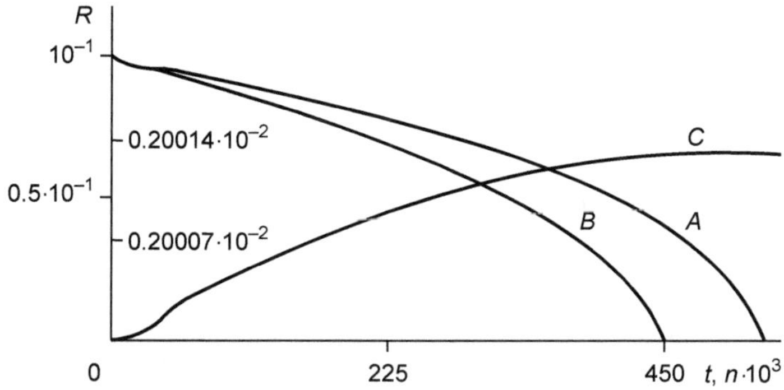

Fig. 4.12 Change with time (t, yr) in the radii (R, cm) of dissolving grains A and B and growing grain C. The values of R for curves A and B are given on the left-hand scale of the y-axis and those for curve C on the right-hand scale. Grains A and B whose number is N, are randomly distributed in a cube with side L_1, while growing grain C is in a cube with side L_2. One cube is placed in another. The parameters: $a_o = 10^{-9}$; $b_o = 2 \times 10^{-9}$ g/cm^3; $k = 10^{-3}$ cm^4/(g s). Data for model calculations: $N = 40$ $(20A + 20B)$; $L_1 = 2$, $L_2 = 0.3$ cm; $a_s = 10^{-8}$, $b_s = 2 \times 10^{-8}$ g/cm^3; $D_a = 9 \times 10^{-8}$, $D_b = 0.5 \times 10^{-7}$ cm^2/s; $R_a^\circ = 10^{-1}$, $R_b^\circ = 10^{-1}$, $R_c^\circ = 0.2 \times 10^{-2}$ cm; $V_a = 3.5 \times 10^{-1}$, $V_b = 4 \times 10^{-1}$, $V_c = 3 \times 10^{-1}$ cm^3/g. The curves A and B relate to the left-hand ordinate scale, and the curve C—to the right-hand one

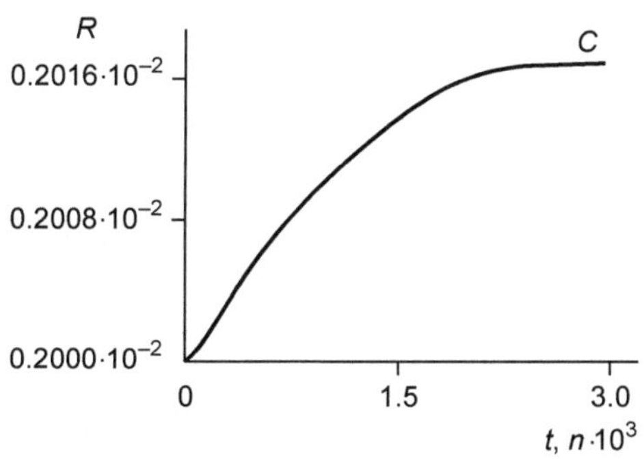

Fig. 4.13 Change with time (t, yr) in the radius (R, cm) of growing grain C. The conditions for distribution of grains are the same as in Fig. 4.12. The parameters: $a_o = 10^{-9}$; $b_o = 2 \times 10^{-9}$ g/cm^3; $k = 10^{-3}$ cm^4/(g s). Data for model calculations: $N = 20$ $(10A + 10B)$; $L_1 = 1$, $L_2 = 0.3$ cm; $a_s = 2 \times 10^{-6}$, $b_s = 10^{-6}$ g/cm^3; $D_a = 2.5 \times 10^{-8}$, $D_b = 2 \times 10^{-8}$ cm^2/s; $R_a^\circ = 5 \times 10^{-2}$, $R_b^\circ = 5 \times 10^{-2}$, $R_c^\circ = 0.2 \times 10^{-2}$ cm; $V_a = 3.5 \times 10^{-1}$, $V_b = 4 \times 10^{-1}$, $V_c = 3 \times 10^{-1}$ cm^3/g

intrusion (Reverdatto et al. 1979). As grains A and B undergo dissolution, the composition of grain C (the ratio of concentration a and b) can vary significantly.

Figure 4.14 shows the parameters for the composition of a part of the growing zone in grain C. Such variation in the grain composition can be expected for grain growth at the expense of a depleting source (Sheplev et al. 1990).

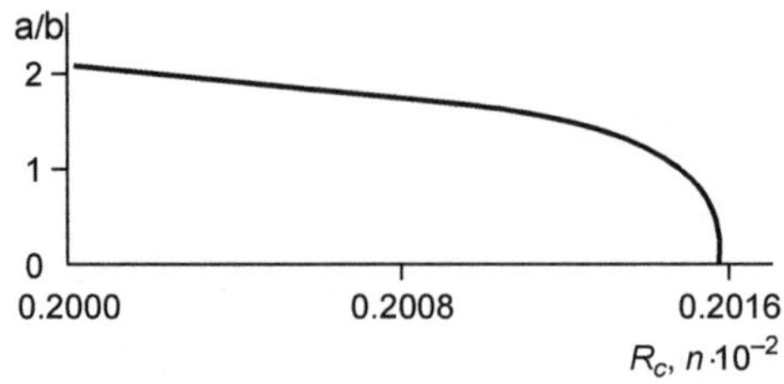

Fig. 4.14 Change in the composition (*a/b*) of growing grain *C* with increasing radius (cm). The conditions for distribution of grains *A, B,* and *C* are the same as in Fig. 4.13. Data for model calculations: $N = 20$ $(10A + 10B)$; $L_1 = 1$, $L_2 = 0.3$ cm; $a_o = 10^{-9}$; $b_o = 2 \times 10^{-9}$ g/cm^3; $k = 10^{-3}$ cm^4/(g s); $a_s = 2 \times 10^{-6}$, $b_s = 10^{-6}$ g/cm^3; $D_a = 2.5 \times 10^{-8}$, $D_b = 2 \times 10^{-8}$ cm^2/s; $R_a^\circ = 5 \times 10^{-2}$, $R_b^\circ = 5 \times 10^{-2}$, $R_c^\circ = 0.2 \times 10^{-2}$ cm; $V_a = 3.5 \times 10^{-1}$, $V_b = 4 \times 10^{-1}$, $V_c = 3 \times 10^{-1}$ cm^3/g

Changes in the size and composition of grains in Figs. 4.12, 4.13 and 4.14 are only illustrative because of the assumptions listed and conditional parameter values used for calculations. The latter is mainly true for the solute diffusion coefficients and reaction rates. Since only part of the surface of mineral grains in the rock participates in the reactions, the rate constants are much smaller compared to the data for free aqueous solutions (Fisher 1978; Fisher and Lasaga 1981; Lasaga 1981; Walther and Wood 1984). The coefficients of diffusive mass transfer are highly dependent on the thickness of intergranular fluid films (the width of intergranular channels filled with fluid) and are less than diffusion coefficients in free aqueous solution (see Sect. 1.2, and also Balashov et al. 1983; Fisher 1978; Lasaga 1981; Walther and Wood 1984), but greater than those of lattice diffusion (Freer 1981). Although the solubility of minerals in water and the saturation concentrations are approximately known (Lasaga 1981; Walther and Wood 1984, and others), the properties of the intergranular fluid can differ considerably from those of free aqueous solution. The initial concentrations of substances a_o and b_o in the solution are of no particular importance, but the saturation concentrations $a_s > a_o$ and $b_s > b_o$ need to be known more accurately because the dissolution rates, and hence the dissolution time of grains *A* and *B* depend strongly on these parameters: an order of magnitude increase in a_s and b_s results in an order of magnitude decrease in the dissolution time and vice versa.

In summary, we now turn to the above-mentioned simplifying assumptions. The first assumption of a distance between grains exceeding the grain size is very difficult to weaken or relinquish. The case of a short distance between grains requires the numerical solution of a three-dimensional problem with very complicated geometry for each specific case. The second and third assumptions can be weakened by introducing small parameters and constructing a first approximation for the unperturbed case, for which the above results may serve as a zeroth approximation. Abandoning the assumption of unbounded volume (fourth assumption) indicates loss of spherical symmetry; however, some special cases can be considered with taking account of the existing solutions.

4.2.2.2 Intergranular Mass Transfer in the Matrix and Metamorphic Reactions

As discussed in the Chaps. 1 and 2, metamorphic mineral transformations are driven by changes in *P-T* conditions of the Earth's crust. Three types of such transformations can be distinguished: (1) polymorphic transformations in minerals, which involve no change in chemical composition but only in crystal structure, (2) reactions involved the appearance or disappearance of new phases, and (3) diffusion-controlled exchange reactions leading to changes in the composition of minerals but not their modal proportions. Metamorphic reactions of the second type are (1) congruent dissolution of the early phases with complete removal of the dissolved material into a fluid and/or (2) incongruent dissolution of the early phases with partial removal of the dissolved material into a fluid and the preservation of the undissolved crystalline part of a mineral. The topochemical transformations that involve the formation of new phases and the inheritance of fragments of crystal structure of some early phase can play an important role, in metamorphic exchange reactions. The reaction generally involves dissolution of the old (early) phases, nucleation of new phases, material transport to nucleation sites, growth of new phases, and changes in in their composition by means of diffusion (Walther and Wood 1984). In the sequence of steps of mineral formation, the slowest one is rate-controlling. In practice, the dissolution and growth of new minerals during metamorphism are usually faster than material transport.

As discussed in Sect. 1.2, there are three mechanisms of mass transfer in metamorphic rocks: by fluid, by diffusion through the intergranular fluid or by diffusion along grain boundaries or through deformed crystal lattice (Walther and Wood 1984). The low permeability of the rocks with narrow intergranular channels (up to $\sim$100 Å) makes fluid infiltration impossible, while diffusion of dissolved substances through intergranular and intragranular fluid films becomes the predominant mass transfer mechanism. If film thickness does not exceed 10 Å, diffusion through the fluid film becomes impossible and mass transfer in this case is controlled by diffusion along the grain boundaries. In general, the flux of diffusing material in low-permeability rocks does not exceed 10^{-11}–10^{-12} g cm^{-2} s^{-1}; the flux along the grain boundaries is less than 10^{-15} g cm^{-2} s^{-1} (Fig. 1.19).

A variety of diffusion-controlled processes within a rock during metamorphism can be classified into the following types: (1) diffusion-controlled bimetasomatism at the boundaries between rocks or mineral grains of different composition leading to the formation of coronites and other zoned reaction structures; and (2) mineral transformations in the matrix minerals, involving their interaction with porphyroblasts.

From a thermodynamic point of view, a metamorphic reaction starts when a mineral or mineral assemblage intersects the boundaries of its *P-T* stability field. However, taking into account the kinetics, since the dissolution and growth of minerals occur over a certain period of time, both dissolving and growing mineral phases commonly coexist on metamorphic isograds. The interaction of mineral grains in the matrix during metamorphism occurs by means of a complex back-and-forth diffusive exchange of chemical components via a fluid through the network of intergranular channels. The transport of dissolved substances is controlled by concentration gradients. All minerals and all chemical elements participate in the reaction to varying degrees. The contents of most chemical elements are negligible, the compositions of minerals and their transformations in a metamorphic rock can be accounted for by a small number (usually up to 10–12) of the major chemical components. This enables the interaction of minerals to be adequately studied in terms of the Gibbs phase rule and material balance during reactions. The compositions of newly-formed minerals are continuously changing during mineral growth in response to a change in *P-T* conditions.

Exchange diffusion of components takes place at the contact between chemically contrasting mineral grains or rocks undergoing metamorphism, which leads to the development of reaction zoning as a result of bimetasomatism. This type of interaction results in the formation of coronites (see Sect. 4.2.1) or a linear zonal sequence of layers (see, for example, Thompson 1975). Mineral zoning is not always commonly developed, and contrasting media tend to interact by exchange diffusion accompanied only by a change in the mineral composition at the contact of rocks (Lepezin et al. 1990; Seroglazov 1992; Lepezin and Khlestov 2009; Lepezin 2015). In this case, original textures are often preserved in metamorphic rocks. However, poikilo- and porphyroblasts, regions of pseudomorphic replacement of minerals, development of corona, fibroblastic and similar structures may develop in lepidogranoblastic (in general case) metapelites. Unlike coronites, these rocks does not exhibit replacement by successive multiple mineral layers, which can be explained by only a minor increase in concentrations of diffusing chemical components in the reaction zone.

Reaction zoning may have not been developed in the rock matrix due to the absence of contrasting compositional heterogeneities in the protolith. The latter can interact with minerals around the matrix by mutual diffusion of components, forming zonal segregations, which, as noted above, represent frozen reaction textures (see Sect. 4.2.1). The reaction zoning in segregations, such as coronites, reflects variations in the concentration and mobility of the components. For example, in metapelites, the melanocratic cores of segregations are usually composed (in different combinations) of andalusite, sillimanite, kyanite, biotite, ilmenite, spinel, cordierite, staurolite, garnet and other minerals with high contents of Al, Ti, Mg, and Fe, while the leucocratic rims are largely composed of quartz, feldspars, white mica, etc., i.e., the concentrators of alkalis, calcium and other mobile components.

The structural and textural homogeneity of the protolith in the absence of deformation implies a similar homogeneity of isochemically metamorphosed rocks. The primary structural and compositional inhomogeneity of the protolith is inherited to some extent by the metamorphic rock. The distribution of newly formed minerals is largely controlled by the primary compositional inhomogeneity of the protolith, the initial concentrations and mobilities of chemical components, of which aluminum appears to be the least mobile (Carmichael 1969; Fisher 1973). The distribution of zones with the highest contents of Al and Ti (another least mobile element) in the metamorphic rock can be used primarily to deduce the textural and structural features of the protolith. The chemical inhomogeneities due to variations in the concentrations of the more mobile elements, such as potassium, sodium, calcium, magnesium, iron, etc., are less stable during metamorphism and can be partly destroyed during recrystallization of minerals (often with a change in composition). However, relatively large inhomogeneities, even being modified, are usually preserved in the protolith, even if they are represented by more mobile components than Al and Ti. These are, for example, siliceous concretions in limestones, ferro-manganese, phosphorite, clay nodules in sedimentary rocks, etc.

During the subsequent retrograde metamorphism, rock undergo recrystallization, which, due to a decrease in surface energy, leads to the formation of large mineral grains at the expense of small ones. Syn- or post-metamorphic deformations may cause changes in textures of the rocks, leading to the development of schistosity, gneissic banding, crenulation cleavage, etc., and the production of blastic and clastic structures.

According to the Korzhinskii principle of local equilibrium, thermodynamic equilibrium can be attained in each elementary volume of the rock over time. A progressive increase in the size of local equilibrium domains results in a transition to disequilibrium. Obviously, it will take a longer time to equilibrate the chemical potentials and reach the new equilibrium state in a larger volume. There should be a minimum (elementary equilibrium) volume, in which a redistribution of matter takes place during the metamorphic reaction with preservation of the material balance. This local volume is limited by the distance over which the most mobile of the components participating in the reaction would have had to move (except for volatiles). The estimation of the minimum volumes in which metamorphic reactions take place can be made on the basis of microtextural observations and the spatial distribution of minerals participating in the reaction, with taking into account their composition, grain size and distances between them in the rock.

As discussed in Sect. 1.2, Carmichael (1969) was the first to notice that mineral transformations at isograds can be described as metamorphic reactions balanced for the reacting components, except for volatiles. He studied samples of metapelite crystalline schists from the area in the Whetstone Lake area, south-eastern Ontario, Canada, exhibiting a metamorphic zoning similar to that described by Barrow (1893, 1912) in Scotland (see Sect. 1.2). The studied thin sections of pelitic schists were composed of quartz, felsic plagioclase, muscovite, biotite, garnet, staurolite, kyanite, and sillimanite. Two isograds were identified in the zoning: sillimanite and sillimanite/kyanite-garnet-biotite. The polymorphic transition Ky $\rightarrow$ Sil has traditionally been used to locate the sillimanite isograd in Scotland. However, in the case of a pseudomorphic reaction, sillimanite would have to grow near or directly above the kyanite relics. In reality, the appearance of sillimanite is mainly confined to biotite and quartz. Based on these microtextural observations, Carmichael suggested the following possible reactions for the formation of sillimanite: reaction (1) (Fig. 4.15a) produces muscovite in place of kyanite, and then muscovite is converted to sillimanite by reaction (2). If these reactions occur simultaneously in a small volume of rock, the final reaction will be a simple transformation Ky $\rightarrow$ Sil.

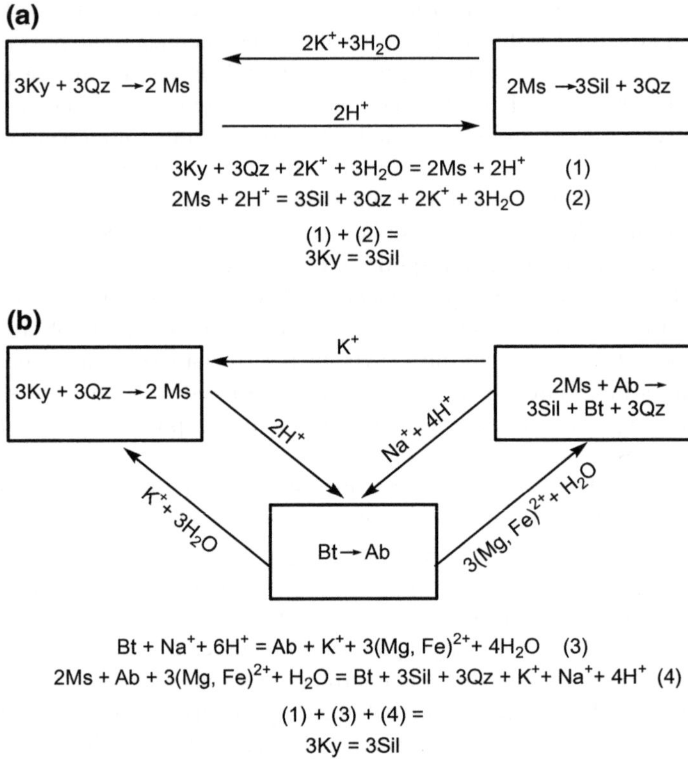

Fig. 4.15 Mechanism of a common cyclic reactions accompanied by local metasomatism without changing the bulk rock composition at the sillimanite isograd. The net reactions for the whole system are 3 Kyanite = 3 Sillimanite. The exchange of ions among the local systems is shown schematically (Carmichael 1969)

Figure 4.15 a show the schematic relationships reflecting the actual numbers of ions that would have to diffuse along the grain boundaries between the two reaction domains (1) and (2). These two reaction domains can be considered as local systems, in which kyanite is converted to sillimanite.

Further, Carmichael showed that this net reaction can proceed even in a more complex way (Fig. 4.15b). Based on microtextural observations, he written reactions (3) and (4), the reality of which is confirmed by the observed replacement of biotite by plagioclase and the simultaneous formation of biotite, sillimanite and quartz in the same rock. By adding these reactions with reaction (1), he obtained the same net reaction Ky → Sil, as was obtained by adding together reactions (1) and (2). In this case, the net reaction is carried out without change in the bulk chemical composition of the rock, but each particular reaction is not isochemical and thus requires that hypothetical ions must enter the "dispersed phase" (Carmichael 1969).

He regarded the net metamorphic reaction as a combination of local metasomatic cation-exchange reactions, which proceed simultaneously in different domains of the rock. Local systems of particular phase interactions in metapelites are limited to a small volume of rock, each of which is controlled by the limit of migration for aluminum, while other chemical components are more mobile. Local systems can exchange components with each other by diffusion through the fluid, and this defines the volume of the net reaction.

Carmichael (1969) used a simple approach: a study of thin sections, analysis of textural and structural relationships in the metamorphic rock, spatial distribution and approximate estimates of the volumes and relationships of minerals, idealized mineral compositions and the use of schematic chemical reactions of mineral interaction. Despite the rough approximation, his conclusions were generally valid, although they were not justified quantitatively.

Carmichael postulated that aluminum is the least mobile component in the metamorphic reaction and has the very small limit of migration of order of 0.2 mm, while that for potassium, sodium, calcium, magnesium and iron is of the order of 2–4 mm, and that for titanium is of the order of 0.5 mm. The distances between the local systems are of the order of several millimeters, ranging up to a few centimeters in very coarse-grained rock. In fact, each local system is closed to aluminum, and the net reaction is closed to all components, except for volatiles, implies that metamorphism is isochemical with respect to the rock as a whole.

Carmichael's ideas about metamorphic reactions in the rock matrix were later elaborated by a number of researchers. The precise stoichiometric coefficients for the overall reaction among the phases in a rock were calculated using the principle of the conservation of matter, provided that an estimate can be made of the components that enter or leave the system. If matter is conserved, the material lost or gained by the system must equal a linear combination of the compositions of the reacting phases. If the chemical analyses of the phases are error-free, the stoichiometric coefficient of each phase in the reaction may be obtained by Foster (1977):

$$[\gamma]^{-1}\cdot[F] = [S], \tag{4.72}$$

where

$$[\gamma] = \begin{bmatrix} \gamma_{11} & \gamma_{12} & \cdots & \gamma_{1p} \\ \gamma_{21} & \gamma_{22} & \cdots & \gamma_{2p} \\ \cdots & \cdots & \cdots & \cdots \\ \gamma_{p1} & \gamma_{p2} & \cdots & \gamma_{pp} \end{bmatrix}, \tag{4.73}$$

$$[F] = \begin{bmatrix} f_1 \\ f_2 \\ \cdots \\ f_p \end{bmatrix}, \tag{4.74}$$

$$[S] = \begin{bmatrix} S_1 \\ S_2 \\ \dots \\ S_p \end{bmatrix}, \tag{4.75}$$

γ_{ij} is the number of moles of component i in one mole of phase j; p is the total number of phases present; f_i is the number of moles of component i lost (+) or gained (−) by the system; S_j is the number of moles of phase j taking part in the reaction; $[\gamma]^{-1}$ is the inverse of $[\gamma]$. If there are more than p independent components in the phases taking part in the reaction, any p of them can be used and will give the same reaction provided the chemical analyses are perfect.

Unfortunately, because this approach balances the reaction exactly, even small analytical errors produce incompatible results when different components are used as basis vectors. To permit the "best fit" of the data using n components, so that the analytical errors tend to balance out, the following matrix equation was used:

$$\left([\gamma]^T[\gamma]\right)^{-1}[\gamma]^T[F] = [S], \tag{4.76}$$

where $[\gamma]$ is $n \times p$; $[\gamma]^T$ is the $p \times n$ transpose of $[\gamma]$; $[F]$ is $n \times 1$. This solution results for Eq. (4.76) in a least squares fit of the phase compositions to the known gain or loss of components from the system.

The method of studying mineral transformations during metamorphism in the matrix was applied to sillimanite-containing metapelites in the Rangeley-Oquossoc region, northwestern Maine, USA (Foster 1977). Metapelites were formed from Ordovician and Silurian-Devonian sediments that have undergone two periods of deformation, followed by intrusion of quartz diorite at 500–650 °C and 4–5 kbar (Guidotti 1970, 1974). The metapelites with lepidogranoblastic texture are composed of staurolite, garnet, sillimanite, biotite, muscovite, plagioclase, ilmenite and quartz. Foster (1977) noted that the prograde metamorphic processes that took place in metapelites produced tree types of mineral heterogeneities: (1) sillimanite segregations composed of sillimanite cores and biotite rims; (2) staurolite segregations composed of poikilitic staurolite porphyroblasts and muscovite rims; and (3) garnet segregations consisting of garnet porphyroblast embedded in a matrix composed of muscovite, biotite, plagioclase, quartz, and ilmenite (Fig. 4.16). Garnet has two distinct morphologies. Garnets in or near sillimanite segregations occur as xenomorphic grains. Garnets in the matrix tend to be isometric up to 1.5 mm in diameter. The modal amount and grain size of garnet tend increase close to the intrusion contact. Foster distinguished two garnet generations, old garnet and new garnet; old garnets usually form cores of porphyroblasts. The staurolite and sillimanite segregations are distributed throughout the rocks in a nonuniform manner, which is caused by reactions between minerals. The reactions begin with the nucleation and growth of sillimanite in the biotite-muscovite-plagioclase-ilmenite-quartz matrix, surrounding the garnet and staurolite porphyroblasts. The

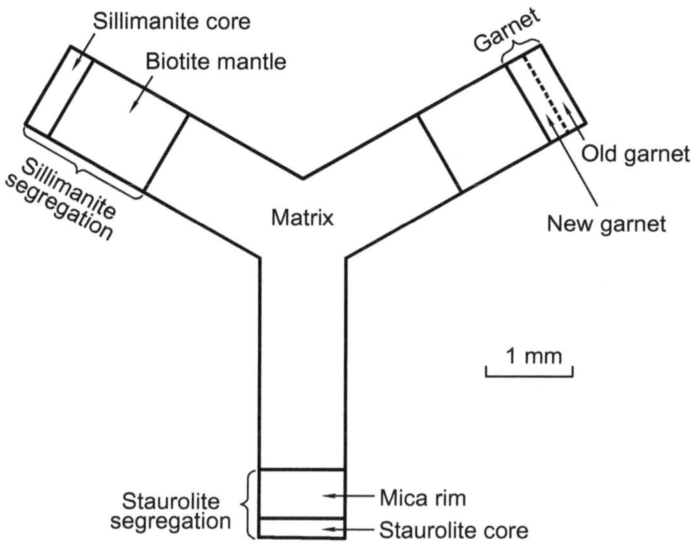

Fig. 4.16 Schematic representation of diffusion system in sillimanite-garnet-staurolite-bearing pelitic schists near Rangeley, Maine (Foster 1977)

overall reaction calculated for the eight-component system using the data on minerals from Foster (1977) can be written as (Reverdatto and Kolobov 1987):

$$0.927\mathrm{St} + 0.057\mathrm{Ms} + 3.712\mathrm{Qz} = 0.554\mathrm{Grt} + 0.054\mathrm{Bt} + 0.043\mathrm{Ilm} + 0.010\mathrm{Pl} + 3.550\mathrm{Sil} + 1.000\mathrm{H_2O}. \quad (4.77)$$

Variations in the composition of minerals produce little changes in the coefficients of each phase involved in the reaction (Foster 1977). However, the calculated volume change is 10–15%. Table 4.1 shows a comparison of the calculated volume fractions of minerals for the net reaction (the molar volumes of each phase were converted to volume fractions using the molar volume values from Robie et al. 1966; Griffen and Ribbe 1973; Holland and Powell 1998). As a result of detailed mineralogical studies and calculations of the volumes of initial and newly formed phases, Foster concluded that the net reaction is produced by the sum of several minor local reactions in the matrix. The local reactions are not isochemical, requiring that mass transfer take place between the staurolite and sillimanite-biotite segregations, garnet porphyroblasts and matrix. The mass transfer occurring in a rock between segregations and garnet grains is shown schematically in Fig. 4.17. Analysis of mass balance in a volume of $\sim 1\ \mathrm{cm}^3$ пof rock shows that the system is left closed to all components except water.

Another example is metapelitic schists found at the contact with a granitoid massif in the Sangilen Mountains, Tuva (Reverdatto and Kolobov 1987). Schists

Table 4.1 Comparison of calculated volume coefficients of minerals for reaction equation and net volume changes (vol.%) in individual segregations during metamorphism for sillimanite-bearing pelitic schists near Rangley, Maine, USA

Mineral	St	Ms	Qz	Grt	Bt	Ilm	Pl	Sil
After Foster (1977)	−1.145	−0.0803	−0.1915	+0.3400	+0.0676	+0.0093	+0.0012	+1.000
According to reaction Eq. (4.77)	−1.167	−0.0453	−0.4754	+0.3593	+0.0465	+0.0077	+0.0057	+1.000

Staurolite segregation	
Volume of mica rims (Foster 1977)[a]	5.36%
Volume of replaced staurolite (Foster 1977)[b]	−5.14%
The same (according to reaction equation) (4.77)[b]	−5.24%
Garnet in matrix	
Volume of newly-formed garnet (Foster 1977)[b]	1.19%
The same (according to reaction equation) (4.77)[b]	1.15%

Volumes produced (+) or consumed (−) per unit volume of sillimanite produced
[a]Measured in rock using drawings of thin-sections and a planimeter
[b]Calculated from volume coefficients in upper part of Table 4.1

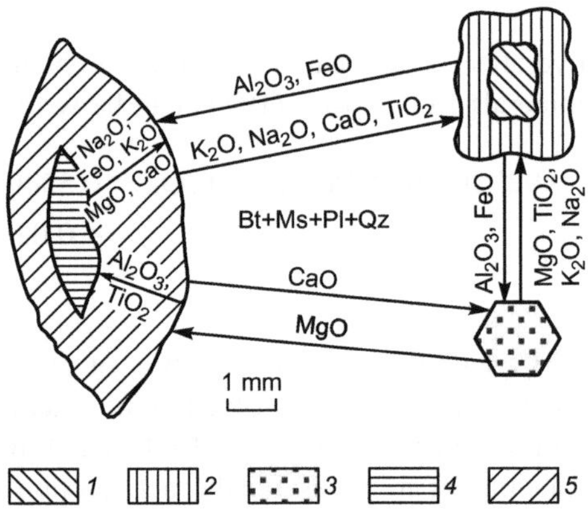

Fig. 4.17 Cation exchange reaction mechanism within segregations and porphyroblasts of garnet for sillimanite-bearing rocks near Rangeley, Maine (Foster 1977). Mineral assemblages: *1*—St + Qz, *2*—Ms + Bt + Pl + Ilm+ Qz, *3*—Grt, *4*—Sil + Pl + Ilm + Qz, *5*—Bt + Pl + Ilm + Qz

are composed of staurolite, cordierite, garnet, biotite, muscovite, plagioclase, ilmenite, and quart. Contact metamorphism resulted in the development of andalusite and other mineral transformations. Staurolite forms porphyroblasts and aggregates. In the outer zone of the contact aureole, the muscovite rims develop around staurolite grains. The degree of replacement increases toward the intrusion, and muscovite pseudomorphs with rare staurolite relics occur at a distance of about 2/3 of the aureole width. The growth of andalusite porphyroblasts with a large number of inclusions of quartz, biotite, and plagioclase occurs simultaneously. Outside the contact aureole garnet occurs as irregular grains with inclusions of quartz, graphite, and ilmenite. Near the granitoid intrusion, garnet becomes well faceted and contains no inclusions. The net reaction involving replacement of staurolite, growth of andalusite and recrystallization of garnet for the ten-component system can be written as (Reverdatto and Kolobov 1987):

$$\begin{aligned}&0.847\mathrm{St}+0.606\mathrm{Crd}+0.744\mathrm{Grt}_1+0.847\mathrm{Ms}+0.019\mathrm{Pl}_1\\&=5.015\mathrm{And}+0.896\mathrm{Bt}+1.012\mathrm{Grt}_2+0.193\mathrm{Pl}_2+0.232\mathrm{Qz}+1.000\mathrm{H_2O}.\end{aligned}\tag{4.78}$$

Garnet-1 and plagioclase-1 represent the compositions of the grain cores, while garnet-2 and plagioclase-2 of the grain rims. The calculate reaction (4.78) can be quantitatively verified against the mineralogical calculations (Table 4.2). The volume ratios of the minerals involved in the reaction are close to the data obtained by recalculating the molar volumes of each phase participating in the reaction (4.78) to volume fractions using the data on molar volumes from Robie et al. (1966), and Griffen and Ribbe (1973). The volume of rock in which this reaction took place is only up to a few cubic centimeters. This results from the distribution of minerals and the distance between grains and their size (Fig. 4.18). This conclusion was drawn based on the results of measurements from differently oriented thin sections as well as quantitative and mineralogical calculations.

In more recent works, mass transfer constrains in the matrix have been further developed. The basis for mass transfer analysis during metamorphism are the mineral reaction and modal calculations. Because of the small grain size, the modes of minerals were calculated from the composition of the minerals in a sample and the sample's whole-rock chemical composition, according to the approach proposed by Ferry (1984). For each sample:

Table 4.2 Comparison of observed mineral volume ratios with values calculated from the mineral reaction (4.78) for rocks from the contact aureole (Sangilen, Tuva) (Reverdatto and Kolobov 1987)

Minerals	V_j, mm^3		V_j, vol.%		V_j	$\frac{V_j}{V_{St}}$	V_j^*	$\frac{V_j^*}{V_{St}^*}$
	Initial	Final	Initial	Final				
St	83.2	0	1.04	0	−1.04	1	−188.88	1
Crd	1019.2	636.8	12.74	7.96	−4.78	4.60	−141.05	0.75
Ms	855.2	800.8	10.69	10.01	−0.68	0.65	−119.18	0.63
Ilm	66.4	65.6	0.83	0.82	−0.01	0.01	−0.60	0.003
And	0	108.0	0	1.35	+1.35	1.30	+258.42	1.37
Bt	1986.4	2054.4	24.83	25.68	+0.85	0.82	+136.68	0.72
Grt	76.0	93.6	0.95	1.17	+0.22	0.21	+30.82	0.16
Pl	417.6	486.4	5.22	6.08	+0.86	0.83	+19.41	0.10
Qz	3488.0	3757.6	43.6	46.97	+3.37	3.24	+5.26	0.03

V_j = a volume of mineral j (in mm^3 and vol.%), V_j = mineral volume calculated from reaction equation. V_j = changes in mineral volumes as a result of mineral reaction (4.78). V_{St} and V^*_{St}—a volume of staurolite relative to which the volume ratios with other mineral phases were determined directly in thin sections and from reaction equation calculations, respectively

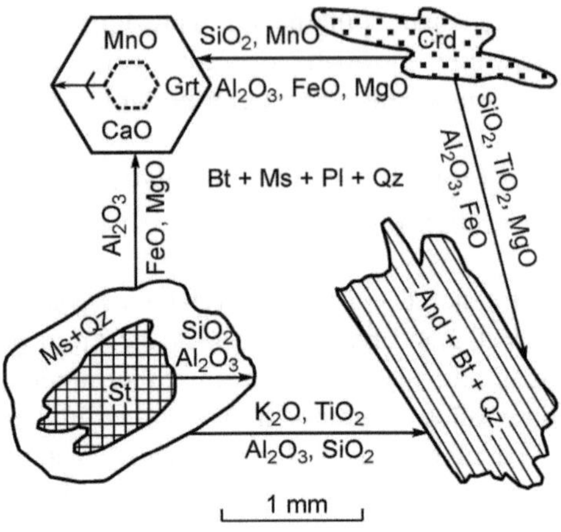

Fig. 4.18 Chemical component redistribution among reacting phases as a result of an overall contact metamorphic reaction (4.78) in pelitic schists of Sangilen area (Tuva) (Reverdatto and Kolobov 1987)

$$i/Al = \left(\sum_j a_{i,j} \cdot n_j\right) / \left(\sum_j a_{Al,j} \cdot n_j\right), \tag{4.79}$$

where i/Al is the whole-rock atomic ratio of element i to Al as a more relatively immobile element (obtained from whole-rock chemical data), n_j represents the number of moles of each mineral species j in a sample, and $a_{i,j}$ is the number of atoms of i per standard formula unit of mineral j (obtained from mineral analyses). The volume of the rock considered was arbitrarily taken as 100 cm^3, i.e.,

$$\sum_j V_j \cdot n_j = 100 \tag{4.80}$$

where V_j is the molar volume of mineral j in cm^3. Numerical values of n_j were calculated for samples by solving the system of Eqs. (4.79) and (4.80); moles mineral j were recalculated as volumes (modes) using molar volume data for minerals from the internally consistent thermodynamic dataset of Holland and Powell (1998).

Chemical reaction equations were calculated using the actual composition of coexisting minerals by means of the algebraic mass balance techniques of the program Mathematica 5.0, using the built-in function, NullSpace for the systems K_2O–FeO–MgO–Al_2O_3–SiO_2–H_2O–CaO±Na_2O±MnO±TiO_2 (KFMASHCa±Na ±Mn±Ti) (Wolfram 2003). To apply this technique, we constructed a composite matrix for each mineral assemblage containing (n) number of columns equal to the number of minerals in the assemblages and (n − 1) rows corresponding to the chemical components. The composition matrix is expressed in terms of cations per formula unit. H_2O was included as a component in the mass balance. For mass-balance calculations, the water contents in minerals were assumed to be stoichiometric in the model formulae inferred from the widely accepted ones in the literature. For each rock, the number of rock-forming components was defined by the actual composition of the system.

In order to test the reliability of the chemical reaction, the calculated mineral volume ratios from reaction equation were compared with the actual mineral volumes in the rock calculated by whole-rock chemical composition. The volumes of reacting minerals (unless H_2O also involved) were estimated by the following relationships:

$$v = n_j V_j, \tag{4.81}$$

where n_j is the number of moles of mineral j from the reaction equation.

An important procedure was the evaluation of the minimum rock volume required for approaching equilibrium among all coexisting phases on the basis of mass balance. Usually these volumes, in which metamorphic reactions took place, are estimated on the basis of observations of spatial relationships between reacting minerals in different-oriented thin sections, taking account of their grain sizes and intergranular distances (e.g., Foster 1977, 1981, 1986). In this study minimum rock volumes were estimated on the basis of the size (thickness) and spatial distribution of different zones of garnet porphyroblasts. The balance of components was estimated for these microdomains of the rock, assuming practically constant volume. The numbers of component moles in such microdomains of the rock both before and after the beginning of the reaction were compared, which, in combination with microstructural interpretations of the mineral relations, enabled to draw conclusions about the mineral transformations.

A number of metamorphic reactions have been studied using the example of contact aureole of the Kharlovo gabbro massif mentioned above (see Fig. 3.17 in Chap. 3). The country rocks consist of psammitic sediments interbedded with rare, thin beds of basic lavas and tuff layers that were regionally metamorphosed to greenschist facies conditions during the Cambrian-Ordovician Caledonian orogeny. Metapsammites consist of chlorite, muscovite, albite and quartz, with sporadic epidote, calcite and ore minerals. The country rocks adjacent to the gabbro pluton are recrystallized to hornfels (Likhanov et al. 1994). Geochemical studies (Reverdatto et al. 1974) on country and contact metamorphic rocks indicate that the contact metamorphism was isochemical. The contact aureole, as defined by the biotite-in isograd, extends up to 1.2 km to the east of massif with distinct zonation due to changing hornfels structure and mineral associations. Mineral assemblages, from lower to higher temperature, are: (1) Chl + Ms + PI + Qz + Ilm + Mag + Ep ± Hem ± Cal (country rocks), (2) Bt + Ms + Chl + PI + Qz + Mag ± Ep (outer biotite zone), (3) Crd + Bt + Ms + Qz + PI + Mag (spotted hornfels zone), (4) Crd + Bt + PI + Qz + And + Mag (middle zone), (5) Crd + Bt + PI + Kfs + Qz + And ± Mag (inner zone), (6) Crd + Bt + Spl + Sil + Pl + Kfs + Mag (thin, quartz-free segregations in the inner zone).

We studied in detail the diabase body (60–80 cm thick) that underwent thermal metamorphism in the middle part of the contact aureole (the zone of cordierite-biotite hornfels) (Likhanov et al. 1995). Rare, basic effusive rocks occur as amygdaloidal diabases. They are massive and greenish to dark-grey in colour, with non-uniformly distributed amygdules that are less than 2 mm in diameter and are filled with calcite and chlorite. The diabases are composed of plagioclase, chlorite, calcite, sphene and ore minerals, and rarely biotite. The contact metamorphism gave rise to hornblende in the central part of the diabase body. In the vicinity of calcite-chlorite amygdules, hornblende grains are commonly rimmed by an alkali-amphibole—arfvedsonite. Based on microtextural and chemical information, the reactions responsible for the formation of biotite, arfvedsonite, and for the epidote breakdown were caculated.

The thermal effects of the gabbro pluton on the country rocks are first perceptible through the appearance of fine biotite flakes in chlorite-muscovite-plagioclase-quartz-ilmenite-magnetite-epidote schists. Biotite grows exclusively from chlorite, replacing it along grain boundaries, fractures and cleavage planes. As a result of this interaction, small aggregates are formed, which are thin chlorite/biotite interstratifications and mixed-layers.

Biotite formation occurred at 400–430 °C in the contact aureole of the Kharlovo massif, in the system SiO_2–TiO_2–Al_2O_3–FeO–MgO–K_2O–H_2O, may be written in terms of the following reaction (Likhanov et al. 1994):

$$\begin{aligned} &1.4\mathrm{Ms}_1 + 0.049\mathrm{Chl} + 0.042\mathrm{Ilm} + 0.0019\mathrm{Mgt} \\ &= 1.1\mathrm{Ms}_2 + 0.27\mathrm{Bt} + 0.7\mathrm{Qz} + 0.72\mathrm{H_2O} \end{aligned} \tag{4.82}$$

where Ms_1 и Ms_2 are the muscovite compositions in country rocks and from the outer biotite zone, respectively. This reaction appears on the A'KF diagram (Likhanov et al. 1994) as the intersection of the more phengitic muscovite-chlorite with the less phengitic muscovite-biotite tie-line. The reaction is also confirmed by mineral abundances in thin section, where the formation of biotite is accompanied by decreasing contents of chlorite, muscovite and ore minerals in parallel with changing white mica composition.

Table 4.3 reports modal abundances under biotite-out and biotite-in conditions. The balance of petrogenetic components was estimated for fine chlorite-biotite-mica segregations. Mass balance analysis showed that the reaction volume in the rock is of the order of 0.01 mm^3 (the boundary of the chlorite-biotite aggregate with other minerals) (Likhanov 1989). Thus, mass transfer of major components for the biotite-forming reaction was limited to very small volumes (microsites). Such scales of mass transfer may indicate that diffusion is main limiting factor for mass exchange during contact metamorphism. A characteristic distance of mass transfer in the metamorphic reaction may therefore be estimated at 0.23 mm. This is the distance the components were transported during diffusion-controlled metamorphic reactions (scale of mass transfer). In our case, it is the radius of a sphere with 0.01 mm^3 volume.

A detailed examination of narrow transition zones between chlorite and newly formed biotite revealed mineral phases corresponding to a thin interlayering in the volume ratio 2:1 and 1:1. During the transition from chlorite to biotite, with an

Table 4.3 Net mass balance of major components before and after biotite formation reaction (4.82) in the spotted hornfelses of Kharlovo contact aureole (Likhanov et al. 1994)

	N	n_{SiO_2}	n_{TiO_2}	$n_{Al_2O_3}$	n_{FeO}	n_{MgO}	n_{K_2O}
Initial concentration of components							
Chl	0.09	0.26	–	0.23	0.19	0.22	–
Ms	0.36	1.23	–	0.81	0.06	0.08	0.28
Mag	0.67	–	0.13	0.01	1.57	0.13	–
Ilm	0.28	–	0.30	–	0.24	–	–
Qz	39.2	39.2	–	–	–	–	–
Total		**40.7**	**0.43**	**1.05**	**2.06**	**0.43**	**0.28**
Final concentration of components							
Chl	0.08	0.21	–	0.23	0.16	0.19	–
Bt	0.04	0.11	0.01	0.07	0.04	0.05	0.03
Ms	0.31	0.98	0.01	0.82	0.03	0.04	0.26
Mag	0.61	–	0.12	0.01	1.43	0.12	–
Ilm	0.22	–	0.23	–	0.19	–	–
Qz	39.7	39.7	–	–	–	–	–
Total		**41.0**	**0.37**	**1.13**	**1.85**	**0.40**	**0.29**

$N = V/V_j$—molar quantity of minerals in estimated volume of rock = 0.1 mm^3 ($N \times 10^{-4}$ mol), $n_k = N \times a_{i,j}$—molar quantity of major (petrogenetic) component i in mineral j ($n_k \times 10^{-8}$ mol). The values to be compared are underlined and bolded

almost constant Si content in the tetrahedral layers, an increase in the amount of K^+ by 0.5–1 cation was accompanied by a gradual decrease by 1.5–3 cations in the octahedra of the "brucite" layer $(Fe + Mg + Al)^{VI}$ (Fig. 4.19). The possible mechanism of this reaction is interpreted as a topochemical mineral transformation with the inheritance of isolated fragments of the chlorite lattice, e.g., the "talc" layer; the octahedral "brucite" layers were replaced by interlayer potassium cations (Fig. 4.20). Such topochemical mineral transformations resulted in the formation of intermediate mixed-layer phases. This mechanism implies the small limits of migration of Al_2O_3 and SiO_2 participating in the reaction, which is also true for a portion of MgO and FeO in the "talc" layer. The limit of migration of the remaining portion of MgO and FeO was larger, since they constituted part of the destroyed "brucite" layer. K_2O as the most mobile of the components involved in the reaction was introduced into the biotite structure from muscovite. This conclusion was confirmed by long-term hydrothermal experiments on the conversion of chlorite in different chemical media at *P-T* parameters of 500 °C and 1 kbar, which allowed the simulation of the biotite isograd conditions for contact metamorphism in metapelites (Likhanov 2003). The experimental results show that the well-ordered mixed-layered aggregates consisting of packets with an inter-planar space distance of 38 Å (near chlorite) and 24 Å (near biotite) and a width of 0.05–0.1 μm are formed at the boundary between the initial chlorite and the newly formed biotite. This is indicative of a gradual replacement of chlorite by biotite due to the

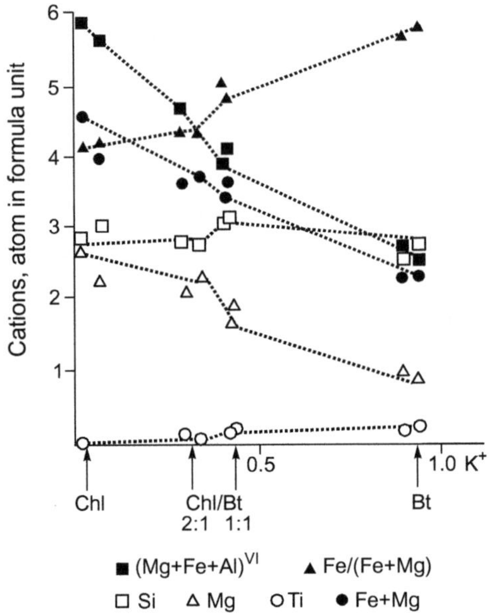

Fig. 4.19 Plot of cation proportion per formula unit versus K^+ content for chlorite, biotite, and 1:1 and 2:1 chlorite/biotite mixed-layer interstratifications (Likhanov et al. 1994)

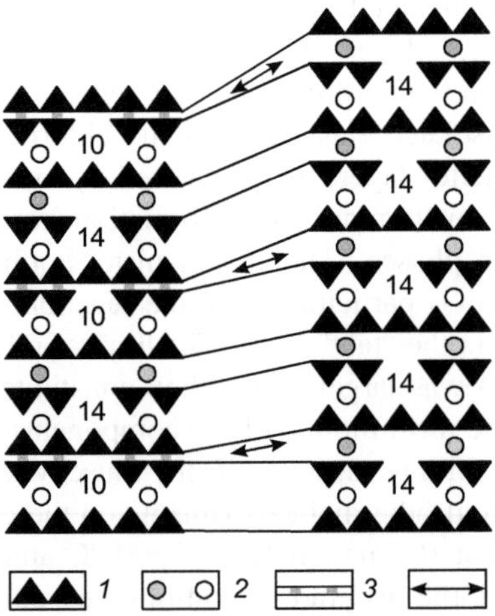

Fig. 4.20 Possible crystal-chemical mechanism for obtaining a regularly interstratified 1:1, 14Å/10Å chlorite/biotite mixed-layer from 14Å chlorite. Model involves the replacement of octahedral brucite-like layers by K cation interlayer sheets. 1—tetrahedral sheet (talc layers); 2—octahedral sheet (dark circles—brucite-like layers in the structure of chlorite); 3—potassium cation interlayer sheet in the biotite structure; 4—interlayer sheet exchange (replacement of brucite-like layers by K cation interlayer sheet

appearance of intermediate mixed-layered phases; otherwise, complete destruction of aluminosilicate phases would take place with the removal of Al and Si into the solution. Atomic absorption spectrometry data for the concentration ratios of the components in solutions before and after the experiments can serve as an indirect proof of the proposed mechanism: the appearance of a large amount of Mg in solution confirms the destruction of the "brucite" layer in the chlorite structure, while a simultaneous decrease in the potassium content of the solution points to substitution of Mg by interlayer potassium cations. Based on the *P*-*T*-*t* parameters of the experiment and the degree of structural transformations in the initial chlorite, the rate of the chlorite-biotite substitution reaction was estimated to be equal to (0.6–1.3) $\times$ 10^{-18} mol cm^{-2} s^{-1}. This value is an order of magnitude lower than the experimental dissolution rate of phlogopite obtained by Walther and Wood (1984). The width of the altered zone and the known duration of the experiment made it possible to estimate the characteristic diffusion coefficient for K_2O, which is 2 $\times$ 10^{-17} cm^{-2} s^{-1} at T = 500 °C and P = 1 kbar.

Many authors regard the plagioclase-epidote equilibrium as a metamorphic grade indicator, which defines upper thermal limit of Pl + Ep stability field in Ms +Bt+Qz-bearing metapelites (Rambaldi 1973; Crawford 1966). Epidote occurs as small (up to 0.2 mm), rare, rounded or irregular grains. In the country rocks it is usually zoned, with rims enriched in Fe and Ca and depleted in Al relative to the core. Near the biotite-in isograd and in the outer biotite zone, the epidote grains are no longer zoned but appear corroded, partially replaced by magnetite and recrystallized into aggregates of smaller prismatic crystals. Moving further towards the contact, epidote disappears completely in the spotted hornfels zone. Near the biotite isograd - where epidote breakdown occurs at 450 °C—the (Ca/(Ca + Na)) ratio increases to 0.05 and reverse zoning is observed.

Taking account of these observations, the epidote consuming reaction in the contact aureole of the Kharlovo massif, in the system SiO_2–Al_2O_3–TiO_2–FeO–MgO–CaO–Na_2O–K_2O–H_2O (Likhanov et al. 1994), is:

$$\begin{aligned} &27.59\mathrm{Pl}_1+0.558\mathrm{Ms}+0.179\mathrm{Chl}+0.222\mathrm{Ep}+0.099\mathrm{Ilm} \\ &=28.37\mathrm{Pl}_2+0.532\mathrm{Bt}+0.035\mathrm{Ti}-\mathrm{Mgt}+\mathrm{Qz}+0.96\mathrm{H_2O}, \end{aligned} \tag{4.83}$$

where Pl_1 и Pl_2 are the plagioclase compositions of X_{An} = 0.01 и 0.05, respectively.

This reaction is supported by the mineral abundances in the rock, where the growth of biotite and newly formed plagioclase with a higher anorthite content is accompanied by a decrease in the modal amount of muscovite and chlorite in parallel with changing plagioclase compositions (Table 4.4). The predominant mechanism that controls reaction (4.83) is inferred to be congruent dissolution of epidote with total transfer of all components to the intergranular solution and transport of material by diffusion to the site of new mineral growth. Chemical element redistributions between reacting phases during epidote breakdown are diagrammatically shown in Fig. 4.21.

Table 4.4 Net mass balance of major components during epidote-consuming reaction (4.83) in the spotted hornfelses of Kharlovo contact aureole (Likhanov 1990)

	N	n_{SiO_2}	n_{CaO}	$n_{Al_2O_3}$	n_{FeO}	n_{MgO}	n_{K_2O}	n_{Na_2O}
Initial concentration of components								
Ms	0.90	2.87	0	2.28	0.23	0.46	0.76	0.18
Chl	0.23	0.67	0	0.56	0.40	0.62	0.01	0
Bt	0	0	0	0	0	0	0	0
Ep	0.13	0.83	0.49	0.55	0.26	0	0	0.01
Pl	4.19	12.61	0.04	4.19	0	0	0	3.85
Qtz	100.14	100.14	0	0	0	0	0	0
Total		**117.12**	**0.53**	**7.58**	**0.89**	**1.08**	**0.77**	**4.04**
Final concentration of components								
Ms	0.69	2.20	0	1.87	0.02	0.04	0.61	0.05
Chl	0.11	0.31	0	0.28	0.20	0.30	0	0
Bt	0.69	1.86	0	1.16	0.95	0.68	0.64	0.02
Ep	0.03	0.19	0.11	0.13	0.05	0	0	0
Pl	5.44	16.21	0.21	5.80	0	0	0	4.46
Qtz	95.83	95.83	0	0	0	0	0	0
Total		**116.60**	**0.33**	**9.24**	**1.22**	**1.02**	**1.25**	**4.53**

N = V/V_j—molar quantity of minerals in estimated volume of rock = 0.1 mm^3 (N × 10^{-4} mol), n_k = N × $a_{i,j}$—molar quantity of major (petrogenetic) component i in mineral j (n_k × 10^{-8} mol). The values to be compared are underlined and bolded

Although the image is quite schematic, it reflects the trends in the chemical composition of the reacting phases (zoned plagioclase), their volume ratios and possible directions of migration of the components during the reaction. Since all Ca migrated to the newly formed plagioclase with a complete dissolution of the epidote, we built a quantitative model of the redistribution of this component with calculated volume ratios and sizes of the zones composed of the more mafic and felsic plagioclases around epidote grains (Likhanov 1990). A comparison of the model Ca redistribution and volume ratios of the reacting phases shows that the volume bounded by the domain made up of the newly formed plagioclase grains was 0.032–0.027 mm^3. Redistribution of Fe, Si, and Al from epidote into other phases requires a smaller volume. This suggests that Al_2O_3 and SiO_2 were the least mobile, while FeO was more mobile, and CaO was the most mobile during dissolution of epidote.

In the metabasites from the middle part of this aureole, arfvedsonite, a relatively uncommon amphibole, was observed. Arfvedsonite is metastable in normal metabasites. It is never formed by the isochemical metamorphism of normal rocks, as it requires rocks initially rich in Fe, Si and Na and depleted in Al. In all the well-known occurrences, arfvedsonite appears to have formed through hydrothermal-metasomatic processes that supplied Na and Fe. This can account for its formation in the quartzites of the Krivoy Rog district (the Ukrainian shield), in the ultrabasic rocks of the Bazhenov massif (Urals) and in the gneisses of the

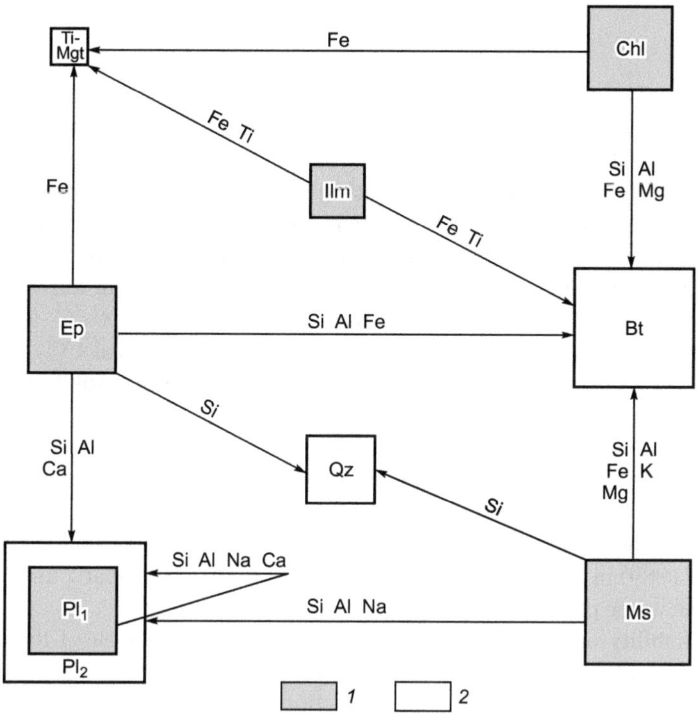

Fig. 4.21 Chemical component redistribution among reacting phases as a result of epidote-consuming reaction (4.83) in the spotted hornfelses of Kharlovo contact aureole (Likhanov 1990). 1—reactants, 2—reaction products

exocontact aureole of the Kave alkaline granites of the Kola Peninsula (Likhanov et al. 1995).

The isochemical contact metamorphism gave rise to hornblende in the central part of the diabase body. In the vicinity of calcite-chlorite amygdules, hornblende grains are commonly rimmed by arfvedsonite, that may form bladed and needle-like crystals varying from several micrometers to a few hundredths of millimeterin size. Near the calcite-chlorite amygdules, albite (Ab) rims are locally replaced by oligoclase (Olg).

Hornblende and arfvedsonite were formed at $T = 525$–550 °C and $P = 1$–1.5 kbar in the middle part of the contact aureole of the Kharlovo gabbro massif according to the following reactions for the system KFMnMASHTiCaNaC (Likhanov et al. 1995):

$$\begin{aligned} &0.8\mathrm{Chl} + 0.27\mathrm{Bt} + 0.83\mathrm{Cal} + 1.38\mathrm{Mgt} + 1.68\mathrm{Spn} + 1.4\mathrm{Ab} \\ &= 1.5\mathrm{Hbl} + 1.58\mathrm{Ilm} + 4.65\mathrm{H_2O} + 0.83\mathrm{CO_2}, \end{aligned} \quad (4.84)$$

and for the system KFMASHTiCaNaC:

$$\begin{aligned}&0.02\text{Chl} + 1.18\text{Ab} + 0.21\text{Cal} + 0.14\text{Mgt}\\&= 0.1\text{Arf} + 1.0\text{Olg} + 0.02\text{H}_2\text{O} + 0.21\text{CO}_2.\end{aligned} \tag{4.85}$$

Equation (4.84) conforms with the experimental study of phase relations carried out by Liou et al. (1974) for the basalt system. Calcite solubility and Ca concentration with decreasing fluid and increasing temperature were studied experimentally by Sharp and Kennedy (1965) and Fein and Walther (1987, 1989); this process could underlie the deceleration and arrest of hornblende growth. Isobaric experiments by the same authors have shown that calcite solubility in terms of calcium molality at P = 1–2 kbar decreased from $10^{-3.2}$ at T = 300 °C to $10^{-5.2}$ at T = 550 °C, with X_{CO_2} between 0.02 and 0.15 in a supercritical CO_2–H_2O fluid.

Equation (4.85) agrees with experimental data on Na and Si solubility in a fluid phase by incongruent dissolution of plagioclase at T = 400–800 °C and P = 1–3 kbar (Kotelnikov and Schekina 1986; Dujon and Lagache 1984; Shvedenkova and Shvedenkov 1990). Presumably higher anorthite content in the near-surface zone of grains is due to reaction of Ca in the fluid with residual aluminum in the plagioclase. The dissolution mechanism seems to be similar to that studied by Johannes (1989) in experiments on the melting of plagioclase-quartz assemblages at P = 2 kbar water pressure and at T = 850 °C.

The reliability of reaction (4.85) is supported by comparison of the calculated volume ratios of mineral with the observed volume ratios of phase contents. The volume ratios of phases participating in reaction (4.85) were first compared with the volume of newly formed arfvedsonite and then with the volume ratios of phases calculated on the basis of quantitative determination of mineral contents in thin sections. Arfvedsonite and oligoclase were found close together only near calcite-chlorite amygdules, where incongruent dissolution of plagioclase took place. During the reaction, Al_2O_3 was almost completely retained in plagioclase, SiO_2 and Na_2O were transferred from plagioclase to arfvedsonite while CaO was transferred from calcite to plagioclase. The volume ratios of phases participating in reaction (4.85) correspond to the observed volume distribution of minerals in the rock. Analysis of petrogenetic component balance confirmed the character of mineral transformations during arfvedsonite growth (Likhanov et al. 1995). The rocks were not altered in this case by metasomatic supply. Balance analysis showed that mass transfer for the main petrogenetic components in the course of the arfvedsonite-forming reaction was limited to very small volumes (microsites) of about 0.07–0.1 mm^3. These volumes were evaluated from examples of specific reactions with specific sizes and compositions of reacting phases. This conclusion may be extended to other microdomains of the rock. Chemical element redistributions during hornblende and arfvedsonite formation are diagrammatically shown in Fig. 4.22.

In fact, due to short term and low-temperature thermal metamorphism, diffusion flows and concentration gradients were so short-ranged that they did not lead to alteration of the whole rock composition. A rare combination of moderate P-T parameters and the short duration of thermal influence on the nonuniform basaltic

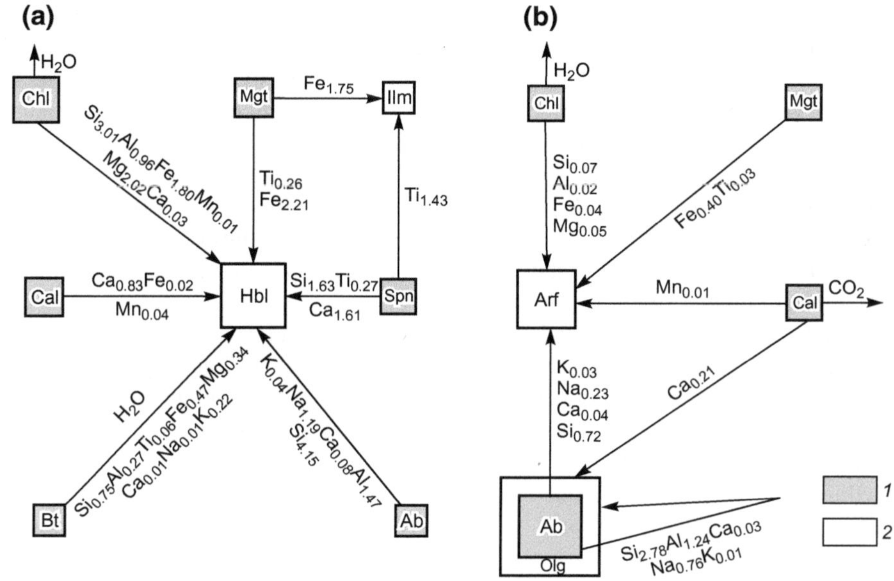

Fig. 4.22 Chemical component redistribution among reacting phases during hornblende (**a**) and arfvedsonite (**b**) formation (4.84) in the middle zone of Kharlovo contact aureole (Likhanov and Reverdatto 1991). 1—reactants, 2—reaction products

rock led to the formation of arfvedsonite, which is unusual in isochemical metamorphism. The formation of arfvedsonite rims around the hornblende grains could be due to the epitaxial nucleation of this mineral. This later growth took place in the stability field of arfvedsonite under the conditions of local saturation of Na, Si and Fe in H_2O–CO_2 solution, as a result of diffusion-controlled reactions coupled with Na and Si production during incongruent dissolution of albitic plagioclase. With higher temperature or longer duration, the reaction would probably have involved larger volume and a greater number of minerals. Under these conditions, further arfvedsonite formation would have been impossible without an additional supply of Na, Si and Fe and a relative deficiency of Ca and Al in the system; arfvedsonite would be unstable and would be replaced by a more stable mineral association.

With the exception of the pioneering work of Carmichael (1969), all of the above studies of mass transfer during mineral transformations are associated with metamorphism near magmatic intrusions. It is also of interest to consider mineral reactions during collision metamorphism in overthrust terranes of the Transangarian Yenisei Ridge, which include the Garevka (GC) and the Teya (TC) metamorphic complexes (Likhanov et al. 2014). In these cases a Neoproterozoic collision-related metamorphism of the kyanite-sillimanite type locally overprinted regionally metamorphosed low-pressure andalusite-bearing rocks. A positive correlation between rock ages and *P-T* estimates for the kyanite-sillimanite metamorphism provides evidence for regional structural and tectonic heterogeneity. The medium-pressure recrystallization was characterized by (1) localized distribution of metamorphic zones in the area directly underlying thrust faults with a measured thickness of 2.5–7 km; (2) syntectonic formation of kyanite-bearing mineral assemblages related to thrusting; (3) gradual increase in metamorphic pressure towards the thrust faults associated with a low metamorphic field gradient (from 7 to 14 °C/km); and (4) equally steep burial *P-T* paths recorded for the highest grade rocks (Likhanov et al. 2006, 2008a; Likhanov and Reverdatto 2014b). These specific features are typical of collisional metamorphism during overthrusting of continental blocks and are evidence of near-isothermal loading in accordance with the transient emplacement of thrust sheets.

The first study area is located between the Yeruda and Tchirimba Rivers in the Teya complex at the eastern margin of the middle part of the Transangarian Yenisei Ridge. During prograde metamorphism, Al- and Fe-rich metapelites of the Yenisei Ridge (Russia) underwent a gradual pressure increase in the vicinity of thrusting, as a result of which, low-pressure metamorphic effects were overprinted by medium-pressure regional metamorphic mineral assemblages and microstructures of kyanite-sillimanite type. An early low-pressure event produced andalusite-bearing assemblages of the andalusite-sillimanite facies series at about 3.5–4 kbar and 540–560 °C. The rocks are mainly characterised by Ms + Chl + Bt + Cld + And + Qz + Ilm ± Crd mineral assemblages. Owing to a gradual pressure increase in the vicinity of thrusting at 4.5–6.7 kbar and 540–600 °C, medium-pressure regional metamorphic sequence of kyanite-sillimanite type was formed, as witnessed by kyanite pseudomorphs after andalusite and by staurolite and garnet formation in the matrix (Likhanov et al. 2000; Likhanov and Reverdatto 2007). The first appearance of kyanite, defining the andalusite-kyanite isograd, marks the boundary between low-pressure and moderate-pressure metamorphic rocks. The pseudomorphs are characterized by the assemblage kyanite + staurolite + muscovite + quartz (no garnet, chlorite, biotite, or plagioclase). The pseudomorph relations suggest a contemporaneous crystallization of all minerals during the early stage of andalusite replacement. Kyanite and andalusite are found only within the pseudomorphs and never occur in the matrix. Near the margins of the pseudomorphs, various assemblages locally containing eight or nine minerals may be found. They are represented by Ms + Bt + Chl + Qtz + Grt + St + Ky ± Pl ± Cld ± Sil ± And (relict) mineral assemblages.

In order to evaluate the effect of additional minor components on the possible mass-balance relations, all phases have been involved in the calculations (Likhanov and Reverdatto 2002). On this basis, the local mineral reaction responsible for the early stage of kyanite-staurolite-muscovite-quartz replacement of andalusite in the outer zone can be written in the KFMnMASHCNT system as:

$$\begin{aligned} &0.096\mathrm{Cld} + 0.811\mathrm{And} + 0.001\mathrm{Bt} + 0.00012\mathrm{Pl} + 0.0008\mathrm{Ilm} \\ &= 0.049\mathrm{St} + 0.554\mathrm{Ky} + 0.060\mathrm{Qz} + 0.001\mathrm{Ms} + 0.000012\mathrm{Mrg} + 0.140\mathrm{H_2O}. \end{aligned} \tag{4.86}$$

The reliability of the reaction equations is confirmed by microstructural relationships, in which Ky-St-Qtz-Ms pseudomorphs are developed at the boundary between andalusite and the biotite-chloritoid aggregate of the matrix. The replacement of andalusite by kyanite-staurolite-muscovite-quartz aggregates means that a simple polymorphic transformation did not take place. Polymorphic replacements are uncommon during prograde metamorphic events, because normal 'static' continental geotherms do not pass from the andalusite *P-T* stability field to that of kyanite (Kerrick 1990); nor do the 'clockwise' *P-T* paths (P_{max} before T_{max}) that are typical of many collisional orogens (e.g. Thompson and England 1984). Such pseudomorphs are usually thought to have formed during retrograde metamorphic events as a result of hydration reactions (e.g., Chinner 1980). However, there are several examples in the literature of prograde transformation of andalusite to kyanite, and they have been attributed either to the metastable crystallization of andalusite in the kyanite *P-T* stability (Hollister 1969), or to a pressure increase resulting from tectonic thickening accomplished by thrusting (Crawford and Mark 1982; Baker 1987; Beddoe-Stephens 1990; Spear et al. 1990) or magma loading (Brown and Walker 1993; Brown 1996; Whitney et al. 1999). The calculated mineral reaction (4.86), inferred to have occurred during pressure increase in rocks overlapped by the Panimba thrust confirm the possibility of prograde transformation of andalusite to kyanite.

Given the thickness of a Ky-St-Ms-Qz replacement zone around andalusite crystals, averaging up to 2 mm, and the estimated distances between reactant and product grains (reactions 4.86), the volume involved in the redistribution of chemical components was calculated. It did not exceed $\sim 1.2 \pm 0.4$ cm^3 (Likhanov and Reverdatto 2002). Possible mechanisms for the observed pseudomorph development include topochemical mineral transformations, in which newly formed minerals inherit some crystal-structural elements from the old ones. Microstructural observations suggest that nucleation of the newly formed grains in pseudomorphs is commonly confined to the grain margins of andalusite crystals, which are the most favorable high-energy sites for a phase to nucleate (Vernon 1976). Kyanite and staurolite crystallization was controlled by the presence of andalusite. Owing to the affinity of its crystal lattice (chains of AlO_6 edge-shared octahedral parallel to the *c* crystallographic axis), andalusite may have been replaced by an epitaxial mechanism producing the very distinctive pseudomorphous microstructure.

We infer that mineral transformations in the matrix to produce garnet and staurolite proceeded simultaneously with the pseudomorph development in the outer zone. Taking account of these microstructural observations, local mineral reactions in the matrix involving the simultaneous formation of staurolite and garnet can be written for the KFMnMASHCNT system (Likhanov and Reverdatto 2002) as:

$$\begin{aligned} &0.147\mathrm{Cld} + 0.153\mathrm{Chl} + 0.343\mathrm{Ms} + 0.005\mathrm{Mrg} + 0.046\mathrm{Ilm} \\ &= 0.073\mathrm{St} + 0.003\mathrm{Grt} + 0.347\mathrm{Bt} + 0.088\mathrm{Qz} + 0.045\mathrm{Pl} + 0.836\mathrm{H_2O} \end{aligned} \tag{4.87}$$

Equation (4.87) is consistent with the observed chemographic relations (Fe/(Fe + Mg): Grt > Cld $\geq$ St > Bt > Chl) in the studied metapelites. They are also supported by mineral abundances in the rock, where the growth of staurolite, garnet, biotite, quartz and plagioclase is accompanied by decrease in the amounts of chloritoid, chlorite and muscovite. Additional microstructural evidence is provided by the presence of inclusions of ilmenite, muscovite, chloritoid and quartz in the staurolite and garnet, and by the absence of kyanite and andalusite from the matrix. Simultaneous appearance of staurolite and garnet in the matrix at the expense of chloritoid, chlorite and muscovite agrees with reactions proposed by Whitney et al. (1996) for relatively aluminous, Fe-rich pelitic schists in Dutchess County (USA) at about the same grade.

To estimate the minimum volume involved in local reaction (4.87) in the matrix, we determined the size and distance between the grains of reacting coexisting minerals in thin sections. The maximum distance between all grains determines the minimum volume required for mass balance of reactants and products. Given the average grain size of the order of 0.5–1 mm, the coexistence of all minerals in a thin section occurs within a square $\sim$7.5 mm in side. Based on these estimations and assuming uniform distribution of minerals in the rock, the minimum volume of $\sim$0.4 cm^3 was determined.

The predominant mechanism that controls reaction (4.87) in the matrix region in the outer zone, in response to chemical potential changes, is inferred to be congruent dissolution of reactants with total transfer of all components to the intergranular solution and transport of material by diffusion to the site of new mineral growth.

From the thermodynamic standpoint, large volume and small entropy effects characterize the calculated reactions. The change in volume of the solids in reaction (4.86), calculated using standard molar volume data for minerals from the internally consistent thermodynamic dataset of Holland and Powell (1998), is very large, namely $-\Delta V = 49\%$. For reaction (4.87) it is somewhat lower, ranging within the interval of $-\Delta V = 42–39\%$. The very large $-\Delta V$ (solids) indicates that in a closed system these reactions would be driven to the right by an increase in lithostatic pressure, which is consistent with the thermobarometric results (Likhanov et al. 2001a).

Net mass balance in metapelites from the outer zone at $P = 4.5–5$ kbar and $T = 550–570$ °C is defined as the sum of two local cation-exchange reactions that form pseudomorphs (4.86) and minerals in the adjacent matrix (4.87). This net mass balance can be represented as:

$$\begin{aligned} &0.243\mathrm{Cld} + 0.811\mathrm{And} + 0.153\mathrm{Chl} + 0.342\mathrm{Ms} + 0.005\mathrm{Mrg} + 0.047\mathrm{Ilm} \\ &= 0.122\mathrm{St} + 0.554\mathrm{Ky} + 0.003\mathrm{Grt} + 0.346\mathrm{Bt} + 0.148\mathrm{Qz} + 0.045\mathrm{Pl} + 0.98\mathrm{H_2O} \end{aligned} \tag{4.88}$$

The calculated net mass balance may be one of the variants of an overall reaction. If minor extra components such as Zn in staurolite and chloritoid, Fe^{3+} in ferromagnesian silicate minerals, as well as the components of a complex C–O–H fluid (O_2, CO, CH_4), that is in equilibrium with graphite are considered, the stoichiometric coefficients for the overall reaction among the phases in a rock may vary. However, the calculated net mass balance (4.88) is consistent with limitations imposed by the microstructure of the rocks, the mineral assemblages and mineral abundances, where the first appearance of staurolite, kyanite and garnet in the outer zone being accompanied by increase in modal amount of biotite, plagioclase and quartz and decreasing amounts of chloritoid, andalusite, chlorite, muscovite and ilmenite (Table 4.5). A combined mass balance confirms the possibility that muscovite is consumed in the matrix and produced in the pseudomorphs, in such a way that the net balance is consumption. The reliability of calculations is also supported by a comparison of the calculated mineral volume ratios from reaction Eq. (4.88), with the actual phase volumes in the rock calculated from whole-rock chemical compositions. The formation of new minerals during the net mass balance (4.88) in the outer zone occurs at the expense of minerals of the protolith (zone I), reacting in the stoichiometric ratio (Ilm:Chl:Cld:Ms:And = 1:3:3.5:9:16) that closely corresponds to the net mass balance results (Ilm:Chl:Cld:Ms:And = 1:3:5:8:17). The results obtained from a comparison of volume ratios of the newly formed minerals (Grt < St < Pl < Bt < Ky < Qz) with the actual volume ratios in the rock, as well as quantitative ratios between the reactants and products of reaction (Ky: And = 1:1.5, Ms:Bt = 1:1) are inferred to the satisfactory, the deviation being less than 20%. These and some other deviations (e.g. for quartz and plagioclase) can be explained by slight differences in bulk composition, by errors in molar volume values for minerals of variable composition, and by analytical uncertainty and mineral heterogeneity. A schematic representation of the redistribution of chemical components among all minerals is graphically shown in Fig. 4.23. The above net mass balance is in good agreement with mass balance results by Whitney et al. (1995) reported for metamorphic rocks from the British Columbia (Canada) in which garnet, staurolite, kyanite and biotite formed at the expense of chloritoid, muscovite and ilmenite. The microstructural relationships suggest that the minimum volume in which net mass balance (4.88) between two local volumes in the outer zone metapelites took place is equal to a sum of local reaction volumes (4.86) and (4.87), not exceeding $\sim 1.6 \pm 0.4$ cm^3.

At increased pressure in the inner zone, garnet exhibits an increase in grossular component from core to rim. In a closed system, garnet growth requires consumption of anorthite component and a shift of remaining plagioclase composition toward albite (Ghent and Stout 1981). It is consistent with the observed normal zoning of the plagioclase. The assumption that was made with respect to plagioclase is that the core-to-rim zoning correlates with the growth of the garnet from core to rim. That is, the core of plagioclase represents the plagioclase composition at the early stage of garnet growth. Based on the above considerations, we calculated the following local reaction responsible for mineral transformations in the matrix in the inner zone at $P = 6.2$–6.7 kbar and $T = 550$–580 °C in the KFMnMASHC chemical system: 0.038Chl + 0.040Grt_c + 0.343Qz + 0.038Ms + 0.595Pl_c = 0.045Bt + 0.068Grt_r + 0.704Pl_r + 0.145H_2O, where *c* and *r* are the

Table 4.5 Net mass balance during mineral reaction (4.88) of collisional metamorphism of rocks in the northern segment of Yenisei Ridge (Likhanov and Reverdatto 2002)

	N	Initial concentration of components								
		n_{SiO_2}	$n_{Al_2O_3}$	n_{FeO}	n_{MgO}	n_{K_2O}	n_{MnO}	n_{CaO}	n_{Na_2O}	n_{TiO_2}
Ms	16.7	50.77	47.26	1.34	1.00	14.36	0	0	1.67	0.50
Chl	3.10	7.84	8.87	8.80	5.43	0	0.03	0	0.09	0.06
Bt	14.6	39.13	25.26	20.88	12.12	12.41	0.15	0	0	2.48
Pl	2.20	6.36	2.46	0.02	0	0	0	0.20	1.91	0
Qtz	248	248	0	0	0	0	0	0	0	0
Cld	4.60	9.15	18.40	8.00	1.15	0	0.09	0	0	0
Ilm	8.00	0	0	7.52	0	0	0.16	0	0	8.16
Mrg	0.14	0.28	0.56	0	0	0	0	0.12	0.02	0
And	77.3	75.7	156	0.70	0	0	0	0	0	0
Ky	0	0	0	0	0	0	0	0	0	0
St	0	0	0	0	0	0	0	0	0	0
Grt	0	0	0	0	0	0	0	0	0	0
Total		437	259	47.3	19.7	26.8	0.43	0.32	3.69	11.2
	N	Final concentration of components								
		n_{SiO_2}	$n_{Al_2O_3}$	n_{FeO}	n_{MgO}	n_{K_2O}	n_{MnO}	n_{CaO}	n_{Na_2O}	n_{TiO_2}
Ms	16.3	49.55	46.13	1.30	0.98	14.02	0	0	1.63	0.49
Chl	3.00	7.59	8.58	8.52	5.25	0	0.03	0	0.09	0.06
Bt	15.7	42.08	27.16	22.45	13.03	13.35	0.16	0	0	2.67
Pl	7.70	22.25	8.62	0.08	0	0	0	0.69	6.70	0
Qtz	333	333	0	0	0	0	0	0	0	0
Cld	0	0	0	0	0	0	0	0	0	0
Ilm	7.50	0	0	7.05	0	0	0.15	0	0	7.65
Mrg	0.06	0.17	0.07	0	0	0	0	0.01	0.05	0
And	22.4	21.8	45.2	0.20	0	0	0	0	0	0
Ky	18.4	18.03	36.98	0.18	0	0	0	0	0	0
St	2.10	16.52	39.12	7.41	1.07	0	0.04	0	0	0.04
Grt	0.2	0.60	0.39	0.52	0.05	0	0.03	0.01	0	0
Total		511	212	47.7	20.4	27.4	0.41	0.71	8.47	10.9

$N = V/V_j$—molar quantity of minerals in estimated volume of rock = 1.6 cм^3 ($N \times 10^{-4}$ mol), $n_k = N \times a_{i,j}$—molar quantity of petrogenetic component i in mineral j ($n_k \times 10^{-4}$ mol)

cores and rims of garnet and plagioclase zoned grains. The reliability of this reaction is supported by the aforementioned interrelationship between grossular zoning in garnet and plagioclase composition and by increase in the amount of biotite and decrease in the amount of chlorite and muscovite. Incongruent behavior of reaction in the matrix in the inner zone is supported by outer rim growth of garnet under the conditions of local saturation of Ca as a result of diffusion-controlled reaction, coupled with Ca consumption during incongruent dissolution of plagioclase (Foster 1977, 1986; Likhanov et al. 1994, 1995).

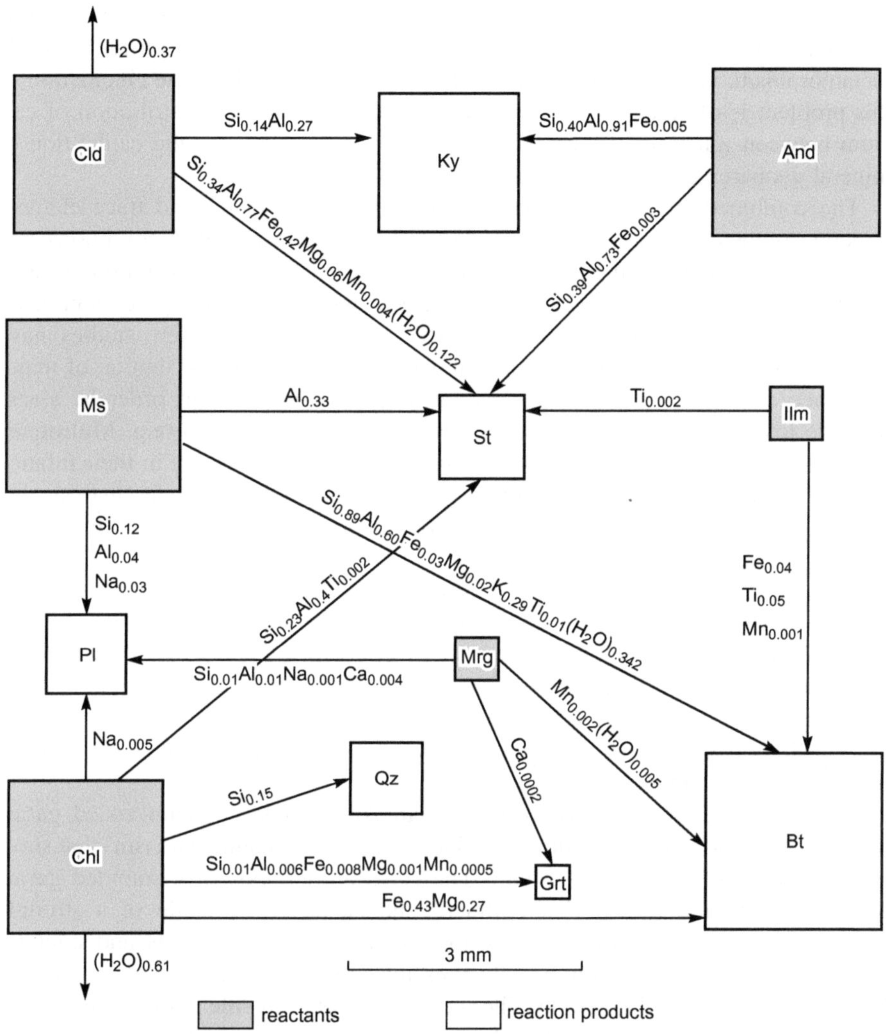

Fig. 4.23 Chemical component redistribution among reacting phases as a result of overall reaction (4.88) that form pseudomorphs and minerals in the adjacent matrix in Fe- and Al-rich metapelites of Yenisei Ridge (Likhanov and Reverdatto 2002). Volumes of all minerals are shown at the figure scale

As noted above, one of the diagnostic features of increase in pressure during prograde metamorphism of metapelites is chemical growth zoning of garnets expressed in a significant increase of grossular component from 1 to 6 wt% with concomitant decrease of spessartine component. Such specific features are typical of collision-related metamorphism during overthrusting of continental blocks. Similar garnet zoning patterns were derived from metapelitic rocks in numerous overthrust areas with multistage history (Spear et al. 2002; Faryad and Chakraborty

2005; Cutts et al. 2010; Likhanov and Reverdatto 2008, 2011a). The nature of this phenomenon, namely a significant calcium enrichment in garnet during prograde metamorphism, is not well understood (Reverdatto et al. 2017). The elucidation of this problem is of significant petrological interest, because a redistribution of calcium between minerals of variable composition is the basis for the calibration of mineral geobarometers (Ghent and Stout 1981).

The combined study of the measurements of in situ major and trace element mineral composition and mineral modes with changing *P* and *T* conditions and metamorphic reactions in rocks is an effective approach to the solution of such problems. This approach allows us to reconstruct the metamorphic process in detail and to carry out quantitative mass-transfer analysis. Quite a few studies have attempted to correlate a mass-balanced model reactions and redistribution of major and trace elements across widespread metamorphic isograds in order to assess possible losses or gains of these elements during metamorphism (e.g. Mulrooney and Rivers 2005; Nehring et al. 2010). But such investigations are in their infancy in metamorphic petrology. This study involves a detailed mass transfer analysis that contributes to our understanding of metamorphic reactions responsible for calcium enrichment in garnet.

Metapelite samples used in this study were collected within the Garevka (GC) metamorphic complex at the western margin of the Transangarian Yenisei Ridge in the lower course of the Garevka River (sample 27) and from outcrops on the right bank of the Yenisei River (sample 56). The metapelitic rocks of the study area are Grt + Bt + Ms + Pl + Qz ± St ± Ilm ± Ky ± Chl ± Ep intensely deformed gneisses and schists crosscut by steeply-dipping, sheeted, coarse-grained felsic dykes (Likhanov and Santosh 2017).

Pelitic gneisses and schists from the Garevka Complex contain zoned garnet porphyroblasts with three texturally distinct zones (core, mantle, and rim) that show clear evidence of multistage growth. The cores are composed of rounded garnet admixed with randomly oriented mineral inclusions. The mantle of a strongly deformed garnet grain is crowded with oriented matrix inclusions and contains abundant ilmenite and graphite inclusions that are concentrated at the boundary between the core and the mantle. The inclusions within mantle define a straight S_1 fabric. Garnets of sub-idiomorphic shapes exhibits thin rims (up to 0.4 mm). These kinds of garnets often have discontinuous edges and contain recrystallized quartz-rich strain shadows. Along with abundant inclusions of the matrix minerals, almost all garnet zones contain monazite, zircon, apatite, and tourmaline as well as rare epidote and xenotime.

The three discrete stages define a counter-clockwise *P-T* path involving initial prograde low pressure heating followed by near isothermal medium-pressure compression and post-peak retrograde decompression and cooling. The first stage occurred as a result of the Grenville-age orogeny during late Meso-early Neoproterozoic (1050–850 Ma) and was marked by low-pressure zoned metamorphism at *c.* 4.8–5.0 kbar and 565–580 °C with a metamorphic field gradient of $dT/dZ = 20–30$ °C/km. At the second stage, the rocks experienced middle Neoproterozoic (~800 Ma) collision-related medium-pressure metamorphism at *c.*

7.7–7.9 kbar and 630–660 °C with *dT/dZ* ≤ 10 °C/km. The final stage evolved as a synexhumation retrograde metamorphism (785–776 Ma) at *c.* 4.8–5.4 kbar and 500 °C with *dT/dZ* ≤ 14 °C/km and recorded uplift of the rocks to upper crustal levels in extensional tectonic settings (Likhanov et al. 2015b).

The garnets are characterized by significant depletion in light REEs (LREE) and enrichment in heavy REEs (HREE), which exceeds the chondrite Yb/La ratio by 10,000 times. Peak concentrations for all HREEs are located in the core and rim zones of each garnet porphyroblasts and HREE concentrations decrease toward the mantle. The cores of the garnet are characterized by a strongly differentiated chondrite-normalized REEs pattern with a regular increase from LREEs to HREEs, somewhat enrichment in middle REEs, La, and Ce, and depletion in HREEs from the core to the mantle of the garnet (Fig. 4.24a). The decrease in the content of HREEs, MREEs, and total REEs along with the increase in LREEs content from the mantle to the rim zone in garnets (Yenisei area, sample 56) is correlated with an increase in the content of all REEs in the newly formed plagioclase (Fig. 4.24a). In contrast, garnet of the Garevka area (sample 27) is characterized by an increase in the content of most REEs (except for Y, Dy, Gd) correlated with a coupled decrease in the REE content in the newly formed plagioclase (Fig. 4.24a). In both cases, the REE contents in analyzed garnet exhibit systematic negative correlations with Ca content in garnet. Another revealed feature is that the increase in temperature and pressure led to a decrease in the HREE and Y content in garnet, which is consistent with similar garnet REE zoning pattern of the Lapland–White Sea belt of the Baltic Shield (Skublov 2005) and other world collisional orogens (Otamendi et al. 2002; Pyle et al. 2001; Franceschelli et al. 2002).

Epidote, known carrier of the REEs, especially the LREEs (Frei et al. 2004) is characterized by a flat REE pattern, which exceeds that of chondrite by 10–30 times, and by a positive Eu anomaly (Fig. 4.24). It is remarkable that the flat spectrum of epidote almost corresponds to the REE pattern of the rock and is somewhat lower and slightly higher in the LREE and HREE areas, respectively. In contrast to garnet, monazite is characterized by the opposite (steep negative) REE patterns caused by strong enrichment in LREEs relative to HREEs (Fig. 4.24b), which suggest equilibrium between monazite and garnet (Lanari and Engi 2017). Other minerals are depleted in almost all REEs in relation to their bulk-rock concentrations (Fig. 4.24b). The high LREE concentrations in quartz are most likely explained by fluid and mineral inclusions.

Because isochemical suites of the metapelitic rocks are available, differences in mineralogy between two assemblages can be attributed to changes in intensive variables rather than to changes in bulk composition. Our previous studies on these rocks (Likhanov and Reverdatto 2016a) have shown that the prograde evolution of chemical and modal compositions of minerals during collisional metamorphism was controlled by gradual pressure increase at nearly constant temperature. Hence, it is possible to write balanced chemical reactions among mineral associations accounting for prograde and retrograde mineral transformations that are consistent with mineral chemistry and microtextural relationships.

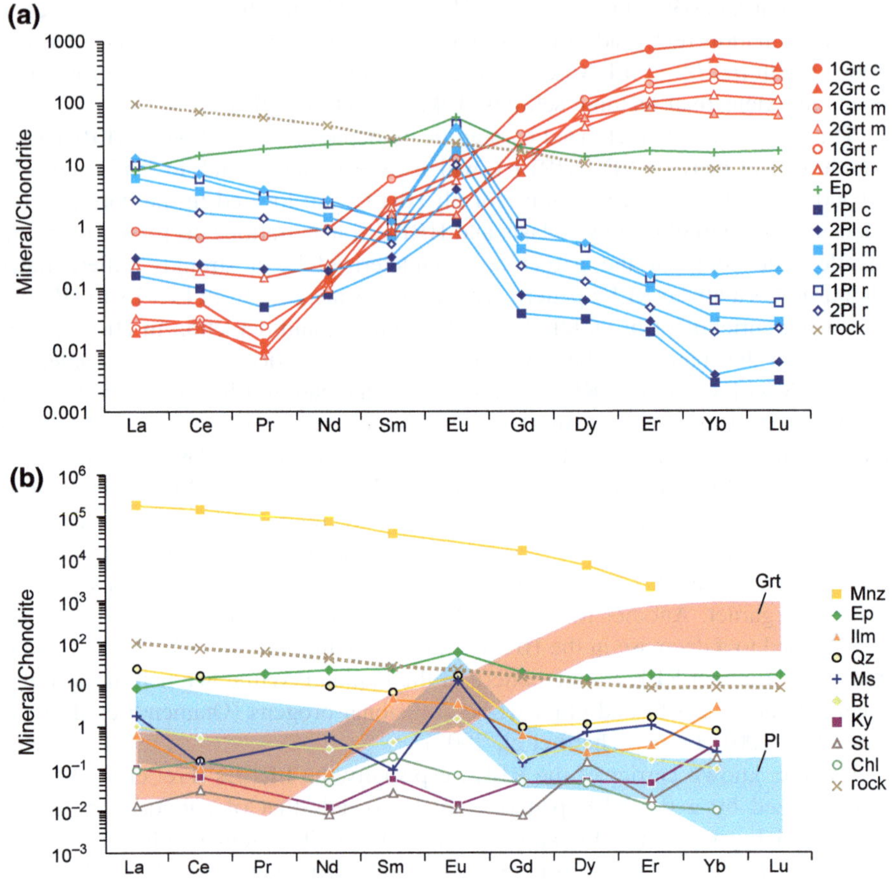

Fig. 4.24 Chondrite-normalized REE patterns (Boynton 1984) of zoned garnet and plagioclase, and epidote (**a**) and of all minerals, which participate in collision-related metamorphic reactions (**b**), compared with the average REE pattern for study rocks. 1*Grt* and 1*Pl*, sample 56; 2*Grt* and 2*Pl*, sample 27. C, m, and r correspond to core, mantle and rim garnet zones, respectively (**a**). Fields of garnet and plagioclase compositions is indicated by red and blue colors, respectively (**b**)

The core mineral assemblages at P = 4.5–5 kbar and T = 560–570 °C define the boundary between the garnet-chlorite and staurolite-biotite zones (e.g., Carmichael 1970) in many low-pressure metamorphic complexes of the andalusite-sillimanite facies series (Miyashiro 1973). It is in good agreement with the *P-T* stability fields of the key assemblages predicted by the petrogenetic grid of Likhanov et al. (2005) for Fe- and Al-rich metapelites in the KFMASH system. The mineral transformations of two rocks from the studied areas are distinct only by the presence of epidote in the newly formed assemblages (Yenisei area, sample 56) at almost equal proportions between the stoichiometric coefficients of reacting and newly formed minerals.

At increased pressure during collisional metamorphism, garnet exhibits an increase in grossular component from core to mantle. In a closed system, garnet growth requires consumption of anorthite component and a shift of remaining (re-equilibrated) plagioclase composition toward albite (e.g., Ghent and Stout 1981). It is consistent with the observed decrease in anorthite component of the plagioclase from core to mantle in rocks of the Garevka area (sample 27). Based on the above considerations, we calculated the following local reaction responsible for mineral transformations during collisional metamorphism of rocks in the Garevka area at P = 9.3 kbar and T = 660 °C in the KFMASHNaCaTiMn chemical system:

$$\begin{aligned} &0.16\mathrm{Grt_c} + 0.69\mathrm{Pl_c} + 0.98\mathrm{Qz} + 0.11\mathrm{Ms} + 0.22\mathrm{St} \\ &\quad = 0.42\mathrm{Grt_m} + 0.67\mathrm{Pl_m} + 0.05\mathrm{Bt} + 0.07\mathrm{Ilm} + 0.221\mathrm{Ky} + 0.35\mathrm{H_2O} \end{aligned} \quad (4.89)$$

where c and m are the garnet and plagioclase compositions in core and mantle of garnet porphyroblasts. The calculated net mass balance may be one of the variants of an overall reaction. If minor extra components such as Zn in staurolite, Fe^{3+} in ferromagnesian silicate minerals, as well as the components of a complex C–O–H fluid (O_2, CO, CH_4) that are in equilibrium with graphite are considered, the stoichiometric coefficients for the overall reaction among the phases in a rock may vary. However, the calculated net mass balance (4.89) is consistent with limitations imposed by the microtexture of the rocks, the mineral assemblages, and mineral abundances, where the first appearance of kyanite and the growth of biotite and ilmenite at the boundary between core and mantle are accompanied by decrease in modal amount of muscovite and staurolite. The reliability of calculations is also supported by a comparison of the calculated mineral volume ratios from reaction Eq. (4.89), with the actual phase volumes in the rock calculated from whole-rock chemical compositions. The formation of new minerals in the mantle occurs at the expense of minerals of the core zone, reacting in the stoichiometric ratio (Grt:Pl:Ms:St = 1:5.5:0.75:1.75) (Table 4.6) that closely corresponds to the net mass balance results for reaction (4.89) (Grt:Pl:Ms:St = 1:4.3:0.69:1.38). The results obtained from a comparison of volume ratios of the newly formed minerals in the given relationships between the stoichiometric coefficients (Grt:Pl:Bt:Ky = 1:1.6:0.12:0.53) with the actual volume ratios in the rock (Grt:Pl:Bt:Ky = 1:1.8:0.19:0.84) (Table 4.6), may be considered satisfactorily consistent; the deviation being less than 20%. These and some other deviations (e.g. for quartz) can be explained by slight differences in bulk composition, by uncertainties in molar volume values for minerals of variable composition, and by analytical uncertainty and mineral heterogeneity. Generally, this reaction is consistent with the mineral transformations in medium-pressure Barrovian-type metapelites (Whitney et al. 1996).

Rocks of the Yenisei area (Sample 56) are characterized by simultaneous increase in Grs component in garnet and the An component in plagioclase with pressure increase. In a closed system, such coupled increase in the CaO content cannot be the result of Ca and Al redistribution between garnet and plagioclase

Table 4.6 Calculated mineral abundances for rocks in different zones in the Garevka area (Likhanov and Reverdatto 2016b)

Mineral	Content, moles				Content, vol.%			
	Sample 56		Sample 27		Sample 56		Sample 27	
	c	m	c	m	c	m	c	m
Muscovite	0.018	0.008	0.03	0.02	2.5	1.1	4.2	3.2
Garnet	0.15	0.21	0.04	0.10	17.3	20.7	7.2	5.8
Epidote	0.015	0	0	0	2.1	0	0	0
Biotite	0.008	0.01	0.015	0.019	1.2	1.5	2.3	2.9
Plagioclase	0.045	0.06	0.22	0.18	4.5	6	22.1	17.8
Quartz	2.91	2.76	2.69	2.80	66.0	62.7	61.1	63.6
Staurolite	0.14	0.09	0.07	0.06	6.4	4.0	3.1	2.9
Ilmenite	0	0.002	0	0.002	0	0.06	0	0.05
Kyanite	0	0.09	0	0.084	0	3.9	0	3.7

Number of moles of each mineral calculated in volume of rock = 100 см^3. C and m correspond to modal compositions of core and mantle, respectively

alone. With account for the data on the presence of epidote in regionally metamorphosed rocks (core zone) and its absence as a product of collisional metamorphism (mantle), the following epidote-consuming mineral reaction at $P = 7.9$ kbar and $T = 625$ °C may be written:

$$\begin{aligned} &0.134\text{Grt}_\text{c} + 0.040\text{Pl}_\text{c} + 0.144\text{Qz} + 0.011\text{Ms} + 0.017\text{Ep} + 0.102\text{St} \\ &= 0.248\text{Grt}_\text{m} + 0.064\text{Pl}_\text{m} + 0.010\text{Bt} + 0.02\text{Ilm} + 0.089\text{Ky} + 0.136\text{H}_2\text{O} \end{aligned} \tag{4.90}$$

The equation of mass balance obtained indicates the proportions of mineral phases at the final stage, in which last traces of relict epidote disappears in the rock. According to experimental data from Liou (1973), Liou et al. (1983), and Poli and Schmidt (2004), in contrast to the wide range of *P-T* stability and occurrence of epidote in metabasites, in metapelites its stability is strongly restricted to a limited *P-T* range, which is mainly controlled by oxygen fugacity (Engi et al. 2017). At low oxygen fugacity close to the QFM buffer, the upper thermal stability limit of epidote in metapelites is less than $T = 600$ °C at $P = 7$ kbar. The presence of graphite and nearly pure ilmenite in studied rocks indicates reduced conditions (lower than the QFM buffer) (Likhanov and Reverdatto 2014a).

This reaction is supported by the mineral abundances in the rock, where the simultaneous growth of garnet, biotite, plagioclase, and kyanite is accompanied by a decrease in the modal amount of muscovite, staurolite, and quartz (Table 4.7). The formation of the reaction products occurred at the expense of the minerals of the core zone reacting in stoichiometric proportions (Grt:Pl:Ms:Ep:St = 1:0.3:0.12:0.10:0.9), that closely corresponds to the net mass balance results for reaction (4.90): Grt:Pl:Ms:Ep:St = 1:0.3:0.10:0.13:0.8. The results obtained from a comparison of volume ratios of the reaction products in the given relationships between the stoichiometric coefficients (Grt:Pl:Bt:Ky = 1:0.26:0.07:0.36) with the actual volume ratios in the rock (Grt:Pl:Bt:Ky = 1:0.28:0.05:0.40) are very similar (Table 4.7).

The predominant mechanism that may control reaction (4.90) is inferred to be congruent dissolution of reactants with transfer of all components through the intergranular medium and to the site of mineral growth. It is supported by growth of garnet in mantle under the conditions of local saturation of Ca as a result of equilibrium-controlled reaction, coupled with Ca consumption during dissolution of epidote (Likhanov et al. 1994).

During the late stage, which is accompanied by a total pressure and temperature decrease, garnet exhibits a decrease in grossular component from mantle to rim zone, whereas the composition of plagioclase are characterized by an increase of the anorthite component in the same direction. Taking account of these observations, local mineral reaction modelling the major element redistribution among coexisting phases in sample 56 during retrograde synexhumation metamorphism at P = 4.8–4.9 kbar and T = 515–560 °C can be represented in the KFMASHCa system as: $0.584Grt_m + 0.221Pl_m + 0.328Qz + 0.148Bt + 0.034H_2O$ $= 0.494Grt_r + 0.467Pl_r + 0.131Ms$, where m and r are the garnet and plagioclase compositions in mantle and rim zones of garnet porphyroblasts. The stoichiometric coefficients for possible mass-balance among reactant and product phases of the Garevka area (sample 27) are similar.

The formation of garnet porphyroblasts with three texturally distinct zones allows us to define the minimum rock volume. This local volume is limited by the distance to which the most mobile of the components participating in the reaction migrates. The actual volumes of mass transfer should exceed the calculated ones, since the newly formed minerals may have received substances not only from the neighbouring within the enclosing them zones, but also from more distant sources in the adjacent zones of garnet (Reverdatto and Kolobov 1987). These volumes in which metamorphic reactions took place were estimated on the basis of the size (thickness) and spatial distribution of different zones of garnet porphyroblasts. The lateral dimensions across mantle were determined. Given the thickness of mantle (Fig. 2.1 in Chap. 2), the volume involved in the redistribution of chemical components was calculated. It did not exceed $\sim$0.4 mm^3 for sample 56 and $\sim$1 mm^3 for sample 27. These values are in good agreement with estimations made by other authors for the scales of mass transfer as a result of metamorphic reactions in metapelites (Carmichael 1969; Foster 1977, 1981, 1986; Dipple et al. 1990; Reverdatto and Kolobov 1987; Likhanov 2018; Likhanov and Reverdatto 2002; Likhanov et al. 1995).

If all reactions occurring in the rock have been taken into account and the system is closed to all components except water, the sum of the mass transfer should be zero for all components. Analysis of the balance of major components (Table 4.7) shows that their redistribution among reacting phases during mineral reactions (4.89, 4.90) of collisional metamorphism may be limited by the mentioned local volumes. This estimation seems to be close to correct, because a deviation for all components with the exception of TiO_2 (Sample 56) does not exceed 20%, which

Table 4.7 Net mass balance of major components before and after mineral reactions of collisional metamorphism (4.89) and (4.90)

	N	Initial concentration of components (sample 56)								
		n_{SiO_2}	$n_{Al_2O_3}$	n_{FeO}	n_{MgO}	n_{K_2O}	n_{MnO}	n_{CaO}	n_{Na_2O}	n_{TiO_2}
Ms	7.2	22.3	20.1	0.48	0.45	5.97	0	0.01	1.00	0.09
Grt	60	180	119	141	14	0	20	5	0.42	0
Ep	6	38	29	8	0.1	0.02	0.12	23	0.55	0.06
Bt	3.2	8.82	5.22	4.42	2.66	3.07	0.04	0.01	0.05	0.52
Pl	18	52	20	0	0	0.3	0	2.48	15	0
Qz	1160	1160	0	0	0	0	0	0	0	0
St	5.6	46	102	17	3	0.01	0.01	0	0.22	0.40
Ilm	0	0	0	0	0	0	0	0	0	0
Ky	0	0	0	0	0	0	0	0	0	0
Total		1507	295	171	20	9	20	30	17	1.1
	N	Final concentration of components (sample 56)								
		n_{SiO_2}	$n_{Al_2O_3}$	n_{FeO}	n_{MgO}	n_{K_2O}	n_{MnO}	n_{CaO}	n_{Na_2O}	n_{TiO_2}
Ms	3.2	9.93	8.77	0.17	0.20	2.73	0	0.01	0.71	0.04
Grt	76	229	150	175	16	0	15	23	0.5	0.3
Ep	0	0	0	0	0	0	0	0	0	0
Bt	4	11	6.36	5.92	3.08	3.85	0.05	0	0.10	0.74
Pl	24	68	28	0.05	0.02	0.07	0	4.56	18	0
Qz	1100	1100	0	0	0	0	0	0	0	0
St	3.6	30	65	11	2.1	0.01	0.01	0	0.14	0.26
Ilm	0.8	0.01	0.01	0.8	0.01	0	0.01	0	0	0.8
Ky	36	36	72	0.36	0	0	0	0	0	0
Total		1483	330	210	21	7	15	28	19	2.1
	N	Initial concentration of components (sample 27)								
		n_{SiO_2}	$n_{Al_2O_3}$	n_{FeO}	n_{MgO}	n_{K_2O}	n_{MnO}	n_{CaO}	n_{Na_2O}	n_{TiO_2}
Ms	30	93.5	82.2	2.43	2.16	24.5	0	0.03	4.53	1.14
Grt	60	180	118	141	15	0	14	7	0.3	0
Bt	15	40	27	19	13	14	0.15	0.02	0.15	2.88
Pl	220	622	259	0.4	0.2	0.7	0	40.2	170	0
Qz	2690	2690	0	0	0	0	0	0	0	0
St	70	567	1308	213	27	0	7	0	0	1.4
Ilm	0	0	0	0	0	0	0	0	0	0
Ky	0	0	0	0	0	0	0	0	0	0
Total		4192	1794	376	57	39	21	47	175	5.4
	N	Final concentration of components (sample 27)								
		n_{SiO_2}	$n_{Al_2O_3}$	n_{FeO}	n_{MgO}	n_{K_2O}	n_{MnO}	n_{CaO}	n_{Na_2O}	n_{TiO_2}
Ms	20	62.2	54	1.5	0	16.4	0	0	3.64	0.18
Grt	50	150	148	117	13	0	9	13	0.1	0.3
Bt	19	51	32	28	13	18.4	0.23	0.02	0.32	2.74

(continued)

Table 4.7 (continued)

	N	Final concentration of components (sample 27)								
		n_{SiO_2}	$n_{Al_2O_3}$	n_{FeO}	n_{MgO}	n_{K_2O}	n_{MnO}	n_{CaO}	n_{Na_2O}	n_{TiO_2}
Pl	180	520	200	0.2	0	1.8	0	31.1	144	0
Qz	2800	2800	0	0	0	0	0	0	0	0
St	60	486	1121	182	23	0	6	0	0	1.2
Ilm	1.6	0.02	0.02	25	0.02	0.04	0.12	0	0.12	1.5
Ky	84	84	168	0.84	0	0	0	0	0	0
Total		4153	1723	354	49	37	15	44	148	5.9

N = V/V$_j$—molar quantity of minerals in estimated volume of rock = 0.4 mm^3 (N $\times$ 10^{-4} mol) for sample 56 and 1 mm^3 (N $\times$ 10^{-4} mol) for sample 27, n_k = N $\times$ $a_{i,j}$—molar quantity of major (petrogenetic) component i in mineral j (n_k $\times$ 10^{-8} mol)

could be related to one of the aforementioned factors. Some deficiency of the initial concentrations of TiO_2 relative to those final concentrations in Sample 56 may be explained by the abundance of ilmenite microinclusions in the mantle of garnet. With the error of 20%, the volume won't exceed 1.2 mm^3.

These local reactions were isochemical except for water, not requiring that mass transfer take place between adjacent zones. The distance of material transport for the most mobile component participating in the reactions does not exceed ~0.2 mm (Sample 56) and ~0.5 mm (Sample 27), respectively. These values are thought (e.g., Ridley 1985) to depend on the rate of diffusive mass transport, pressure and temperature parameters and duration of metamorphism. Thus, diffusional flow and concentration gradients were so short-ranged that they did not lead to appreciable matter redistribution within the rock. This small reaction volume in which mass transfer occurred and limited diffusion may have been due to the short period of collisional metamorphism.

Analysis of the balance of some large ion lithophile elements (Rb, Ba) and some high field strength elements (Th, U) which is characterized by relatively high migration mobility during metamorphism (Nozhkin and Turkina 1993) shows that their redistribution among reacting phases during mineral reactions (4.89, 4.90) may be limited by the mentioned small volumes (Table 4.8).

Rare earth elements also diffuse through garnet and demonstrate more complex behaviour. The HREE contents in analyzed garnet exhibit clear dependence on the Ca content of garnet. In all cases, an abrupt increase in Ca content in garnets during prograde collision-related metamorphism of metapelites is accompanied by a significant decrease in HREE contents 2–3 times as compared to the initial concentrations (Table 4.9). It is consistent with the observed deficit in Y contents between reactant and product assemblages (25 ppm vs. 15 ppm и 17 ppm vs. 6 ppm), determined with using of calculated stoichiometric coefficients from mineral reaction Eq. (4.89, 4.90), respectively.

The systematic negative correlation between HREE and Ca contents in garnet shows that variations in major-element composition affect HREE incorporation at eightfold-coordinated site typically filled by a solid solution of divalent cations in

Table 4.8 Net mass balance of trace elements before and after mineral reactions of collisional metamorphism

	N	Initial contents (sample 56)				N	Final contents (sample 56)			
		n_{Rb}	n_{Ba}	n_{Th}	n_U		n_{Rb}	n_{Ba}	n_{Th}	n_U
Ms	7.2	1166	12,434	0.04	0.19	3.2	518	5526	0.02	0.08
Grt	60	1922	25.1	0.3	1.74	76	2995	21.5	0.62	0.41
Ep	6	57.11	32.48	0.13	1.35	0				
Bt	3.2	1424	1830	0.18	0.10	4	1780	2288	0.22	0.12
Pl	18	1.67	1854	0.11	0.05	24	3.22	1251	1.00	0.07
Qz	1160	977	13,862	319	499	1100	926	13,145	303	473
St	5.6	47.6	1.58	0.02	–	3.6	30.6	1.02	0.01	–
Ilm						0.8	–	–	–	–
Ky						36	8.75	9.32	1.15	1.94
Total		5595	29,039	319	502		6262	26,242	306	475
	N	Initial contents (sample 56)				N	Final contents (sample 56)			
		n_{Rb}	n_{Ba}	n_{Th}	n_U		n_{Rb}	n_{Ba}	n_{Th}	n_U
Ms	30	4860	51,810	0.15	0.78	20	3240	34,540	0.1	0.52
Grt	60	2408	4.8	0.48	0.48	50	1838	2.8	0.5	0.25
Ep										
Bt	15	6677	8580	0.825	0.45	19	8457	10,868	1.05	0.57
Pl	220	21.6	23,980	5.28	0.88	180	23.4	25,056	3.96	0.36
Qz	2690	2265	32,145	740	1157	2800	2358	33,460	770	1204
St	70	595	19.7	0.21	–	60	510	16.9	0.18	–
Ilm						1.6	–	–	–	–
Ky						84	20.4	21.8	2.69	4.53
Total		16,827	116,540	747	1160		16,447	103,966	778	1210

N = V/V_j—molar quantity of minerals in estimated volume of rock = 0.4 mm^3 (N × 10^{-4} mol) for sample 56 and 1 mm^3 (N × 10^{-4} mol) for sample 27, n_k = N× $a_{i,j}$—molar quantity of trace element i in mineral j (n_k × 10^{-8} mol)

garnet (Skublov 2005; van Westrenen et al. 2003. Because the ionic radii of HREEs (1.01–1.05 Å) and Y (1.04 Å) are similar with the ionic radius of Ca^{2+} (1.04 Å) in appropriate coordination they are expected to favor for concomitant replacement of HREEs for Ca into the eightfold sites in garnet (Ague and Carlson 2013). A critical consideration, however, is that the REEs and Y are trivalent, but they substitute onto sites normally occupied by divalent cations. The dominant mechanism for maintaining local neutrality is likely to be coupled substitution of divalent cations like Mg and Fe for trivalent Al on octahedral sites, simultaneously with substitution of trivalent REEs and Y for divalent Ca cations on eightfold site (Carlson 2012).

These relationships may provide evidence for removal of HREEs during the formation of the progressive zoning of garnets outside the system limited by the volumes of matter redistribution during collisional metamorphism. These observed losses of HREEs are balanced by further gains HREEs during retrograde

Table 4.9 Net mass balance of REE before and after mineral reactions of collisional metamorphism (4.89) and (4.90)

	N	Initial concentration of components (sample 56)								
		n_{La}	n_{Ce}	n_{Nd}	n_{Sm}	n_{Eu}	n_{Dy}	n_{Er}	n_{Yb}	n_Y
Ms	7.2	4.00	0.78	3.85	0.13	8.23	1.76	2.10	0.25	0.32
Grt	60	1.14	2.82	4.8	30.9	32	8088	9060	10,980	72,720
Ep	6	15.2	69.1	78.06	26.9	25.7	25.5	20.60	19.16	187.3
Bt	3.2	1.01	1.39	0.57	0.27	0.38	0.38	0.11	0.06	0.14
Pl	18	0.9	1.44	0.85	0.76	1.53	0.18	0.07	0.018	4.34
Qz	1160	8409	14,187	6373	1460	1826	416	390	186	1241
St	5.6	0.02	0.14	0.039	0.03	0.01	0.26	0.02	0.51	0.017
Ilm	0	0	0	0	0	0	0	0	0	0
Ky	0	0	0	0	0	0	0	0	0	0
Total		8431	14,263	6461	1519	1894	8532	9473	11,186	74,153
	N	Final concentration of components (sample 56)								
		n_{La}	n_{Ce}	n_{Nd}	n_{Sm}	n_{Eu}	n_{Dy}	n_{Er}	n_{Yb}	n_Y
Ms	3.2	1.78	0.35	1.72	0.06	3.66	0.78	0.93	0.11	0.14
Grt	76	19.5	39.4	41.8	89	69.8	2654	3177	4676	24,442
Ep	0	0	0	0	0	0	0	0	0	0
Bt	4	1.27	1.73	0.72	0.33	0.47	0.48	0.14	0.08	0.18
Pl	24	45.1	71.8	20.4	3.26	29.6	1.78	0.50	0.17	13.13
Qz	1100	7974	13,453	6043	1385	1731	395	370	176	1177
St	3.6	0.02	0.090	0.03	0.02	0.01	0.17	0.02	0.33	0.011
Ilm	0.8	0.16	0.069	0.04	0.73	0.20	0.06	0.06	0.47	0.01
Ky	36	1.15	1.76	2.66	0.40	0.04	0.58	0.32	2.81	2.84
Total		8043	13,568	6100	1479	1835	3053	3549	4856	26,207
	N	Initial concentration of components (sample 27)								
		n_{La}	n_{Ce}	n_{Nd}	n_{Sm}	n_{Eu}	n_{Dy}	n_{Er}	n_{Yb}	n_Y
Ms	30	16.7	3.24	16.1	0.54	34.3	7.35	8.73	1.05	1.32
Grt	60	0.36	1.08	5.76	9.78	3.3	1644	3671	6330	20,184
Bt	15	4.76	6.50	2.69	1.25	1.77	1.8	0.51	0.3	0.66
Pl	220	21.3	43.6	25.3	13.6	64.6	4.4	1.32	0.22	59.8
Qz	2690	19,500	32,899	14,779	3387	4234	966	904	430	2878
St	70	0.28	1.75	0.49	0.35	0.07	3.22	0.28	6.37	0.21
Ilm	0	0	0	0	0	0	0	0	0	0
Ky	0	0	0	0	0	0	0	0	0	0
Total		19,543	32,955	14,829	3412	4338	2627	4586	6768	23,124
	N	Final concentration of components (sample 27)								
		n_{La}	n_{Ce}	n_{Nd}	n_{Sm}	n_{Eu}	n_{Dy}	n_{Er}	n_{Yb}	n_Y
Ms	20	11.1	2.16	10.7	0.36	22.9	4.9	5.82	0.7	0.88
Grt	50	0.5	1.1	2.95	15.7	5.7	924	867	674	9930
Bt	19	6.02	8.23	3.40	1.58	2.24	2.28	0.65	0.38	0.84

(continued)

Table 4.9 (continued)

	N	Final concentration of components (sample 27)								
		n_{La}	n_{Ce}	n_{Nd}	n_{Sm}	n_{Eu}	n_{Dy}	n_{Er}	n_{Yb}	n_{Y}
Pl	180	717	1022	289	41.6	525	30.2	6.12	6.12	128
Qz	2800	20,297	34,244	15,383	3525	4407	1005	941	448	2996
St	60	0.24	1.5	0.42	0.30	0.06	2.76	0.24	5.46	0.18
Ilm	1.6	0.31	0.14	0.08	1.46	0.41	0.13	0.12	0.94	0.02
Ky	84	2.67	4.12	6.22	0.92	0.08	1.34	0.76	6.55	6.64
Total		21,035	35,283	15,694	3587	4963	1971	1821	1142	13,063

N = V/V_j—molar quantity of minerals in estimated volume of rock = 0.4 mm^3 (N × 10^{-4} mol) for sample 56 and 1 mm^3 (N × 10^{-4} mol) for sample 27, n_k = N× $a_{i,j}$—molar quantity of rare-earth element i in mineral j (n_k × 10^{-8} mol)

synexhumation metamorphism associated with increase in HREEs content and decrease in the concentration of Ca in garnets from the mantle to the rim during reaction of retrograde metamorphism. It seems that the total HREEs balance requires a rather greater reaction volume involved in the redistribution of HREEs than that for redistribution of other trace elements, which provides evidence for their differential migration mobility. The scales of mass transfer for HREEs increase up to ~0.5 mm (Sample 56) and ~1 mm (Sample 27) (Fig. 2.1 in Chap. 2), which results in proportional increase in the overall volumes of HREEs redistribution from 3 to 8 mm^3. The balancing of the HREEs concentrations between garnet, the principle carrier of the HREEs in rock, and all other phases is evident from the mirror shape and REE spectrum pattern of rock-forming minerals in relation to the bulk-rock composition (Fig. 4.24b). Nevertheless, it should be noted that in most of the other studies, REE have smaller equilibrium volume than major elements.

A LREE and MREE mass balance with the exception of Gd yields very good matches between the measured and reconstructed REE abundances in all cases, indicating that metamorphism was essentially isochemical with respect to LREEs and MREEs, with the exception of Gd (Table 4.9). Mass transfer analysis shows that LREEs and MREEs distribution among reacting phases during mineral reactions (4.89, 4.90) of collisional metamorphism may be limited by the local volumes of the order of 0.4–1 mm^3, which provide evidence for their limited migration mobility during collisional metamorphism. It is also confirmed by the calculated balance in Sm (0.33 ppm vs. 0.31 ppm), Eu (0.37 ppm vs. 0.32 ppm), and Nd (0.23 ppm vs. 0.20 ppm) abundances between reactants and products of mineral reactions (4.89) within 5–15% error. Thus, in all cases, the REEs, except for HREEs, act as relatively "immobile" elements, despite major redistribution among the coexisting metamorphic phases.

Thus, the main causes for the sharp increase in Ca content in garnets during collisional metamorphism are the redistribution between garnet and plagioclase, which became more sodic in composition, and/or as a result of diffusion-controlled reaction, coupled with Ca consumption during epidote breakdown.

The data presented here reveal previously unrecognized aspects in behavior of major and trace elements during multistage metamorphism. Further study in this direction will focus on accessory REE-bearing minerals (monazite, xenotime, allanite, bastnaesite, synchysite) which actively participate in metamorphic reactions (e.g., Janots et al. 2008; Kohn and Malloy 2004; Savko and Bazikov 2011; Savko et al. 2012) and include them in overall mass balance calculations.

In general, the obtained data allow us to conclude that mass transfer in metamorphic reactions occurs with the preservation of the balance of matter within very small local volumes of the rock, which increase from a few hundredths of cubic millimeters to a few cubic centimeters depending on the duration (in order of increasing duration of metamorphism: contact metamorphism → medium-pressure zonal metamorphism → collisional metamorphism), *P-T* parameters, strain intensity and a degree of fluid saturation of rocks.

4.2.3 Estimation of Rates of Metamorphic Front Migration

Section 1.2 has already dealt with the kinetics of metamorphic reactions. However, further exploration of this issue is required.

The process of prograde metamorphism is caused by the increased temperature and pressure in the Earth's crust. The pressure acting on a rock mass generates stresses almost instantaneously (in a geological sense) in response to a loading applied. At the same time, the amount of load during collision may vary in time. It can be said that the primary cause of metamorphic reactions is the migration of fronts of elevated T and P. When the conditions for the phase transition are reached, the rate of the metamorphic reaction will depend on the rates of nucleation of new minerals and diffusion-controlled mass transfer in the rocks.

The heating of rocks leading to their metamorphism is a non-stationary process and depends on geological conditions, thermophysical characteristics, the presence of phase transitions, etc. Using model representations, it is possible to estimate the average migration rate of a temperature front in a certain time interval. This is the simplest way to solve a problem when no special accuracy is needed.

England and Thompson (1984) provided theoretical estimates of the heating rate of continental crust during collision and orogeny. The calculations were performed for the case of one-dimensional heat transfer, accounting for the supply of heat from the mantle and radiogenic heat production; before heating, an instantaneous doubling of the crustal thickness (initial thickness of 35 km) was assumed. Modeling evolving geotherms in thickened crust showed that the temperature of 400 °C has migrated at a rate of $n \times 10^{-9}$ cm s^{-1} in a few million years. The other temperature fronts migrated at approximately the same rate.

The heating rate during contact metamorphism can be easily estimated in one-dimensional approximation using mathematical modeling. As discussed in Sect. 3.2.1.1, the phase transition in the country rocks is assumed to be instantaneous, and the model calculations yielded the maximum limiting width of a metamorphic zoning and a distance of corresponding isograds from the magmatic contact. The migration rate of the metamorphic front was assumed to be equal to the rate of conductive heat transfer. For a granite dike with an initial magma temperature of 750 °C and a thickness of 100 m (see Fig. 3.3), the rate of the migration of a temperature of 200 °C in the contact aureole was determined to be $n \times 10^{-6}$ cm s^{-1}. For an ultramafic dike with the same thickness and the initial magma temperature of 1300 °C (see Fig. 3.3), the migration rate of the temperature front of 500 °C was estimated to range between $n \times 10^{-7}$ and $n \times 10^{-6}$ cm s^{-1}; the migration rate of the temperature front of 200 °C was 9.8×10^{-7} cm s^{-1}. The migration rate was higher at the onset of heating (with a total duration of ~300 years) and decreased at the end of heating.

Model calculations of heat transfer near the lateral vertical contact of a large mafic magma body intruded into the mid-crustal depths were used to estimate the rate of heating of rocks in two-dimensional approximation. For example, let us determine the rate of propagation of the 600 °C isotherm. The calculations were made to clarify the origin of metamorphic zoning of the Connemara complex of western Ireland (Reverdatto and Polyansky 2004; see also Sect. 3.2.1.2). Figure 3.22a shows the thermal evolution of the country rocks near vertical lateral contact of a 6 km thick sheet-like mafic magmatic body that was emplaced at a depth of 20 km at an initial temperature of 500 °C. The calculations show that the temperature front of 600 °C has migrated horizontally a distance of 2.5 km in a time period of ~1 Myr after intrusion emplacement. Thus, the heating rate was estimated to be $\sim 8 \times 10^{-9}$ cm s^{-1}. The thermal evolution of the country rocks above the upper boundary of the same sheet-like intrusion after emplacement allows us to determine the heating rate at its roof (Reverdatto and Polyansky 2004). The temperature front of 600 °C has migrated a distance of ~1.7 km over ~0.3 Myr. Thus, the heating rate is estimated to be $\sim 1.7 \times 10^{-8}$ cm s^{-1}. Figure 3.22c shows the thermal evolution of the rocks above the upper boundary of the same sheet-like intrusion tilted at an angle of 35°. The temperature front of 600 °C has migrated horizontally (at a depth of 20 km) over a distance of ~7.5 km over ~1 Myr, and the heating rate was $\sim 2.3 \times 10^{-8}$ cm s^{-1}.

We can very roughly estimate the migration rate of the 350 °C isotherm during the development of a metamorphic zoning near the Kharlovo cylindrical gabbro intrusion (see Sect. 3.2.1.2. and Fig. 3.17). The initial temperature of mafic magma is assumed to be 1100 °C, the depth of emplacement is taken to be 7–8 km. At a geothermal gradient of ~25 °C/km, the initial temperature of the country rocks at this depth should reach 175–200 °C. The 530 °C isotherm within metamorphic zoning is located at a distance of about 1 km from the intrusive contact. Using the temperature dependence of $\frac{x}{\sqrt{t}}$, similar to that shown in Figs. 3.5 and 3.6, where x is the distance (in meters) and t is the heating time (in years), for $\frac{x}{\sqrt{t}}$ lying in the interval between 3 and 4, it can be calculated that the migration rate of the temperature front is approximately equal to $n \times 10^{-8}$ cm s^{-1} (e.g., Reverdatto et al. 2018).

It is interesting to assess the rate of heating observed in sediments filling a rift basin. Figure 3.43 illustrates the modeling results of the evolution of a rift basin, sediment accumulation and distribution of paleoisotherms in the rocks of the Yenisei-Khatanga trough (see Sect. 3.2.2.1). In these models, the evolution of a basin is reconstructed by the backstripping method using stratigraphic-lithological data on the structure and composition of sedimentary columns. The dynamics of basin subsidence is calculated using the Mackenzie's model and geophysical data. These two versions were compared and the discrepancy was estimated. The modeling was performed by the identification of the best fit between the calculated values and the observed dynamics of basin subsidence using different values of physical parameters, extension coefficients for the crust and subcrustal lithosphere. Knowing the conductive heat flow and thermal conductivity of sedimentary rocks, it is possible to calculate the temperature at a given depth and at any given time. We carried out modeling of heating of sedimentary rocks during the formation of the Ust-Yenisei depression in western part of the Yenisei-Khatanga trough (Fig. 3.43). The calculations show that the average migration rate of the temperature front of 400 °C was 1.5×10^{-10} cm s^{-1} over a period of 100 Myr. Basin subsidence took place at approximately the same rate. In the central part of Yenisei-Khatanga trough, the subsidence within the Balakhna depression occurred over a period of 85 Myr following termination of the spreading episode of rifting. The migration rate of the temperature front of 300 °C was estimated at $\sim 7.5 \times 10^{-11}$ cm s^{-1} (Polyanskii et al. 2000). Such a slow rate of migration can be explained by the weak heating of sedimentary rocks during basin subsidence. The period after the active spreading stage of rifting was characterized by a slow cooling of the asthenosphere accompanied by a weak heating of sediments filling the depression.

The above results show that the rate of heating, i.e. the migration rate of a given temperature front during metamorphism ranges from $n \times 10^{-7}$ to $n \times 10^{-11}$ cm s^{-1}. If we discard the extreme values characteristic of contact and burial metamorphism, then the heating rate will be equal to $\sim n \times 10^{-8}$ cm s^{-1}. It should be noted that the models for contact and burial metamorphism were realized in one-dimensional approximation; for three-dimensional cases the heating rates should be lower.

The rate of heating can on average be expressed as $n \times 10^{-13}$ °C s^{-1} (England and Thompson 1984). For contact metamorphism, this rate would be somewhat higher.

It is equally important to determine the rate of lithostatic pressure growth during a collisional event leading to metamorphism. Collision of continental slabs leads to horizontal shortening in the rock strata along the strike as a result of folding, faulting and overthrusting. The rate of folding over a long period is either equal to or lower than that of overthrusting (for example, Khain and Lomize 1995; Biot 1961; Suppe et al. 1992; Livio et al. 2007, and others). This allows us to assume that the overall rate of shortening along the strike depends mainly on the shear rate.

The geodetically measured rate of the modern Indo-Asian convergence is currently about 40 km/Myr, i.e. $\sim 1.3 \times 10^{-7}$ cm s^{-1} (Bilham et al. 1997; Wang et al. 2001). Thermal and kinematic modeling performed (Brewer and Burbank 2006) shows a shear rate of $n \times 10^{-8}$ cm s^{-1} for the Asian plate and Indian plates with respect to the Main Himalayan Thrust in Nepal. In the Himalayan section, a shortening rate of $\sim 6 \times 10^{-8}$ cm s^{-1} was documented over the Miocene (Mugnier et al. 2004). Tectonic processes in the Alpine-Himalayan fold belt resulted in the formation of Pamir-Punjab collision structure. In the Pamir, thrusting and overthrusting rates varied from $\sim 6.3 \times 10^{-9}$ to $\sim 1.3 \times 10^{-7}$ cm s^{-1} over the Cenozoic (Burtman 2013). In the western Alps, overthrusting rates varied from 9.5×10^{-12} to 6.3×10^{-10} cm s^{-1} over the Eocene-Oligocene (Artoni and Meckel 1998). Based on different tectonic models, Tozer et al. (2002) provided the most likely shortening rate of $<1.9 \times 10^{-8}$ cm s^{-1} for the Central Apennines in the Pliocene-Pleistocene. Following the early Paleozoic collision in northern China, shortening rates of $\sim 1.9 \times 10^{-7}$ cm s^{-1} was documented in Inner Mongolia and 1.5×10^{-7}–2.7×10^{-7} cm s^{-1} in other regions (Wan 2010). The rate of movement along the Lesnovsky overthrust in northern Kamchatka was 1.6×10^{-7} cm s^{-1} over a period of ~ 1 Myr (Soloviev et al. 2001).

In the Transangarian part of the Yenisei Ridge, manifestations of the collision-related metamorphism are attributed to ancient regional overthrusts. Using the Mayakon area in the interfluve of the Yeruda and Chirimba Rivers as an example, it was shown that overthrusting during the Late Riphean was the main cause of the kyanite-sillimanite metamorphism (Likhanov et al. 2004, 2008b). Overthrusting led to a 1.5–3 kbars increase in lithostatic pressure in the Precambrian metapetites forming a strip >20 km long and up to 5–7 km wide. It is assumed that the underlying rocks were overlain by 7 km of overburden. Mineralogical and petrological data suggests a low geothermal gradient ranging from 1–7 to 12 °C/km in the zone of collision-related metamorphism. The absence of any noticeable rise in temperature is explained by the overthrusting rate of $n \times 10^{-9}$ cm s^{-1} with subsequent rapid exhumation and erosion.

The rates of subduction of lithospheric plates (Khain and Lomize 1995) are typically between 3×10^{-7} and 3×10^{-8} cm s^{-1}), with rare exceptions. Gutscher and Westbrook (2009) provided slightly different values ranging from 4.7×10^{-7} to 1.6×10^{-8} cm s^{-1} (based on 57 examples of subduction). The maximum rate of subduction hardly exceeds 6.3×10^{-7} cm s^{-1} (Staudigel and King 1992), while the minimum rate is loosely estimated to be equal to or less than 3×10^{-8} cm s^{-1} (Goes et al. 2008).

The calculated rates of lithostatic pressure increase during continental collision (as a result of overthrusting) and subduction of lithospheric plates vary from 10^{-7} to 10^{-12} cm $\times$ s^{-1}, being on average $n \times 10^{-8}$ cm s^{-1}.

Models predict that the subducting slab-mantle interface temperatures are relatively low to great depths (for example, Peacock 2003; Syracuse et al. 2010). Calculations using the MacKenzie's model (McKenzie 1969) show that it takes ~ 40 Myr for the subducting slab to reach the core-mantle boundary after the onset of subduction, while the minimum slab temperature would differ from that of the surrounding mantle (Stein and Stein 1996). In the case of thermomechanical (model) parameters typical of a subsiding slab, the temperature of the coldest part of the slab at a depth of 670 km would be one-half as high as that of the mantle. Since it takes ~8 Myr for the slab to reach this temperature, the rate of heating of the subducting slab would be equal to $\sim 2.6 \times 10^{-7}$ cm s^{-1}.

Based on the above results, the following conclusion can be made. The calculated rates of heating and lithostatic pressure increase during metamorphism in response to magmatic and tectonic processes range from 10^{-7} to 10^{-12} cm s^{-1}, being on average $n \times 10^{-8}$ cm s^{-1}.

The rate of lithostatic pressure increase (in bars per second) can be exemplified by thermal models constructed for subduction zones. In the Izu-Bonin subduction zone (Peacock 2003), the pressure at the top of the subducting slab should be ~30 kbar at a depth of 100 km and ~60 kbar a depth of 200 km. The distance along the top of the slab between the points where such pressures are reached is 130 km. It takes ~2.3 Myr for the slab to move this distance, given that the subduction rate is $\sim 1.8 \times 10^{-7}$ cm s^{-1}. Hence, the rate of lithostatic pressure increase is $\sim 4.2 \times 10^{-10}$ bar s^{-1}. Almost identical estimates were obtained for the NE Japan subduction zone (Peacock 2003). It seems likely that the calculated rate of pressure increase will be similar in all other subduction zones. In the Mediterranean orogens, the calculated exhumation rates during decompression in subduction zones are roughly equal to $n \times 10^{-11}$ bar s^{-1} (Jolivet et al. 2003).

Numerical modeling of the effects of overthrusting using the prescribed values of all input parameters, such as thickness, distance of overthrusting, heat flow and the initial temperature distribution in the rock sequence, allows us to investigate the history of temperature distributions in the overthrusting sheet and underlying bedrocks, depending on the rate of movement (Yalcin et al. 1997). By assuming a thickness of an overthrusting sheet of 5 km that moved a distance of 10 km at a thrusting rate of 1.6×10^{-7} cm s^{-1} the calculated rate of lithostatic pressure increase would be approximately equal to $\sim 2.4 \times 10^{-10}$ bar s^{-1}. Such thrusting rate caused an S-shaped temperature profile, a decrease in the thrusting rate by an order of magnitude or so did not cause an inversion in the temperature trend.

It is of interest to compare the rates of increase in *P-T* parameters in response to tectonic and magmatic processes with the rates of metamorphic front migration. The kinetics of metamorphism has already been discussed above (see Sect. 1.2 and others). It was emphasized that the metamorphic reaction is a complex process, which involves the nucleation of minerals, diffusion transport of matter and the growth of mineral grains. Diffusion transfer is the slowest process controlling the overall reaction rate.

Using experimentally determined coefficients of diffusion through fluid films along the grain boundaries, Fisher (1978) gave the approximate estimates of growth rates of mineral segregations in metamorphic rocks. A typical spherical segregation with a radius of 0.5 cm, composed of ~0.2 mm mineral grains grew at a rate of about 0.3×10^{-12} cm s^{-1} for 50,000 years. Variations of the reasonable durations of metamorphism and size of segregations give a wider range of calculated growth rates between 10^{-11} and 10^{-14} cm s^{-1}. The calculated growth rates of the reaction rim in a spherical segregation with a radius of 1 cm range from 10^{-13} to 10^{-14} cm s^{-1}.

The growth rate of biotite can be approximately calculated for metapelites of the Kharlovo gabbro massif (Fig. 3.17 in Chap. 3). Likhanov et al. (2001b) proposed an approach for deriving kinetic parameters of diffusion-controlled metamorphic reactions based on comparison of the calculated temperature evolution and the observed variations in the chemical composition of this mineral. Unlike other methods used to construct quantitative models of reactions which produce distinctly expressed zoned microtextures, the proposed approach allows better understanding of the processes occurring in texture-homogeneous rocks, where complete consumption of some reactant mineral phases prevent accurate reconstruction of the metamorphic process.

Biotite was chosen as the main mineral characterizing the ``reaction product'', for the following reasons. First, biotite is ubiquitous in all samples from the study area and is the most abundant and widespread ferromagnesian phase in all contact metamorphic zones and mineral assemblages. Second, it displays a clear trend of increasing TiO_2 content with increasing metamorphic temperature (e.g. Miyashiro 1958). An abrupt change in TiO_2 content (from 2.32 to 3.38 wt%) coincides with the andalusite-in isograd of the middle zone of the contact aureole. All these data confirm that biotite took part in practically all local metamorphic reactions in the contact aureole and can be used in evaluating the rate of metamorphic reactions in which biotite participated directly.

The study area previously well investigated (Reverdatto et al. 1974; Likhanov 1989, 1990) is the contact aureole of the Kharlovo gabbro massif. Emplacement of a composite stock of titanomagnetitic gabbro, forming a distinct "cylindric-shaped" structure with a radius of 1950 m, in Cambrian-Ordovician metapsammites of the Altay Mountains, regionally metamorphosed to greenschist facies conditions, produced an extensive contact aureole with distinct zonation characterized by muscovite and amphibole hornfels facies.

We modeled contact metamorphism by two concurrent processes. The first process was cooling and solidification of an igneous body due to heat conductivity, which necessarily involves a substantial temperature change in the surrounding rocks. This temperature change causes the second process which is alteration of the surrounding rock. The rock responds to heat with redistribution of chemical components among the mineral phases (assuming a closed system). If the rate of these changes is high enough, the new component distribution achieves an equilibrium state at a new temperature. Otherwise, the distribution is frozen in a non-equilibrium state or the rocks remain unaltered.

Metamorphic zoning is typically observed around an intrusion. Zoning of this kind implies relatively gradual changes within zones and abrupt transitions between them. Since metamorphism is a set of chemical reactions, one may expect reaction rates to depend on temperature exponentially. This relation implies that under the wide range of temperature conditions, metamorphic changes near an igneous contact should be step-like (e.g. Clayton and Bostick 1986). As shown in Fig. 4.25, a compositional profile has two comparatively homogeneous zones with a narrow transition zone between them.

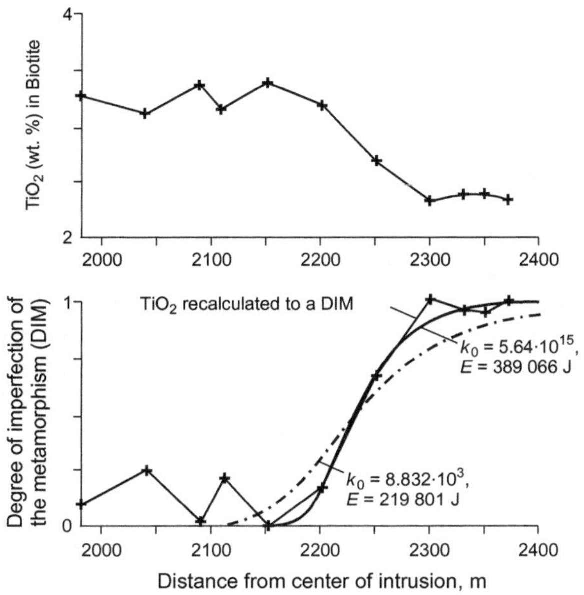

Fig. 4.25 Profile of TiO_2 content in biotite and model profiles. Upper panel: wt% TiO_2 versus distance from center of intrusion. Lower panel: recalculated TiO_2 content result in the observed DIM profile. Two model calculations show good agreement with observed DIM

Step-like behavior is supported by another observation. Within a definite metamorphic zone, a mineral assemblage was stable. This fact implies that compositional variation in a particular mineral occurs mainly due to temperature variation across the zone. However, in an adjoining zone (with higher or lower temperatures) some reactions stop due to extremely low reaction rates, while others proceed due to a shift in chemical equilibrium. The combination of different temperatures and the shift in chemical equilibrium results in a different mineral assemblage (metamorphic zone). The shift of chemical equilibrium across the transition zone has a much more powerful effect on mineral compositions than that due to temperature variation within the zone (Ferry 1983).

To represent metamorphic changes quantitatively, we need to introduce a formal criterion to characterize a metamorphic process. The extent (ξ) of chemical changes during metamorphism at a given moment may be described in dimensionless, normalized form, $d\xi$—called degree of advancement of reaction (Helgeson 1968) or reaction progress variable (Fitts 1962), which, for a mineral, may be defined as: current quantity of a component/quantity of a component at equilibrium.

The degree of advancement of reaction is a reasonable formal parameter for describing the dynamics of a metamorphic system. For the purposes of this paper, however, we used an inverse criterion: ($y = 1 - d\xi$), where y is the degree of incompleteness (or imperfection) of reaction (DIM). The degree of incompleteness of reaction may also be interpreted as concentration (e.g. Bolton et al. 1999).

This dimensionless parameter is equal to 1 for unaltered rock (before the intrusion) and equal to 0 for rock that has reached its equilibrium state for the

P-T conditions of the contact metamorphism. The rock changes that are identified as metamorphism may be converted to DIM.

The upper panel of Fig. 4.25 shows TiO_2 variations versus distance from the center of the intrusion. A sharp drop between two relatively steady intervals is observed. A respectively uniform percentage of TiO_2 at distances between 1980 and 2160 m suggests that changes in biotite due to temperature reached the metamorphic equilibrium state (y = 0), whereas at distances beyond 2300 m TiO_2 is uniform (y = 1). After smoothing and scale attribution, this data could express DIM (lower panel of Fig. 4.25). The same method may be applied to any mineral. Less formal observations may also be quantified with this method. For instance, changes in modes or rock volumes which a rock undergoes could be plotted against metamorphic temperature or against distance in the field perpendicular to isograds.

By taking into account the shape of profile and the temperature-dependence of the reaction rate, it was assumed that the dynamics of biotite formation is governed by the first-order equation:

$$\frac{dy(r,t)}{dt} = -y(r,t) \cdot k_0 \cdot \exp\left(\frac{-E}{RT(r,t)}\right), \tag{4.91}$$

where E is the activation energy, R is the universal gas constant, T is the absolute temperature, k_o is the frequency factor, t is the time and r is the distance. The temperature evolution around the magmatic body was obtained by solving the heat transfer equation for a cylinder. The kinetic parameters of the biotite formation reaction (E and k_o) are unknown, but we have the observed compositional profile expressed through the degree of incompleteness of reaction ($y_{observed}$) and the temperature distribution. In the simplest case, the kinetic parameters E and k_o can be written in such a way that the difference between observed DIM and calculated DIM is minimum. Because of non-linearity of Eq. (4.91), E and k_o turn out to be mutually related, which makes the inverse problem ambiguous. The resulting solutions of the inverse problem are equivalent along the line $k_o \cdot \exp(-E/RT) = const$. There is an infinite set of combinations of E and k_o lying on above line, which being substituted into Eq. (4.91) will give results similar to the observed TiO_2 profile. Although there is a large spread in the kinetic parameters, the model profile of the degree of incompleteness (y) coincides with observed data within analytical errors in the determination of TiO_2 content and an error of model calculations. Using the mean value theorem (Bronshtein and Semendyayev 1979) allows us to extract an effective reaction rate, instead of finding E and k_o. In doing so, the effective rate will be the main factor controlling the slope in compositional profiles. Different combinations of E and k_o affect the slope and curvatures of that change. The effective rate is expressed as: $k_{eff} = k_0 \cdot \exp\left(\frac{-E}{RT^*}\right)$, where T^* is the effective temperature. The effective rate is the main factor responsible for the position of steep change in a composition profile. Different combinations of E and k_0 affect the slope and curvatures of that change. As a result, we may reasonably assume that the effective rate is approximately the same for all our calculations.

Since the effective rate is the same for both sets of kinetic parameters, the intersection point of rates in Fig. 4.26 determines the effective temperature, and, accordingly, the rate. The condition for intersection of curves is $k_{0,1} \cdot \exp\left(\frac{-E_1}{RT^*}\right) = k_{0,2} \cdot \exp\left(\frac{-E_2}{RT^*}\right)$, where the additional lower index 1 or 2 belongs to the first or second set of parameters. Thus, the effective temperature is determined by the formula $T^* = \frac{E_2 - E_1}{R \ln\left(\frac{k_{0,2}}{k_{0,1}}\right)}$. The coincidence of profiles with a relatively large spread of kinetic parameters indicates that the reaction was carried out at a temperature of ~475 °C (Fig. 4.26). For this temperature, all combinations of E and k_O are in good approximation with the observed profile and are located on the line $k_O \cdot \exp\left(\frac{-E}{RT^*}\right) = 10^{-12}$. Thus, the effective rate of metamorphic reactions involving biotite in metapelitic rocks of the Kharlovo gabbro massif turns out to be $\sim 10^{-12}\ \mathrm{s}^{-1}$.

Thus, in terms of the overall process, our study seems to be adequate and will certainly facilitate the choice of specific isograds for further study. Theoretically, the method described in this chapter can be used to determine the relationship between time, temperature and mineral composition changes for a wide range of contact metamorphic reactions. The approach is, perhaps, a unique tool for the determination of the kinetic parameters of metamorphism, and could be utilized for a better understanding of the processes occurring in texture-homogeneous rocks, in which chemical reactions often reach completion, making it impossible to establish the ratio between reactant and product minerals.

Baxter (2003) summarized the estimated metamorphic reaction rates based on natural observations and justified the diffusion-controlled growth of mineral grains. He suggested that observed difference in the reaction rates between contact and regional metamorphism can be attributed to higher temperatures attained during contact metamorphism. Baxter expressed the overall reaction rate in units of reactant mass divided by area and time, assuming that only 20% of mineral grain surfaces participated in the reaction. Hence, the rates of metamorphic front migration can be obtained by means of simple transformation of values of overall reaction rates. The units used by Baxter should also be changed in order to compare

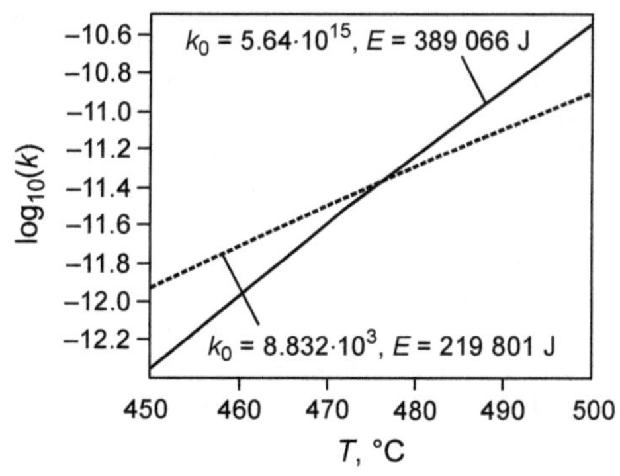

Fig. 4.26 Rate of reaction (k) as a function of temperature for two sets of kinetic parameters—E and k_O

the rates of metamorphic fronts with the rates (in cm s^{-1}) of change in metamorphic *P-T* parameters. Based on the data of Baxter, the rates of reaction front advancement can be estimated in the range of $n \times 10^{-14}$ to $n \times 10^{-18}$ cm s^{-1} for regional metamorphism (in metapelites) and $n \times 10^{-10}$ to $n \times 10^{-16}$ cm s^{-1} for contact metamorphism.

These model estimates are only tentative, because of the uncertainties in the diffusion rates of mass transfer during mineral growth. Nevertheless, the rates of prograde metamorphic reactions are lower than the rates of the increase in pressure and temperature in response to tectono-magmatic events. The rate of most mineral transformations in rocks lags behind that of the onset of *P-T* metamorphic conditions. However, these *P-T* conditions can persist for a long time in a stable geodynamic setting. The exception is a magmatic process, in which the thermal resources of an intrusive body act as a limiting factor for the duration of heating.

In general, the low rates of metamorphic reactions lead to the fact that the changes in *P-T* metamorphic parameters are recorded in mineral equilibria with a time lag, and the compositions of newly-formed coexisting minerals during their evolution may not reflect the peak *T* and/or *P* conditions. However, this problem requires further investigation.

To summarize, it should be noted that metamorphic reactions rates increase considerably with increasing rates of metamorphic strain usually accompanying regional metamorphism. A possible reason for this may be an increase in the reacting surface area of mineral grains.

References

Abu-Alam TS, Stuwe K (2009) Exhumation during oblique transpression: the Feiran-Solaf region, Egypt. J Metamorph Geol 27:439–459

Ague JJ, Carlson WD (2013) Metamorphism as garnet sees it: the kinetics of nucleation and growth, equilibration, and diffusional relaxation. Elements 9:439–445

Artoni A, Meckel LD (1998) History and deformation rates of a thrust sheet top basin: the Barreme basin, western Alps, SE France. Geol Soc Spec Publ 134:213–237 (Blackwell, London)

Ashworth JR (1993) Fluid-absent diffusion kinetics of Al inferred from retrograde meta- morphic coronas. Am Mineral 78:331–337

Ashworth JR, Birdi JJ (1990) Diffusion modeling of coronas around olivine in an open system. Geochim Cosmochim Acta 54:2389–2401

Ashworth JR, Sheplev VS (1997) Diffusion modelling of metamorphic layered coronas with stability criterion and consideration of affinity. Geochim Cosmochim Acta 61:3671–3689

Ashworth JR, Sheplev VS, Bryxina NA et al (1998) Diffusion-controlled corona reaction and overstepping of equilibrium in a garnet granulite, Yenisei ridge, Siberia. J Metamorph Geol 16:231–246

Baker AJ (1987) Models for the tectonothermal evolution of the eastern Dalradian of Scotland. J Metamorph Geol 5:101–118

Balashov VN, Zaraisky GP, Tikhomirova VI et al (1983) Diffusion of rock-forming components in pore solutions at T = 250 °C and P = 100 MPa. Geochem Int 20(1):28–40

Balen D, Massonne H-J, Petrinec Z (2015) Collision-related Early Paleozoic evolution of a crustal fragment from the northern Gondwana margin (Slavonian Mountains, Tisia Mega Unit,

Croatia): reconstruction of the *P-T* path, timing and paleotectonic implications. Lithos 232:211–228
Barrow G (1893) On an intrusion of muscovite-biotite gneiss in the south-eastern Highlands of Scotland, and its accompanying metamorphism. Q J Geol Soc Lond 49:330–388
Barrow G (1912) On the geology of the Lower Deeside and the southern Highland border. Proc Geol Assoc 23:268–284
Baxter EF (2003) Natural constraints on metamorphic reaction rates. In: Vance D, Muller W, Villa IM (eds) Geochronology: linking the isotopic record with petrology and textures. Geological Society Special Publication, vol 220. Blackwell, London, pp 183–202
Beaumont C, Jamieson RA, Nguyen MH et al (2001) Hymalayan tectonics explained by extrusion of a low-viscosity crustal channel coupled to focused surface denudation. Nature 414:738–742
Beddoe-Stephens B (1990) Pressures and temperatures of Dalradian metamorphism and the andalusite-kyanite transformation in the northeast Grampians. Scott J Geol 26:3–14
Berman RG (1991) Thermobarometry using multi-equilibrium calculations: a new technique, with petrological applications. Can Mineral 29:833–856
Bilham R, Larson K, Freymueller J et al (1997) GPS measurements of present-day convergence across the Nepal Himalaya. Nature 386:61–64
Biot MA (1961) Theory of folding of stratified viscoelastic media and its implications in tectonics and orogenesis. Geol Soc Am Bull 72:1595–1620
Bohlen SR (1991) On the formation of granulites. J Metamorph Geol 9:223–229
Boltaks BI (1961) Diffuziya v poluprovodnikakh (Diffusion in semiconductors). Fizmathgis, Moscow
Bolton EW, Lasaga AC, Rye DM (1999) Long-term flow/chemistry feedback in a porous medium with heterogeneous permeability: kinetic control of dissolution and precipitation. Am J Sci 299:1–68
Boynton WV (1984) Cosmochemistry of the rare earth elements: meteorite studies. In: Henderson P (ed) Rare earth element geochemistry. Elsevier, Amsterdam, pp 63–114
Brewer ID, Burbank DW (2006) Thermal and kinematic modeling of bedrock and detrital cooling ages in the central Himalaya. J Geophys Res 111:B09409
Bronshtein IN, Semendyayev KA (1979) Handbook of mathematics. Verlag Harri Deutsch, Berlin
Brown EH (1996) High-pressure metamorphism caused by magma loading in Fiordland, New Zealand. J Metamorph Geol 14:441–452
Brown M (2001) From microscope to mountain belt: 150 years of petrology and its contribution to understanding geodynamics, particularly the tectonics of orogens. J Geodynamics 32:115–164
Brown M (2007) Metamorphic conditions in orogenic belts: a record of secular change. Int Geol Rev 49:193–234
Brown EH, Walker NW (1993) A magma-loading model for Barrovian metamorphism in the Southeast Coast Plutonic Complex, British Columbia and Washington. Geol Soc Am Bull 105:479–500
Bucher K, Grapes R (2011) Petrogenesis of metamorphic rocks, 8th edn. Springer-Verlag, Berlin-Heidelberg
Burtman VS (2013) The geodynamics of the Pamir-Punjab syntaxis. Geotectonics 47(1):31–51
Cagnard F, Barbey P, Gapais D (2011) Transition between "Archaean-type" and "modern-type" tectonics: insights from the Finnish Lapland Granulite Belt. Precambr Res 187:127–142
Cai J, Liu F, Liu P et al (2014) Metamorphic *P-T* path and tectonic implications of pelitic granulites from the Daqingshan Complex of the Khondalite Belt, North China Craton. Precambr Res 241:161–184
Carlson WD (2012) Rates and mechanism of Y, REE, and Cr diffusion in garnet. Am Mineral 97:1598–1618
Carmichael DM (1969) On the mechanism of prograde metamorphic reactions in quartz-bearing pelitic rocks. Contr Mineral Petrol 20:244–267
Carmichael DM (1970) Intersecting isograds in the Whetstone Lake Area, Ontario. J Petrol 11:147–181
Chinner GA (1980) Kyanite isograds of Grampian metamorphism. J Geol Soc Lond 137:35–39

Clayton JL, Bostick NH (1986) Temperature effect on kerogen and on molecular and isotopic composition of organic matter in Pierre Shale near an igneous dike. Org Geochem 10:135–143
Cloos M (1982) Flow melanges: numerical modelling and geological constraints on their origin in the Franciscan subduction complex. Geol Soc Am Bull 93:330–345
Collins WJ, Vernon RH (1991) Orogeny associated with anticlockwise P-T-t paths: Evidence from low-P, high-T metamorphic terranes in the Arunta inlier, Central Australia. Geology 19:835–838
Crawford ML (1966) Composition of plagioclase and associated minerals in some schists of Vermont, USA and South Westland, New Zealand. Contrib Mineral Petrol 13:269–294
Cruciani G, Franceschelli M, Groppo C (2011) *P-T* evolution of eclogite-facies metabasite from NE Sardinia, Italy: Insights into the prograde evolution of Variscan eclogites. Lithos 121:135–150
Cutts KA, Kinny PD, Strachan RA et al (2010) Three metamorphic events recorded in a single garnet: Integrated phase modelling, in situ LA-ICPMS and SIMS geochronology from the Moine Supergroup, NW Scotland. J Metamorph Geol 28:249–267
De Groot SR, Mazur P (1962) Nonequilibrium thermodynamics. North Holland Publishing Co., Amsterdam
Denbig K (1966) The principles of chemical equilibrium. Cambridge University Press, Cambridge
Dipple GM, Wintsch RP, Andrews MS (1990) Identification of the scales of different mobility in a ductile fault zone. J Metamorph Geol 8:645–661
Droop GTR, Bucher-Nurminen K (1984) Reaction textures and metamorphic evolution of sapphirine-bearing granulites from the Gruf Complex, Italian central Alps. J Petrol 25:766–803
Dujon SC, Lagache M (1984) Echanges entre plagioclases et solutions aqueuses de chlorures sodi-calciques á différentes pressions et temperature (400 á 800 °C, 1 a 3 kilobars). Bull Mineral 197:553–569
Engi M, Lanari P, Kohn MJ (2017) Significant ages—an Introduction to Petrochronology. Rev Mineral Geochem 83:1–12
England P, Molnar P (1993) The interpretation of inverted metamorphic isograds using simple physical calculation. Tectonics 12:145–157
England PC, Thompson AB (1984) Pressure-temperature-time paths of regional metamorphism: heat transfer during the evolution of regions of thickened continental crust. J Petrol 25:894–928
Ernst WG (1988) Tectonic history of subduction zones inferred from retrograde blueschist P-T paths. Geology 16:1081–1084
Ernst RE, Wingate MTD, Buchan KL et al (2008) Global record of 1600–700 Ma Large Igneous Provinces (LIPs): implications for the reconstruction of the proposed Nuna (Columbia) and Rodinia supercontinents. Precambr Res 160:159–178
Escuder-Viruete J, Pérez-Estaún A (2013) Contrasting exhumation P-T paths followed by high-P rocks in the northern Caribbean subduction–accretionary complex: insights from the structural geology, microtextures and equilibrium assemblage diagrams. Lithos 160–161:117–144
Faryad SW, Chakraborty S (2005) Duration of Eo-Alpine metamorphic events obtained from multicomponent diffusion modeling of garnet: a case study from the Eastern Alps. Contrib Mineral Petrol 150:306–318
Fein JB, Walther JV (1987) Calcite solubility in supercritical CO_2–H_2O fluids. Geochim Cosmochim Acta 51:1665–1673
Fein JB, Walther JV (1989) Calcite solubility and speciation in supercritical NaCl–HCl aqueous fluids. Contrib Mineral Petrol 103:317–324
Ferry JM (1983) Applications of the reaction progress variable in Metamorphic Petrology. J Petrol 24:343–376
Fisher GW (1973) Nonequilibrium thermodynamics as a model for diffusion-controlled metamorphic processes. Am J Sci 273:897–924
Fisher GW (1977) Nonequilibrium thermodynamics in metamorphism. In: Fraser DG (ed) Thermodynamics in geology. NATO advanced study institutes series—C, vol 30. Reidel Publishing Co., Dordrecht, pp 381–403
Fisher GW (1978) Rate laws in metamorphism. Geochim Cosmochim Acta 42:1035–1050

Fisher GW, Elliott D (1974) Criteria for quasisteady diffusion and local equilibrium in metamorphism. In: Hofmann et al (eds) Geochemical transport and kinetics. Carnegie Institution of Washington, Publ 634, pp 231–241

Fisher GW, Lasaga AC (1981) Irreversible thermodynamics in petrology. In: Lasaga AC, Kirkpatrick RJ (eds) Kinetics of geochemical processes. Reviews in mineralogy, vol 8. Mineral Society of America, Book Crafter Inc., Chelsea, Michigan, pp 171–209

Fitts DD (1962) Nonequilibrium thermodynamics: a phenomenological theory of irreversible processes in fluid system. McGraw-Hill, New York

Foster CT Jr (1977) Mass transfer in sullimanite-bearing pelitic schists near Rangeley, Maine. Am Mineral 62:727–746

Foster CT Jr (1981) A thermodynamic model of mineral segregations in the lower sillimanite zone near Rangeley, Maine. Am Miner 66:260–277

Foster CT Jr (1986) Thermodynamic models of reactions involving garnet in a sillimanite/staurolite schists. Mineral Mag 50:427–439

Franceschelli M, Memmi I, Ottolini L et al (2002) Trace- and major-element zoning in garnet: a case study in the pelitic schists of NE Sardinia (Italy). Neues Jahrb Mineral Monatsh 8:337–351

Frantz JD, Mao HK (1979) Bimetasomatism resulting from intergranular diffusion: II. Prediction of multimineralic zone sequences. Am J Sci 279:302–323

Freer R (1981) Diffusion in silicate minerals and glasses: a data digest and guide to the literature. Contrib Mineral Petrol 76:440–454

Frei D, Liebscher A, Franz G et al (2004) Trace element geochemistry of epidote minerals. Rev Mineral Geochem 56:553–605

Gao J, Klemdt R (2003) Formation of HP-LT rocks and their tectonic implications in the western Tianshan Orogen, NW China: geochemical and age constraints. Lithos 66:1–22

Gerya TV (2002) P-T-trendy i model' formirovaniya granulitovykh kompleksov dokembriya (P-T trends and model of formation of Precambrian granulite complexes). Doctor of Science Dissertation, Moscow State University, Moscow

Gerya TV (2010) Introduction to numerical geodynamic modelling. Cambridge University Press, Cambridge

Gerya TV (2014) Precambrian geodynamics: concepts and models. Gondwana Res 25:442–463

Gerya TV, Maresch WV (2004) Metapelites of the Kanskiy granulite complex: kinked P-T path and geodynamic model. J Petrol 45:1393–1412

Ghent ED, Stout MZ (1981) Geobarometry and geothermometry of plagioclase-biotite-garnet-muscovite assemblages. Contrib Mineral Petrol 76:92–97

Gladkochub DP, Pisarevsky SA, Donskaya TV et al (2010) Proterozoic mafic magmatism in Siberian craton: an overview and implications for paleocontinental reconstruction. Precambr Res 183:660–668

Goes S, Capitanio FA, Morra G (2008) Evidence of lower-mantle slab penetration phases in plate motions. Nature 451:981–984

Griffen DT, Ribbe PH (1973) The crystal chemistry of staurolite. Am J Sci 273A:479–495

Groppo C, Rolfo F (2008) Counterclockwise *P-T* evolution of the Aghil Range: metamorphic record of an accretionary melange between Kunlun and Karakorum (SW Sinkiang, China). Lithos 105:365–378

Guidotti CV (1970) The mineralogy and petrology of the transition from the lower to upper sillimanite zone in the Oquossoc area, Maine. J Petrol 11:277–336

Guidotti CV (1974) Transition from staurolite to sillimanite zone, Rangeley quadrangle, Maine. Geol Soc Am Bull 85:475–490

Gutscher M-A, Westbrook GK (2009) Great earthquakes in slow-subduction, low-taper margins. In: Lallemand S, Funiciello F (eds) Subduction zone geodynamics. Springer, Berlin, pp 119–133

Hand M, Dirks PHGM, Powell R et al (1992) How well established is isobaric cooling in Proterozoic orogenic belts? An example from the Arunta inlier, central Australia. Geology 20:649–652

Harley SL (1989) The origin of granulites: metamorphic perspective. Geol Mag 126:215–247
He Z, Zhang Z, Zong K (2014) Metamorphic *P-T-t* evolution of mafic HP granulites in the northeastern segment of the Tarim Craton (Dunhuang block): evidence for early Paleozoic continental subduction. Lithos 196–197:1–13
Helgeson HC (1968) Evaluation of irreversible reactions in geochemical processes involving minerals and aqueous solutions. I. Thermodynamic relations. Geochim Cosmochim Acta 32:853–877
Holland TJB, Powell R (1990) An enlarged and updated internally consistent thermodynamic dataset with uncertainties and correlations: the system K_2O–Na_2O–CaO–MgO–MnO–FeO–Fe_2O_3–Al_2O_3–TiO_2–SiO_2–C–H_2–O_2. J Metamorph Geol 8:89–124
Holland TJB, Powell R (1998) An internally consistent thermodynamic data set for phases of petrological interest. J Metamorph Geol 16:309–343
Hollister LS (1969) Metastable paragenetic sequence of andalusite, kyanite, and sillimanite, Kwoeik area, British Columbia. Am J Sci 267:352–370
Huerta AD, Royden LH, Hodges KV (1999) The effects of accretion, erosion and radiogenic heat on the metamorphic evolution of collisional orogens. J Metamorph Geol 17:349–366
Jamieson RA, Beaumont C, Nguyen MH et al (2002) Interaction of metamorphism, deformation and exhumation in large convergent orogens. J Metamorph Geol 20:9–24
Janots E, Engi M, Berger A (2008) Prograde metamorphic sequence of REE minerals in pelitic rocks of the Central Alps: implications for allanite-monazite-xenotime phase relations from 250 to 610 °C. J Metamorph Geol 26:509–526
Joesten R (1977) Evolution of mineral assemblage zoning in diffusion metasomatism. Geochim Cosmochim Acta 41:649–670
Joesten R (1991) Grain-boundary diffusion kinetics in silicate and oxide minerals. In: Ganguly J (ed) Diffusion, atomic ordering and mass transport. Advances in physical geochemistry, vol 8. Springer, New York, pp 345–395
Joesten R, Fisher G (1988) Kinetics of diffusion controlled mineral growth in the Christmas Mountains (Texas) contact aureole. Geol Soc Am Bull 100:714–732
Johannes W (1989) Melting of plagioclase-quartz assemblages at 2 kbar water pressure. Contrib Mineral Petrol 103:270–276
Johnson MRW, Harley SL (2012) Orogenesis: the making of mountains. Cambridge University Press, Cambridge
Johnson SE, Vernon RH (1995) Stepping stones and pitfalls in the determination of an anti-clockwise *P-T-t* deformation path: the low-P, high-T Cooma Complex, Australia. J Metamorph Geol 13:165–183
Johnson SE, Brown M, Solar GS (2003) Low-pressure subsolidus and suprasolidus phase equilibria in the MnNCKFMASH system: constraints on conditions of regional metamorphism in western Maine, northern Appalachians. Am Mineral 88:624–638
Jolivet L, Faccenna C, Goffe B et al (2003) Subduction tectonics and exhumation of high-pressure metamorphic rocks in the Mediterranean orogens. Am J Sci 303:353–409
Katchalsky A, Curran PF (1965) Nonequilibrium thermodynamics in biophysics. Harvard University Press, Cambridge, MA
Kerrick DM (1990) The Al_2SiO_5 polymorphs. Mineralogical Society of America, Washington DC
Khain VE, Lomize MG (1995) Geotektonika s osnovami geodinamiki (Geotectonics with the basics of geodynamics). Moscow State University, Moscow
Kohn MJ, Malloy MA (2004) Formation of monazite via prograde metamorphic reactions among common silicates: Implications for age determinations. Geochim Cosmochim Acta 68:101–113
Korobeinikov SN, Polyansky OP, Likhanov II et al (2006) Mathematical modeling of overthrusting fault as a cause of andalusite–kyanite metamorphic zoning in the Yenisei Ridge. Dokl Earth Sci 408(4):652–656
Korzhinskii DS (1955) Ocherk metasomaticheskikh protsessov (An overview of metasomatic processes). In: Betekhtin AG (ed) Key problems in the theory of magmatic mineral deposits, 2nd edn. Izd Acad Nauk SSSR, Moscow, pp 335–456

Korzhinskii DS (1957) Fiziko-khimicheskiye osnovy analiza paragenezisov mineralov (Physico-chemical basis for the analysis of mineral parageneses). Izd Acad Nauk SSSR, Moscow

Korzhinskii DS (1962) Teoriya protsessov mineraloobrazovaniya (The theory of mineral formation processes). Izd Acad Nauk SSSR, Moscow

Korzhinskii DS (1973) Teoreticheskiye osnovy analiza paragenezisov mineralov (Theoretical bases of the analysis of mineral parageneses). Nauka, Moscow

Korzhinskii DS (1982) Teoriya metasomaticheskoy zonal'nosti (The theory of metasomatic zoning). Nauka, Moscow

Kotelnikov AR, Schekina TI (1986) Eksperimental'noye izucheniye kinetiki vzaimo-deystviya plagioklazov s vodno-solevym flyuidom pri 500 °C i P = 1 kbar (Experimental investigation of kinetic interaction of plagioclases with water-salt fluid at 500 °C and P = 1 kbar). Geokhimiya 9:1233–1244

Krebs M, Maresch WV, Schertl H-P et al (2008) The dynamics of intra-oceanic subduction zones: a direct comparison between fossil petrological evidence (Rio San Juan Complex, Dominican Republic) and numerical simulation. Lithos 103:106–137

Kuznetsova RP, Sheplev VS, Kolobov VY (1992) Analiz rosta zonal'nykh mineral'nykh segregatsiy i polucheniye kharakteristik massoperenosa pri metamorfizme. 4. Issledovaniye sistemy SiO_2–MgO–CaO (Analysis of growth of zoned mineral segregation and characteristics of mass transfer under metamorphism. 4. The SiO_2–MgO–CaO system). Geol Geofiz 33 (9):42–50

Kuznetsova RP, Kolobov VY, Sheplev VS (1994) Analiz rosta zonal'nykh mineral'nykh segregatsiy i polucheniye kharakteristik massoperenosa pri metamorfizme. 5. Issledovaniye sistemy SiO_2–Al_2O_3–FeO–MgO–K_2O–(Na_2O) (Analysis of growth of zoned mineral segregation and characteristics of mass transfer during metamorphism. 5. The SiO_2–Al_2O_3–FeO–MgO–K_2O–(Na_2O) system). Geol Geofiz 35(10):87–96

Lanari P, Engi M (2017) Local bulk compositional effect on mineral assemblages. Rev Mineral Geochem 83:55–102

Lasaga AC (1981) Rate laws of chemical reactions. In: Lasaga AC, Kirkpatrick RJ (eds) Kinetics of geochemical processes. Reviews in mineralogy, vol 8. Book Crfter Inc, Chelsea, Michigan, pp 1–68

Lazaro C, Garcia-Casco A, Rojas Agramonte Y et al (2009) Fifty-five-million-year history of oceanic subduction and exhumation at the northern edge of the Caribbean plate (Sierra del Convento mélange, Cuba). J Metamorph Geol 27:19–40

Lepezin GG (2015) Material transfer through the interface between peraluminous metapelite and gedrite-bearing gneiss at high temperatures and moderate pressures. Geochem Int 53(1):39–59

Lepezin GG, Khlestov VV (2009) Mass transfer at the contact of high-Al metapelites: an example of the high-temperature Sharyzhalgai Complex, eastern Sayan. Geochem Int 47(3):244–259

Lepezin GG, Seroglazov VV, Usova LV (1990) Masshtaby massoperenosa na kontakte metapelitov i metabazitov (The scale of mass transfer at the contact of metapelites and metabasites). Dokl Akad Nauk 314:1218–1222

Li J, Klemd R, Gao J et al (2016) Polycyclic metamorphic evolution of eclogite: Evidence for multistage burial–exhumation cycling in a subduction channel. J Petrol 57:119–146

Lichtner PC (1988) The quasi-stationary state approximation to coupled mass transport and fluid-rock interaction in a porous medium. Geochim Cosmochim Acta 52:143–165

Likhanov II (1989) Nizkotemperaturnaya biotitovaya izograda v kontaktovom oreole Kharlovskogo gabbrovogo massiva (Severo-Zapadnyy Altay) (Low-grade biotite isograde within the contact aureole of Kharlovo gabbro massif). Geol Geofis 30(7):46–54

Likhanov II (1990) Razlozheniye epidota pri nizkotemperaturnom kontaktovom meta-morfizme metapelitov (Epidote consuming reaction in low-temperature contact metamorphism of metapelites). In: Zapiski (Proceedings) of VMO, vol 119, pp 40–48

Likhanov II (2003) Mineral'nyye reaktsii i massoperenos pri metamorfizme nizkikh i umerennykh davleniy (Mineral reactions and mass-transfer during low- and medium-pressure metamorphism). Doctor of Sciences Dissertation. GEO Press, Novosibirsk

Likhanov II (2018) Mass-transfer and differential element mobility in metapelites during multistage metamorphism of Yenisei Ridge, Siberia. In: Ferrero S, Labari P, Gonsalses P, Grosch EG (eds) Metamorphic geology: microscale to mountain belts, vol 478. Geological Society Special Publication, Blackwell, London. https://doi.org/10.1144/sp478.11

Likhanov II, Reverdatto VV (1991) Arfvedsonite formed as a diabase-metapelite reaction product in contact metamorphism. Trans (Dokl) USSR Acad Sci Earth Sci Sect 317A(3):154–158

Likhanov II, Reverdatto VV (2002) Mass transfer during andalusite replacement by kyanite in Al- and Fe-rich metapelites in the Yenisei Range. Petrology 10(5):479–494

Likhanov II, Reverdatto VV (2007) Provenance of Precambrian Fe- and Al-rich metapelites in the Yenisei Ridge and Kuznetsk Alatau, Siberia: geochemical signatures. Acta Geol Sini Engl Edn 81(3):409–423

Likhanov II, Reverdatto VV (2008) Precambrian Fe- and Al-rich pelites from the Yenisey Ridge, Siberia: geochemical signatures for protolith origin and evolution during metamorphism. Int Geol Rev 50:597–623

Likhanov II, Reverdatto VV (2011a) Lower Proterozoic metapelites in the northern Yenisei Range: nature and age of protolith and the behaviour of material during collisional metamorphism. Geochem Int 49(3):224–252

Likhanov II, Reverdatto VV (2011b) Neoproterozoic collisional metamorphism in overthrust terranes of the Transangarian Yenisei Ridge, Siberia. Int Geol Rev 53:802–845

Likhanov II, Reverdatto VV (2014a) Geochemistry, age and petrogenesis of rocks from the Garevka metamorphic complex, Yenisei Ridge. Geochem Int 52(1):1–21

Likhanov II, Reverdatto VV (2014b) P-T-t constraints on the metamorphic evolution of the Transangarian Yenisei Ridge: geodynamic and petrological implications. Russ Geol Geophys 55(3):299–322

Likhanov II, Reverdatto VV (2016a) Geochemistry, petrogenesis and age of metamorphic rocks of the Angara complex at the junction of South and North Yenisei Ridge. Geochem Int 54 (2):127–148

Likhanov II, Reverdatto VV (2016b) Quantitative analysis of mass-transfer during polymetamorphism in pelites of the Transangarian Yenisei Ridge. Russ Geol Geophys 57(8):1204–1220

Likhanov II, Santosh M (2017) Neoproterozoic intraplate magmatism along the western margin of the Siberian Craton: implications for breakup of the Rodinia supercontinent. Precambr Res 300:315–331

Likhanov II, Reverdatto VV, Memmi I (1994) Short-range mobilization of elements in the biotite zone of contact aureole of the Kharlovo gabbro intrusion (Russia). Eur J Mineral 6:133–144

Likhanov II, Reverdatto VV, Memmi I (1995) The origin of arfvedsonite in metabasites from the contact aureole of the Kharlovo gabbro intrusion (Russia). Eur J Mineral 7:379–389

Likhanov II, Polyanskii OP, Kozlov PS et al (2000) Replacement of andalusite by kyanite with increasing pressure at a low geothermal gradient in metapelites of the Enisei Ridge. Dokl Earth Sci 375(9):1411–1416

Likhanov II, Polyanskii OP, Reverdatto VV et al (2001a) Metamorphic evolution of high-alumina metapelites near the Panimba overthrust (Yenisei Range): mineral associations, P-T conditions, and tectonic model. Geol Geofiz 42(8):1205–1220

Likhanov II, Ten AA, Reverdatto VV et al (2001b) Inverse modeling approach for obtaining kinetic parameters of diffusion-controlled metamorphic reactions in the Kharlovo contact aureole (South Siberia, Russia). Mineral Petrol 71:51–65

Likhanov II, Polyansky OP, Reverdatto VV et al (2004) Evidence from Fe- and Al-rich metapelites for thrust loading in the Transangarian Region of the Yenisei Ridge, eastern Siberia. J Metamorph Geol 22:743–762

Likhanov II, Reverdatto VV, Selyatizkii AY (2005) Mineral equilibria and P-T diagram for Fe- and Al-rich metapelites in the KFMASH system (K_2O-FeO-MgO-Al_2O_3-SiO_2-H_2O). Petrology 13(1):73–83

Likhanov II, Kozlov PS, Popov NV et al (2006) Collision metamorphism as a result of thrusting in the Transangara region of the Yenisei Ridge. Dokl Earth Sci 411(1):1313–1317

Likhanov II, Reverdatto VV, Kozlov PS et al (2008a) Collision metamorphism of Precambrian complexes in the Transangarian Yenisei Range. Petrology 16(2):136–160

Likhanov II, Reverdatto VV, Verschinin AE (2008b) Fe- and Al-rich metapelites of the Teya sequence, Yenisei Range: geochemistry, protoliths and the behavior of their matter during metamorphism. Geochem Int 46(1):17–36

Likhanov II, Reverdatto VV, Kozlov PS (2009) Kyanite-sillimanite metamorphism of the Precambrian complexes, Transangarian region of the Yenisei Ridge. Russ Geol Geophys 50 (12):1034–1051

Likhanov II, Reverdatto VV, Kozlov PS et al (2010a) Upper Riphean age of kyanite-sillimanite metamorphism in Transangarian Yenisei Ridge: evidence from ^{40}Ar-^{39}Ar data. Dokl Earth Sci 433(2):1108–1113

Likhanov II, Reverdatto VV, Travin AV (2010b) Exhumation rate of rocks from Neoproterozoic collisional metamorphic complexes of the Yenisei Ridge. Dokl Earth Sci 435(1):1518–1523

Likhanov II, Reverdatto VV, Kozlov PS (2011a) Collision-related metamorphic complexes of the Yenisei Ridge: their evolution, ages, and exhumation rate. Russ Geol Geophys 52(10):1256–1269

Likhanov II, Reverdatto VV, Kozlov PS (2011b) The Teya polymetamorphic complex in the Transangarian Yenisei Ridge: an example of metamorphic superimposed zoning of low- and medium-pressure facies series. Dokl Earth Sci 436(2):213–218

Likhanov II, Reverdatto VV, Kozlov PS et al (2011c) New evidence for Grenville events on the western margin of the Siberian craton: the example of the Garevka metamorphic complex in the Transangarian Yenisei Ridge. Dokl Earth Sci 438(2):782–787

Likhanov II, Reverdatto VV, Kozlov PS et al (2013a) Neoproterozoic metamorphic evolution in the Transangarian Yenisei Ridge: evidence from monazite and xenotime geochronology. Dokl Earth Sci 450(1):556–561

Likhanov II, Reverdatto VV, Kozlov PS et al (2013b) Three metamorphic events in Precambrian P-T-t history of the Transangarian Yenisei Ridge recorded in garnet grains in metapelites. Petrology 21(6):561–578

Likhanov II, Nozhkin AD, Reverdatto VV et al (2014) Grenville tectonic events and evolution of the Yenisei Ridge at the western margin of the Siberian craton. Geotectonics 48(5):371–389

Likhanov II, Nozhkin AD, Reverdatto VV et al (2015a) P-T evolution of ultrahigh temperature metamorphism: evidence for a late Paleoproterozoic intraplate extension at the southwestern margin of the Siberian Craton. Dokl Earth Sci 465(1):1139–1142

Likhanov II, Reverdatto VV, Kozlov PS et al (2015b) *P-T-t* constraints on polymetamorphic complexes in the Yenisei Ridge, East Siberia: implications for Neoproterozoic paleocontinental reconstructions. J Asian Earth Sci 113:391–410

Likhanov II, Nozhkin AD, Reverdatto VV et al (2016) Metamorphic evolution of ultrahigh-temperature Fe- and Al-rich granulites in the South Yenisei Ridge and tectonic implications. Petrology 24(4):392–408

Likhanov II, Régnier J-L, Santosh M (2018) Blueschist facies fault tectonites from the western margin of the Siberian Craton: implications for subduction and exhumation associated with early stages of the Paleo-Asian Ocean. Lithos 304–307:468–488

Liou JG (1973) Synthesis and stability relations of epidote, $Ca_2Al_3FeSi_3O_{12}(OH)$. J Petrol 14:381–413

Liou JG, Kuniyoshi S, Ito K (1974) Experimental studies of the phase relations between greenschist and amphibolite in a basaltic system. Am J Sci 274:613–632

Liou JG, Kim HS, Maruyama S (1983) Prehnite-epidote equilibria and their petrologic applications. J Petrol 24:321–342

Liu M, Peterson JC, Yund RA (1997) Diffusion-controlled growth of albite and pyroxene reaction rim. Contrib Mineral Petrol 126:217–223

Livio F, Sileo G, Michetti AM et al (2007) Pleistocene compressive tectonics in Central Southern Alps (Italy): rates of folding determined from growth strata. Geophys Res Abstr 9:02740

McKenzie DP (1969) Speculations on the consequences and causes of plate motions. Geophys J Roy Astron Soc 18:1–32

Miyashiro A (1958) Regional metamorphism of the Gosaisyo-Takanuki district in the central Abakuma Plateau. J Fac Sci Univ Tokyo 11:211–229
Miyashiro A (1973) Metamorphism and metamorphic belts. Allen and Unwin, London
Mugnier J-L, Huyghe P, Leturmy P et al (2004) Episodicity and rates of thrust-sheet motion in the Himalayas (western Nepal). In: McClay KR (ed) Thrust tectonics and hydrocarbon systems, vol 82. American Association Petroleum Geologist Memoir, AAPG, Boulder, pp 93–116
Mulrooney D, Rivers T (2005) Redistribution of the rare-earth elements among coexisting minerals in metamorphic rocks across the epidote-out isograd: an example from the St. Anthony Complex, northern Newfoundland, Canada. Can Mineral 43:263–294
Nehring F, Foley SF, Holtta P (2010) Trace element partitioning in the granulite facies. Contrib Mineral Petrol 159:493–519
Nozhkin AD, Turkina OM (1993) Geochimiya granulitov iz kanskogo i sharyzhalgaiskogo kompleksov (Geochemistry of granulites from kansk and sharyzhalgay complexes). UIGGM Press, Novosibirsk
Nozhkin AD, Likhanov II, Savko KA et al (2018) Sapphirine-bearing ultrahigh-temperature granulites of the Anabar shield: chemical composition, U-Pb zircon ages and P-T conditions of metamorphism. Doklady Earth Sci 479(1):347–351
Oelkers EH (1996) Physical and chemical properties of rocks and fluids for chemical mass transport calculations. In: Lichtner PC, Steefel CI, Oelkers EH (eds) Reactive transport in porous media. Review in mineralogy, vol 34. Mineralogical Society of America, Washington, pp 131–191
Otamendi JE, de la Rosa JD, Patino Douce AE et al (2002) Rayleigh fractionation of heavy rare earths and yttrium during metamorpfic garnet growth. Geology 30:159–162
Pattison DRM (2001) Instability of Al_2SiO_5 “triple point” assemblages in muscovite+biotite +quartz-bearing metapelites, with implications. Am Mineral 86:1414–1422
Peacock SM (2003) Thermal structure and metamorphic evolution of subducting slabs. In: Eiler J, Abers GA (eds) Inside the subduction factory. Geophysical monograph series, vol 238. AGU, Washington, pp 7–22
Perchuk LL (1989) Vzaimosoglasovanie nekotorykh Fe-Mg geotermomotrov na osnove zakona | Nernsta: revisiya (Mutual consistence between some Fe-Mg-geothermometers based on the Nernst law: revision). Geokhimiya 27(5):611–622
Perchuk LL, Gerya TV, van Reenen DD et al (2001) Formation and dynamics of granulite complexes within cratons. Gondwana Res 4:729–732
Perchuk LL, Gerya TV, van Reenen DD et al (2006) P-T paths and problems of high-temperature polymetamorphism. Petrology 14:117–153
Perchuk AL, Safonov OG, Sazonova LV et al (2015) Osnovy petrologii magmaticheskih processov (Basics of petrology of magmatic and metamorphic processes). Universitetskaya Kniga, Moscow
Poli S, Schmidt MW (2004) Experimental subsolidus studies on epidote minerals. Rev Mineral Geochem 56:171–195
Polyanskii OP, Reverdatto VV, Anan’ev VA (2000) Evolution of the rift sedimentary basin as an indicator of geodynamic setting (on example of the Enisei-Khatanga depression). Dokl Akad Nauk 370(1):71–75
Prigogine I (1961) Introduction to thermodynamics of irreversible processes, 2nd edn. Interscience, New York
Prigogine I, Defay R (1962) Chemische thermodynamic. Verlag für Grundstoffindustrie, Leipzig
Pyle JM, Spear FS, Rudnick RL et al (2001) Monazite-xenotine-garnet equilibrium in metapelites and a new monazite-garnet thermometer. J Petrol 42:2083–2107
Rambaldi ER (1973) Variation in the composition of plagioclase and epidote in some metamorphic rocks near Bancroft, Ontario. Can J Earth Sci 10:852–868
Raychenko AI (1981) Matematicheskaya teoriya diffusii v prilozheniyah (Mathematical theory of diffusion in applications). Naukova dumka, Kiev

Reinhardt J, Rubenach MJ (1989) Temperature-time relationships across metamorphic zones: evidence from porphyroblast-matrix relationships in progressively deformed metapelites. Tectonophysics 158:141–161

Reverdatto VV, Kolobov VY (1987) Mass transportation in metamorphism. Geol Geofiz 28(3):1–9

Reverdatto VV, Polyansky OP (1992) Evolution of PT-parameters in the alternative models of metamorphism. Dokl Akad Nauk 325(5):1017–1020

Reverdatto VV, Polyansky OP (2004) Modelling of the thermal history of metamorphic zoning in the Connemara region (western Ireland). Tectonophysics 379:77–91

Reverdatto VV, Sharapov VN, Lavrent'ev YG et al (1974) Investigations in isochemical contact metamorphism. Contrib Mineral Petrol 48:287–299

Reverdatto VV, Pertsev NN, Korolyuk VN (1979) P_{CO2}-T-evolution and origin of zoning in melilite during the regressive stage of contact metamorphism in carbonate-bearing rocks. Contrib Mineral Petrol 70:203–208

Reverdatto VV, Likhanov II, Polyansky OP et al (2017) Priroda i modeli metamorfizma (Nature and models of metamorphism). Publishing House SB RAS, Novosibirsk

Reverdatto VV, Babichev AV, Likhanov II et al (2018) Movement rates of metamorphic fronts in tocks near magmatic intrusive bodies. Dokl Earth Sci 480(2):751–753

Ridley J (1985) The effect of reaction enthalpy on the progress of a metamorphic reaction. In: Thompson AB, Rubie DC (eds) Advances in physical geochemistry, vol 4. Springer, Berlin, pp 80–97

Robie RA, Bethke PM, Toulmin MS et al (1966) X-ray crystallographic data, densities and molar volumes of minerals. In: Clark SP (ed) Handbook of physical constants. Geological Society of America, New York, pp 29–73

Robinson D, Beavins RE (1989) Diastathermal (extensional) metamorphism at very low grades and possible high grade analogues. Earth Planet Sci Lett 92:81–88

Rubenach MJ (1992) Proterozoic low-pressure/high-temperature metamorphism and an anticlockwise *P-T-t* path for the Hazeldene area, Mount Isa Inlier, Queensland, Australia. J Metamorph Geol 10:333–346

Sandiford M, Powell R (1991) Some remarks on high-temperature-low-pressure metamorphism in convergent orogens. J Metamorph Geol 9:333–340

Sarkar T, Schenk V (2014) Two-stage granulite formation in a Proterozoic magmatic arc (Ongole domain of the Eastern Ghats Belt, India): Part 1. Petrology and pressure–temperature evolution. Precambr Res 255:485–509

Savko KA, Bazikov NS (2011) Phase Equilibria of bastnaesite, allanite, and monazite: bastnaesite-out isograde in metapelites of the Vorontsovskaya Group, Voronezh Crystalline Massif. Petrology 19:445–469

Savko KA, Bazikov NS, Korish EKh et al (2012) Accessory rare earths-bearing minerals in Palaeoproterozoic schists of Voronezh crystalline massif. Zap RMO 141:107–128

Schenk V (1984) Petrology of felsic granulites, metapelites, metabasites and metacarbonates from southern Calabria (Italy): prograde metamorphism, uplift and cooling of a former lower crust. J Petrol 25:255–298

Seroglazov VV (1992) Mass transfer at the contact of high-Al gneisses and metaultrabasites (Sharyzhalgai Complex, eastern Sayan). Dokl Akad Nauk 323(5):925–929

Sharp WE, Kennedy GC (1965) The system $CaO–CO_2–H_2O$ in two-phase region calcite—aqueous solution. J Geol 73(2):391–403

Sheplev VS (1998) Matematicheskoe modelirovanie chimitcheskoy zonalnosti v metamorfitcheskih reakcionnyh strukturah (Mathematical modeling of chemical zoning in metamorphic reaction structures). Doctor of Science Dissertation, OIGGM SO RAN, Novosibirsk

Sheplev VS, Reverdatto VV (1998) Mineralogical geothermobarometry under unsteady equilibrium conditions. Dokl Akad Nauk 361(3):392–396

Sheplev VS, Korneeva LV, Reverdatto VV (1990) A simple model of dissolution and growth of scattered mineral grains during metamorphism. Geokhimiya 28(7):954–961

Sheplev VS, Kolobov VY, Kuznetsova RN et al (1991) Analysis of growth of zonated mineral segregation and characteristics of mass transfer during metamorphism. 1. Theoretical model in a quasi-stationary approximation. Sov Geol Geophy 32(12):1–12
Sheplev VS, Kuznetsova RP, Kolobov VY (1992a) Analysis of growth of zonated mineral segregation and characteristics of mass transfer during metamorphism. 2. The system SiO_2–Al_2O_3–MgO–$NaCa_2O_{5/2}$). Geol Geofiz 33(2):73–80
Sheplev VS, Kuznetsova RP, Kolobov VY (1992b) Analysis of growth of zonal mineral segregation and characteristics of mass transfer under metamorphism 3 The model of steady diffusion. Geol Geofiz 33(6):46–52
Sheplev VS, Kolobov VY, Kuznetsova RP (1998) Analysis of growth of zonated mineral associations. In: Augustithis SS (ed) Theophrastus' contributions to advanced studies in geology, vol II. Theophrastus Publications, Athens, pp 223–247
Shmonov VM, Vitovtova VM, Zharikov AV (2002) Fluidnaya pronixaemost' porod zemnoy kory (Fluid permeability of rocks of the Earth's crust). Nauchniy Mir, Moscow
Shvedenkov GY, Reverdatto VV, Bul'bak TA et al (2006) Bimetasomatic zoning in the CaO-MgO-SiO_2-H_2O-CO_2 system: experiments with the use of natural rock samples. Petrology 14(5):515–527
Shvedenkova SV, Shvedenkov GY (1990) Experimental investigation of calcium and sodium distribution between plagioclase and solution at 350 °C and 100 MPa. Geol Geofis 2:75–80
Sills JD, Rollinson HR (1987) The metamorphic evolution of the Lewisian Complex. In: Park RG, Tarney J (eds) Evolution of the Lewisian and comparable Precambrian high grade terrains, vol 28. Geological Society, London, Special Publications, pp 81–92
Skippen C (1974) An experimental model for low pressure metamorphism of siliceous dolomitic marble. Am J Sci 274:487–509
Sklyarov EV (2006) Exhumation of metamorphic complexes: basic mechanisms. Russ Geol Geophys 47(1):68–72
Sklyarov EV, Gladkochub DP, Donskaya TV et al (2001) Metamorfizm i tektonika (Metamorphism and Tectonics). Intermet Engineering, Moscow
Skublov SG (2005) Geokhimiya redkozemelnyh elementov v porodoobrazuyushih metamorficheskih mineralah (Geochemistry of rare-earth elements in the rock-forming metamorphic minerals). Nauka, St. Petersburg
Sobolev SV, Babeyko AYu (1994) Modeling of mineralogical composition, density and elastic-wave velocities in anhydrous magmatic rocks. Surv Geophys 15(5):515–544
Soloviev AV, Shapiro MN, Garver DI (2001) O skorostyah formirovaniya kollizionnyih nadvigov (On the rates of formation of collision thrusts (Lesnovsky overthrust, northern Kamchatka)). Bull MOIP Geol Dept 76(5):29–32
Spear FS (1989) Relative thermobarometry and metamorphic *P-T* paths. In: Daly JS, Cliff RA, Yardley BWD (eds) Evolution of metamorphic belts. Geological Society Special Publication, Blackwell, Oxford, pp 63–82
Spear FS, Peacock SM (1989) Metamorphic pressure-temperature-time paths. American Geophysical Union, Washington
Spear FS, Hickmott DD, Selverstone J (1990) Metamorphic consequences of thrust emplacement, Fall Mountain, New Hampshire. Geol Soc Am Bull 102:1344–1360
Spear FS, Kohn MJ, Cheney JT et al (2002) Metamorphic, thermal and tectonic evolution of Central New England. J Petrol 43:2097–2120
Staudigel H, King SD (1992) Ultrafast subduction—the key to slab recycling efficiency and mantle differentiation. Earth Planet Sci Lett 109:517–530
Stein S, Stein CA (1996) Thermo-mechanical evolution of oceanic lithosphere: implications for the subduction process and deep earthquakes. In: Bebout GE, Scholl DW, Kirby SH et al (eds) Subduction: top to bottom. Geophysical monograph, vol 96. American Geophysical Union, New York, pp 1–17
Suppe J, Chou GT, Hook SC (1992) Rates of folding and faulting determined from growth strata. In: McClay KR (ed) Thrust tectonics. Springer, Dordrecht, pp 105–121

Syracuse EM, van Keken PE, Abers GA (2010) The global range of subduction zone thermal models. Phys Earth Planet Inter 183:73–90

Tam PY, Zhao G, Sun M et al (2012a) Petrology and metamorphic *P-T* path of high-pressure mafic granulites from the Jiaobei massif in the Jiao-Liao-Ji Belt, North China Craton. Lithos 155:94–109

Tam PY, Zhao G, Zhou X et al (2012b) Metamorphic *P-T* path and implications of high-pressure pelitic granulites from the Jiaobei massif in the Jiao-Liao-Ji Belt, North China Craton. Gondwana Res 22:104–117

Teyssier C, Whitney DL (2002) Gneiss domes and orogeny. Geology 30:1139–1142

Thompson AB (1975) Calc-silikate diffusion zones between marble and pelitic schists. J Petrol 16 (2):314–346

Thompson AB, England PC (1984) Pressure-temperature-paths of regional metamorphism, II: their influence and interpretation using mineral assemblages in metamorphic rocks. J Petrol 25:929–954

Tong L, Xu Y-G, Cawood PA et al (2014) Anticlockwise P-T evolution at ~280 Ma recorded from ultrahigh-temperature metapelitic granulite in the Chinese Altai orogenic belt, a possible link with the Tarim mantle plume? J Asian Earth Sci 94:1–11

Tozer RSJ, Butler RWH, Corrado S (2002) Comparing thin- and thick-skanned thrust tectonic models of the Central Apenines, Italy. In: Europ Geosci Union, Stephan Mueller Spec Publ Ser, vol 1. pp 181–194. https://doi.org/10.5194/smsps-1-181-2002

Trommsdorff V (1966) Progressive metamorphose kieseliger korbonatgesteine in den Zentralalpen zwischen Bernina und Simplon. Schweiz Mineral Petrogr Mitt 46(2):431–460

Trommsdorff V (1972) Change in T-X during metamorphism of siliceous dolomitic rocks of the Central Alps. Schweiz Mineral Petrogr Mitt 52(3):567–571

Tsunogae T, van Reenen DD (2006) Corundum + quartz and Mg-staurolite bearing granulite from the Limpopo Belt, southern Africa: implications for a *P-T* path. Lithos 92:576–587

Van Westrenen W, Allan NL, Blundy JD et al (2003) Trace element incorporation into pyrope-grossular solid solutions: an atomistic simulation study. Phys Chem Miner 30:217–229

Vernon RH (1976) Metamorphic processes, reactions and microstructure development. George Allen and Unwin, London

Walther JV, Wood BJ (1984) Rate and mechanism in prograde metamorphism. Contrib Mineral Petrol 88:246–259

Wan T (2010) The tectonics of China: data, maps and evolution. Springer, Heidelberg

Wan B, Windley BF, Xiao W et al (2015) Paleoproterozoic high-pressure metamorphism in the northern North China Craton and implications for the Nuna supercontinent. Nat Commun 6:8344

Wang Q, Zhang PZ, Freymueller JT et al (2001) Present-day crustal deformation in China constrained by Global Positioning System measurements. Science 294:574–577

Waters DJ (1986) Metamorphic history of sapphirine-bearing and related magnesian gneisses from Namaqualand, South Africa. J Petrol 27:541–565

Weare JH, Stephens JR, Eugster HP (1976) Diffusion metasomatism and mineral reaction zones: general principles and application to feldspar alteration. Am J Sci 276:767–816

Wernicke B (1985) Uniform-sense normal simple shear of the continental lithosphere. Can J Earth Sci 22:108–125

White RW, Powell R, Clarke GL (2002) The interpretation of reaction textures in Fe-rich metapelitic granulites of the Musgrave Block, central Australia: constraints from mineral equilibria calculations in the system K_2O–FeO–MgO–Al_2O_3–SiO_2–H_2O–TiO_2–Fe_2O_3. J Metamorph Geol 20:41–55

Whitney DL, Lang HM, Ghent ED (1995) Quantitative determination of metamorphic reaction history: mass balance between groundmass and mineral inclusion assemblages in metamorphic rocks. Contrib Mineral Petrol 120:404–411

Whitney DL, Mechum TA, Kuehner SM et al (1996) Progressive metamorphism of pelitic rocks from protolith to granulite facies, Dutchess County, New York, USA: Constraints on the timing of fluid infiltration during regional metamorphism. J Metamorph Geol 14:163–181

Whitney DL, Miller RB, Paterson SR (1999) *P-T-t* evidence for mechanisms of vertical tectonic motion in a contractional orogen: north-western US and Canadian Cordillera. J Metamorph Geol 17:75–90

Will TM, Schmadicke E (2003) Isobaric cooling and anti-clockwise *P-T* paths in the Variscan Odenwald crystalline complex, Germany. J Metamorph Geol 21:469–480

Wolfram S (2003) The mathematica book, 5th edn. Wolfram Media Inc., Champaign

Xiang H, Zhang L, Zhong ZQ et al (2012) Ultrahigh-temperature metamorphism and anticlockwise P-T–t path of Paleozoic granulites from north Qinling-Tongbai orogen, Central China. Gondwana Res 21:559–576

Yalcin MN, Littke R, Sachsenhofer RF (1997) Thermal history of sedimentary basins. In: Welte DH, Horsfield B, Baker DR (eds) Petroleum and basin evolution. Insights from petroleum geochemistry, geology and basin modeling. Springer, Heidelberg, pp 71–168

Yu S, Zhang J, Real PGD (2011) Petrology and P-T path of high-pressure granulite from the Dulan area, North Qaidam Mountains, northwestern China. J Asian Earth Sci 42:641–660

Yund RA (1997) Rates of grain boundary diffusion through enstatite and forsterite reaction rims. Contrib Mineral Petrol 126:224–226

Zhai QG, Zhang RY, Jahn BM et al (2011) Triassic eclogites from central Qiangtang, northern Tibet, China: petrology, geochronology and metamorphic *P-T* path. Lithos 125:173–189

Conclusions

In this book, we attempted to establish a causal relationship between geodynamics and magmatism, on the one hand, and metamorphism, on the other hand. Analysis of the problem reveals that metamorphism is a consequence and an indicator of geodynamics and magmatism.

The study of metamorphism can be used to draw several important conclusions. Metamorphism at shallow levels commonly results from the igneous intrusion. Melting of the lithosphere is accompanied by gravitational instability, diapiric ascent of magmas, and elevated heat flows from the earth's interior. The process of extension and compression of the crust and lithosphere is manifested in burial and collision-related types of metamorphism. High-pressure and low-temperature metamorphism is caused by rapid subsidence of shallow rocks into the lithosphere followed by rapid exhumation. Large masses of granulites in shield areas were formed under specific conditions in the Archean and Paleoproterozoic.

A better understanding of causal relationships requires a study of *P*-*T* metamorphic parameters and, especially, *P*-*T* trends. The relationship between peak *T* and peak *P* in the metamorphic history of the rock should be used as one of the key parameters for elucidating geodynamic and magmatic causes of different types of metamorphism. Their interpretation using thermomechanical numerical models accounting for variable rates and mechanisms of subsidence and exhumation can be used to solve many geodynamical problems.

As regards the kinetics of prograde metamorphism, the first steps were made to estimate the reaction rates and sizes of local-equilibrium elementary volumes in the rocks where these reactions take place. It was concluded that the mass transfer in metamorphic reactions occurred with the preservation of the matter balance within very small volumes of the rocks and increased from a few hundredths of cubic millimeters to a few cubic centimeters, depending on the duration of a metamorphic event. The smallest local volumes are characteristic of contact metamorphism, while the largest ones are typical of collisional metamorphism. Further progress can be expected in a quantitative analysis of mass transfer involving trace elements. It

V. V. Reverdatto et al., *The Nature and Models of Metamorphism*,
Springer Geology, https://doi.org/10.1007/978-3-030-03029-2

seems likely that in the future, the kinetic features will help establish links between metamorphism, geodynamics, and magmatism.

Rates of prograde metamorphic reactions are likely to be slower than the rate of the increase in *P-T* parameters in response to tectonomagmatic events. Therefore, the rate of most mineral transformations in rocks lags behind that of the onset of *P-T* metamorphic conditions and the compositions of newly-formed coexisting minerals during their evolution may not reflect the peak *T* and/or *P* conditions.

We believe that metamorphic petrology has a very promising future. Advances in this science could be expected not only in traditional fields, but also in many related disciplines: solid-state physics and physical chemistry. We hope that the new scientific results will help elucidate the causes leading to generation of zones of superlithostatic pressure in rocks undergoing deformation. Metamorphism will play an ever-increasing role as an indicator of geodynamic conditions, while a better understanding of the nature of protoliths will provide new knowledge on the Archean-Proterozoic geology and will be helpful in mineral exploration.

MIX
Papier aus verantwortungsvollen Quellen
Paper from responsible sources
FSC® C105338

If you have any concerns about our products, you can contact us on
ProductSafety@springernature.com

In case Publisher is established outside the EU, the EU authorized representative is:
Springer Nature Customer Service Center GmbH
Europaplatz 3, 69115 Heidelberg, Germany

Printed by Libri Plureos GmbH
in Hamburg, Germany